Applied Probability and Statistics

BAILEY · The Elements of Stochastic Processes wi̶ to the Natural Sciences

BAILEY · Mathematics, Statistics and Systems for ̶

BARTHOLOMEW · Stochastic Models for Social Pr̶ *Edition*

BECK and ARNOLD · Parameter Estimation in Engineering and Science

BENNETT and FRANKLIN · Statistical Analysis in Chemistry and the Chemical Industry

BHAT · Elements of Applied Stochastic Processes

BLOOMFIELD · Fourier Analysis of Time Series: An Introduction

BOX · The Life of a Scientist

BOX and DRAPER · Evolutionary Operation: A Statistical Method for Process Improvement

BOX, HUNTER, and HUNTER · Statistics for Experimenters: An Introduction to Design, Data Analysis and Model Building

BROWN and HOLLANDER · Statistics: A Biomedical Introduction

BROWNLEE · Statistical Theory and Methodology in Science and Engineering, *Second Edition*

BURY · Statistical Models in Applied Science

CHAMBERS · Computational Methods for Data Analysis

CHATTERJEE and PRICE · Regression Analysis by Example

CHERNOFF and MOSES · Elementary Decision Theory

CHOW · Analysis and Control of Dynamic Economic Systems

CLELLAND, deCANI, BROWN, BURSK, and MURRAY · Basic Statistics with Business Applications, *Second Edition*

COCHRAN · Sampling Techniques, *Third Edition*

COCHRAN and COX · Experimental Designs, *Second Edition*

COX · Planning of Experiments

COX and MILLER · The Theory of Stochastic Processes, *Second Edition*

DANIEL · Application of Statistics to Industrial Experimentation

DANIEL and WOOD · Fitting Equations to Data

DAVID · Order Statistics

DEMING · Sample Design in Business Research

DODGE and ROMIG · Sampling Inspection Tables. *Second Edition*

DRAPER and SMITH · Applied Regression Analysis

DUNN · Basic Statistics: A Primer for the Biomedical Sciences, Second Edition

DUNN and CLARK · Applied Statistics: Analysis of Variance and Regression

ELANDT-JOHNSON · Probability Models and Statistical Methods in Genetics

FLEISS · Statistical Methods for Rates and Proportions

GIBBONS, OLKIN and SOBEL · Selecting and Ordering Populations: A New Statistical Methodology

GNANADESIKAN · Methods for Statistical Data Analysis of Multivariate Observations

GOLDBERGER · Econometric Theory

GROSS and CLARK · Survival Distributions: Reliability Applications in the Biomedical Sciences

GROSS and HARRIS · Fundamentals of Queueing Theory

GUTTMAN, WILKS and HUNTER · Introductory Engineering Statistics, *Second Edition*

continued on back

Sampling Techniques

A Wiley Publication in Applied Statistics

Sampling Techniques

third edition

WILLIAM G. COCHRAN

Professor of Statistics, Emeritus
Harvard University

John Wiley & Sons

New York · Santa Barbara · London · Sydney · Toronto

Library of Congress Cataloging in Publication Data:

Cochran, William Gemmell, 1909–
  Sampling techniques.

  (Wiley series in probability and mathematical statistics)
  Includes bibliographical references and index.
  1. Sampling (Statistics) I. Title.
QA276.6.C6   1977   001.4'222   77–728
ISBN 0–471–16240–X

Printed in the United States of America

10 9 8 7 6 5 4 3 2 1

to Betty

Preface

As did the previous editions, this textbook presents a comprehensive account of sampling theory as it has been developed for use in sample surveys. It contains illustrations to show how the theory is applied in practice, and exercises to be worked by the student. The book will be useful both as a text for a course on sample surveys in which the major emphasis is on theory and for individual reading by the student.

The minimum mathematical equipment necessary to follow the great bulk of the material is a familiarity with algebra, especially relatively complicated algebraic expressions, plus a knowledge of probability for finite sample spaces, including combinatorial probabilities. The book presupposes an introductory statistics course that covers means and standard deviations, the normal, binomial, hypergeometric, and multinomial distributions, the central limit theorem, linear regression, and the simpler types of analyses of variance. Since much of classical sample survey theory deals with the distributions of estimators over the set of randomizations provided by the sampling plan, some knowledge of nonparametric methods is helpful.

The topics in this edition are presented in essentially the same order as in earlier editions. New sections have been included, or sections rewritten, primarily for one of three reasons: (1) to present introductions to topics (sampling plans or methods of estimation) relatively new in the field; (2) to cover further work done during the last 15 years on older methods, intended either to improve them or to learn more about the performance of rival methods; and (3) to shorten, clarify, or simplify proofs given in previous editions.

New topics in this edition include the approximate methods developed for the difficult problem of attaching standard errors or confidence limits to nonlinear estimates made from the results of surveys with complex plans. These methods will be more and more needed as statistical analyses (e.g., regressions) are performed on the results. For surveys containing sensitive questions that some respondents are unlikely to be willing to answer truthfully, a new device is to present the respondent with either the sensitive question or an innocuous question; the specific choice, made by randomization, is unknown to the interviewer. In some sampling problems it may seem economically attractive, or essential in countries without full sampling resources, to use two overlapping lists (or frames, as they are called) to cover the complete population. The method of double sampling has been extended to cases where the objective is to compare the means of a number of subgroups within the population. There has been interesting work on the attractive properties that the ratio and regression estimators have if it can be assumed that the finite population is itself a random sample from an infinite superpopulation in which a mathematical model appropriate to the ratio or regression estimator holds. This kind of assumption is now new—I noticed recently that Laplace used it around 1800 in a sampling problem—but it clarifies the relation between sample survey theory and standard statistical theory.

An example of further work on topics included in previous editions is Chapter 9A, which has been written partly from material previously in Chapter 9; this was done mainly to give a more adequate account of what seem to me the principal methods produced for sampling with unequal probabilities without replacement. These include the similar methods given independently by Brewer, J. N. K. Rao, and Durbin, Murthy's method, the Rao, Hartley, Cochran method, and Madow's method related to systematic sampling, with comparisons of the performances of the methods on natural populations. New studies have been done of the sizes of components of errors of measurement in surveys by repeat measurements by different interviewers, by interpenetrating subsamples, and by a combination of the two approaches. For the ratio estimator, data from natural populations have been used to appraise the small-sample biases in the standard large-sample formulas for the variance and the estimated variance. Attempts have also been made to create less biased variants of the ratio estimator itself and of the formula for estimating its sampling variance. In stratified sampling there has been additional work on allocating sample sizes to strata when more than one item is of importance and on estimating sample errors when only one unit is to be selected per stratum. Some new systematic sampling methods for handling populations having linear trends are also of interest.

Alva L. Finkner and Emil H. Jebe prepared a large part of the lecture notes from which the first edition of this book was written. Some investigations that provided background material were supported by the Office of Naval Research, Navy Department. From discussions of recent developments in sampling or suggestions about this edition, I have been greatly helped by Tore Dalenius, David J. Finney, Daniel G. Horvitz, Leslie Kish, P. S. R. Sambasiva Rao, Martin Sandelius, Joseph Sedransk, Amode R. Sen, and especially Jon N. K. Rao, whose painstaking reading of the new and revised sections of this edition resulted in many constructive suggestions about gaps, weaknesses, obscurities, and selection of topics. For typing and other work involved in production of a typescript I am indebted to Rowena Foss, Holly Grano, and Edith Klotz. My thanks to all.

William G. Cochran

South Orleans, Massachusetts
February, 1977

Contents

CHAPTER   PAGE

1   INTRODUCTION   1

1.1   Advantages of the Sampling Method   1
1.2   Some Uses of Sample Surveys   2
1.3   The Principal Steps in a Sample Survey   4
1.4   The Role of Sampling Theory   8
1.5   Probability Sampling   9
1.6   Alternatives to Probability Sampling   10
1.7   Use of the Normal Distribution   11
1.8   Bias and Its Effects   12
1.9   The Mean Square Error   15

     Exercises   16

CHAPTER

2   SIMPLE RANDOM SAMPLING   18

2.1   Simple Random Sampling   18
2.2   Selection of a Simple Random Sample   19
2.3   Definitions and Notation   20
2.4   Properties of the Estimates   21
2.5   Variances of the Estimates   23
2.6   The Finite Population Correction   24
2.7   Estimation of the Standard Error from a Sample   25
2.8   Confidence Limits   27
2.9   An Alternative Method of Proof   28
2.10   Random Sampling with Replacement   29
2.11   Estimation of a Ratio   30
2.12   Estimates of Means Over Subpopulations   34
2.13   Estimates of Totals Over Subpopulations   35
2.14   Comparisons Between Domain Means   39
2.15   Validity of the Normal Approximation   39
2.16   Linear Estimators of the Population Mean   44

     Exercises   45

CHAPTER

3 SAMPLING PROPORTIONS AND PERCENTAGES   50

3.1 Qualitative Characteristics   50
3.2 Variances of the Sample Estimates   50
3.3 The Effect of P on the Standard Errors   53
3.4 The Binomial Distribution   55
3.5 The Hypergeometric Distribution   55
3.6 Confidence Limits   57
3.7 Classification into More than Two Classes   60
3.8 Confidence Limits with More than Two Classes   60
3.9 The Conditional Distribution of p   61
3.10 Proportions and Totals Over Subpopulations   63
3.11 Comparisons Between Different Domains   64
3.12 Estimation of Proportions in Cluster Sampling   64

     Exercises   68

CHAPTER

4 THE ESTIMATION OF SAMPLE SIZE   72

4.1 A Hypothetical Example   72
4.2 Analysis of the Problem   73
4.3 The Specification of Precision   74
4.4 The Formula for n in Sampling for Proportions   75
4.5 Rare Items—Inverse Sampling   76
4.6 The Formula for n with Continuous Data   77
4.7 Advance Estimates of Population Variances   78
4.8 Sample Size with More than One Item   81
4.9 Sample Size when Estimates Are Wanted for Subdivisions of the Population   82
4.10 Sample Size in Decision Problems   83
4.11 The Design Effect (*Deff*)   85

     Exercises   86

CHAPTER

5 STRATIFIED RANDOM SAMPLING   89

5.1 Description   89
5.2 Notation   90
5.3 Properties of the Estimates   91
5.4 The Estimated Variance and Confidence Limits   95
5.5 Optimum Allocation   96

5.6 Relative Precision of Stratified Random and Simple Random Sampling 99
5.7 When Does Stratification Produce Large Gains in Precision? . . . 101
5.8 Allocation Requiring More than 100 Per Cent Sampling 104
5.9 Estimation of Sample Size with Continuous Data 105
5.10 Stratified Sampling for Proportions 107
5.11 Gains in Precision in Stratified Sampling for Proportions 109
5.12 Estimation of Sample Size with Proportions 110

Exercises 111

CHAPTER
5A FURTHER ASPECTS OF STRATIFIED SAMPLING 115

5A.1 Effects of Deviations from the Optimum Allocation 115
5A.2 Effects of Errors in the Stratum Sizes 117
5A.3 The Problem of Allocation with More than One Item 119
5A.4 Other Methods of Allocation with More than One Item 121
5A.5 Two-Way Stratification with Small Samples 124
5A.6 Controlled Selection 126
5A.7 The Construction of Strata 127
5A.8 Number of Strata 132
5A.9 Stratification After Selection of the Sample (Poststratification) . . 134
5A.10 Quota Sampling 135
5A.11 Estimation from a Sample of the Gain Due to Stratification . . . 136
5A.12 Estimation of Variance with One Unit per Stratum 138
5A.13 Strata as Domains of Study 140
5A.14 Estimating Totals and Means Over Subpopulations 142
5A.15 Sampling from Two Frames 144

Exercises 146

CHAPTER
6 RATIO ESTIMATORS 150

6.1 Methods of Estimation 150
6.2 The Ratio Estimate 150
6.3 Approximate Variance of the Ratio Estimate 153
6.4 Estimation of the Variance from a Sample 155
6.5 Confidence Limits 156
6.6 Comparison of the Ratio Estimate with Mean per Unit 157
6.7 Conditions Under Which the Ratio Estimate Is a Best Linear Unbiased Estimator 158
6.8 Bias of the Ratio Estimate 160

6.9 Accuracy of the Formulas for the Variance and Estimated
Variance 162
6.10 Ratio Estimates in Stratified Random Sampling 164
6.11 The Combined Ratio Estimate 165
6.12 Comparison of the Combined and Separate Estimates 167
6.13 Short-Cut Computation of the Estimated Variance 169
6.14 Optimum Allocation with a Ratio Estimate 172
6.15 Unbiased Ratio-type Estimates 174
6.16 Comparison of the Methods 177
6.17 Improved Estimation of Variance 178
6.18 Comparison of Two Ratios 180
6.19 Ratio of Two Ratios 183
6.20 Multivariate Ratio Estimates 184
6.21 Product Estimators 186

Exercises   186

CHAPTER

7 REGRESSION ESTIMATORS

7.1 The Linear Regression Estimate 189
7.2 Regression Estimates with Preassigned b 190
7.3 Regression Estimates when b Is Computed from the Sample . . . 193
7.4 Sample Estimate of Variance 195
7.5 Large-Sample Comparison with the Ratio Estimate and the Mean
per Unit 195
7.6 Accuracy of the Large-Sample Formulas for $V(\bar{y}_{lr})$ and $v(\bar{y}_{lr})$ 197
7.7 Bias of the Linear Regression Estimate 198
7.8 The Linear Regression Estimator Under a Linear Regression
Model 199
7.9 Regression Estimates in Stratified Sampling 200
7.10 Regression Coefficients Estimated from the Sample 201
7.11 Comparison of the Two Types of Regression Estimate 203

Exercises   203

CHAPTER

8 SYSTEMATIC SAMPLING

8.1 Description 205
8.2 Relation to Cluster Sampling 207
8.3 Variance of the Estimated Mean 207
8.4 Comparison of Systematic with Stratified Random Sampling . . . 212
8.5 Populations in "Random" Order 212

8.6   Populations with Linear Trend   214
8.7   Methods for Populations with Linear Trends   216
8.8   Populations with Periodic Variation   217
8.9   Autocorrelated Populations   219
8.10 Natural Populations   221
8.11 Estimation of the Variance from a Single Sample   223
8.12 Stratified Systematic Sampling   226
8.13 Systematic Sampling in Two Dimensions   227
8.14 Summary   229

      Exercises   231

CHAPTER
9   SINGLE-STAGE CLUSTER SAMPLING: CLUSTERS OF EQUAL SIZES

233

9.1   Reasons for Cluster Sampling   233
9.2   A Simple Rule   234
9.3   Comparisons of Precision Made from Survey Data   238
9.4   Variance in Terms of Intracluster Correlation   240
9.5   Variance Functions   243
9.6   A Cost Function   244
9.7   Cluster Sampling for Proportions   246

      Exercises   247

CHAPTER
9A   SINGLE-STAGE CLUSTER SAMPLING: CLUSTERS OF UNEQUAL SIZES

249

9A.1   Cluster Units of Unequal Sizes   249
9A.2   Sampling with Probability Proportional to Size   250
9A.3   Selection with Unequal Probabilities with Replacement . .   252
9A.4   The Optimum Measure of Size   255
9A.5   Relative Accuracies of Three Techniques   255
9A.6   Sampling with Unequal Probabilities Without Replacement .   258
9A.7   The Horvitz-Thompson Estimator   259
9A.8   Brewer's Method   261
9A.9   Murthy's Method   263
9A.10 Methods Related to Systematic Sampling   265
9A.11 The Rao, Hartley, Cochran Method   266
9A.12 Numerical Comparisons   267
9A.13 Stratified and Ratio Estimates   270

       Exercises   272

CHAPTER

10   SUBSAMPLING WITH UNITS OF EQUAL SIZE 274

10.1	Two-Stage Sampling	274
10.2	Finding Means and Variances in Two-Stage Sampling	275
10.3	Variance of the Estimated Mean in Two-Stage Sampling	276
10.4	Sample Estimation of the Variance	278
10.5	The Estimation of Proportions	279
10.6	Optimum Sampling and Subsampling Fractions	280
10.7	Estimation of m_{opt} from a Pilot Survey	283
10.8	Three-Stage Sampling	285
10.9	Stratified Sampling of the Units	288
10.10	Optimum Allocation with Stratified Sampling	289
	Exercises	290

CHAPTER

11   SUBSAMPLING WITH UNITS OF UNEQUAL SIZES 292

11.1	Introduction	292
11.2	Sampling Methods when $n = 1$	293
11.3	Sampling with Probability Proportional to Estimated Size	297
11.4	Summary of Methods for $n = 1$	299
11.5	Sampling Methods When $n > 1$	300
11.6	Two Useful Results	300
11.7	Units Selected with Equal Probabilities: Unbiased Estimator	303
11.8	Units Selected with Equal Probabilities: Ratio to Size Estimate	303
11.9	Units Selected with Unequal Probabilities with Replacement: Unbiased Estimator	306
11.10	Units Selected Without Replacement	308
11.11	Comparison of the Methods	310
11.12	Ratios to Another Variable	311
11.13	Choice of Sampling and Subsampling Fractions. Equal Probabilities	313
11.14	Optimum Selection Probabilities and Sampling and Subsampling Rates	314
11.15	Stratified Sampling. Unbiased Estimators	316
11.16	Stratified Sampling. Ratio Estimates	317
11.17	Nonlinear Estimators in Complex Surveys	318
11.18	Taylor Series Expansion	319
11.19	Balanced Repeated Replications	320
11.20	The Jackknife Method	321
11.21	Comparison of the Three Approaches	322
	Exercises	324

CHAPTER

12   DOUBLE SAMPLING   327

    12.1   Description of the Technique   327
    12.2   Double Sampling for Stratification   327
    12.3   Optimum Allocation   331
    12.4   Estimated Variance in Double Sampling for Stratification   333
    12.5   Double Sampling for Analytical Comparisons   335
    12.6   Regression Estimators   338
    12.7   Optimum Allocation and Comparison with Single Sampling . . .   341
    12.8   Estimated Variance in Double Sampling for Regression   343
    12.9   Ratio Estimators   343
    12.10 Repeated Sampling of the Same Population   344
    12.11 Sampling on Two Occasions   346
    12.12 Sampling on More than Two Occasions   348
    12.13 Simplifications and Further Developments   351

           Exercises   355

CHAPTER

13   SOURCES OF ERROR IN SURVEYS   359

    13.1   Introduction   359
    13.2   Effects of Nonresponse   359
    13.3   Types of Nonresponse   364
    13.4   Call-backs   365
    13.5   A Mathematical Model of the Effects of Call-backs   367
    13.6   Optimum Sampling Fraction Among the Nonrespondents   370
    13.7   Adjustments for Bias Without Call-backs   374
    13.8   A Mathematical Model for Errors of Measurement   377
    13.9   Effects of Constant Bias   379
    13.10 Effects of Errors that Are Uncorrelated Within the Sample . . .   380
    13.11 Effects of Intrasample Correlation Between Errors of Measure-
           ment   383
    13.12 Summary of the Effects of Errors of Measurement   384
    13.13 The Study of Errors of Measurement   384
    13.14 Repeated Measurement of Subsamples   386
    13.15 Interpenetrating Subsamples   388
    13.16 Combination of Interpenetration and Repeated Measurement . .   391
    13.17 Sensitive Questions: Randomized Responses   392
    13.18 The Unrelated Second Question   393
    13.19 Summary   395

           Exercises   396

References   400

Answers to Exercises   412

Author Index   419

Subject Index   422

CHAPTER 1

Introduction

1.1 ADVANTAGES OF THE SAMPLING METHOD

Our knowledge, our attitudes, and our actions are based to a very large extent on samples. This is equally true in everyday life and in scientific research. A person's opinion of an institution that conducts thousands of transactions every day is often determined by the one or two encounters he has had with the institution in the course of several years. Travelers who spend 10 days in a foreign country and then proceed to write a book telling the inhabitants how to revive their industries, reform their political system, balance their budget, and improve the food in their hotels are a familiar figure of fun. But in a real sense they differ from the political scientist who devotes 20 years to living and studying in the country only in that they base their conclusions on a much smaller sample of experience and are less likely to be aware of the extent of their ignorance. In science and human affairs alike we lack the resources to study more than a fragment of the phenomena that might advance our knowledge.

This book contains an account of the body of theory that has been built up to provide a background for good sampling methods. In most of the applications for which this theory was constructed, the aggregate about which information is desired is finite and delimited—the inhabitants of a town, the machines in a factory, the fish in a lake. In some cases it may seem feasible to obtain the information by taking a complete enumeration or census of the aggregate. Administrators accustomed to dealing with censuses were at first inclined to be suspicious of samples and reluctant to use them in place of censuses. Although this attitude no longer persists, it may be well to list the principal advantages of sampling as compared with complete enumeration.

Reduced Cost

If data are secured from only a small fraction of the aggregate, expenditures are smaller than if a complete census is attempted. With large populations, results accurate enough to be useful can be obtained from samples that represent only a small fraction of the population. In the United States the most important recurrent surveys taken by the government use samples of around 105,000 persons, or about one person in 1240. Surveys used to provide facts bearing on sales and advertising policy in market research may employ samples of only a few thousand.

Greater Speed

For the same reason, the data can be collected and summarized more quickly with a sample than with a complete count. This is a vital consideration when the information is urgently needed.

Greater Scope

In certain types of inquiry highly trained personnel or specialized equipment, limited in availability, must be used to obtain the data. A complete census is impracticable: the choice lies between obtaining the information by sampling or not at all. Thus surveys that rely on sampling have more scope and flexibility regarding the types of information that can be obtained. On the other hand, if accurate information is wanted for many subdivisions of the population, the size of sample needed to do the job is sometimes so large that a complete enumeration offers the best solution.

Greater Accuracy

Because personnel of higher quality can be employed and given intensive training and because more careful supervision of the field work and processing of results becomes feasible when the volume of work is reduced, a sample may produce more accurate results than the kind of complete enumeration that can be taken.

1.2 SOME USES OF SAMPLE SURVEYS

To an observer of developments in sampling over the last 25 years the most striking feature is the rapid increase in the number and types of surveys taken by sampling. The Statistical Office of the United Nations publishes reports from time to time on "Sample Surveys of Current Interest" conducted by member countries. The 1968 report lists surveys from 46 countries. Many of these surveys seek information of obvious importance to national planning on topics such as agricultural production and land use, unemployment and the size of the labor force, industrial production, wholesale and retail prices, health status of the people, and family incomes and expenditures. But more specialized inquiries can also be found: for example, annual leave arrangements (Australia), causes of divorce (Hungary), rural debt and investment (India), household water consumption (Israel), radio listening (Malaysia), holiday spending (Netherlands), age structure of cows (Czechoslovakia), and job vacancies (United States).

Sampling has come to play a prominent part in national decennial censuses. In the United States a 5% sample was introduced into the 1940 Census by asking extra questions about occupation, parentage, fertility, and the like, of those persons whose names fell on two of the 40 lines on each page of the schedule. The use of sampling was greatly extended in 1950. From a 20% sample (every fifth line) information was obtained on items such as income, years in school, migration, and service in armed forces. By taking every sixth person in the 20% sample, a further sample of $3\frac{1}{3}$% was created to give information on marriage and fertility. A series of questions dealing with the condition and age of housing was split into five sets, each set being filled in at every fifth house. Sampling was also employed to speed up publication of the results. Preliminary tabulations for many important items, made on a sample basis, appeared more than a year and half before the final reports.

This process continued in the 1960 and 1970 Censuses. Except for certain basic information required from every person for constitutional or legal reasons, the whole census was shifted to a sample basis. This change, accompanied by greatly increased mechanization, resulted in much earlier publication and substantial savings.

In addition to their use in censuses, continuing samples are employed by government bureaus to obtain current information. In the United States, examples are the Current Population Survey, which provides monthly data on the size and composition of the labor force and on the number of unemployed, the National Health Survey, and the series of samples needed for the calculation of the monthly Consumer Price Index.

On a smaller scale, local governments—city, state, and county—are making increased use of sample surveys to obtain information needed for future planning and for meeting pressing problems. In the United States most large cities have commercial agencies that make a business of planning and conducting sample surveys for clients.

Market research is heavily dependent on the sampling approach. Estimates of the sizes of television and radio audiences for different programs and of newspaper and magazine readership (including the advertisements) are kept continually under scrutiny. Manufacturers and retailers want to know the reactions of people to new products or new methods of packaging, their complaints about old products, and their reasons for preferring one product to another.

Business and industry have many uses for sampling in attempting to increase the efficiency of their internal operations. The important areas of quality control and acceptance sampling are outside the scope of this book. But, obviously, decisions taken with respect to level or change of quality or to acceptance or rejection of batches are well grounded only if results obtained from the sample data are valid (within a reasonable tolerance) for the whole batch. The sampling of records of business transactions (accounts, payrolls, stock, personnel)—usually much easier than the sampling of people—can provide serviceable information quickly and economically. Savings can also be made through sampling in the estimation of inventories, in studies of the condition and length of the life of equipment, in the inspection of the accuracy and rate of output of clerical work, in investigating how key personnel distribute their working time among different tasks, and, more generally, in the field known as operations research. The books by Deming (1960) and Slonim (1960) contain many interesting examples showing the range of applications of the sampling method in business.

Opinion, attitude, and election polls, which did much to bring the technique of sampling before the public eye, continue to be a popular feature of newspapers. In the field of accounting and auditing, which has employed sampling for many years, a new interest has arisen in adapting modern developments to the particular problems of this field. Thus, Neter (1972) describes how airlines and railways save money by using samples of records to apportion income from freight and passenger service. The status of sample surveys as evidence in lawsuits has also been subject to lively discussion. Gallup (1972) has noted the major contribution that sample surveys can make to the process of informed government by determining quickly people's opinions on proposed or new government programs and has stressed their role as sources of information in social science.

Sample surveys can be classified broadly into two types—*descriptive* and *analytical.* In a descriptive survey the objective is simply to obtain certain information about large groups: for example, the numbers of men, women, and children who view a television program. In an analytical survey, comparisons are made between different subgroups of the population, in order to discover whether differences exist among them and to form or to verify hypotheses about the reasons for these differences. The Indianapolis fertility survey, for instance, was an attempt to determine the extent to which married couples plan the number and spacing of children, the husband's and wife's attitudes toward this planning, the reasons for these attitudes, and the degree of success attained (Kiser and Whelpton, 1953).

The distinction between descriptive and analytical surveys is not, of course, clear-cut. Many surveys provide data that serve both purposes. Along with the rise in the number of descriptive surveys, there has, however, been a noticeable increase in surveys taken primarily for analytical purposes, particularly in the study of human behavior and health. Surveys of the teeth of school children before and after fluoridation of water, of the death rates and causes of death of people who smoke different amounts, and the huge study of the effectiveness of the Salk polio vaccine may be cited. The study by Coleman (1966) on equality of educational opportunity, conducted on a national sample of schools, contained many regression analyses that estimated the relative contributions of school characteristics, home background, and the child's outlook to variations in exam results.

1.3 THE PRINCIPAL STEPS IN A SAMPLE SURVEY

As a preliminary to a discussion of the role that theory plays in a sample survey, it is useful to describe briefly the steps involved in the planning and execution of a survey. Surveys vary greatly in their complexity. To take a sample from 5000 cards, neatly arranged and numbered in a file, is an easy task. It is another matter to sample the inhabitants of a region where transport is by water through the forests, where there are no maps, where 15 different dialects are spoken, and where the inhabitants are very suspicious of an inquisitve stranger. Problems that are baffling in one survey may be trivial or nonexistent in another.

The principal steps in a survey are grouped somewhat arbitrarily under 11 headings.

Objectives of the Survey

A lucid statement of the objectives is most helpful. Without this, it is easy in a complex survey to forget the objectives when engrossed in the details of planning, and to make decisions that are at variance with the objectives.

Population to be Sampled

The word *population* is used to denote the aggregate from which the sample is chosen. The definition of the population may present no problem, as when sampling a batch of electric light bulbs in order to estimate the average length of life of a bulb. In sampling a population of farms, on the other hand, rules must be set up to define a farm, and borderline cases arise. These rules must be usable in practice: the enumerator must be able to decide in the field, without much hesitation, whether or not a doubtful case belongs to the population.

The population to be sampled (the *sampled* population) should coincide with the population about which information is wanted (the *target* population). Sometimes, for reasons of practicability or convenience, the sampled population is more restricted than the target population. If so, it should be remembered that conclusions drawn from the sample apply to the sampled population. Judgment about the extent to which these conclusions will also apply to the target population must depend on other sources of information. Any supplementary information that can be gathered about the nature of the differences between sampled and target population may be helpful.

Data to be Collected

It is well to verify that all the data are relevant to the purposes of the survey and that no essential data are omitted. There is frequently a tendency, particularly with human populations, to ask too many questions, some of which are never subsequently analyzed. An overlong questionnaire lowers the quality of the answers to important as well as unimportant questions.

Degree of Precision Desired

The results of sample surveys are always subject to some uncertainty because only part of the population has been measured and because of errors of measurement. This uncertainty can be reduced by taking larger samples and by using superior instruments of measurement. But this usually costs time and money. Consequently, the specification of the degree of precision wanted in the results is an important step. This step is the responsibility of the person who is going to use the data. It may present difficulties, since many administrators are unaccustomed to thinking in terms of the amount of error that can be tolerated in estimates, consistent with making good decisions. The statistician can often help at this stage.

Methods of Measurement

There may be a choice of measuring instrument and of method of approach to the population. Data about a person's state of health may be obtained from statements that he or she makes or from a medical examination. The survey may employ a self-administered questionnaire, an interviewer who reads a standard set of questions with no discretion, or an interviewing process that allows much latitude in the form and ordering of the questions. The approach may be by mail, by telephone, by personal visit, or by a combination of the three. Much study has been made of interviewing methods and problems (see, e.g., Hyman, 1954 and Payne, 1951).

A major part of the preliminary work is the construction of record forms on which the questions and answers are to be entered. With simple questionnaires, the answers can sometimes be precoded—that is, entered in a manner in which they can be routinely transferred to mechanical equipment. In fact, for the construction of good record forms, it is necessary to visualize the structure of the final summary tables that will be used for drawing conclusions.

The Frame

Before selecting the sample, the population must be divided into parts that are called *sampling units*, or *units*. These units must cover the whole of the population and they must not overlap, in the sense that every element in the population belongs to one and only one unit. Sometimes the appropriate unit is obvious, as in a population of light bulbs, in which the unit is the single bulb. Sometimes there is a choice of unit. In sampling the people in a town, the unit might be an individual person, the members of a family, or all persons living in the same city block. In sampling an agricultural crop, the unit might be a field, a farm, or an area of land whose shape and dimensions are at our disposal.

The construction of this list of sampling units, called a *frame*, is often one of the major practical problems. From bitter experience, samplers have acquired a critical attitude toward lists that have been routinely collected for some purpose. Despite assurances to the contrary, such lists are often found to be incomplete, or partly illegible, or to contain an unknown amount of duplication. A good frame may be hard to come by when the population is specialized, as in populations of bookmakers or of people who keep turkeys. Jessen (1955) presents an interesting method of constructing a frame from the branches of a fruit tree.

Selection of the Sample

There is now a variety of plans by which the sample may be selected. For each plan that is considered, rough estimates of the size of sample can be made from a knowledge of the degree of precision desired. The relative costs and time involved for each plan are also compared before making a decision.

The Pretest

It has been found useful to try out the questionnaire and the field methods on a small scale. This nearly always results in improvements in the questionnaire and may reveal other troubles that will be serious on a large scale, for example, that the cost will be much greater than expected.

Organization of the Field Work

In extensive surveys many problems of business administration are met. The personnel must receive training in the purpose of the survey and in the methods of measurement to be employed and must be adequately supervised in their work. A procedure for *early* checking of the quality of the returns is invaluable. Plans must be made for handling nonresponse, that is, the failure of the enumerator to obtain information from certain of the units in the sample.

Summary and Analysis of the Data

The first step is to edit the completed questionnaires, in the hope of amending recording errors, or at least of deleting data that are obviously erroneous. Decisions about computing procedure are needed in cases in which answers to certain questions were omitted by some respondents or were deleted in the editing process. Thereafter, the computations that lead to the estimates are performed. Different methods of estimation may be available for the same data.

In the presentation of results it is good practice to report the amount of error to be expected in the most important estimates. One of the advantages of probability sampling is that such statements can be made, although they have to be severely qualified if the amount of nonresponse is substantial.

Information Gained for Future Surveys

The more information we have initially about a population, the easier it is to devise a sample that will give accurate estimates. Any completed sample is potentially a guide to improved future sampling, in the data that it supplies about the means, standard deviations, and nature of the variability of the principal measurements and about the costs involved in getting the data. Sampling practice advances more rapidly when provisions are made to assemble and record information of this type.

There is another important respect in which any completed sample facilitates future samples. Things never go exactly as planned in a complex survey. The alert sampler learns to recognize mistakes in execution and to see that they do not occur in future surveys. ⸣

1.4 THE ROLE OF SAMPLING THEORY

This list of the steps in a sample survey has been given in order to emphasize that sampling is a practical business, which calls for several different types of skill. In some of the steps—the definition of the population, the determination of the data to be collected and of the methods of measurement, and the organization of the field work—sampling theory plays at most a minor role. Although these topics are not discussed further in this book, their importance should be realized. Sampling demands attention to all phases of the activity: poor work in one phase may ruin a survey in which everything else is done well.

The purpose of sampling theory is to make sampling more efficient. It attempts to develop methods of sample selection and of estimation that provide, at the lowest possible cost, estimates that are precise enough for our purpose. This principle of specified precision at minimum cost recurs repeatedly in the presentation of theory.

In order to apply this principle, we must be able to predict, for any sampling procedure that is under consideration, the precision and the cost to be expected. So far as precision is concerned, we cannot foretell exactly how large an error will be present in an estimate in any specific situation, for this would require a knowledge of the true value for the population. Instead, the precision of a sampling procedure is judged by examining the frequency distribution generated for the estimate if the procedure is applied again and again to the same population. This is, of course, the standard technique by which precision is judged in statistical theory.

A further simplification is introduced. With samples of the sizes that are common in practice, there is often good reason to suppose that the sample estimates are approximately normally distributed. With a normally distributed estimate, the whole shape of the frequency distribution is known if we know the mean and the standard deviation (or the variance). A considerable part of sample survey theory is concerned with finding formulas for these means and variances.

There are two differences between standard sample survey theory and the classical sampling theory as taught in books on mathematical statistics. In classical theory the measurements that are made on the sampling units in the population are usually assumed to follow a frequency distribution, for example, the normal distribution, of known mathematical form apart from certain population parameters such as the mean and variance whose values have to be estimated from the sample data. In sample survey theory, on the other hand, the attitude has been to assume only very limited information about this frequency distribution. In particular, its mathematical form is not assumed known, so that the approach might be described as model-free or distribution-free. This attitude is natural for large surveys in which many different measurements with differing frequency distributions are made on the units. In surveys in which only a few measurements per unit are made, studies of their frequency distributions may justify the assumption of known mathematical forms, permitting the results from classical theory to be applied.

A second difference is that the populations in survey work contain a *finite* number of units. Results are slightly more complicated when sampling is from a finite instead of an infinite population. For practical purposes these differences in results for finite and infinite populations can often be ignored. Cases in which this is not so will be pointed out.

1.5 PROBABILITY SAMPLING

The sampling procedures considered in this book have the following mathematical properties in common.

1. We are able to define the set of distinct samples, $S_1, S_2, \cdots, S_v$, which the procedure is capable of selecting if applied to a specific population. This means that we can say precisely what sampling units belong to S_1, to S_2, and so on. For example, suppose that the population contains six units, numbered 1 to 6. A common procedure for choosing a sample of size 2 gives three possible candidates—$S_1 \sim (1, 4); S_2 \sim (2, 5); S_3 \sim (3, 6)$. Note that not all possible samples of size 2 need be included.

2. Each possible sample S_i has assigned to it a known probability of selection π_i.

3. We select one of the S_i by a random process in which each S_i receives its appropriate probability π_i of being selected. In the example we might assign equal probabilities to the three samples. Then the draw itself can be made by choosing a random number between 1 and 3. If this number is j, S_j is the sample that is taken.

4. The method for computing the estimate from the sample must be stated and must lead to a unique estimate for any specific sample. We may declare, for example, that the estimate is to be the average of the measurements on the individual units in the sample.

For any sampling procedure that satisfies these properties, we are in a position to calculate the frequency distribution of the estimates it generates if repeatedly applied to the same population. We know how frequently any particular sample S_i will be selected, and we know how to calculate the estimate from the data in S_i. It is clear, therefore, that a sampling theory can be developed for any procedure of this type, although the details of the development may be intricate. The term *probability sampling* refers to a method of this type.

In practice we seldom draw a probability sample by writing down the S_i and π_i as outlined above. This is intolerably laborious with a large population, where a sampling procedure may produce billions of possible samples. The draw is most commonly made by specifying probabilities of inclusion for the individual units and drawing units, one by one or in groups until the sample of desired size and type is constructed. For the purposes of a theory it is sufficient to know that we could write down the S_i and π_i if we wanted to and had unlimited time.

1.6 ALTERNATIVES TO PROBABILITY SAMPLING

The following are some common types of nonprobability sampling.

1. The sample is restricted to a part of the population that is readily accessible. A sample of coal from an open wagon may be taken from the top 6 to 9 in.

2. The sample is selected haphazardly. In picking 10 rabbits from a large cage in a laboratory, the investigator may take those that his hands rest on, without conscious planning.

3. With a small but heterogeneous population, the sampler inspects the whole of it and selects a small sample of "typical" units—that is, units that are close to his impression of the average of the population.

4. The sample consists essentially of volunteers, in studies in which the measuring process is unpleasant or troublesome to the person being measured.

Under the right conditions, any of these methods can give useful results. They are not, however, amenable to the development of a sampling theory that is model-free, since no element of random selection is involved. About the only way of examining how good one of them may be is to find a situation in which the results are known, either for the whole population or for a probability sample, and make comparisons. Even if a method appears to do well in one such comparison, this does not guarantee that it will do well under different circumstances.

In this connection, some of the earliest uses of sampling by country and city governments from 1850 onward were intended to save money in making estimates from the results of a Census. For the most important items in the Census, the country or city totals were calculated from the complete Census data. For other items a sample of say 15 or 25% of the Census returns was selected in order to lighten the work of estimating country or city totals for these items. Two rival methods of sample selection came into use. One, called *random selection*, was an application of probability sampling in which each unit in the population (e.g., each Census return) had an equal chance of being included in the sample. For this method it was realized that by use of sampling theory and the normal distribution, as noted previously, the sampler could predict approximately from the sample data the amount of error to be expected in the estimates made from the sample. Moreover, for the most important items for which complete Census data were available, he could check to some extent the accuracy of the predictions.

The other method was *purposive selection*. This was not specifically defined in detail but usually had two common features. The sampling unit consisted of groups of returns, often relatively large groups. For example, in the 1921 Italian

Census the country consisted of 8354 communes grouped into 214 districts. In drawing a 14% sample, the Italian statisticians Gini and Galvani selected 29 districts purposively rather than 1250 communes. Second, the 29 districts were chosen so that the sample gave accurate estimates for 7 important control variables for which results were known for the whole country. The hope was that this sample would give good estimates for other variables highly correlated with the control variables.

In the 1920s the International Statistical Institute appointed a commission to report on the advantages and disadvantages of the two methods. The report, by Jensen (1926), seemed on balance to favor purposive selection. However, purposive selection was abandoned relatively soon as a method of sampling for obtaining national estimates in surveys in which many items were measured. It lacked the flexibility that later developments of probability sampling produced, it was unable to predict from the sample the accuracy to be expected in the estimates, and it used sampling units that were too large. Gini and Galvani concluded that the probability method called stratified random sampling (Chapter 5), with the commune as a sampling unit, would have given better results than their method.

1.7 USE OF THE NORMAL DISTRIBUTION

It is sometimes useful to employ the word *estimator* to denote the rule by which an estimate of some population characteristic μ is calculated from the sample results, the word *estimate* being applied to the value obtained from a specific sample. An estimator $\hat{\mu}$ of μ given by a sampling plan is called unbiased if the mean value of $\hat{\mu}$, taken over all possible samples provided by the plan, is equal to μ. In the notation of section 1.5, this condition may be written

$$E(\hat{\mu}) = \sum_{i=1}^{v} \pi_i \hat{\mu}_i = \mu$$

where $\hat{\mu}_i$ is the estimate given by the ith sample. The symbol E, which stands for "the expected value of," is used frequently.

As mentioned in section 1.4, the samples in surveys are often large enough so that estimates made from them are approximately normally distributed. Furthermore, with probability sampling, we have formulas that give the mean and variance of the estimates. Suppose that we have taken a sample by a procedure known to give an unbiased estimator and have computed the sample estimate $\hat{\mu}$ and its standard deviation $\sigma_{\hat{\mu}}$ (often called, alternatively, its standard error). How good is the estimate? We cannot know the exact value of the error of estimate $(\hat{\mu} - \mu)$ but, from the properties of the normal curve, the chances are

0.32 (about 1 in 3) that the absolute error $|\hat{\mu} - \mu|$ exceeds $\sigma_{\hat{\mu}}$
0.05 (1 in 20) that the absolute error $|\hat{\mu} - \mu|$ exceeds $1.96\sigma_{\hat{\mu}} \doteq 2\sigma_{\hat{\mu}}$
0.01 (1 in 100) that the absolute error $|\hat{\mu} - \mu|$ exceeds $2.58\sigma_{\hat{\mu}}$

For example, if a probability sample of the records of batteries in routine use in a large factory shows an average life $\hat{\mu} = 394$ days, with a standard error $\sigma_{\hat{\mu}} = 4.6$ days, the chances are 99 in 100 that the average life in the population of batteries lies between

$$\hat{\mu} = 394 - (2.58)(4.6) = 382 \text{ days}$$

and

$$\hat{\mu}_U = 394 + (2.58)(4.6) = 406 \text{ days}$$

The limits, 382 days and 406 days, are called lower and upper *confidence limits*. With a single estimate from a single survey, the statement "μ lies between 382 and 406 days" is not certain to be correct. The "99% confidence" figure implies that if the same sampling plan were used many times in a population, a confidence statement being made from each sample, about 99% of these statements would be correct and 1% wrong. When sampling is being introduced into an operation in which complete censuses have previously been used, a demonstration of this property is sometimes made by drawing repeated samples of the type proposed from a population for which complete records exist, so that μ is known (see, e.g., Trueblood and Cyert, 1957). The practical verification that approximately the stated proportion of statements is correct does much to educate and reassure administrators about the nature of sampling. Similarly, when a single sample is taken from each of a series of different populations, about 95% of the 95% confidence statements are correct.

The preceding discussion assumes that $\sigma_{\hat{\mu}}$, as computed from the sample, is known exactly. Actually, $\sigma_{\hat{\mu}}$, like $\hat{\mu}$, is subject to a sampling error. With a normally distributed variable, tables of Student's t distribution are used instead of the normal tables to calculate confidence limits for μ when the sample is small. Replacement of the normal table by the t table makes almost no difference if the number of degrees of freedom in $\sigma_{\hat{\mu}}$ exceeds 50. With certain types of stratified sampling and with the method of replicated sampling (section 11.19) the degrees of freedom are small and the t table is needed.

1.8 BIAS AND ITS EFFECTS

In sample survey theory it is necessary to consider biased estimators for two reasons.

1. In some of the most common problems, particularly in the estimation of ratios, estimators that are otherwise convenient and suitable are found to be biased.

2. Even with estimators that are unbiased in probability sampling, errors of measurement and nonreponse may produce biases in the numbers that we are able to compute from the data. This happens, for instance, if the persons who refuse to be interviewed are almost all opposed to some expenditure of public funds, whereas those who are interviewed are split evenly for and against.

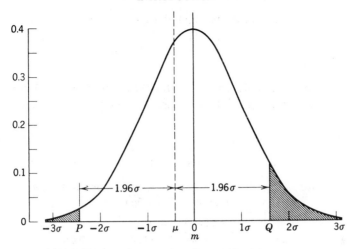

Fig. 1.1 Effect of bias on errors of estimation.

To examine the effect of bias, suppose that the estimate $\hat{\mu}$ is normally distributed about a mean m that is a distance B from the true population value μ, as shown in Fig. 1.1. The amount of bias is $B = m - \mu$. Suppose that we do not know that any bias is present. We compute the standard deviation σ of the frequency distribution of the estimate—this will, of course, be the standard deviation about the mean m of the distribution, not about the true mean μ. We are using σ in place of $\sigma_{\hat{\mu}}$. As a statement about the accuracy of the estimate, we declare that the probability is 0.05 that the estimate $\hat{\mu}$ is in error by more than 1.96σ.

We will consider how the presence of bias distorts this probability. To do this, we calculate the true probability that the estimate is in error by more than 1.96σ, where error is measured from the true mean μ. The two tails of the distribution must be examined separately. For the upper tail, the probability of an error of more than $+1.96\sigma$ is the shaded area above Q in Fig. 1.1. This area is given by

$$\frac{1}{\sigma\sqrt{2\pi}} \int_{\mu+1.96\sigma}^{\infty} e^{-(\hat{\mu}-m)^2/2\sigma^2} \, d\hat{\mu}$$

Put $\hat{\mu} - m = \sigma t$. The lower limit of the range of integration for t is

$$\frac{\mu - m}{\sigma} + 1.96 = 1.96 - \frac{B}{\sigma}$$

Thus the area is

$$\frac{1}{\sqrt{2\pi}} \int_{1.96-(B/\sigma)}^{\infty} e^{-t^2/2} \, dt$$

Similarly, the lower tail, that is, the shaded area below P, has an area

$$\frac{1}{\sqrt{2\pi}} \int_{-\infty}^{-1.96-(B/\sigma)} e^{-t^2/2} \, dt$$

From the form of the integrals it is clear that the amount of disturbance depends solely on the ratio of the bias to the standard deviation. The results are shown in Table 1.1.

TABLE 1.1

EFFECT OF A BIAS B ON THE PROBABILITY OF AN ERROR
GREATER THAN 1.96σ

| B/σ | Probability of Error | | Total |
	$< -1.96\sigma$	$>1.96\sigma$	
0.02	0.0238	0.0262	0.0500
0.04	0.0228	0.0274	0.0502
0.06	0.0217	0.0287	0.0504
0.08	0.0207	0.0301	0.0508
0.10	0.0197	0.0314	0.0511
0.20	0.0154	0.0392	0.0546
0.40	0.0091	0.0594	0.0685
0.60	0.0052	0.0869	0.0921
0.80	0.0029	0.1230	0.1259
1.00	0.0015	0.1685	0.1700
1.50	0.0003	0.3228	0.3231

For the total probability of an error of more than 1.96σ, the bias has little effect provided that it is less than one tenth of the standard deviation. At this point the total probability is 0.0511 instead of the 0.05 that we think it is. As the bias increases further, the disturbance becomes more serious. At $B = \sigma$, the total probability of error is 0.17, more than three times the presumed value.

The two tails are affected differently. With a positive bias, as in this example, the probability of an underestimate by more than 1.96σ shrinks rapidly from the presumed 0.025 to become negligible when $B = \sigma$. The probability of the corresponding overestimate mounts steadily. In most applications the total error is the primary interest, but occasionally we are particularly interested in errors in one direction.

As a working rule, the effect of bias on the accuracy of an estimate is negligible if the bias is less than one tenth of the standard deviation of the estimate. If we have a biased method of estimation for which $B/\sigma < 0.1$, where B is the absolute value of the bias, it can be claimed that the bias is not an appreciable disadvantage of the method. Even with $B/\sigma = 0.2$, the disturbance in the probability of the total error is modest.

In using these results, a distinction must be made between the two sources of bias mentioned at the beginning of this section. With biases of the type that arise in estimating ratios, an upper limit to the ratio B/σ can be found mathematically. If the sample is large enough, we can be confident that B/σ will not exceed 0.1. With biases caused by errors of measurement or nonresponse, on the other hand, it is usually impossible to find a guaranteed upper limit to B/σ that is small. This troublesome problem is discussed in Chapter 13.

1.9 THE MEAN SQUARE ERROR

In order to compare a biased estimator with an unbiased estimator, or two estimators with different amounts of bias, a useful criterion is the mean square error (MSE) of the estimate, measured from the population value that is being estimated. Formally,

$$\begin{aligned}
\text{MSE}(\hat{\mu}) = E(\hat{\mu} - \mu)^2 &= E[(\hat{\mu} - m) + (m - \mu)]^2 \\
&= E(\hat{\mu} - m)^2 + 2(m - \mu)E(\hat{\mu} - m) + (m - \mu)^2 \\
&= (\text{variance of } \hat{\mu}) + (\text{bias})^2
\end{aligned}$$

the cross-product term vanishing since $E(\hat{\mu} - m) = 0$.

Use of the MSE as a criterion of the accuracy of an estimator amounts to regarding two estimates that have the same MSE as equivalent. This is not strictly correct because the frequency distributions of errors $(\hat{\mu} - \mu)$ of different sizes will not be the same for the two estimates if they have different amounts of bias. It has been shown, however, by Hansen, Hurwitz, and Madow (1953) that if B/σ is less than about one half, the two frequency distributions are almost identical in regard to *absolute* errors $|\hat{\mu} - \mu|$ of different sizes. Table 1.2 illustrates this result.

TABLE 1.2

PROBABILITY OF AN ABSOLUTE ERROR $\geq 1\sqrt{\text{MSE}}$,
$1.96\sqrt{\text{MSE}}$ AND $2.576\sqrt{\text{MSE}}$

Probability

B/σ	$1\sqrt{\text{MSE}}$	$1.96\sqrt{\text{MSE}}$	$2.576\sqrt{\text{MSE}}$
0	0.317	0.0500	0.0100
0.2	0.317	0.0499	0.0100
0.4	0.319	0.0495	0.0095
0.6	0.324	0.0479	0.0083

Even at $B/\sigma = 0.6$, the changes in the probabilities as compared with those for $B/\sigma = 0$ are slight.

Because of the difficulty of ensuring that no unsuspected bias enters into estimates, we will usually speak of the *precision* of an estimate instead of its *accuracy*. Accuracy refers to the size of deviations from the true mean μ, whereas precision refers to the size of deviations from the mean m obtained by repeated application of the sampling procedure.

EXERCISES

1.1   Suppose that you were using sampling to estimate the total number of words in a book that contains illustrations.

(*a*) Is there any problem of definition of the population? (*b*) What are the pros and cons of (1) the page, (2) the line, as a sampling unit?

1.2   A sample is to be taken from a list of names that are on cards (one name to a card) numbered consecutively in a file. Each name is to have an equal chance of being drawn in the sample. What problems arise in the following common situations? (*a*) Some of the names do not belong to the target population, although this fact cannot be verified for any name until it has been drawn. (*b*) Some names appear on more than one card. All cards with the same name bear consecutive numbers and therefore appear together in the file. (*c*) Some names appear on more than one card, but cards bearing the same name may be scattered anywhere about the file.

1.3   The problem of finding a frame that is complete and enables the sample to be drawn is often an obstacle. What kinds of frames might be tried for the following surveys? Have the frames any serious weaknesses? (*a*) A survey of stores that sell luggage in a large city. (*b*) A survey of the kinds of articles left behind in subways or buses. (*c*) A survey of persons bitten by snakes during the last year. (*d*) A survey to estimate the number of hours per week spent by family members in watching television.

1.4   A city directory, 4 years old, lists the addresses in order along each street, and gives the names of the persons living at each address. For a current interview survey of the people in the city, what are the deficiencies of this frame? Can they be remedied by the interviewers during the course of the field work? In using the directory, would you draw a list of addresses (dwelling places) or a list of persons?

1.5   In estimating by sampling the actual value of the small items in the inventory of a large firm, the actual and the book value were recorded for each item in the sample. For the total sample, the ratio of actual to book value was 1.021, this estimate being approximately normally distributed with a standard error of 0.0082. If the book value of the inventory is $80,000, compute 95% confidence limits for the actual value.

1.6   Frequently data must be treated as a sample, although at first sight they appear to be a complete enumeration. A proprietor of a parking lot finds that business is poor on Sunday mornings. After 26 Sundays in operation, his average receipts per Sunday morning are exactly $10. The standard error of this figure, computed from week-to-week variations, is $1.2. The attendant costs $7 each Sunday. The proprietor is willing to keep the lot open at this time if his expected future profit is $5 per Sunday morning. What is the confidence probability that the long-term profit rate will be at least $5? What assumption must be made in order to answer this question?

1.7   In Table 1.2, what happens to the probability of exceeding $1\sqrt{\text{MSE}}$. $1.96\sqrt{\text{MSE}}$, and $2.576\sqrt{\text{MSE}}$ when B/σ tends to infinity, that is, when the MSE is due entirely to bias? Do your results agree with the directions of the changes noted in Table 1.2 as B/σ moves from 0 to 0.6?

1.8   When it is necessary to compare two estimates that have different frequency distributions of errors $(\hat{\mu} - \mu)$, it is occasionally possible, in specialized problems, to compute the cost or loss that will result from an error $(\hat{\mu} - \mu)$ of any given size. The estimate that gives the smaller expected loss is preferred, other things being equal. Show that if the loss is a quadratic function $\lambda(\hat{\mu} - \mu)^2$ of the error, we should choose the estimate with the smaller mean square error.

CHAPTER 2

Simple Random Sampling

2.1 SIMPLE RANDOM SAMPLING

Simple random sampling is a method of selecting n units out of the N such that every one of the $_NC_n$ distinct samples has an equal chance of being drawn. In practice a simple random sample is drawn unit by unit. The units in the population are numbered from 1 to N. A series of random numbers between 1 and N is then drawn, either by means of a table of random numbers or by means of a computer program that produces such a table. At any draw the process used must give an equal chance of selection to any number in the population *not already drawn*. The units that bear these n numbers constitute the sample.

It is easily verified that all $_NC_n$ distinct samples have an equal chance of being selected by this method. Consider one distinct sample, that is, one set of n specified units. At the first draw the probability that some one of the n specified units is selected is n/N. At the second draw the probability that some one of the remaining $(n-1)$ specified units is drawn is $(n-1)/(N-1)$, and so on. Hence the the probability that all n specified units are selected in n draws is

$$\frac{n}{N} \cdot \frac{(n-1)}{(N-1)} \cdot \frac{(n-2)}{(N-2)} \cdots \frac{1}{(N-n+1)} = \frac{n!(N-n)!}{(N)!} = \frac{1}{_NC_n} \qquad (2.1)$$

Since a number that has been drawn is removed from the population for all subsequent draws, this method is also called random sampling *without replacement*. Random sampling *with replacement* is entirely feasible: at any draw, all N members of the population are given an equal chance of being drawn, no matter how often they have already been drawn. The formulas for the variances and estimated variances of estimates made from the sample are often simpler when sampling is with replacement than when it is without replacement. For this reason sampling with replacement is sometimes used in the more complex sampling plans, although at first sight there seems little point in having the same unit two or more times in the sample.

2.2 SELECTION OF A SIMPLE RANDOM SAMPLE

Tables of random numbers are tables of the digits $0, 1, 2, \ldots 9$, each digit having an equal chance of selection at any draw. Among the larger tables are those published by the Rand Corporation (1955)—1 million digits—and by Kendall and Smith (1938)—100,000 digits. Numerous tables are available, many in standard statistical texts. Table 2.1 shows 1000 random digits for illustration, from Snedecor and Cochran (1967).

TABLE 2.1

ONE THOUSAND RANDOM DIGITS

	00–04	05–09	10–14	15–19	20–24	25–29	30–34	35–39	40–44	45–49
00	54463	22662	65905	70639	79365	67382	29085	69831	47058	08186
01	15389	85205	18850	39226	42249	90669	96325	23248	60933	26927
02	85941	40756	82414	02015	13858	78030	16269	65978	01385	15345
03	61149	69440	11286	88218	58925	03638	52862	62733	33451	77455
04	05219	81619	10651	67079	92511	59888	84502	72095	83463	75577
05	41417	98326	87719	92294	46614	50948	64886	20002	97365	30976
06	28357	94070	20652	35774	16249	75019	21145	05217	47286	76305
07	17783	00015	10806	83091	91530	36466	39981	62481	49177	75779
08	40950	84820	29881	85966	62800	70326	84740	62660	77379	90279
09	82995	64157	66164	41180	10089	41757	78258	96488	88629	37231
10	96754	17676	55659	44105	47361	34833	86679	23930	53249	27083
11	34357	88040	53364	71726	45690	66334	60332	22554	90600	71113
12	06318	37403	49927	57715	50423	67372	63116	48888	21505	80182
13	62111	52820	07243	79931	89292	84767	85693	73947	22278	11551
14	47534	09243	67879	00544	23410	12740	02540	54440	32949	13491
15	98614	75993	84460	62846	59844	14922	48730	73443	48167	34770
16	24856	03648	44898	09351	98795	18644	39765	71058	90368	44104
17	96887	12479	80621	66223	86085	78285	02432	53342	42846	94771
18	90801	21472	42815	77408	37390	76766	52615	32141	30268	18106
19	55165	77312	83666	36028	28420	70219	81369	41943	47366	41067

In using these tables to select a simple random sample, the first step is to number the units in the population from 1 to N. If the first digit of N is a number between 5 and 9, the following method of selection is adequate. Suppose $N = 528$, and we want $n = 10$. Select three columns from Table 2.1, say columns 25 to 27. Go down the three columns, selecting the first 10 *distinct* numbers between 001 and 528. These are 36, 509, 364, 417, 348, 127, 149, 186, 290, and 162. For the last two numbers we jumped to columns 30 to 32. In repeated selections it is advisable to vary the starting point in the table.

The disadvantage of this method is that the three-digit numbers 000 and 529 to 999 are not used, although skipping numbers does not waste much time. When the first digit of N is less than 5, some may still prefer this method if n is small and a large table of random digits is available.

With $N = 128$, for example, a second method that involves less rejection and is easily applied is as follows. In a series of three-digit numbers, subtract 200 from all numbers between 201 and 400, 400 from all numbers between 401 and 600, 600 from all numbers between 601 and 800, 800 from all numbers between 801 and 999 and, of course, 000 from all numbers between 000 and 200. All remainders greater than 129 and the numbers 000, 200, and so forth, are rejected. Using columns 05 to 07 in Table 2.1, we get 26, 52, 7, 94, 16, 48, 41, 80, 128, and 92, the draw requiring 15 three-digit numbers for $n = 10$. In this sample the rejection rate $5/15 = 33\%$ is close to the probability of rejection $72/200 = 36\%$ for this method. In using this method with a number N like 384, note that one subtracts 400 from a number between 401 and 800, but automatically rejects all numbers greater than 800. Subtraction of 800 from numbers between 801 and 999 would give a higher probability of acceptance to remainders between 001 and 199 than to remainders between 200 and 384.

Other methods of sampling are often preferable to simple random sampling on the grounds of convenience or of increased precision. Simple random sampling serves best to introduce sampling theory.

2.3 DEFINITIONS AND NOTATION

In a sample survey we decide on certain properties that we attempt to measure and record for every unit that comes into the sample. These properties of the units are referred to as *characteristics* or, more simply, as *items*.

The values obtained for any specific item in the N units that comprise the population are denoted by $y_1, y_2, \ldots, y_N$. The corresponding values for the units in the sample are denoted by $y_1, y_2, \ldots, y_n$, or, if we wish to refer to a typical sample member, by y_i $(i = 1, 2, \ldots, n)$. Note that the sample will not consist of the *first n* units in the population, except in the instance, usually rare, in which these units happen to be drawn. If this point is kept in mind, my experience has been that no confusion need result.

Capital letters refer to characteristics of the *population* and *lowercase* letters to those of the *sample*. For totals and means we have the following definitions.

	Population	Sample
Total:	$Y = \sum\limits_{}^{N} y_i = y_1 + y_2 + \cdots + y_N$	$\sum\limits_{}^{n} y_i = y_1 + y_2 + \cdots + y_n$
Mean:	$\bar{Y} = \dfrac{y_1 + y_2 + \cdots + y_N}{N} = \dfrac{\sum\limits_{}^{N} y_i}{N}$	$\bar{y} = \dfrac{y_1 + y_2 + \cdots + y_n}{n} = \dfrac{\sum\limits_{}^{n} y_i}{n}$

Although sampling is undertaken for many purposes, interest centers most frequently on four characteristics of the population.

1. Mean $= \bar{Y}$ (e.g., the average number of children per school).
2. Total $= Y$ (e.g., the total number of acres of wheat in a region).
3. Ratio of two totals or means $R = Y/X = \bar{Y}/\bar{X}$ (e.g., ratio of liquid assets to total assets in a group of families).
4. Proportion of units that fall into some defined class (e.g., proportion of people with false teeth).

Estimation of the first three quantities is discussed in this chapter.

The symbol $\hat{\ }$ denotes an estimate of a population characteristic made from a sample. In this chapter only the simplest estimators are considered.

Estimator

Population mean $\bar{Y}$	$\hat{\bar{Y}} = \bar{y} =$ sample mean
Population total Y	$\hat{Y} = N\bar{y} = N\sum_{}^{n} y_i/n$
Population ratio R	$\hat{R} = \bar{y}/\bar{x} = \sum_{}^{n} y_i \Big/ \sum_{}^{n} x_i$

In $\hat{Y}$ the factor N/n by which the sample total is multiplied is sometimes called the *expansion* or *raising* or *inflation* factor. Its inverse n/N, the ratio of the size of the sample to that of the population, is called the *sampling fraction* and is denoted by the letter f.

2.4 PROPERTIES OF THE ESTIMATES

The precision of any estimate made from a sample depends both on the method by which the estimate is calculated from the sample data and on the plan of sampling. To save space we sometimes write of "the precision of the sample mean" or "the precision of simple random sampling," without specifically mentioning the other fundamental factor. This has been done, we hope, only in instances in which it is clear from the context what the missing factor is. When studying any formula that is presented, the reader should make sure that he or she knows the specific method of sampling and method of estimation for which the formula has been established.

In this book a method of estimation is called *consistent* if the estimate becomes exactly equal to the population value when $n = N$, that is, when the sample consists of the whole population. For simple random sampling it is obvious that $\bar{y}$ and $N\bar{y}$ are consistent estimates of the population mean and total, respectively. Consistency is a desirable property of estimators. On the other hand, an inconsistent estimator is not necessarily useless, since it may give satisfactory precision when n is small compared to N. Its utility is likely to be confined to this situation.

Hansen, Hurwitz, and Madow (1953) and Murthy (1967) give an alternative definition of consistency, similar to that in classical statistics. An estimator is consistent if the probability that it is in error by more than any given amount tends to zero as the sample becomes large. Exact statement of this definition requires care with complex survey plans.

As we have seen, a method of estimation is *unbiased* if the average value of the estimate, taken over all possible samples of given size n, is exactly equal to the true population value. If the method is to be unbiased without qualification, this result must hold for any population of finite values y_i and for any n. To investigate whether $\bar{y}$ is unbiased with simple random sampling, we calculate the value of $\bar{y}$ for all $_NC_n$ samples and find the average of the estimates. The symbol E denotes this average over all possible samples.

Theorem 2.1.   The sample mean $\bar{y}$ is an unbiased estimate of $\bar{Y}$.
Proof.   By its definition

$$E\bar{y} = \frac{\sum \bar{y}}{_NC_n} = \frac{\sum (y_1 + y_2 + \cdots + y_n)}{n[N!/n!(N-n)!]} \tag{2.2}$$

where the sum extends over all $_NC_n$ samples. To evaluate this sum, we find out in how many samples any specific value y_i appears. Since there are $(N-1)$ other units available for the rest of the sample and $(n-1)$ other places to fill in the sample, the number of samples containing y_i is

$$_{N-1}C_{n-1} = \frac{(N-1)!}{(n-1)!(N-n)!} \tag{2.3}$$

Hence

$$\sum (y_1 + y_2 + \cdots + y_n) = \frac{(N-1)!}{(n-1)!(N-n)!}(y_1 + y_2 + \cdots + y_N)$$

From (2.2) this gives

$$E\bar{y} = \frac{(N-1)!}{(n-1)!(N-n)!} \frac{n!(N-n)!}{nN!}(y_1 + y_2 + \cdots + y_N)$$

$$= \frac{(y_1 + y_2 + \cdots + y_N)}{N} = \bar{Y} \tag{2.4}$$

Corollary. $\hat{Y} = N\bar{y}$ is an unbiased estimate of the population total Y.

A less cumbersome proof of theorem 2.1 is obtained as follows. Since every unit appears in the same number of samples, it is clear that

$$E(y_1 + y_2 + \cdots + y_n) \text{ must be some multiple of } y_1 + y_2 + \cdots + y_N \tag{2.5}$$

The multiplier must be n/N, since the expression on the left has n terms and that on the right has N terms. This leads to the result.

2.5 VARIANCES OF THE ESTIMATES

The variance of the y_i in a finite population is usually defined as

$$\sigma^2 = \frac{\sum_{1}^{N} (y_i - \bar{Y})^2}{N} \qquad (2.6)$$

As a matter of notation, results are presented in terms of a slightly different expression, in which the divisor $(N-1)$ is used instead of N. We take

$$S^2 = \frac{\sum_{1}^{N} (y_i - \bar{Y})^2}{N-1} \qquad (2.7)$$

This convention has been used by those who approach sampling theory by means of the analysis of variance. Its advantage is that most results take a slightly simpler form. Provided that the same notation is maintained consistently, all results are equivalent in either notation.

We now consider the variance of $\bar{y}$. By this we mean $E(\bar{y} - \bar{Y})^2$ taken over all $_NC_n$ samples.

Theorem 2.2. The variance of the mean $\bar{y}$ from a simple random sample is

$$V(\bar{y}) = E(\bar{y} - \bar{Y})^2 = \frac{S^2}{n} \frac{(N-n)}{N} = \frac{S^2}{n}(1-f) \qquad (2.8)$$

where $f = n/N$ is the sampling fraction.
Proof.

$$n(\bar{y} - \bar{Y}) = (y_1 - \bar{Y}) + (y_2 - \bar{Y}) + \cdots + (y_n - \bar{Y}) \qquad (2.9)$$

By the argument of symmetry used in relation (2.5), it follows that

$$E[(y_1 - \bar{Y})^2 + \cdots + (y_n - \bar{Y})^2] = \frac{n}{N}[(y_1 - \bar{Y})^2 + \cdots + (y_N - \bar{Y})^2] \qquad (2.10)$$

and also that

$$E[(y_1 - \bar{Y})(y_2 - \bar{Y}) + (y_1 - \bar{Y})(y_3 - \bar{Y}) + \cdots + (y_{n-1} - \bar{Y})(y_n - \bar{Y})]$$
$$= \frac{n(n-1)}{N(N-1)}[(y_1 - \bar{Y})(y_2 - \bar{Y}) + (y_1 - \bar{Y})(y_3 - \bar{Y})$$
$$+ \cdots + (y_{N-1} - \bar{Y})(y_N - \bar{Y})] \qquad (2.11)$$

In (2.11) the sums of products extend over all pairs of units in the sample and population, respectively. The sum on the left contains $n(n-1)/2$ terms and that on the right contains $N(N-1)/2$ terms.

Now square (2.9) and average over all simple random samples. Using (2.10) and (2.11), we obtain

$$n^2 E(\bar{y} - \bar{Y})^2 = \frac{n}{N} \Big\{ (y_1 - \bar{Y})^2 + \cdots + (y_N - \bar{Y})^2$$

$$+ \frac{2(n-1)}{N-1} [(y_1 - \bar{Y})(y_2 - \bar{Y}) + \cdots + (y_{N-1} - \bar{Y})(y_N - \bar{Y})] \Big\}$$

Completing the square on the cross-product term, we have

$$n^2 E(\bar{y} - \bar{Y})^2 = \frac{n}{N} \Big\{ \Big(1 - \frac{n-1}{N-1}\Big) [(y_1 - \bar{Y})^2 + \cdots + (y_N - \bar{Y})^2]$$

$$+ \frac{n-1}{N-1} [(y_1 - \bar{Y}) + \cdots + (y_N - \bar{Y})]^2 \Big\}$$

The second term inside the curly bracket vanishes, since the sum of the y_i equals $N\bar{Y}$. Division by n^2 gives

$$V(\bar{y}) = E(\bar{y} - \bar{Y})^2 = \frac{N-n}{nN(N-1)} \sum_{i=1}^{N} (y_i - \bar{Y})^2 = \frac{S^2}{n} \frac{(N-n)}{N}$$

This completes the proof.

Corollary 1. The standard error of $\bar{y}$ is

$$\sigma_{\bar{y}} = \frac{S}{\sqrt{n}} \sqrt{(N-n)/N} = \frac{S}{\sqrt{n}} \sqrt{1-f} \qquad (2.12)$$

Corollary 2. The variance of $\hat{Y} = N\bar{y}$, as an estimate of the population total Y, is

$$V(\hat{Y}) = E(\hat{Y} - Y)^2 = \frac{N^2 S^2}{n} \frac{(N-n)}{N} = \frac{N^2 S^2}{n} (1-f) \qquad (2.13)$$

Corollary 3. The standard error of $\hat{Y}$ is

$$\sigma_{\hat{Y}} = \frac{NS}{\sqrt{n}} \sqrt{(N-n)/N} = \frac{NS}{\sqrt{n}} \sqrt{1-f} \qquad (2.14)$$

2.6 THE FINITE POPULATION CORRECTION

For a random sample of size n from an infinite population, it is well known that the variance of the mean is σ^2/n. The only change in this result when the population is finite is the introduction of the factor $(N-n)/N$. The factors $(N-n)/N$ for the variance and $\sqrt{(N-n)/N}$ for the standard error are called the *finite population corrections* (fpc). They are given with a divisor $(N-1)$ in place of N by writers who present results in terms of σ. Provided that the sampling fraction n/N remains low, these factors are close to unity, and the size of the population as such has no direct effect on the standard error of the sample mean. For instance, if S is the same in the two populations, a sample of 500 from a population of 200,000 gives almost as precise an estimate of the population mean as a sample of 500 from a population of 10,000. Persons unfamiliar with sampling often find this result difficult to believe and, indeed, it is remarkable. To them it seems intuitively obvious that if information has been obtained about only a very small fraction of the population, the sample mean cannot be accurate. It is instructive for the reader to consider why this point of view is erroneous.

In practice the fpc can be ignored whenever the sampling fraction does not exceed 5% and for many purposes even if it is as high as 10%. The effect of ignoring the correction is to overestimate the standard error of the estimate $\bar{y}$.

The following theorem, which is an extension of theorem 2.2, is not required for the discussion in this chapter, but it is proved here for later reference.

Theorem 2.3. If y_i, x_i are a pair of variates defined on every unit in the population and $\bar{y}$, $\bar{x}$ are the corresponding means from a simple random sample of size n, then their *covariance*

$$E(\bar{y} - \bar{Y})(\bar{x} - \bar{X}) = \frac{N-n}{nN} \frac{1}{N-1} \sum_{i=1}^{N} (y_i - \bar{Y})(x_i - \bar{X}) \qquad (2.15)$$

This theorem reduces to theorem 2.2 if the variates y_i, x_i are equal on every unit.

Proof. Apply theorem 2.2 to the variate $u_i = y_i + x_i$. The population mean of u_i is $\bar{U} = \bar{Y} + \bar{X}$, and theorem 2.2 gives

$$E(\bar{u} - \bar{U})^2 = \frac{N-n}{nN} \frac{1}{N-1} \sum_{i=1}^{N} (u_i - \bar{U})^2$$

that is

$$E[(\bar{y} - \bar{Y}) + (\bar{x} - \bar{X})]^2 = \frac{N-1}{nN} \frac{1}{N-1} \sum_{i=1}^{N} [(y_i - \bar{Y}) + (x_i - \bar{X})]^2 \qquad (2.16)$$

Expand the quadratic terms on both sides. By theorem 2.2,

$$E(\bar{y} - \bar{Y})^2 = \frac{N-n}{nN} \frac{1}{N-1} \sum_{i=1}^{N} (y_i - \bar{Y})^2$$

with a similar relation for $E(\bar{x} - \bar{X})^2$. Hence these two terms cancel on the left and right sides of (2.16). The result of the theorem (equation 2.15) follows from the cross-product terms.

2.7 ESTIMATION OF THE STANDARD ERROR FROM A SAMPLE

The formulas for the standard errors of the estimated population mean and total are used primarily for three purposes: (1) to compare the precision obtained by simple random sampling with that given by other methods of sampling, (2) to estimate the size of the sample needed in a survey that is being planned, and (3) to estimate the precision actually attained in a survey that has been completed. The formulas involve S^2, the population variance. In practice this will not be known, but it can be estimated from the sample data. The relevant result is stated in theorem 2.4.

Theorem 2.4. For a simple random sample

$$s^2 = \frac{\sum\limits_{1}^{n} (y_i - \bar{y})^2}{n-1}$$

is an unbiased estimate of

$$S^2 = \frac{\sum\limits_{1}^{N} (y_i - \bar{Y})^2}{N-1}$$

Proof. We may write

$$s^2 = \frac{1}{n-1} \sum_{i=1}^{n} [(y_i - \bar{Y}) - (\bar{y} - \bar{Y})]^2 \qquad (2.17)$$

$$= \frac{1}{n-1} \left[\sum_{i=1}^{n} (y_i - \bar{Y})^2 - n(\bar{y} - \bar{Y})^2 \right] \qquad (2.18)$$

Now average over all simple random samples of size n. By the argument of symmetry used in theorem 2.2,

$$E\left[\sum_{i=1}^{n} (y_i - \bar{Y})^2 \right] = \frac{n}{N} \sum_{i=1}^{N} (y_i - \bar{Y})^2 = \frac{n(N-1)}{N} S^2$$

by the definition of S^2. Furthermore, by theorem 2.2,

$$E[n(\bar{y} - \bar{Y})^2] = \frac{N-n}{N} S^2$$

Hence

$$E(s^2) = \frac{S^2}{(n-1)N} [n(N-1) - (N-n)] = S^2 \qquad (2.19)$$

Corollary. Unbiased estimates of the variances of $\bar{y}$ and $\hat{Y} = N\bar{y}$ are

$$v(\bar{y}) = s_{\bar{y}}^2 = \frac{s^2}{n} \left(\frac{N-n}{N} \right) = \frac{s^2}{n} (1-f) \qquad (2.20)$$

$$v(\hat{Y}) = s_{\hat{Y}}^2 = \frac{N^2 s^2}{n} \left(\frac{N-n}{N} \right) = \frac{N^2 s^2}{n} (1-f) \qquad (2.21)$$

For the standard errors we take

$$s_{\bar{y}} = \frac{s}{\sqrt{n}}\sqrt{1-f}, \qquad s_{\hat{Y}} = \frac{Ns}{\sqrt{n}}\sqrt{1-f} \qquad (2.22)$$

These estimates are slightly biased: for most applications the bias is unimportant.

The reader should note the symbols employed for true and estimated variances of the estimates. Thus, for $\bar{y}$, we write

True variance: $V(\bar{y}) = \sigma_{\bar{y}}^2$

Estimated variance: $v(\bar{y}) = s_{\bar{y}}^2$

2.8 CONFIDENCE LIMITS

It is usually assumed that the estimates $\bar{y}$ and $\hat{Y}$ are normally distributed about the corresponding population values. The reasons for this assumption and its limitations are considered in section 2.15. If the assumption holds, lower and upper confidence limits for the population mean and total are as follows:
Mean:

$$\hat{\bar{Y}}_L = \bar{y} - \frac{ts}{\sqrt{n}}\sqrt{1-f}, \qquad \hat{\bar{Y}}_U = \bar{y} + \frac{ts}{\sqrt{n}}\sqrt{1-f} \qquad (2.23)$$

Total:

$$\hat{Y}_L = N\bar{y} - \frac{tNs}{\sqrt{n}}\sqrt{1-f}, \qquad \hat{Y}_U = N\bar{y} + \frac{tNs}{\sqrt{n}}\sqrt{1-f} \qquad (2.24)$$

The symbol t is the value of the normal deviate corresponding to the desired confidence probability. The most common values are

Confidence probability (%)	50	80	90	95	99
	0.67	1.28	1.64	1.96	2.58

If the sample size is less than 50, the percentage points may be taken from Student's t table with $(n-1)$ degrees of freedom, these being the degrees of freedom in the estimated variance s^2. The t distribution holds exactly only if the observations y_i are themselves normally distributed and N is infinite. Moderate departures from normality do not affect it greatly. For small samples with very skew distributions, special methods are needed.

Example.   Signatures to a petition were collected on 676 sheets. Each sheet had enough space for 42 signatures, but on many sheets a smaller number of signatures had been collected. The numbers of signatures per sheet were counted on a random sample of 50 sheets (about a 7% sample), with the results shown in Table 2.2.

Estimate the total number of signatures to the petition and the 80% confidence limits.

The sampling unit is a sheet, and the observations y_i are the numbers of signatures per sheet. Since about half the sheets had the maximum number of signatures, 42, the data are presented as a frequency distribution. Note that the original distribution appears to be far from normal, the greatest frequency being at the upper end. Nevertheless, there is reason to believe from experience that the means of samples of 50 are approximately normally distributed.

We find

$$n = \Sigma f_i = 50, \qquad y = \Sigma f_i y_i = 1471, \qquad \Sigma f_i y_i^2 = 54{,}497$$

Hence the estimated total number of signatures is

$$\hat{Y} = N\bar{y} = \frac{(676)(1471)}{50} = 19{,}888$$

For the sample variance s^2 we have

$$s^2 = \frac{1}{n-1}[\Sigma f_i(y_i - \bar{y})^2] = \frac{1}{n-1}\left[\Sigma f_i y_i^2 - \frac{(\Sigma f_i y_i)^2}{\Sigma f_i}\right]$$

$$= \frac{1}{49}\left[54{,}497 - \frac{(1471)^2}{50}\right] = 229.0$$

From (2.22) the 80% confidence limits are

$$19{,}888 \pm \frac{tNs}{\sqrt{n}}\sqrt{1-f} = 19{,}888 \pm \frac{(1.28)(676)(15.13)\sqrt{1-0.0740}}{\sqrt{50}}$$

This gives 18,107 and 21,669 for the 80% limits. A complete count showed 21,045 signatures.

TABLE 2.2

RESULTS FOR A SAMPLE OF 50 PETITION SHEETS y_i = NUMBER OF SIGNATURES: f_i = FREQUENCY

y_i	42	41	36	32	29	27	23	19	16	15
f_i	23	4	1	1	1	2	1	1	2	2

y_i	14	11	10	9	7	6	5	4	3	Total
f_i	1	1	1	1	1	3	2	1	1	50

2.9   AN ALTERNATIVE METHOD OF PROOF

Cornfield (1944) suggested a method of proving the principal results for simple random sampling without replacement that enables us to use standard results from infinite population theory. Let a_i be a random variate that takes the value 1 if the ith unit is in the sample and the value 0 otherwise. The sample mean $\bar{y}$ may be written

$$\bar{y} = \frac{1}{n}\sum_{i=1}^{N} a_i y_i \qquad (2.25)$$

where the sum extends over all N units in the population. In this expression the a_i are random variables and the y_i are a set of fixed numbers.

Clearly

$$\Pr(a_i = 1) = \frac{n}{N}, \qquad \Pr(a_i = 0) = 1 - \frac{n}{N}$$

Thus a_i is distributed as a binomial variate in a single trial, with $P = n/N$. Hence

$$E(a_i) = P = \frac{n}{N}, \qquad V(a_i) = PQ = \frac{n}{N}\left(1 - \frac{n}{N}\right) \tag{2.26}$$

To find $V(\bar{y})$ we need also the covariance of a_i and a_j. The product $a_i a_j$ is 1 if the ith and jth unit are both in the sample and is zero otherwise. The probability that two specific units are both in the sample is easily found to be $n(n-1)/N(N-1)$. Hence

$$\text{Cov}(a_i a_j) = E(a_i a_j) - E(a_i)E(a_j)$$

$$= \frac{n(n-1)}{N(N-1)} - \left(\frac{n}{N}\right)^2 = -\frac{n}{N(N-1)}\left(1 - \frac{n}{N}\right) \tag{2.27}$$

Applying this approach to find $V(\bar{y})$, we have, from (2.25),

$$V(\bar{y}) = \frac{1}{n^2}\left[\sum_{i=1}^{N} y_i^2 V(a_i) + 2 \sum_{i<j}^{N} y_i y_j \, \text{Cov}(a_i a_j)\right] \tag{2.28}$$

$$= \frac{1-f}{nN}\left(\sum y_i^2 - \frac{2}{N-1}\sum y_i y_j\right) \tag{2.29}$$

using (2.26) and (2.27). Completing the square on the cross-product term gives

$$V(\bar{y}) = \frac{1-f}{nN}\left(\frac{N}{N-1}\sum y_i^2 - \frac{1}{N-1}Y^2\right) \tag{2.30}$$

$$= \frac{1-f}{n(N-1)}\sum (y_i - \bar{Y})^2 = \frac{(1-f)S^2}{n} \tag{2.31}$$

The method gives easy proofs of theorems 2.3 and 2.4. It may be used to find higher moments of the distribution of $\bar{y}$, although for this purpose a method given by Tukey (1950), with further development by Wishart (1952), is more powerful.

2.10 RANDOM SAMPLING WITH REPLACEMENT

A similar approach applies when sampling is with replacement. In this event the ith unit may appear $0, 1, 2, \ldots, n$ times in the sample. Let t_i be the number of times that the ith unit appears in the sample. Then

$$\bar{y} = \frac{1}{n}\sum_{i=1}^{N} t_i y_i \tag{2.32}$$

Since the probability that the ith unit is drawn is $1/N$ at each draw, the variate t_i distributed as a binomial number of successes out of n trials with $p = 1/N$. Hence

$$E(t_i) = \frac{n}{N}, \qquad V(t_i) = n\left(\frac{1}{N}\right)\left(1 - \frac{1}{N}\right) \tag{2.33}$$

Jointly, the variates t_i follow a multinomial distribution. For this,

$$\text{Cov}\,(t_i t_j) = -\frac{n}{N^2} \tag{2.34}$$

Using (2.32), (2.33), and (2.34), we have, for sampling with replacement,

$$V(\bar{y}) = \frac{1}{n^2}\left[\sum_{i=1}^{N} y_i^2 \frac{n(N-1)}{N^2} - 2 \sum_{i<j}^{N} y_i y_j \frac{n}{N^2} \right] \tag{2.35}$$

$$= \frac{1}{nN} \sum_{i=1}^{N} (y_i - \bar{Y})^2 = \frac{\sigma^2}{n} = \frac{N-1}{N}\frac{S^2}{n} \tag{2.36}$$

Consequently, $V(\bar{y})$ in sampling without replacement is only $(N-n)/(N-1)$ times its value in sampling with replacement. If instead of $\bar{y}$ the mean $\bar{y}_d$ of the different or distinct units in the sample is used as an estimate when sampling is with replacement, Murthy (1967) has shown that the leading term in the average variance of $\bar{y}_d$ is $(1-f/2)S^2/n$, following work by Basu (1958) and Des Raj and Khamis (1958). In some applications the cost of measuring the distinct units in the sample may be predominating, so that the cost of the sample is proportional to the number of distinct units. In this situation, Seth and J. N. K. Rao (1964) showed that for given average cost, $V(\bar{y})$ in sampling without replacement is less than $V(\bar{y}_d)$ in sampling with replacement. They also prove the more general result that if $\bar{y}_d' = f(\nu)\bar{y}_d/Ef(\nu)$, where ν is the number of distinct units in the sample and $f(\nu)$ is a function of ν, then $V(\bar{y}) < V(\bar{y}_d')$ if $S^2 < N\bar{Y}^2$, a condition satisfied by nearly all populations encountered in sample surveys.

2.11   ESTIMATION OF A RATIO

Frequently the quantity that is to be estimated from a simple random sample is the ratio of two variables both of which vary from unit to unit. In a household survey examples are the average number of suits of clothes per adult male, the average expenditure on cosmetics per adult female, and the average number of hours per week spent watching television per child aged 10 to 15. In order to estimate the first of these items, we would record for the ith household ($i = 1, 2, \ldots, n$) the number of adult males x_i who live there and the total number of suits y_i that they possess. The population parameter to be estimated is the ratio

$$R = \frac{\text{total number of suits}}{\text{total number of adult males}} = \frac{\sum\limits_{1}^{N} y_i}{\sum\limits_{1}^{N} x_i} \qquad (2.37)$$

The corresponding sample estimate is

$$\hat{R} = \frac{\sum\limits_{1}^{n} y_i}{\sum\limits_{1}^{n} x_i} = \frac{\bar{y}}{\bar{x}} \qquad (2.38)$$

Examples of this kind occur frequently when the sampling unit (the household) comprises a group or cluster of elements (adult males) and our interest is in the population mean *per element*. Ratios also appear in many other applications, for example, the ratio of loans for building purposes to total loans in a bank or the ratio of acres of wheat to total acres on a farm.

The sampling distribution of $\hat{R}$ is more complicated than that of $\bar{y}$ because both the numerator $\bar{y}$ and the denominator $\bar{x}$ vary from sample to sample. In small samples the distribution of $\hat{R}$ is skew and $\hat{R}$ is usually a slightly biased estimate of R. In large samples the distribution of $\hat{R}$ tends to normality and the bias becomes negligible. The following approximate result will serve for most purposes: the distribution of $\hat{R}$ is studied in more detail in Chapter 6.

Theorem 2.5. If variates y_i, x_i are measured on each unit of a simple random sample of size n, assumed large, the MSE and variance of $\hat{R} = \bar{y}/\bar{x}$ are each approximately

$$\text{MSE}(\hat{R}) \doteq V(\hat{R}) \doteq \frac{1-f}{n\bar{X}^2} \frac{\sum\limits_{i=1}^{N} (y_i - Rx_i)^2}{N-1} \qquad (2.39)$$

where $R = \bar{Y}/\bar{X}$ is the ratio of the population means and $f = n/N$.
Proof.

$$\hat{R} - R = \frac{\bar{y}}{\bar{x}} - R = \frac{\bar{y} - R\bar{x}}{\bar{x}} \qquad (2.40)$$

If n is large, $\bar{x}$ should not differ greatly from $\bar{X}$. In order to avoid having to work out the distribution of the ratio of two random variables $(\bar{y} - R\bar{x})$ and $\bar{x}$, we replace $\bar{x}$ by $\bar{X}$ in the denominator of (2.40) as an approximation. This gives

$$\hat{R} - R \doteq \frac{\bar{y} - R\bar{x}}{\bar{X}} \qquad (2.41)$$

Now average over all simple random samples of size n.

$$E(\hat{R} - R) \doteq \frac{E(\bar{y} - R\bar{x})}{\bar{X}} = \frac{\bar{Y} - R\bar{X}}{\bar{X}} = 0 \qquad (2.42)$$

since $R = \bar{Y}/\bar{X}$. This shows that to the order of approximation used here $\hat{R}$ is an unbiased estimate of R.

From (2.41) we also obtain the result

$$\text{MSE}(\hat{R}) = E(\hat{R} - R)^2 \doteq \frac{1}{\bar{X}^2} E(\bar{y} - R\bar{x})^2 \qquad (2.43)$$

The quantity $\bar{y} - R\bar{x}$ is the sample mean of the variate $d_i = y_i - Rx_i$, whose population mean $\bar{D} = \bar{Y} - R\bar{X} = 0$. Hence we can find $V(\hat{R})$ by applying theorem 2.2 for the variance of the mean of a simple random sample to the variate d_i and dividing by $\bar{X}^2$. This gives

$$V(\hat{R}) \doteq \frac{1}{\bar{X}^2} E(\bar{y} - R\bar{x})^2 = \frac{1}{\bar{X}^2} \frac{S_d^2}{n}(1 - f) \qquad (2.44)$$

$$= \frac{1 - f}{n\bar{X}^2} \frac{\sum_{i=1}^{N}(d_i - \bar{D})^2}{(N - 1)} = \frac{1 - f}{n\bar{X}^2} \frac{\sum_{i=1}^{N}(y_i - Rx_i)^2}{N - 1} \qquad (2.45)$$

This completes the proof.

The way in which theorem 2.5 was proved is worth noting. It was shown that the formula in theorem 2.2 for the variance of the sample mean $\bar{y}$ gives the formula for the approximate variance of the ratio $\bar{y}/\bar{x}$, if the variate y_i is replaced by the variate $(y_i - Rx_i)/\bar{X}$. The same result, or its natural extension, holds also in more complex sampling situations and is used frequently later in this book.

As a sample estimate of

$$\frac{\sum_{i=1}^{N}(y_i - Rx_i)^2}{N - 1}$$

it is customary to take

$$\frac{\sum_{i=1}^{n}(y_i - \hat{R}x_i)^2}{n - 1}$$

This estimate can be shown to have a bias of order $1/n$.
For the estimated standard error of $\hat{R}$, this gives

$$s(\hat{R}) = \frac{\sqrt{1 - f}}{\sqrt{n}\bar{X}} \sqrt{\frac{\sum(y_i - \hat{R}x_i)^2}{n - 1}} \qquad (2.46)$$

If $\bar{X}$ is not known, the sample estimate $\bar{x}$ is substituted in the denominator. One way to compute $s(\hat{R})$ is to express it as

$$s(\hat{R}) = \frac{\sqrt{1-f}}{\sqrt{n}\bar{X}} \sqrt{\frac{\sum y_i^2 - 2\hat{R} \sum y_i x_i + \hat{R}^2 \sum x_i^2}{n-1}} \tag{2.47}$$

Example. Table 2.3 shows the number of persons (x_1), the weekly family income (x_2), and the weekly expenditure on food (y) in a simple random sample of 33 low-income families. Since the sample is small, the data are intended only to illustrate the calculations.

Estimate from the sample (a) the mean weekly expenditure on food per family, (b) the mean weekly expenditure on food per person, and (c) the percentage of the income that is spent on food. Compute the standard errors of these estimates.

Weekly Expenditure on Food per Family. This is the ordinary sample mean

$$\bar{y} = \frac{907.2}{33} = \$27.49$$

By theorem 2.2 (ignoring the fpc), its standard error is

$$s_{\bar{y}} = \frac{1}{\sqrt{n}} \sqrt{\frac{\sum (y_i - \bar{y})^2}{n-1}} = \frac{1}{\sqrt{n(n-1)}} \sqrt{\sum y_i^2 - \frac{(\sum y_i)^2}{n}}$$

$$= \frac{1}{\sqrt{(33)(32)}} \sqrt{28,224 - (907.2)^2/33} = \$1.76$$

(The uncorrected sum of squares 28,224 is given underneath Table 2.3.)

Weekly Expenditure on Food per Person. Since the size of family varies, the estimate is a ratio of two variables,

$$\hat{R}_1 = \frac{\sum y}{\sum x_1} = \frac{907.2}{123} = \$7.38 \text{ per person}$$

The sums of squares and products needed to compute $S(\hat{R})$ by (2.47) are found under Table 2.3. We need in addition

$$2\hat{R}_1 = 14.7512, \qquad \hat{R}_1^2 = 54.3996, \qquad \bar{x}_1 = 3.7273$$

Extra decimals are carried in $\hat{R}_1$, $2\hat{R}_1$, $\hat{R}_1^2$ to preserve accuracy.

Hence, from (2.47),

$$s(\hat{R}_1) = \frac{1}{\sqrt{33}(3.7273)} \sqrt{\frac{(28,224) - (14.7512)(3595.5) + (54.3996)(533)}{32}}$$

$$= \$0.534$$

Percentage of Income Spent on Food. This again is a ratio of two variables

$$\hat{R}_2 = 100 \frac{\sum y}{\sum x_2} = \frac{(100)(907.2)}{2394} = 37.9\%$$

By (2.47) the reader may verify that the standard error is 2.38%.

TABLE 2.3

Size, Weekly Income, and Food Cost of 33 Families

Family Number	Size x_1	Income x_2	Food Cost y	Family Number	Size x_1	Income x_2	Food Cost y
1	2	62	14.3	18	4	83	36.0
2	3	62	20.8	19	2	85	20.6
3	3	87	22.7	20	4	73	27.7
4	5	65	30.5	21	2	66	25.9
5	4	58	41.2	22	5	58	23.3
6	7	92	28.2	23	3	77	39.8
7	2	88	24.2	24	4	69	16.8
8	4	79	30.0	25	7	65	37.8
9	2	83	24.2	26	3	77	34.8
10	5	62	44.4	27	3	69	28.7
11	3	63	13.4	28	6	95	63.0
12	6	62	19.8	29	2	77	19.5
13	4	60	29.4	30	2	69	21.6
14	4	75	27.1	31	6	69	18.2
15	2	90	22.2	32	4	67	20.1
16	5	75	37.7	33	2	63	20.7
17	3	69	22.6				
				Total	123	2394	907.2

$\sum x_1^2 = 533$, $\sum x_2^2 = 177{,}254$, $\sum y^2 = 28{,}224$

$\sum x_1 y = 3595.5$, $\sum x_2 y = 66{,}678$

2.12 ESTIMATES OF MEANS OVER SUBPOPULATIONS

In many surveys, estimates are made for each of a number of classes into which the population is subdivided. In a household survey separate estimates might be wanted for families with $0, 1, 2, \ldots$ children, for owners and renters, or for families in different occupation groups. The term *domains of study* has been given to these subpopulations by the U.N. Subcommission on Sampling (1950).

In the simplest situation each unit in the population falls into one of the domains. Let the jth domain contain N_j units, and let n_j be the number of units in a simple random sample of size n that happen to fall in this domain. If y_{jk} $(k = 1, 2, \ldots, n_j)$ are the measurements on these units, the population mean $\bar{Y}_j$ for the jth domain is estimated by

$$\bar{y}_j = \sum_{k=1}^{n_j} \frac{y_{jk}}{n_j} \tag{2.48}$$

At first sight $\bar{y}_j$ seems to be a ratio estimate as in section 2.11. Although n is fixed, n_j will vary from one sample of size n to another. The complication of a ratio estimate can be avoided by considering the distribution of $\bar{y}_j$ over samples in which both n and n_j are fixed. We assume $n_j > 0$.

In the totality of samples with given n and n_j the probability that any specific set of n_j units from the N_j units in domain j is drawn is

$$\frac{_{N-N_j}C_{n-n_j}}{_{N-N_j}C_{n-n_j} \cdot \,_{N_j}C_{n_j}} = \frac{1}{_{N_j}C_{n_j}}$$

Since each specific set of n_j units from domain j can appear with all selections of $(n - n_j)$ units from the $(N - n_j)$ that are not in domain j, the numerator above is the number of samples containing a specified set of n_j, and the denominator is the total number of samples. It follows that theorems 2.1, 2.2, and 2.4 apply to the y_{jk} if we put n_j for n and N_j for N.

From theorem 2.1: $\bar{y}_j$ is an unbiased estimate of $\bar{Y}_j$ (2.49)

From theorem 2.2: the standard error of $\bar{y}_j$ is $\dfrac{S_j}{\sqrt{n_j}}\sqrt{1 - (n_j/N_j)}$ (2.50)

where

$$S_j^2 = \sum_{k=1}^{N_j} \frac{(y_{jk} - \bar{Y}_j)^2}{N_j - 1} \tag{2.51}$$

From theorem 2.4: An estimate of the standard error of $\bar{y}_j$ is

$$\frac{s_j}{\sqrt{n_j}}\sqrt{1 - (n_j/N_j)} \tag{2.52}$$

where

$$s_j^2 = \sum_{k=1}^{n_j} \frac{(y_{jk} - \bar{y}_j)^2}{n_j - 1} \tag{2.53}$$

If the value of N_j is not known, the quantity n/N may be used in place of n_j/N_j when computing the fpc. (With simple random sampling, n_j/N_j is an unbiased estimate of n/N.)

2.13 ESTIMATES OF TOTALS OVER SUBPOPULATIONS

In a firm's list of accounts receivable, in which some accounts have been paid and some not, we might wish to estimate by a sample the total dollar amount of unpaid bills. If N_j (the number of unpaid bills in the population) is known, there is no problem. The sample estimate is $N_j\bar{y}_j$ and its conditional standard error is N_j times expression (2.50).

Alternatively, if the total amount receivable in the list is known, a ratio estimate can be used. The sample gives an estimate of the ratio (total amount of unpaid bills)/(total amount of all bills). This is multiplied by the known total amount receivable in the list.

If neither N_j nor the total receivables is known, these estimates cannot be made. Instead, we multiply the sample total of the y's over units falling in the jth domain by the raising factor N/n. This gives the estimate

$$\hat{Y}_j = \frac{N}{n} \sum_{k=1}^{n_j} y_{jk} \tag{2.54}$$

We will show that $\hat{Y}_j$ is unbiased and obtain its standard error over repeated samples of size n. The device of keeping n_j fixed as well as n does not help in this problem.

In presenting the proof we revert to the original notation, in which y_i is the measurement on the ith unit in the population. Define for every unit in the population a new variate y_i', where

$$y_i' = \begin{cases} y_i & \text{if the unit is in the } j\text{th domain,} \\ 0 & \text{otherwise} \end{cases}$$

The population total of the y_i' is

$$\sum_{i=1}^{N} y_i' = \sum_{j\text{th dom}} y_i = Y_j \tag{2.55}$$

In a simple random sample of size n, $y_i' = y_i$ for each of the n_j units that lie in the jth domain; $y_i' = 0$ for each of the remaining $n - n_j$ units. If $\bar{y}'$ is the ordinary sample mean of the y_i', the quantity

$$N\bar{y}' = \frac{N}{n} \sum_{i=1}^{n} y_i' = \frac{N}{n} \sum_{k=1}^{n_j} y_{jk} = \hat{Y}_j \tag{2.56}$$

This result shows that the estimate $\hat{Y}_j$ as defined in equation (2.54) is N times the sample mean of the y_i'.

In repeated samples of size n we can clearly apply theorems 2.1, 2.2, and 2.4 to the variates y_i'. These show that $\hat{Y}_j$ is an unbiased estimate of Y_j with standard error

$$\sigma(\hat{Y}_j) = \frac{NS'}{\sqrt{n}} \sqrt{1 - (n/N)} \tag{2.57}$$

where S' is the population standard deviation of the y_i'. In order to compute S', we regard the population as consisting of the N_j values y_i that are in the jth domain and of $N - N_j$ zero values. Thus

$$S'^2 = \frac{1}{N-1}\left(\sum_{j\text{th dom}} y_i^2 - \frac{Y_j^2}{N}\right) \tag{2.58}$$

From theorem 2.4 a sample estimate of the standard error of $\hat{Y}_j$ is

$$s(\hat{Y}_j) = \frac{Ns'}{\sqrt{n}}\sqrt{1-(n/N)} \tag{2.59}$$

In computing s', any unit not in the jth domain is given a zero value. Some students seem to have a psychological objection to doing this, but the method is sound.

The methods of this and the preceding section also apply to surveys in which the frame used contains units that do not belong to the population as it has been defined. An example illustrates this application.

Example. From a list of 2422 minor household expenditures a simple random sample of 180 items was drawn in order to estimate the total spent for operation of the household. Certain types of expenditure (on clothing and car upkeep) were not considered relevant. Of the 180 sample items, 152 were relevant. The sum and uncorrected sum of squares of the relevant amounts (in dollars) were as follows.

$$\sum y_i' = 343.5, \qquad \sum y_i'^2 = 1491.38$$

Estimate the total expenditure for household operation and give the standard error of the estimate.

$$\hat{Y}_j = \frac{N}{n}\sum_{i=1}^{n} y_i' = \frac{(2422)(343.5)}{180} = \$4622$$

From (2.59)

$$s(\hat{Y}_j) = \frac{Ns'}{\sqrt{n}}\sqrt{1-(n/N)}$$

In computing s' we regard our sample of 180 items as having 28 zeros. Hence

$$s'^2 = \frac{1}{(179)}\left[\sum y_i'^2 - \frac{(\sum y_i')^2}{180}\right]$$

$$= \frac{1}{(179)}\left[1491.38 - \frac{(343.5)^2}{180}\right] = 4.670$$

Finally,

$$s_{\hat{Y}_j} = (2422)\sqrt{\frac{4.670}{180}\left(1 - \frac{180}{2422}\right)} = \$375$$

The estimate is not precise, its coefficient of variation 375/4622 being about 8%.

In this example expenditures on car upkeep and clothing were excluded as not relevant and therefore were scored as zeros in the sample. In some applications it is known in advance that certain units in the population contribute nothing to the total that is being estimated. For instance, in a survey of stores to estimate total sales of luggage, some stores do not handle luggage; certain area sampling units for farm studies contain no farms. Sometimes it is possible, by expenditure of effort, to identify and count the units that contribute nothing, so that in our notation $(N - N_j)$, hence N_j, is known.

Consequently it is worth examining by how much $V(\hat{Y}_j)$ is reduced when N_j is known. If N_j is not known, (2.57) gives

$$V(\hat{Y}_j) = \frac{N^2 S'^2}{n}\left(1 - \frac{n}{N}\right)$$

If $\bar{Y}_j$ and S_j are the mean and standard deviation in the domain of interest (i.e., among the nonzero units) the reader may verify that

$$(N-1)S'^2 = (N_j - 1)S_j^2 + N_j \bar{Y}_j^2\left(1 - \frac{N_j}{N}\right) \tag{2.60}$$

Since terms in $1/N_j$ and $1/N$ are nearly always negligible,

$$S'^2 \doteq P_j S_j^2 + P_j Q_j \bar{Y}_j^2 \tag{2.61}$$

where $P_j = N_j/N$ and $Q_j = 1 - P_j$. This gives

$$V(\hat{Y}_j) \doteq \frac{N^2}{n}(P_j S_j^2 + P_j Q_j \bar{Y}_j^2)\left(1 - \frac{n}{N}\right) \tag{2.62}$$

If nonzero units are identified, we draw a sample of size n_j from them. The estimate of the domain total is $N_j \bar{y}_j$ with variance

$$V(N_j \bar{y}_j) = \frac{N_j^2}{n_j}S_j^2\left(1 - \frac{n_j}{N_j}\right) = \frac{N^2}{n_j}P_j^2 S_j^2\left(1 - \frac{n_j}{N_j}\right) \tag{2.63}$$

The comparable variances are (2.62) and (2.63). In (2.62) the average number of nonzero units in the sample of size n is nP_j. If we take $n_j = nP_j$ in (2.63), so that the number of nonzeros to be measured is about the same with both methods, (2.63) becomes

$$V(N_j \bar{y}_j) = \frac{N^2}{n}P_j S_j^2\left(1 - \frac{n}{N}\right) \tag{2.64}$$

The ratio of the variances (2.64) to (2.62) is

$$\frac{V(N_j \text{ known})}{V(N_j \text{ not known})} = \frac{S_j^2}{S_j^2 + Q_j \bar{Y}_j^2} = \frac{C_j^2}{C_j^2 + Q_j} \tag{2.65}$$

where $C_j = S_j/\bar{Y}_j$ is the coefficient of variation among the nonzeros. As might be expected, the reduction in variance due to a knowledge of N_j is greater when the proportion of zero units is large and when y_j varies relatively little among the nonzero units. For further study of this problem, see Jessen and Houseman (1944).

2.14 COMPARISONS BETWEEN DOMAIN MEANS

Let $\bar{y}_j$, $\bar{y}_k$ be the sample means in the jth and kth of a set of domains into which the units in a simple random sample are classified. The variance of their difference is

$$V(\bar{y}_j - \bar{y}_k) = V(\bar{y}_j) + V(\bar{y}_k) \tag{2.66}$$

This formula applies also to the difference between two ratios $\hat{R}_j$ and $\hat{R}_k$.

One point should be noted. It is seldom of scientific interest to ask whether $\bar{Y}_j = \bar{Y}_k$, because these means would not be exactly equal in a finite population, except by a rare chance, even if the data in both domains were drawn at random from the same infinite population. Instead, we test the null hypothesis that the two domains were drawn from *infinite* populations having the same mean. Consequently we omit the fpc when computing $V(\bar{y}_j)$ and $V(\bar{y}_k)$, using the formula

$$V(\bar{y}_j - \bar{y}_k) = \frac{S_j^2}{n_j} + \frac{S_k^2}{n_k} \tag{2.67}$$

A formula similar to (2.67) is obtained for tests of significance if one frames the question: Could the samples from the two domains have been drawn at random from the same *finite* population?

Under this null hypothesis it may be proved (see exercise 2.16) that

$$V(\bar{y}_j - \bar{y}_k) = S_{jk}^2 \left(\frac{1}{n_j} + \frac{1}{n_k}\right)$$

where S_{jk}^2 is the variance of the finite population consisting of the combined domains.

2.15 VALIDITY OF THE NORMAL APPROXIMATION

Confidence that the normal approximation is adequate in most practical situations comes from a variety of sources. In the theory of probability much study has been made of the distribution of means of random samples. It has been proved that for any population that has a finite standard deviation the distribution of the sample mean tends to normality as n increases (see, e.g., Feller, 1957). This work relates to infinite populations.

For sampling without replacement from finite populations, Hájek (1960) has given necessary and sufficient conditions under which the distribution of the sample mean tends to normality, following work by Erdös and Rényi (1959) and Madow (1948). Hájek assumes a sequence of values n_ν, N_ν tending to infinity in such a way that $(N_\nu - n_\nu)$ also tends to infinity. The measurements in the νth population are denoted by $y_{\nu i}$, $(i = 1, 2, \ldots, N_\nu)$. For this population, let $S_{\nu \tau}$ be the set of units in the population for which

$$|y_{\nu i} - \bar{Y}_\nu| > \tau \sqrt{n_\nu(1 - f_\nu)} S_\nu$$

where $\bar{Y}_\nu$, S_ν, f_ν are the population mean, s.d., and fpc, and τ is a number >0. Then the Lindeberg-type condition

$$\lim_{\nu \to \infty} \frac{\sum_{S_{\nu\tau}} (y_{\nu i} - \bar{Y}_\nu)^2}{(N_\nu - 1)S_\nu^2} = 0$$

is necessary and sufficient to ensure that $\bar{y}_\nu$ tends to normality with the mean and variance given in theorems 2.1 and 2.2.

This imposing body of knowledge leaves something to be desired. It is not easy to answer the direct question: "For this population, how large must n be so that the normal approximation is accurate enough?" Non-normal distributions vary greatly both in the nature and in the degree of their departure from normality. The distributions of many types of economic enterprise (stores, chicken farms, towns) exhibit a marked positive skewness, with a few large units and many small units. The same kind of skewness is displayed by some biological populations (e.g., the number of rats or flies per city block).

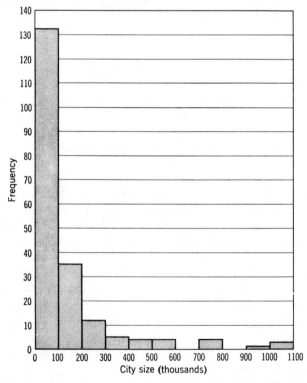

Fig. 2.1. Frequency distribution of sizes of 196 United States Cities in 1920.

As an illustration of a positively skewed distribution, Fig. 2.1 shows the frequency distribution of the numbers of inhabitants in 196 large United States cities in 1920. (The four largest cities, New York, Chicago, Philadelphia, and Detroit, were omitted. Their inclusion would extend the horizontal scale to more than five times the length shown and would, of course, greatly accentuate the skewness.) Figure 2.2 shows the frequency distribution of the total number of

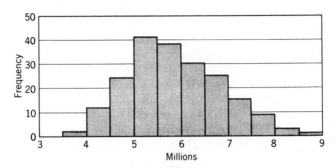

Fig. 2.2 Frequency distribution of totals of 200 simple random samples with $n=49$.

inhabitants in each of 200 simple random samples, with $n = 49$, drawn from this population. The distribution of the sample totals, and likewise of the means, is much more similar to a normal curve but still displays some positive skewness.

From statistical theory and from the results of sampling experiments on skewed populations, some statements can be made about what usually happens to confidence probabilities when we sample from positively skew populations, as follows:

1. The frequency with which the assertion

$$\bar{y} - 1.96 s_{\bar{y}} < \bar{Y} < \bar{y} + 1.96 s_{\bar{y}}$$

is wrong is usually higher than 5%.

2. The frequency with which

$$\bar{Y} > \bar{y} + 1.96 s_{\bar{y}}$$

is *greater* than 2.5%.

3. The frequency with which

$$\bar{Y} < \bar{y} - 1.96 s_{\bar{y}}$$

is *less* than 2.5%.

As an illustration, consider a variate y that is essentially binomially distributed, so that the exact distribution of $\bar{y}$ can be read from the binomial tables. The variate

y takes only two values—the value h with probability P and the value 0 with probability Q. The population mean is $\bar{Y} = Ph$. A simple random sample of size n shows a units that have the value h and $n - a$ units that have the value 0. For the sample,

$$\sum y = ah, \qquad \bar{y} = \frac{ah}{n}$$

$$(n-1)s^2 = \sum y^2 - n\bar{y}^2 = ah^2 - \frac{a^2h^2}{n}$$

$$s_{\bar{y}}^2 = \frac{s^2}{n} = \frac{h^2}{n^2} \frac{a(n-a)}{n-1}$$

Hence 95% normal confidence limits for $\bar{Y}$ are estimated as

$$\bar{y} \pm 1.96 s_{\bar{y}} = \frac{h}{n}\left[a \pm 1.96\sqrt{\frac{a(n-a)}{n-1}}\right] \tag{2.68}$$

Let $n = 400$, $P = 0.1$. Then $\bar{Y} = 0.1h$. By trial we find that if $a = 29$ in expression (2.68) the upper confidence limit is $39.18h/400 = 0.098h$, whereas $a = 30$ gives $40.34h/400 = 0.101h$. Hence any value of $a \le 29$ gives an upper confidence limit that is too low. Similarly we find that if $a \ge 54$ the lower limit is too high.

The variate a follows the binomial distribution with $n = 400$, $P = 0.1$. The tables (Harvard Computation Laboratory, 1955) show that

Pr (stated upper limit too low) $=$ Pr $(a \le 29) = 0.0357$
Pr (stated lower limit too high $=$ Pr $(a \ge 54) = 0.0217$

Pr (confidence statement wrong) $= 0.0574$

The total probability of being wrong is not far from 0.05. In more than 60% of the wrong statements, the true mean is higher than the stated upper limit.

There is no safe general rule as to how large n must be for use of the normal approximation in computing confidence limits. For populations in which the principal deviation from normality consists of marked positive skewness, a crude rule that I have occasionally found useful is

$$n > 25G_1^2 \tag{2.69}$$

where G_1 is Fisher's measure of skewness (Fisher, 1932).

$$G_1 = \frac{E(y_i - \bar{Y})^3}{\sigma^3} = \frac{1}{N\sigma^3}\sum_{i=1}^{N}(y_i - \bar{Y})^3 \tag{2.70}$$

This rule is designed so that a 95% confidence probability statement will be wrong not more than 6% of the time. It is derived mathematically by assuming that any disturbance due to moments of the distribution of $\bar{y}$ higher than the third is negligible. The rule attempts to control only the *total* frequency of wrong statements, ignoring the direction of the error of estimate.

By calculating G_1, or an estimate, for a specific population, we can obtain a rough idea of the sample size needed for application of the normal approximation to compute confidence limits. The result should be checked by sampling experiments whenever possible.

TABLE 2.4

FREQUENCY DISTRIBUTION OF ACRES IN CROPS ON 556 FARMS

Class Intervals (acres)	Coded Scale y_i	Frequency f_i	$f_i y_i$	$f_i y_i^2$	$f_i y_i^3$
0–29	−0.9	47	−42.3	38.1	−34.3
30–63	0	143	0	0	0
64–97	1	154	154	154	154
98–131	2	82	164	328	656
132–165	3	62	186	558	1,674
166–199	4	33	132	528	2,112
200–233	5	13	65	325	1,625
234–267	6	6	36	216	1,296
268–301	7	4	28	196	1,372
302–335	8	6	48	384	3,072
336–369	9	2	18	162	1,458
370–403	10	0	0	0	0
404–437	11	2	22	242	2,662
438–471	12	0	0	0	0
472–505	13	2	26	338	4,394
Totals		556	836.7	3,469.1	20,440.7

$$E(y_i) = \bar{Y} = \frac{836.7}{556} = 1.50486$$

$$E(y_i^2) = \frac{3469.1}{556} = 6.23939$$

$$E(y_i^3) = \frac{20,440.7}{556} = 36.76385$$

$$\sigma^2 = E(y_i^2) - \bar{Y}^2 = 3.97479$$

$$\kappa_3 = E(y_i - \bar{Y})^3 = E(y_i^3) - 3E(y_i^2)\bar{Y} + 2\bar{Y}^3$$

$$= 15.411$$

$$G_1 = \frac{\kappa_3}{\sigma^3} = \frac{15.411}{7.925} = 1.9$$

Example. The data in Table 2.4 show the numbers of acres devoted to crops on 556 farms in Seneca County, New York. The data come from a series of studies by West (1951), who drew repeated samples of size 100 from this population and examined the frequency distributions of $\bar{y}$, s, and Student's t for several items of interest in farm management surveys.

The computation of G_1 is shown under the table. The computations are made on a coded scale, and, since G_1 is a pure number, there is no need to return to the original scale. Note that the first class-interval was slightly different from the others.

Since $G_1 = 1.9$, we take as a suggested minimum n

$$n = (25)(1.9)^2 = 90$$

For samples of size 100, West found with this item (acres in crops) that neither the distribution of $\bar{y}$ nor that of Student's t differed significantly from the corresponding theoretical normal distributions.

Good sampling practice tends to make the normal approximation more valid. Failure of the normal approximation occurs mostly when the population contains some extreme individuals who dominate the sample average when they are present. However, these extremes also have a much more serious effect of increasing the variance of the sample and decreasing the precision. Consequently, it is wise to segregate them and make separate plans for coping with them, perhaps by taking a complete enumeration of them if they are not numerous. This removal of the extremes from the main body of the population reduces the skewness and improves the normal approximation. This technique is an example of stratified sampling, which is discussed in Chapter 5.

2.16   LINEAR ESTIMATORS OF THE POPULATION MEAN

Under simple random sampling, is the sample mean $\bar{y}$ the best estimator of $\bar{Y}$? This question has naturally attracted a good deal of work. The answer depends on the set of competitors to $\bar{y}$ that are allowed and on the definition of "best." For sampling, the units in the population are usually numbered in some way from 1 to N. These numbers that identify the units are often called the *labels* attached to the units.

Early results proved by Horvitz and Thompson (1952) for linear estimators in simple random sampling are as follows. If any y_i always receives the same weight w_i whenever unit labeled i is drawn, the sample mean $\bar{y}$ is the *only* unbiased estimator of $\bar{Y}$ of the form $\sum\limits_{}^{n} w_i y_i$. Since every unit appears in a fraction n/N of simple random samples, $E\left(\sum\limits^{n} w_i y_i\right) = n\left(\sum\limits^{N} w_i y_i\right)\Big/ N \equiv \bar{Y}$ only if every $w_i = 1/n$. If, alternatively, the weight depends only on the *order* in which the unit is drawn into the sample, then $\bar{y}$ has minimum variance among unbiased linear estimators of the form $\sum\limits_{d}^{n} w_d y_{(d)}$, where $y_{(d)}$ is the y-value on the unit that turns up at the dth draw.

A wider class of competitors is the set $\sum\limits^{n} w_{is}y_i$, where w_{is} may depend on the other units that fall into the sample as well as on i. Godambe (1955) showed that in this class no unbiased estimate of $\bar{Y}$ exists with minimum variance for all populations.

Further properties of $\bar{y}$ have been developed by Hartley and J. N. K. Rao (1968, 1969), Royall (1968), and C. R. Rao (1971). In any finite population there will be at most $T \leq N$ distinct values of the y_i. In the sample, let there be n_t values equal to y_t, where $\sum n_t = n$. Hartley and Rao (1968) show that $\bar{y}$ has minimum variance among unbiased estimators of $\bar{Y}$ that are functions only of the n_t and y_t. For random sampling with replacement, they show that the mean of the *distinct* values in the sample is the maximum likelihood estimator of $\bar{Y}$, although it does not have minimum variance in all populations.

C. R. Rao (1971), following work by Kempthorne (1969), considered unbiased estimators $\hat{Y} = \sum w_{is}y_i$ discussed by Godambe. In order to represent the case in which the labels l supply no information about the values y_l, he calculated the average of $V(\hat{Y})$ over all $N!$ permutations of the labels attached to the values, and showed that over these permutations, $\bar{y}$ has minimum average variance. Royall (1970b) has given a more general result.

Godambe's (1955) work has stimulated numerous investigations on sampling design and estimation, including topics such as criteria by which to judge estimators, the role of maximum likelihood, the use of auxiliary information that the labels may carry about the y_i, Bayesian estimators, and methods of estimation when assumptions can be made about the frequency distribution of the y_i. The influence of this work on sampling practice has been limited thus far, but should steadily increase. Some reference to it will be made from time to time. For reviews, see J. N. K. Rao* (1975a) and Smith (1976).

EXERCISES

2.1 In a population with $N = 6$ the values of y_i are 8, 3, 1, 11, 4, and 7. Calculate the sample mean $\bar{y}$ for all possible simple random samples of size 2. Verify that $\bar{y}$ is an unbiased estimate of $\bar{Y}$ and that its variance is as given in theorem 2.2.

2.2 For the same population, calculate s^2 for all simple random samples of size 3 and verify that $E(s^2) = S^2$.

2.3 If random samples of size 2 are drawn with replacement from this population, show by finding all possible samples that $V(\bar{y})$ satifies the equation

$$V(\bar{y}) = \frac{\sigma^2}{n} = \frac{S^2}{n} \frac{(N-1)}{N}$$

2.4 A simple random sample of 30 households was drawn from a city area containing 14,848 households. The numbers of persons per household in the sample were as follows.

5, 6, 3, 3, 2, 3, 3, 3, 4, 4, 3, 2, 7, 4, 3, 5, 4, 4, 3, 3, 4, 3, 3, 1, 2, 4, 3, 4, 2, 4

* Henceforth in this book the surname Rao will refer to J. N. K. Rao unless otherwise noted.

Estimate the total number of people in the area and compute the probability that this estimate is within $\pm 10\%$ of the true value.

2.5   In a study of the possible use of sampling to cut down the work in taking inventory in a stock room, a count is made of the value of the articles on each of 36 shelves in the room. The values to the nearest dollar are as follows.

$$29, 38, 42, 44, 45, 47, 51, 53, 53, 54, 56, 56, 56, 58, 58, 59, 60, 60,$$

$$60, 60, 61, 61, 61, 62, 64, 65, 65, 67, 67, 68, 69, 71, 74, 77, 82, 85.$$

The estimate of total value made from a sample is to be correct within \$200, apart from a 1 in 20 chance. An advisor suggests that a simple random sample of 12 shelves will meet the requirements. Do you agree?

$$\Sigma y = 2138, \quad \Sigma y^2 = 131,682$$

2.6   After the sample in Table 2.2 (p. 28) was taken, the number of completely filled sheets (with 42 signatures each) was counted and found to be 326. Use this information to make an improved estimate of the total number of signatures and find the standard error of your estimate.

2.7   From a list of 468 small 2-year colleges a simple random sample of 100 colleges was drawn. The sample contained 54 public and 46 private colleges. Data for number of students (y) and number of teachers (x) are shown below.

	n	$\Sigma(y)$	$\Sigma(x)$
Public	54	31,281	2,024
Private	46	13,707	1,075
	$\Sigma(y^2)$	$\Sigma(yx)$	$\Sigma(x^2)$
Public	29,881,219	1,729,349	111,090
Private	6,366,785	431,041	33,119

(a) For each type of college in the population, estimate the ratio (number of students)/(number of teachers). (b) Compute the standard errors of your estimates. (c) For the public colleges, find 90% confidence limits for the student/teacher ratio in the whole population.

2.8   In the preceding example test at the 5% level whether the student/teacher ratio is significantly different in the two types of colleges.

2.9   For the public colleges, estimate the total number of teachers (a) given that the total number of public colleges in the population is 251, (b) without knowing this figure. In each case compute the standard error of your estimate.

2.10   The table below shows the numbers of inhabitants in each of the 197 United States cities that had populations over 50,000 in 1940. Calculate the standard error of the estimated total number of inhabitants in all 197 cities for the following methods of sampling: (a) a simple random sample of size 50, (b) a sample that includes the five largest cities and is a simple random sample of size 45 from the remaining 192 cities (c) a sample that includes the nine largest cities and is a simple random sample of size 41 from the remaining cities.

FREQUENCY DISTRIBUTION OF CITY SIZES

Size Class (1000's)	f	Size Class (1000's)	f	Size Class (1000's)	f
50–100	105	550–600	2	....	....
100–150	36	600–650	1	1500–1550	1
150–200	13	650–700	2	....	....
200–250	6	700–750	0	1600–1650	1
250–300	7	750–800	1	....	....
300–350	8	800–850	1	1900–1950	1
350–400	4	850–900	2	....	....
400–450	1	900–950	0	3350–3400	1
450–500	3	950–1000	0	....	....
500–550	0	1000–1050	0	7450–7500	1

Gaps in the intervals are indicated by . . . .

2.11 Calculate the coefficient of skewness G_1 for the original population and the population remaining after removing (a) the five largest cities, (b) the nine largest cities.

2.12 A small survey is to be taken to compare home-owners with renters. In the population about 75% are owners, 25% are renters. For one item the variance is thought to be about 15 for both owners and renters. The standard error of the difference between the two domain means is not to exceed 1. How large a sample is needed (a) if owners and renters can be identified in advance of drawing the sample, (b) if not? (An approximate answer will do in (b); an exact discussion requires binomial tables.)

2.13 A simple random sample of size 3 is drawn from a population of size N with replacement. Show that the probabilities that the sample contains 1, 2, and 3 *different* units (for example, *aaa*, *aab*, *abc*, respectively) are

$$P_1 = \frac{1}{N^2}, \qquad P_2 = \frac{3(N-1)}{N^2}, \qquad P_3 = \frac{(N-1)(N-2)}{N^2}$$

As an estimate of $\bar{Y}$ we take $\bar{y}'$, the unweighted mean over the different units in the sample. Show that the average variance of $\bar{y}'$ is

$$V(\bar{y}') = \frac{(2N-1)(N-1)S^2}{6N^2} \doteq (1-f/2)S^2/3$$

One way to do this is to show that

$$V(\bar{y}') = S^2\left(\frac{N-1}{N}P_1 + \frac{N-2}{2N}P_2 + \frac{N-3}{3N}P_3\right)$$

Hence show that $V(\bar{y}') < V(\bar{y})$, where $\bar{y}$ is the ordinary mean of the n observations in the sample. The result that $V(\bar{y}') < V(\bar{y})$ for any $n > 2$ was proved by Des Raj and Khamis (1958).

2.14 Two dentists A and B make a survey of the state of the teeth of 200 children in a village. Dr. A selects a simple random sample of 20 children and counts the number of decayed teeth for each child, with the following results.

Number of decayed teeth/child	0	1	2	3	4	5	6	7	8	9	10
Number of children	8	4	2	2	1	1	0	0	0	1	1

Dr. B, using the same dental techniques, examines all 200 children, recording merely those who have no decayed teeth. He finds 60 children with no decayed teeth.

Estimate the total number of decayed teeth in the village children, (a) using A's results only, (b) using both A's and B's results. (c) Are the estimates unbiased? (d) Which estimate do you expect to be more precise?

2.15   A company intends to interview a simple random sample of employees who have been with it more than 5 years. The company has \$1000 to spend, and each interview costs \$10. There is no separate list of employees with more than 5 years service, but a list can be compiled from the files at a cost of \$200. The company can either (a) compile the list and interview a simple random sample drawn from the eligible employees or (b) draw a simple random sample of all employees, interviewing only those eligible. The cost of rejecting those not eligible in the sample is assumed negligible.

Show that for estimating a total over the population of eligible employees, plan (a) gives a smaller variance than plan (b) only if $C_j < 2\sqrt{Q_j}$, where C_j is the coefficient of variation of the item among eligible employees and Q_j is the proportion of noneligibles in the company. Ignore the fpc.

2.16   A simple random sample of size $n = n_1 + n_2$ with mean $\bar{y}$ is drawn from a finite population, and a simple random subsample of size n_1 is drawn from it with mean $\bar{y}_1$. Show that (a) $V(\bar{y}_1 - \bar{y}_2) = S^2[(1/n_1) + (1/n_2)]$. where $\bar{y}_2$ is the mean of the remaining n_2 units in the sample, (b) $V(\bar{y}_1 - \bar{y}) = S^2[(1/n_1) - (1/n)]$, (c) Cov$\{\bar{y}, \bar{y}_1 - \bar{y}\} = 0$. Repeated sampling implies repetition of the drawing of both the sample and the subsample.

2.17   The number of distinct simple random samples of size n is of course $N!/n!(N - n)!$. There has been some interest in finding smaller sets of samples of size n that have the same properties as the set of simple random samples. One set is that of *balanced incomplete block* (bib) designs. These are samples of n distinct units out of N such that (i) every unit appears in the same number (r) of samples, (ii) every pair of units appears together in λ samples.

Verify that $\lambda = r(n - 1)/(N - 1)$ and that the number of distinct samples in the set is rN/n. Over the set of *bib* samples, prove in the usual notation that if $\bar{y}$ is the mean of a sample, (a) $V(\bar{y}) = (1 - f)S^2/n$ and (b) $v(\bar{y}) = (1 - f)\sum_{}^{n}(y_i - \bar{y})^2/n(n - 1)$ is an unbiased estimate of $V(\bar{y})$.

Note.   There is no general method for finding the smallest r for which a *bib* can be constructed. Sometimes the smallest known r provides $N!/n!(N - n)!$ samples, bringing us back to simple random samples. But for $N = 91$, $n = 10$, the smallest *bib* set has 91 samples as against over 6 United States trillion SRS. Avadhani and Sukhatme (1973) have shown how *bib* designs may be used in attempting to reduce travel costs between sampling units.

2.18   The following is an illustration by Royall (1968) of the fact that in simple random sampling the sample mean $\bar{y}$ does not have uniformly minimum variance in the class of estimators of the form $\sum_{}^{n} w_{is}y_i$ considered by Godambe (1955), where the weight w_{is} may depend on the other units that fall in the sample. For $N = 3$, $n = 2$, consider the estimator

$$\hat{Y}_{12} = \tfrac{1}{2}y_1 + \tfrac{1}{2}y_2; \qquad \hat{Y}_{13} = \tfrac{1}{2}y_1 + \tfrac{2}{3}y_3; \qquad \hat{Y}_{23} = \tfrac{1}{2}y_2 + \tfrac{1}{3}y_3$$

where $\hat{Y}_{ij}$ is the estimator for the sample that has units (i, j). Prove Royall's results that $\hat{Y}_{ij}$ is unbiased and that $V(\hat{Y}_{ij}) < V(\bar{y})$ if $y_3(3y_2 - 3y_1 - y_3) > 0$. The illustration is taken from an earlier example by Roy and Chakravarti (1960).

2.19   This exercise is another example of estimators geared to particular features of populations. After the decision to take a simple random sample had been made, it was realized that y_1 would be unusually low and y_N would be unusually high. For this situation, Särndal (1972) examined the following unbiased estimator of $\bar{Y}$.

$$\hat{\bar{Y}}_S = \bar{y} + c \qquad \text{if the sample contains } y_1 \text{ but not } y_N$$
$$= \bar{y} - c \qquad \text{if the sample contains } y_N \text{ but not } y_1$$
$$= \bar{y} \qquad \text{for all other samples}$$

where c is a constant. Prove Särndal's result that $\bar{Y}_S$ is unbiased with

$$V(\hat{\bar{Y}}_S) = (1-f)\left[\frac{S^2}{n} - \frac{2c}{(N-1)}(y_N - y_1 - nc)\right]$$

so that $V(\hat{\bar{Y}}_S) < V(\bar{y})$ if $0 < c < (y_N - y_1)/n$.

2.20   For a population with $N = 8$ and values $y_i = 1, 4, 5, 5, 6, 6, 8, 13$, show that with $n = 4$, $V(\bar{y}) = 1.5$, while $V(\hat{\bar{Y}}_S) = 0.214$ when $c = 1.5$ (its best value), and 0.357 when $c = 1$ or 2.

Given the information in exercise 2.19, an alternative sampling plan is to include both y_1 and y_8 in every sample, drawing a simple random sample of size 2 from $y_2, \ldots, y_7$, with mean $\bar{y}_2$. The estimate of $\bar{Y}$ is

$$\hat{\bar{Y}}_{st} = (y_1 + 6\bar{y}_2 + y_8)/8$$

Show that $\hat{\bar{Y}}_{st}$ is unbiased, with variance $9V(\bar{y}_2)/16$. For this population, show that $V(\hat{\bar{Y}}_{st}) = 0.350$. This estimator is an example of stratified sampling (Chapter 5) with three strata: y_1; $y_2 \ldots y_7$; and y_8.

CHAPTER 3

Sampling Proportions
and Percentages

3.1 QUALITATIVE CHARACTERISTICS

Sometimes we wish to estimate the total number, the proportion, or the percentage of units in the population that possess some characteristic or attribute or fall into some defined class. Many of the results regularly published from censuses or surveys are of this form, for example, numbers of unemployed persons, the percentage of the population that is native-born. The classification may be introduced directly into the questionnaire, as in questions that are answered by a simple "yes" or "no." In other cases the original measurements are more or less continuous, and the classification is introduced in the tabulation of results. Thus we may record the respondents' ages to the nearest year but publish the percentage of the population aged 60 and over.

Notation. We suppose that every unit in the population falls into one of the two classes C and C'. The notation is as follows:

Number of units in C in		Proportion of units in C in	
Population	Sample	Population	Sample
A	a	$P = A/N$	$p = a/n$

The sample estimate of P is p, and the sample estimate of A is Np or Na/n. In statistical work the *binomial* distribution is often applied to estimates like a and p. As will be seen, the correct distribution for finite populations is the *hypergeometric*, although the binomial is usually a satisfactory approximation.

3.2 VARIANCES OF THE SAMPLE ESTIMATES

By means of a simple device it is possible to apply the theorems established in Chapter 2 to this situation. For any unit in the sample or population, define y_i as 1 if the unit is in C and as 0 if it is in C'. For this population of values y_i, it is clear that

$$Y = \sum_1^N y_i = A \tag{3.1}$$

50

$$\bar{Y} = \frac{\sum\limits_{1}^{N} y_i}{N} = \frac{A}{N} = P \tag{3.2}$$

Also, for the sample,

$$\bar{y} = \frac{\sum\limits_{1}^{n} y_i}{n} = \frac{a}{n} = p \tag{3.3}$$

Consequently the problem of estimating A and P can be regarded as that of estimating the total and mean of a population in which every y_i is either 1 or 0. In order to use the theorems in Chapter 2, we first express S^2 and s^2 in terms of P and p. Note that

$$\sum_{1}^{N} y_i^2 = A = NP, \qquad \sum_{1}^{n} y_i^2 = a = np$$

Hence

$$S^2 = \frac{\sum\limits_{1}^{N} (y_i - \bar{Y})^2}{N-1} = \frac{\sum\limits_{1}^{N} y_i^2 - N\bar{Y}^2}{N-1}$$

$$= \frac{1}{N-1}(NP - NP^2) = \frac{N}{N-1}PQ \tag{3.4}$$

where $Q = 1 - P$. Similarly

$$s^2 = \frac{\sum\limits_{1}^{n} (y_i - \bar{y})^2}{n-1} = \frac{n}{n-1}pq \tag{3.5}$$

Application of theorems 2.1, 2.2, and 2.4 to this population gives the following results for simple random sampling of the units that are being classified.

Theorem 3.1. The sample proportion $p = a/n$ is an unbiased estimate of the population proportion $P = A/N$.

Theorem 3.2. The variance of p is

$$V(p) = E(p - P)^2 = \frac{S^2}{n}\left(\frac{N-n}{N}\right) = \frac{PQ}{n}\left(\frac{N-n}{N-1}\right) \tag{3.6}$$

using (3.4).

Corollary 1. If p and P are the sample and population *percentages*, respectively, falling into class C, (3.6) continues to hold for the variance of p.

Corollary 2. The variance of $\hat{A} = Np$, the estimated total number of units in class C, is

$$V(\hat{A}) = \frac{N^2 PQ}{n}\left(\frac{N-n}{N-1}\right) \tag{3.7}$$

Theorem 3.3. An unbiased estimate of the variance of p, derived from the sample, is

$$v(p) = s_p^2 = \frac{N-n}{(n-1)N}pq \tag{3.8}$$

Proof. In the corollary of theorem 2.4 it was shown that for a variate y_i an unbiased estimate of the variance of the sample mean $\bar{y}$ is

$$v(\bar{y}) = \frac{s^2}{n}\frac{(N-n)}{N} \tag{3.9}$$

For proportions, P takes the place of $\bar{y}$, and in (3.5) we showed that

$$s^2 = \frac{n}{n-1}pq \tag{3.10}$$

Hence

$$v(p) = s_p^2 = \frac{N-n}{(n-1)N}pq \tag{3.11}$$

It follows that if N is very large relative to n, so that the fpc is negligible, an unbiased estimate of the variance of p is

$$\frac{pq}{n-1}$$

The result may appear puzzling to some readers, since the expression pq/n is almost invariably used in practice for the estimated variance. The fact is that pq/n is not unbiased even with an infinite population.

Corollary. An unbiased estimate of the variance of $\hat{A} = Np$, the estimated total number of units in class C in the population, is

$$v(\hat{A}) = s_{Np}^2 = \frac{N(N-n)}{n-1}pq \tag{3.12}$$

Example. From a list of 3042 names and addresses, a simple random sample of 200 names showed on investigation 38 wrong addresses. Estimate the total number of addresses needing correction in the list and find the standard error of this estimate. We have

$$N = 3042, \qquad n = 200, \qquad a = 38, \qquad p = 0.19$$

The estimated total number of wrong addresses is

$$\hat{A} = Np = (3042)(0.19) = 578$$

$$S_{\hat{A}} = \sqrt{[(3042)(2842)(0.19)(0.81)/199]} = \sqrt{6686} = 81.8$$

Since the sampling ratio is under 7%, the fpc makes little difference. To remove it, replace the term $N-n$ by N. If, in addition, we replace $n-1$ by n, we have the simpler formula

$$s_{Np} = N\sqrt{pq/n} = (3042)\sqrt{(0.19)(0.81)/200} = 84.4$$

This is in fairly close agreement with the previous result, 81.8.

The preceding formulas for the variance and the estimated variance of p hold only if the *units* are classified into C or C' so that p is the ratio of the number of units in C in the sample to the total number of units in the sample. In many surveys each unit is composed of a group of elements, and it is the elements that are classified. A few examples are as follows:

Sampling Unit	Elements
Family	Members of the family
Restaurant	Employees
Crate of eggs	Individual eggs
Peach tree	Individual peaches

If a simple random sample of units is drawn in order to estimate the proportion P of *elements* in the population that belong to class C, the preceding formulas do not apply. Appropriate methods are given in section 3.12.

3.3   THE EFFECT OF P ON THE STANDARD ERRORS

Equation (3.6) shows how the variance of the estimated percentage changes with P, for fixed n and N. If the fpc is ignored, we have

$$V(p) = \frac{PQ}{n}$$

The function PQ and its square root are shown in Table 3.1. These functions may be regarded as the variance and standard deviation, respectively, for a sample of size 1.

The functions have their greatest values when the population is equally divided between the two classes, and are symmetrical about this point. The standard error of p changes relatively little when P lies anywhere between 30 and 70%. At the maximum value of $\sqrt{PQ}$, 50, a sample size of 100 is needed to reduce the standard

SAMPLING TECHNIQUES

TABLE 3.1

VALUES OF PQ AND $\sqrt{PQ}$

P = Population percentage in class C

P	0	10	20	30	40	50	60	70	80	90	100
PQ	0	900	1600	2100	2400	2500	2400	2100	1600	900	0
$\sqrt{PQ}$	0	30	40	46	49	50	49	46	40	30	0

error of the estimate to 5%. To attain a 1% standard error requires a sample size of 2500.

This approach is not appropriate when interest lies in the total *number* of units in the population that are in class C. In this event it is more natural to ask: Is the estimate likely to be correct to within, say, 7% of the true total? Thus we tend to think of the standard error expressed as a fraction or percentage of the true value, NP. The fraction is

$$\frac{\sigma_{Np}}{NP} = \frac{N\sqrt{PQ}}{\sqrt{n}NP}\sqrt{\frac{N-n}{N-1}} = \frac{1}{\sqrt{n}}\sqrt{\frac{Q}{P}}\sqrt{\frac{N-n}{N-1}} \tag{3.13}$$

This quantity is called the *coefficient of variation* of the estimate. If the fpc is ignored, the coefficient is $\sqrt{Q/nP}$. The ratio $\sqrt{Q/P}$, which might be considered the coefficient of variation for a sample of size 1, is shown in Table 3.2.

TABLE 3.2

VALUES OF $\sqrt{Q/P}$ FOR DIFFERENT VALUES OF P

P = Population percentage in class C

P	0	0.1	0.5	1	5	10	20
$\sqrt{Q/P}$	∞	31.6	14.1	9.9	4.4	3.0	2.0

P	30	40	50	60	70	80	90
$\sqrt{Q/P}$	1.5	1.2	1.0	0.8	0.7	0.5	0.3

For a fixed sample size, the coefficient of variation of the estimated total in class C decreases steadily as the true percentage in C increases. The coefficient is high when P is less than 5%. Very large samples are needed for precise estimates of the total number possessing any attribute that is rare in the population. For $P = 1\%$, we must have $\sqrt{n} = 99$ in order to reduce the coefficient of variation of the estimate to 0.1 or 10%. This gives a sample size of 9801. Simple random sampling, or any method of sampling that is adapted for general purposes, is an expensive method of estimating the total number of units of a scarce type.

3.4 THE BINOMIAL DISTRIBUTION

Since the population is of a particularly simple type, in which the y_i are either 1 or 0, we can find the actual frequency distribution of the estimate p and not merely its mean and variance.

The population contains A units that are in class C and $N-A$ units in C', where $P = A/N$. If the first unit that is drawn happens to be in C, there will remain in the population $A-1$ units in C and $N-A$ in C'. Thus the proportion of units in C, after the first draw, changes slightly to $(A-1)/(N-1)$. Alternatively, if the first unit drawn is in C', the proportion in C changes to $A/(N-1)$. In sampling without replacement, the proportion keeps changing in this way throughout the draw. In the present section these variations are ignored, that is, P is assumed constant. This amounts to assuming that A and $N-A$ are both large relative to the sample size n, or that sampling is with replacement.

With this assumption, the process of drawing the sample consists of a series of n trials, in each of which the probability that the unit drawn is in C is P. This situation gives rise to the familiar binomial frequency distribution for the number of units in C in the sample. The probability that the sample contains a units in C is

$$\Pr(a) = \frac{n!}{a!(n-a)!}P^a Q^{n-a} \tag{3.14}$$

From this expression we may tabulate the frequency distribution of a, of $p = a/n$, or of the estimated total Np.

There are three comprehensive sets of tables. All give P by intervals of 0.01. The ranges for n are as follows.

U.S. Bureau of Standards (1950):

$n = 1(1)49$, (*i.e., goes from* 1 *to* 49 *by intervals of* 1).

Romig (1952): $n = 50(5)100$.

Harvard Computation Laboratory (1955):

$n = 1(1)50(2)100(10)200(20)500(50)1000$

3.5 THE HYPERGEOMETRIC DISTRIBUTION

The distribution of p can be found without the assumption that the population is large in relation to the sample. The numbers of units in the two classes C and C' in the population are A and A', respectively. We will calculate the probability that the corresponding numbers in the sample are a and a', where

$$a + a' = n, \qquad A + A' = N$$

In simple random sampling each of the $\binom{N}{n}$ different selections of n units out of N has an equal chance of being drawn. To find the probability wanted, we count how many of these samples contain exactly a units from C and a' from C'. The number of different selections of a units among the A that are in C is $\binom{A}{a}$, whereas the number of different selections of a' among A' is $\binom{A'}{a'}$. Each selection of the first type can be combined with any one of the second to give a different sample of the required type. The total number of samples of the required type is therefore

$$\binom{A}{a} \cdot \binom{A'}{a'}$$

Hence, if a simple random sample of size n is drawn, the probability that it is of the required type is

$$\Pr(a, a' | A, A') = \binom{A}{a} \cdot \binom{A'}{a'} \Big/ \binom{N}{n} \tag{3.15}$$

This is the frequency distribution of a or np, from which that of p is immediately derivable. The distribution is called the *hypergeometric* distribution.

For computing purposes the hypergeometric probability (3.15) may be written as follows.

$$\frac{n!}{a!(n-a)!} \cdot \frac{A(A-1)\ldots(A-a+1)(A')(A'-1)\ldots(A'-a'+1)}{N(N-1)\ldots(N-n+1)} \tag{3.16}$$

Example. A family of eight contains three males and five females. Find the frequency distribution of the number of males in a simple random sample of size 4. In this case

$$A = 3; \quad A' = 5, \quad N = 8; \quad n = 4$$

From (3.16) the distribution of the number of males, a, is as follows:

a	Probability
0	$\dfrac{4!}{0!4!} \cdot \dfrac{5.4.3.2}{8.7.6.5} = \dfrac{1}{14}$
1	$\dfrac{4!}{1!3!} \cdot \dfrac{3.5.4.3}{8.7.6.5} = \dfrac{6}{14}$
2	$\dfrac{4!}{2!2!} \cdot \dfrac{3.2.5.4}{8.7.6.5} = \dfrac{6}{14}$
3	$\dfrac{4!}{3!1!} \cdot \dfrac{3.2.1.5}{8.7.6.5} = \dfrac{1}{14}$
4	Impossible $= 0$

The reader may verify that the mean number of males is $\frac{3}{2}$ and the variance is $\frac{15}{28}$. These results agree with the formulas previously established in section 3.2, which give

$$E(np) = nP = \frac{nA}{N} = \frac{(4)(3)}{8} = \frac{3}{2}$$

$$V(np) = nPQ\frac{N-n}{N-1} = 4 \cdot \frac{3}{8} \cdot \frac{5}{8} \cdot \frac{4}{7} = \frac{15}{28}$$

3.6 CONFIDENCE LIMITS

We first discuss the meaning of confidence limits in the case of qualitative characteristics. In the sample, a out of n fall in class C. Suppose that inferences are to be made about the number A in the population that fall in class C. For an upper confidence limit to A, we compute a value $\hat{A}_U$ such that for this value the probability of getting a or less falling in C in the sample is some small quantity α_U, for example, 0.025. Formally, $\hat{A}_U$ satisfies the equation

$$\sum_{j=0}^{a} \Pr(j, n-j|\hat{A}_U, N-\hat{A}_U) = \alpha_U U \tag{3.17}$$

where Pr is the probability term for the hypergeometric distribution, as defined in (3.15).

When α_U is chosen in advance, (3.17) requires in general a nonintegral value of $\hat{A}_U$ to satisfy it, whereas conceptually $\hat{A}_U$ should be a whole number. In practice we choose $\hat{A}_U$ as the smallest integral value of A such that the left side of (3.17) is less than or equal to α_U. Similarly, the lower confidence limit $\hat{A}_L$ is the largest integral value such that

$$\sum_{j=a}^{n} \Pr(j, n-j|\hat{A}_L, N-\hat{A}_L) \leq \alpha_L \tag{3.18}$$

Confidence limits for P are then found by taking $\hat{P}_U = \hat{A}_U/N$, $\hat{P}_L = \hat{A}_L/N$.

Numerous methods are available for computing confidence limits.

Exact Methods

Chung and DeLury (1950) present charts of the 90, 95, and 99% limits for P for $N = 500$, 2500, and 10,000. Values for intermediate population sizes are obtainable by interpolation. Lieberman and Owen (1961) give tables of individual and cumulative terms of the hypergeometric distribution, but N extends only to 100.

The Normal Approximation

From (3.8) for the estimated variance of p, one form of the normal approximation to the confidence limits for P is

$$p \pm \left[t\sqrt{1-f}\sqrt{pq/(n-1)} + \frac{1}{2n} \right] \tag{3.19}$$

where $f = n/N$ and t is the normal deviate corresponding to the confidence probability. Use of the more familiar term $\sqrt{pq/n}$ seldom makes an appreciable difference. The last term on the right is a correction for continuity. This produces only a slight improvement in the approximation. However, without the correction, the normal approximation usually gives too narrow a confidence interval.

TABLE 3.3

SMALLEST VALUES OF np FOR USE OF THE NORMAL
APPROXIMATION

p	np = Number Observed in the *Smaller* Class	n = Sample Size
0.5	15	30
0.4	20	50
0.3	24	80
0.2	40	200
0.1	60	600
0.05	70	1400
~0*	80	∞

* This means that p is extremely small, so that np follows the Poisson distribution.

The error in the normal approximation depends on all the quantities n, p, N, α_U, and α_L. The quantity to which the error is most sensitive is np or more specifically the number observed in the *smaller* class. Table 3.3 gives working rules for deciding when the normal approximation (3.19) may be used.

The rules in Table 3.3 are constructed so that with 95% confidence limits the true frequency with which the limits fail to enclose P is not greater than 5.5%. Furthermore, the probability that the upper limit is below P is between 2.5 and 3.5%, and the probability that the lower limit exceeds P is between 2.5 and 1.5%.

Example 1. In a simple random sample of size 100, from a population of size 500, there are 37 units in class C. Find the 95% confidence limits for the proportion and for the total number in class C in the population. In this example

$$n = 100, \quad N = 500, \quad p = 0.37$$

The example lies in the range in which the normal approximation is recommended. The estimated standard error of p is

$$\sqrt{(1-f)pq/(n-1)} = \sqrt{(0.8)(0.37)(0.63)/99} = 0.0434$$

The correction for continuity, $1/2n$, equals 0.005. Hence the 95% limits for P are estimated as

$$0.37 \pm (1.96 \times 0.0434 + 0.005) = 0.37 \pm 0.090$$

$$\hat{P}_L = 0.280, \qquad \hat{P}_U = 0.460$$

The limits as read from the charts by Chung and DeLury are 0.285 and 0.462, respectively.

To find limits for the total number in class C in the population, we multiply by N, obtaining 140 and 230, respectively.

Binomial Approximations

When the normal approximation does not apply, limits for P may be found from the binomial tables (section 3.4) and adjusted, if necessary, to take account of the fpc. Table VIII₁ in Fisher and Yates' *Statistical Tables* (1957) gives binomial confidence limits for P for any value of n, and is a useful alternative to the ordinary binomial tables. Example 2 shows how the binomial approximation is computed.

Example 2. For another item in the sample in example 1, nine of the 100 units fall in class C. From Romig's table for $n = 100$ the 95% limits for P are found to be 0.041 and 0.165. (The Fisher-Yates tables give 0.042 and 0.164.) If f, the sampling fraction, is less than 5%, limits found in this way are close enough for most purposes. In this example, $f = 0.2$ and adjustment is needed.

To apply the adjustment, we shorten the interval between p and each limit by the factor $\sqrt{1-f} = \sqrt{0.8} = 0.894$. The adjusted limits are as follows:

$$\hat{P}_L = 0.090 - (0.894)(0.090 - 0.041) = 0.046$$

$$\hat{P}_U = 0.090 + (0.894)(0.165 - 0.090) = 0.157$$

The limits read from the charts by Chung and DeLury are 0.045 and 0.157, respectively.

Burstein (1975) has produced a variant of this calculation that is slightly more accurate. Suppose that a units out of n are in class C (in this example, $a = 9$, $n = 100$). In $\hat{P}_L$, replace $a/n = 0.090$ by $(a - 0.5)/n = 0.085$. In $\hat{P}_U$, replace a/n by $(a + a/n)/n = 0.0909$. Also, $(1-f)$ is taken as $(N-n)/(N-1)$. Thus, by Burstein's method,

$$\hat{P}_L = 0.085 - (0.895)(0.085 - 0.041) = 0.046$$

$$\hat{P}_U = 0.0909 + (0.895)(0.165 - 0.0909) = 0.157$$

there being no change in the limits in this example.

Example 3. In auditing records in which a very low error rate is demanded, the upper confidence limit for A is primarily of interest. Suppose that 200 of 1000 records are verified and that the batch of 1000 is accepted if no errors are found. Special tables have been constructed to give the upper confidence limit for the number of errors in the batch. A good approximation results from the following relation. The probability that no errors are found in n when A errors are present in N is, from the hypergeometric distribution,

$$\frac{(N-A)(N-A-1)\ldots(N-A-n+1)}{N(N-1)\ldots(N-n+1)} \doteq \left(\frac{N-A-u}{N-u}\right)^n$$

where $u = (n-1)/2$. For example, with $n = 200$, $A = 10$, $N = 1000$, the approximation gives $(890.5/900.5)^{200}$, which is found by logs to be 0.107. Thus $A = 10$ (a 1% error rate) is approximately the 90% upper confidence limit for the number of errors in the batch.

3.7 CLASSIFICATION INTO MORE THAN TWO CLASSES

Frequently, in the presentation of results, the units are classified into more than two classes. Thus a sample from a human population may be arranged in 15 five-year age groups. Even when a question is supposed to be answered by a simple "yes" or "no," the results actually obtained may fall into four classes: "yes," "no," "don't know," and "no answer." The extension of the theory to such cases is illustrated by the situation in which there are three classes.

We suppose that the number falling in the ith class is A_i in the population and a_i in the sample, where

$$N = \sum A_i, \qquad n = \sum a_i, \qquad P_i = \frac{A_i}{N}, \qquad p_i = \frac{a_i}{n}$$

When the sample size n is small in relation to all the A_i, the probabilities P_i may be considered effectively constant throughout the drawing of the sample. The probability of drawing the observed sample is given by the *multinomial* expression

$$\Pr(a_i) = \frac{n!}{a_1! a_2! a_3!} P_1{}^{a_1} P_2{}^{a_2} P_3{}^{a_3} \tag{3.20}$$

This is the appropriate extension of the binomial distribution and is a good approximation when the sampling fraction is small.

The correct expression for the probability of drawing the observed sample is

$$\Pr(a_i | A_i) = \binom{A_1}{a_1}\binom{A_2}{a_2}\binom{A_3}{a_3} \Big/ \binom{N}{n} \tag{3.21}$$

This expression is the natural extension of (3.15), section 3.5, for the hypergeometric distribution. The numerator is the number of distinct samples of size n that can be formed with a_1 units in class 1, a_2 in class 2, and a_3 in class 3.

3.8 CONFIDENCE LIMITS WITH MORE THAN TWO CLASSES

Two different cases must be distinguished.

Case 1. We calculate

$$p = \frac{\text{number in any one class in sample}}{n} = \frac{a_1}{n}$$

or

$$p = \frac{\text{total number in a group of classes}}{n} = \frac{a_1 + a_2 + a_3}{n}$$

In either of these situations, although the original classification contains more than two classes, p itself is obtained from a subdivision of the n units into only two classes. The theory already presented applies to this case. Confidence limits are calculated as described in section 3.6.

Case 2. Sometimes certain classes are omitted, p being computed from a breakdown of the remaining classes into two parts. For example, we might omit persons who did not know or gave no answer and consider the ratio of number of "yes" answers to "yes" plus "no" answers. Ratios that are structurally of this type are often of interest in sample surveys. The denominator of such a ratio is not n but some smaller number n'.

Although n' varies from sample to sample, previous results can still be used by considering the conditional distribution of p in samples in which both n and n' are fixed. This device was already employed in section 2.12. Suppose that

$$p = \frac{a_1}{a_1 + a_2}, \qquad n' = a_1 + a_2, \qquad n = a_1 + a_2 + a_3$$

so that a_3 is the number in the sample falling in classes in which we are not at the moment interested. Then, as shown in the next section, the conditional distribution of a_1 and a_2 is the hypergeometric distribution obtained when the sample is of size n' and the population of size $N' = A_1 + A_2$. Hence, from (3.19), the normal approximations to conditional confidence limits for $P = A_1/(A_1 + A_2)$ are

$$p \pm \left[t \sqrt{\left(1 - \frac{n'}{N'}\right) \frac{pq}{(n'-1)}} + \frac{1}{2n'} \right] \tag{3.22}$$

If the value of N' is not known, n/N may be substituted for n'/N' in the fpc term in (3.22).

3.9 THE CONDITIONAL DISTRIBUTION OF p

To find this distribution, we restrict our attention to samples of size n in which $n' = a_1 + a_2$ fall in classes 1 and 2. The number of distinct samples of this type is

$$\binom{N'}{n'}\binom{N-N'}{n-n'} = \binom{A_1+A_2}{a_1+a_2}\binom{A_3}{a_3} \tag{3.23}$$

Among these samples, the number that have a_1 in class 1 and a_2 in class 2 has already been given as the numerator in (3.21), section 3.7. Dividing this numerator by (3.23), we have

$$\Pr(a_1 | A_1, A_2, n, n') = \binom{A_1}{a_1}\binom{A_2}{a_2} \bigg/ \binom{A_1+A_2}{a_1+a_2} \tag{3.24}$$

This is an ordinary hypergeometric distribution for a sample of size n' from a population of size $N' = A_1 + A_2$.

Example. Consider a population that consists of the five units, b, c, d, e, f, that fall in three classes.

Class	A_i	Units Denoted By
1	1	b
2	2	c, d
3	2	e, f

With random samples of size 3, we wish to estimate $P = A_1/(A_1 + A_2)$ or, in this case, $\frac{1}{3}$. Thus $N = 5$ and $N' = 3$.

There are 10 possible samples of size 3, all with equal initial probabilities. These are grouped according to the value of n'.

$$n' = 1$$

Sample	a_1	a_2	p	Conditional Probability	$(p - P)$
bef	1	0	1	$\frac{1}{3}$	$\frac{2}{3}$
cef or def	0	1	0	$\frac{2}{3}$	$-\frac{1}{3}$

If samples are specified by the values of a_1, a_2, only two types are obtainable: $a_1 = 1$, $a_2 = 0$; $a_1 = 0$, $a_2 = 1$. Their conditional probabilities, $\frac{1}{3}$ and $\frac{2}{3}$, respectively, agree with the general expression (3.24). Furthermore,

$$E(p) = \tfrac{1}{3}$$

$$\sigma_p^2 = \left(\frac{1}{3}\right)\left(\frac{4}{9}\right) + \left(\frac{2}{3}\right)\left(\frac{1}{9}\right) = \frac{6}{27} = \frac{2}{9}$$

The estimate p is unbiased, and its variance agrees with the general formula

$$\sigma_p^2 = \left(\frac{N' - n'}{N' - 1}\right)\frac{PQ}{n'} = \left(\frac{3 - 1}{3 - 1}\right)\left(\frac{1}{3}\right)\left(\frac{2}{3}\right) = \frac{2}{9}$$

For $n' = 2$ there are six possible samples, which give only two sets of values of a_1, a_2.

$$n' = 2$$

Sample	a_1	a_2	p	Conditional Probability	$(p - P)$
$bce, bcf, bde,$ or bdf	1	1	$\frac{1}{2}$	$\frac{2}{3}$	$\frac{1}{6}$
cde or cdf	0	2	0	$\frac{1}{3}$	$-\frac{1}{3}$

The estimate is again unbiased and its variance is

$$\sigma_p^2 = \left(\frac{2}{3}\right)\left(\frac{1}{36}\right) + \left(\frac{1}{3}\right)\left(\frac{1}{9}\right) = \frac{1}{18}$$

which may be verified from the general formula. Note that the variance is only one fourth of that obtained when $n' = 1$. In a conditional approach the variance changes with the configuration of the sample that was drawn.

For $n' = 3$, there is only one possible sample, *bcd*. This gives the correct population fraction, $\frac{1}{3}$. The conditional variance of p is zero, as indicated by the general formula, which reduces to zero when $N' = n'$.

3.10 PROPORTIONS AND TOTALS OVER SUBPOPULATIONS

If separate estimates are to be made for each of a number of subpopulations or domains of study to which the units in the sample are allotted, the results in sections 3.8 and 3.9 are applicable. The sample data may be presented as follows.

Class	Domain 1 C	C'	Domain 2 C	C'	...	Domain k C	C'	Total
Number of units	a_1	a_1'	a_2	a_2'	...	a_k	a_k'	n

Of the n units, $(a_1 + a_1')$ are found to fall in domain 1 and of these a_1 fall in class C. The proportion falling in class C in domain 1 is estimated by $p_1 = a_1/(a_1 + a_1')$. The frequency distribution and confidence limits for p_1 were discussed under Case 2 in sections 3.8 and 3.9.

For estimating the *total* number A_1 of units in class C in domain 1, there are two possibilities. If N_1, the total number of units in domain 1 in the population, is known, we may use the conditional estimate

$$\hat{A}_1 = N_1 p_1 = \frac{N_1 a_1}{a_1 + a_1'} \tag{3.25}$$

Its standard error is computed as

$$s(\hat{A}_1) = N_1 \sqrt{1 - (n_1/N_1)} \sqrt{p_1 q_1/(n_1 - 1)} \tag{3.26}$$

where $n_1 = a_1 + a_1'$.

If N_1 is not known, the estimate is

$$\hat{A}_1' = \frac{N a_1}{n} \tag{3.27}$$

with estimated standard error

$$s(\hat{A}_1') = N \sqrt{1 - (n/N)} \sqrt{pq/(n - 1)} \tag{3.28}$$

where $p = a_1/n$.

3.11 COMPARISONS BETWEEN DIFFERENT DOMAINS

Since proportions are estimated independently in different domains, comparisons between such proportions are made by standard elementary methods. For example, to test whether the proportion $p_1 = a_1/(a_1 + a_1')$ differs significantly from the proportion $p_2 = a_2/(a_2 + a_2')$, we form the usual 2×2 table.

	Domain	
	1	2
C	a_1	a_2
C'	a_1'	a_2'
Total	n_1	n_1'

The ordinary χ^2 test (Fisher, 1958) or the normal approximation to the distribution of $(p_1 - p_2)$ is appropriate. Similarly, comparisons among proportions for more than two domains are made by the methods for a $2 \times k$ contingency table.

Occasionally it is desired to test whether a_1 differs significantly from a_2; for example, whether the *number* of Republicans who favor some proposal is greater than the *number* of Democrats in favor. On the null hypothesis that these two numbers are equal in the population, the total $n' = a_1 + a_2$ in the two classes in question should divide with equal probability between the two classes. Consequently we may regard a_1 as a binomial number of successes in n' trials, with probability of success $\frac{1}{2}$ on the null hypothesis. It may be verified that the normal deviate (corrected for continuity) is

$$\frac{2(|a_1 - \frac{1}{2}n'| - \frac{1}{2})}{\sqrt{n'}}$$

3.12 ESTIMATION OF PROPORTIONS IN CLUSTER SAMPLING

As mentioned in section 3.2, the preceding methods are not valid if each unit is a cluster of elements and we are estimating the proportion of elements that fall into class C.

If each unit contains the same number m of elements, let $p_i = a_i/m$ be the proportion of elements in the ith unit that fall into class C. The proportion falling in C in the sample is

$$p = \frac{\sum a_i}{nm} = \frac{1}{n} \sum^n p_i$$

that is, the estimate p is the unweighted mean of the quantities p_i. Consequently, if y_i is replaced by p_i, the formulas in Chapter 2 may be applied directly to give the true and estimated variance of p.

$$V(p) = \frac{1-f}{n} \frac{\sum\limits_{}^{N}(p_i - P)^2}{N-1} \qquad (3.29)$$

An unbiased sample estimate of this variance is

$$v(p) = \frac{1-f}{n} \frac{\sum\limits_{}^{n}(p_i - p)^2}{n-1} \qquad (3.30)$$

Example 1. A group of 61 leprosy patients were treated with a drug for 48 weeks. To measure the effect of the drug on the leprosy bacilli, the presence of bacilli at six sites on the body of each patient was tested bacteriologically. Among the 366 sites, 153, or 41.8%, were negative. What is the standard error of this percentage?

This example comes from a controlled experiment rather than a survey, but it illustrates how erroneous the binomial formula may be. By the binomial formula, we have $n = 366$, and

$$\text{s.e. } (p) = \sqrt{pq/(n-1)} = \sqrt{(41.8)(58.2)/365} = 2.58\%$$

Each patient is a cluster unit with $m = 6$ elements (sites). To find the standard error by the correct formula, we need the frequency distribution of the 61 values of p_i. It is more convenient to tabulate the distribution of y_i, the number of negative sites per patient. With p_i expressed in percents, $p_i = 100 y_i/6$. From the distribution in Table 3.4 we find $\sum fy^2 = 669$ and

$$\text{s.e. } (\bar{y}) = \sqrt{\frac{\sum f_i(y_i - \bar{y})^2}{n(n-1)}} = \sqrt{\frac{669 - [(153)^2/61]}{(61)(60)}} = 0.279$$

Hence

$$\text{s.e. } (p) = \frac{100}{6}\text{s.e. } (\bar{y}) = 4.65\%$$

This figure is about 1.8 times the value given by the binomial formula. The binomial formula requires the assumption that results at different sites on the same patient are independent, although actually they have a strong positive correlation. The last line of Table 3.4 shows the expected number of patients with 0, 1, 2, ... negative sites, computed from the binomial $(0.58 + 0.42)^6$. Note the marked excesses of observed frequencies f of patients with zero negatives and with five and six negatives.

TABLE 3.4

NUMBER OF NEGATIVE SITES PER PATIENT

$y_i = 6p_i/100$	0	1	2	3	4	5	6	Total
f	17	11	4	4	7	14	4	61
fy_i	0	11	8	12	28	70	24	153
f_{exp}	2.3	10.1	18.3	17.6	9.6	2.8	0.3	61.0

If the size of cluster is not constant, let m_i be the number of elements in the ith cluster unit and let $p_i = a_i/m_i$. The proportion of units falling in class C in the sample is

$$p = \frac{\sum\limits_{}^{n} a_i}{\sum\limits_{}^{n} m_i} \tag{3.31}$$

Structurally, this is a typical ratio estimate, discussed in section 2.11 and later in Chapter 6. It is slightly biased, although the bias is seldom likely to be of practical importance.

If we put a_i for y_i and m_i for x_i in (2.39), the approximate variance of p is

$$V(p) \doteq \frac{1-f}{n\bar{M}^2} \frac{\sum (a_i - Pm_i)^2}{N-1} \tag{3.32}$$

where P is the proportion of elements in C in the population and $\bar{M} = \sum\limits_{}^{N} m_i/N$ is the average number of elements per cluster. An alternative expression is

$$V(p) \doteq \frac{1-f}{n} \sum\limits_{}^{N} \left(\frac{m_i}{\bar{M}}\right)^2 \frac{(p_i - P)^2}{N-1} \tag{3.33}$$

This form shows that the approximate variance involves a weighted sum of squares of deviations of the p_i from the population value P.

For the estimated variance we have

$$v(p) = \frac{1-f}{n\bar{m}^2} \frac{\sum a_i^2 - 2p \sum a_i m_i + p^2 \sum m_i^2}{n-1} \tag{3.34}$$

where $\bar{m} = \sum m_i/n$ is the average number of elements per cluster in the sample.

Example 2. A simple random sample of 30 households was drawn from a census taken in 1947 in wards 6 and 7 of the Eastern Health District of Baltimore. The population contains about 15,000 households. In Table 3.5 the persons in each household are classified (a) according to whether they had consulted a doctor in the last 12 months, (b) according to sex.

Our purpose is to contrast the ratio formula with the inappropriate binomial formula. Consider first the proportion of people who had consulted a doctor. For the binomial formula, we would take

$$n = 104, \qquad p = \frac{30}{104} = 0.2885$$

Hence

$$v_{bin}(p) = \frac{pq}{n} = \frac{(0.2885)(0.7115)}{104} = 0.00197$$

TABLE 3.5

DATA FOR A SIMPLE RANDOM SAMPLE OF 30 HOUSEHOLDS

Household Number	Number of Persons m_i	Number of		Doctor Seen in Last Year	
		Males	Females a_i	Yes	No a_i
1	5	1	4	5	0
2	6	3	3	0	6
3	3	1	2	2	1
4	3	1	2	3	0
5	2	1	1	0	2
6	3	1	2	0	3
7	3	1	2	0	3
8	3	1	2	0	3
9	4	2	2	0	4
10	4	3	1	0	4
11	3	2	1	0	3
12	2	1	1	0	2
13	7	3	4	0	7
14	4	3	1	4	0
15	3	2	1	1	2
16	5	3	2	2	3
17	4	3	1	0	4
18	4	3	1	0	4
19	3	2	1	1	2
20	3	1	2	3	0
21	4	1	3	2	2
22	3	2	1	0	3
23	3	2	1	0	3
24	1	0	1	0	1
25	2	1	1	2	0
26	4	3	1	2	2
27	3	1	2	0	3
28	4	2	2	2	2
29	2	1	1	0	2
30	4	2	2	1	3
Totals	104	53	51	30	74

For the ratio formula, we note that there are 30 clusters and take

$n = 30$

m_i = total number in ith household

a_i = number in ith household who had seen a doctor

$p = 0.2885$, as before

$$m = \frac{104}{30} = 3.4667$$

$\sum a_i^2 = 86; \sum m_i^2 = 404; \sum a_i m_i = 113$

The fpc may be ignored. Hence, from (3.34),

$$v(p) = \frac{(86) - 2(0.2885)(113) + (0.2885)^2(404)}{(30)(29)(3.4667)^2} = 0.00520$$

The variance given by the ratio method, 0.00520, is much larger than that given by the binomial formula, 0.00197. For various reasons, families differ in the frequency with which their members consult a doctor. For the sample as a whole, the proportion who consult a doctor is only a little more than one in four, but there are several families in which every member has seen a doctor. Similar results would be obtained for any characteristic in which the members of the same family tend to act in the same way.

In estimating the proportion of males in the population, the results are different. By the same type of calculation, we find

binomial formula: $v(p) = 0.00240$
ratio formula $v(p) = 0.00114$

Here the binomial formula *overestimates* the variance. The reason is interesting. Most households are set up as a result of a marriage, hence contain at least one male and one female. Consequently the proportion of males per family varies less from one half than would be expected from the binomial formula. None of the 30 families, except one with only one member, is composed entirely of males, or entirely of females. If the binomial distribution were applicable, with a true P of approximately one half, households with all members of the same sex would constitute one quarter of the households of size 3 and one eighth of the households of size 4. This property of the sex ratio has been discussed by Hansen and Hurwitz (1942). Other illustrations of the error committed by improper use of the binomial formula in sociological investigations have been given by Kish (1957).

EXERCISES

3.1   For a population with $N = 6$, $A = 4$, $A' = 2$, work out the value of a for all possible simple random samples of size 3. Verify the theorems given for the mean and variance of $p = a/n$. Verify that

$$\frac{N-n}{(n-1)N} pq$$

is an unbiased estimate of the variance of p.

3.2   In a simple random sample of 200 from a population of 2000 colleges, 120 colleges were in favor of a proposal, 57 were opposed, and 23 had no opinion. Estimate 95%

confidence limits for the number of colleges in the population that favored the proposal.

3.3 Do the results of the previous sample furnish conclusive evidence that the majority of the colleges in the population favored this proposal?

3.4 A population with $N = 7$ consists of the elements $B_1, C_1, C_2, C_3, D_1, D_2$, and D_3. A simple random sample of size 4 is taken in order to estimate the proportion of C's to C's + D's. Work out the conditional distributions of this proportion, p, and verify the formula for its conditional variance.

3.5 In the preceding exercise, what is the probability that a sample of size 4 contains B_1? Find the average variance of p in exercise 3.4 over all simple random samples of size 4. This is 0.0393 as against 0.025 with $N = 6$, $n = 4$, and B_1 absent.

3.6 A simple random sample of 290 households was chosen from a city area containing 14,828 households. Each family was asked whether it owned or rented the house and also whether it had the exclusive use of an indoor toilet. Results were as follows.

	Owned		Rented		Total
Exclusive use of toilet	Yes	No	Yes	No	
	141	6	109	34	290

(a) For families who rent, estimate the percentage in the area with exclusive use of an indoor toilet and give the standard error of your estimate; (b) estimate the total number of renting families in the area who do not have exclusive indoor toilet facilities and give the standard error of this estimate.

3.7 If, in example 3.6, the total number of renting families in the city area is 7526, make a new estimate of the number of renters without exclusive toilet facilities and give the standard error of this estimate.

3.8 For estimating the total number of units in class C in domain 1 (section 3.10), the estimate $\hat{A}_1 = N_1 p_1$ was recommended if N_1 were known, as against $\hat{A}_1' = Na_1/n$ if N_1 were not known. Ignoring the fpc, show that in large samples the ratio of the variance of $\hat{A}_1$ to that of $\hat{A}_1'$ is approximately $Q_1/(Q_1 + P_1\pi)$, where π is the proportion of the population that is not in domain 1, and P_1, as in section 3.10, is the proportion of the units in domain 1 that fall in class C. State the conditions under which knowledge of N_1 produces large reductions in variance.

3.9 In a simple random sample of size 5 from a population of size 30, no units in the sample were in class C. By the hypergeometric distribution, find the upper limit to the number A of units in class C in the population, corresponding to a one-tailed confidence probability of 95%. Find also the approximation to A_U obtained by computing the upper 95% binomial limit P_U and shortening the interval as described in section 3.6. Try also the method on p. 59, Example 3.

3.10 A student health service has a record of the total number of eligible students N and of the total number of visits Y made by students during a year. Some students made no visits. The service wishes to estimate the mean number of visits Y/N_1 for the N_1 students who made at least one visit, but does not know the value of N_1. A simple random sample of n eligible students is taken. In it n_1 students out of the n made at least one visit and their total number of visits was y. Ignore the fpc in this question. (a) Show that y/n_1 is an unbiased estimate of Y/N_1 and that its conditional variance is S^2/n_1, where S^2 is the variance of the number of visits among students making at least one visit. (b) A second method of estimating Y/N_1 is to use $\hat{N}_1 = Nn_1/n$ as an estimate of N_1 and hence Yn/Nn_1 as an estimate of Y/N_1. Show that this estimate is biased and that the ratio of the bias to the true value Y/N_1 is approximately $(N-N_1)/nN_1$. Find an approximate expression for the variance of the estimate Y_n/Nn_1 and show that the estimate in (a) has a higher variance if

$$S^2 > \frac{(N-N_1)n_1}{N_1 n}\left(\frac{Y}{N_1}\right)^2$$

Hint. If p is a binomial estimate of P, based on n trials, then approximately

$$E\left(\frac{1}{p}\right) = \frac{1}{P} + \frac{Q}{nP^2}, \qquad V\left(\frac{1}{p}\right) = \frac{Q}{nP^3}$$

3.11   Which of the two previous estimates seems more precise in the following circumstances? $N = 2004$, $Y = 3011$. The sample with $n = 100$ showed that 73 students made at least one visit. Their total number of visits was 152 and the estimated variance s^2 was 1.55.

3.12   A simple random sample of n cluster units, each with m elements, is taken from a population in which the proportion of elements in class C is P. As the intracluster correlation varies, what are the highest and lowest possible values of the true variance of p (the sample estimate of P) and how do they compare with the binomial variance? Ignore the fpc.

3.13   For the sample of 30 households in Table 3.5, the data shown below refer to visits to the dentist in the last year. Estimate the variance of the proportion of persons who saw a dentist, and compare this with the binomial estimate of the variance.

3.14   In sampling for a rare attribute, one method is to continue drawing a simple random sample until m units that possess the rare attribute have been found (Haldane, 1945) where m is chosen in advance. If the fpc is ignored, prove that the probability that the

Number of Persons	Dentist Seen		Number of Persons	Dentist Seen	
	Yes	No		Yes	No
5	1	4	5	1	4
6	0	6	4	4	0
3	1	2	4	1	3
3	2	1	3	1	2
2	0	2	3	0	3
3	0	3	4	1	3
3	1	2	3	0	3
3	1	2	3	1	2
4	1	3	1	0	1
4	0	4	2	0	2
3	1	2	4	0	4
2	0	2	3	1	2
7	2	5	4	1	3
4	1	3	2	0	2
3	0	3	4	0	4

total sample required is of size n is

$$\frac{(n-1)!}{(m-1)!(n-m)!}P^m Q^{n-m} \qquad (n \geq m)$$

where P is the frequency of the rare attribute. Find the average size of the total sample and show that if $m > 1$, $p = (m-1)/(n-1)$ is an unbiased estimate of P. (For further discussion, see Finney, 1949, and Sandelius, 1951, who considers a plan in which sampling continues until either m have been found or the total sample size has reached a preassigned limit n_0.) See also section 4.5.

CHAPTER 4

The Estimation of Sample Size

4.1 A HYPOTHETICAL EXAMPLE

In the planning of a sample survey, a stage is always reached at which a decision must be made about the size of the sample. The decision is important. Too large a sample implies a waste of resources, and too small a sample diminishes the utility of the results. The decision cannot always be made satisfactorily; often we do not possess enough information to be sure that our choice of sample size is the best one. Sampling theory provides a framework within which to think intelligently about the problem.

A hypothetical example brings out the steps involved in reaching a solution. An anthropologist is preparing to study the inhabitants of some island. Among other things, he wishes to estimate the percentage of inhabitants belonging to blood group O. Cooperation has been secured so that it is feasible to take a simple random sample. How large should the sample be?

This equation cannot be discussed without first receiving an answer to another question. How accurately does the anthropologist wish to know the percentage of people with blood group O? In reply he states that he will be content if the percentage is correct within ±5% in the sense that, if the sample shows 43% to have blood group O, the percentage for the whole island is sure to lie between 38 and 48.

To avoid misunderstanding, it may be advisable to point out to the anthropologist that we cannot absolutely guarantee accuracy within 5% except by measuring everyone. However large n is taken, there is a chance of a very unlucky sample that is in error by more than the desired 5%. The anthropologist replies coldly that he is aware of this, that he is willing to take a 1 in 20 chance of getting an unlucky sample, and that all he asks for is the value of n instead of a lecture on statistics.

We are now in a position to make a rough estimate of n. To simplify matters, the fpc is ignored, and the sample percentage p is assumed to be normally distributed. Whether these assumptions are reasonable can be verified when the initial n is known.

In technical terms, p is to lie in the range $(P \pm 5)$, except for a 1 in 20 chance. Since p is assumed normally distributed about P, it will lie in the range $(P \pm 2\sigma_p)$, apart from a 1 in 20 chance. Furthermore,

$$\sigma_p \doteq \sqrt{PQ/n}$$

Hence, we may put

$$2\sqrt{PQ/n} = 5 \quad \text{or} \quad n = \frac{4PQ}{25}$$

At this point a difficulty appears that is common to all problems in the estimation of sample size. A formula for n has been obtained, but n depends on some property of the population that is to be sampled. In this instance the property is the quantity P that we would like to measure. We therefore ask the anthropologist if he can give us some idea of the likely value of P. He replies that from previous data on other ethnic groups, and from his speculations about the racial history of this island, he will be surprised if P lies outside the range 30 to 60%.

This information is sufficient to provide a usable answer. For any value of P between 30 and 60, the product PQ lies between 2100 and a maximum of 2500 at $P = 50$. The corresponding n lies between 336 and 400. To be on the safe side, 400 is taken as the initial estimate of n.

The assumptions made in this analysis can now be reexamined. With $n = 400$ and a P between 30 and 60, the distribution of p should be close to normal. Whether the fpc is required depends on the number of people on the island. If the population exceeds 8000, the sampling fraction is less than 5% and no adjustment for fpc is called for. The method of applying the readjustment, if it is needed, is discussed in section 4.4.

4.2 ANALYSIS OF THE PROBLEM

The principal steps involved in the choice of a sample size are as follows.

1. There must be some statement concerning what is expected of the sample. This statement may be in terms of desired limits of error, as in the previous example, or in terms of some decision that is to be made or action that is to be taken when the sample results are known. The responsibility for framing the statement rests primarily with the persons who wish to use the results of the survey, although they frequently need guidance in putting their wishes into numerical terms.

2. Some equation that connects n with the desired precision of the sample must be found. The equation will vary with the content of the statement of precision and with the kind of sampling that is contemplated. One of the advantages of probability sampling is that it enables this equation to be constructed.

3. This equation will contain, as parameters, certain unknown properties of the population. These must be estimated in order to give specific results.

4. It often happens that data are to be published for certain major subdivisions of the population and that desired limits of error are set up for each subdivision. A separate calculation is made for the n in each subdivision, and the total n is found by addition.

5. More than one item or characteristic is usually measured in a sample survey: sometimes the number of items is large. If a desired degree of precision is prescribed for each item, the calculations lead to a series of conflicting values of n, one for each item. Some method must be found for reconciling these values.

6. Finally, the chosen value of n must be appraised to see whether it is consistent with the resources available to take the sample. This demands an estimation of the cost, labor, time, and materials required to obtain the proposed size of sample. It sometimes becomes apparent that n will have to be drastically reduced. A hard decision must then be faced—whether to proceed with a much smaller sample size, thus reducing precision, or to abandon efforts until more resources can be found.

In succeeding sections some of these questions are examined in more detail.

4.3   THE SPECIFICATION OF PRECISION

The statement of precision desired may be made by giving the amount of error that we are willing to tolerate in the sample estimates. This amount is determined, as best we can, in the light of the uses to which the sample results are to be put. Sometimes it is difficult to decide how much error *should* be tolerated, particularly when the results have several different uses. Suppose that we asked the anthropologist why he wished the percentage with blood group O to be correct to 5% instead of, say, 4 or 6%. He might reply that the blood group data are to be used primarily for racial classification. He strongly suspects that the islanders belong either to a racial type with a P of about 35% or to one with a P of about 50%. An error limit of 5% in the estimate seemed to him small enough to permit classification into one of these types. He would, however, have no violent objection to 4 or 6% limits of error.

Thus the choice of a 5% limit of error by the anthropologist was to some extent arbitrary. In this respect the example is typical of the way in which a limit of error is often decided on. In fact, the anthropologist was more certain of what he wanted than many other scientists and administrators will be found to be. When the question of desired degree of precision is first raised, such persons may confess that they have never thought about it and have no idea of the answer. My experience has been, however, that after discussion they can frequently indicate at least roughly the size of a limit of error that appears reasonable to them.

Further than this we may not be able to go in many practical situations. Part of the difficulty is that not enough is known about the consequences of errors of different sizes as they affect the wisdom of practical decisions that are made from survey results. Even when these consequences are known, however, the results of many important surveys are used by different people for different purposes, and some of the purposes are not foreseen at the time when the survey is planned. Therefore, an element of guesswork is likely to be prominent in the specification of precision for some time to come.

If the sample is taken for a very specific purpose, (e.g., for making a single "yes" or "no" decision or for deciding how much money to spend on a certain venture), the precision needed can usually be stated in a more definite manner, in terms of the consequences of errors in the decision. A general approach to problems of this type is given in section 4.10, which, although in need of amplification, offers a logical start on a solution.

4.4   THE FORMULA FOR n IN SAMPLING FOR PROPORTIONS

The units are classified into two classes, C and C'. Some margin of error d in the estimated proportion p of units in class C has been agreed on, and there is a small risk α that we are willing to incur that the actual error is larger than d; that is, we want

$$\Pr\left(|p - P| \ge d\right) = \alpha$$

Simple random sampling is assumed, and p is taken as normally distributed. From theorem 3.2, section 3.2,

$$\sigma_p = \sqrt{\frac{N-n}{N-1}}\sqrt{\frac{PQ}{n}}$$

Hence the formula that connects n with the desired degree of precision is

$$d = t\sqrt{\frac{N-n}{N-1}}\sqrt{\frac{PQ}{n}}$$

where t is the abscissa of the normal curve that cuts off an area of α at the tails. Solving for n, we find

$$n = \frac{\dfrac{t^2 PQ}{d^2}}{1 + \dfrac{1}{N}\left(\dfrac{t^2 PQ}{d^2} - 1\right)} \qquad (4.1)$$

For practical use, an advance estimate p of P is substituted in this formula. If N is large, a first approximation is

$$n_0 = \frac{t^2 pq}{d^2} = \frac{pq}{V} \qquad (4.2)$$

where

$$V = \frac{pq}{n_0} = \text{desired variance of the sample proportion}$$

In practice we first calculate n_0. If n_0/N is negligible, n_0 is a satisfactory approximation to the n of (4.1). If not, it is apparent on comparison of (4.1) and (4.2) that n is obtained as

$$n = \frac{n_0}{1 + (n_0 - 1)/N} \doteq \frac{n_0}{1 + (n_0/N)} \tag{4.3}$$

Example. In the hypothetical blood groups example we had

$$d = 0.05, \qquad p = 0.5, \qquad \alpha = 0.05, \qquad t = 2$$

Thus

$$n_0 = \frac{(4)(0.5)(0.5)}{(0.0025)} = 400$$

Let us assume that there are only 3200 people on the island. The fpc is needed, and we find

$$n = \frac{n_0}{1 + (n_0 - 1)/N} = \frac{400}{1 + \frac{399}{3200}} = 356$$

The formula for n_0 holds also if d, p, and q are all expressed as percentages instead of proportions. Since the product pq increases as p moves toward $\frac{1}{2}$, or 50%, a conservative estimate of n is obtained by choosing for p the value nearest to $\frac{1}{2}$ in the range in which p is thought likely to lie. If p seems likely to lie between 5 and 9%, for instance, we assume 9% for the estimation of n.

Sometimes, particularly when estimating the total number NP of units in class C, we wish to control the *relative* error r instead of the absolute error in Np; for example, we may wish to estimate NP with an error not exceeding 10%. That is, we want

$$\Pr\left(\frac{|Np - NP|}{NP} \geq r\right) = \Pr\left(|p - P| \geq rP\right) = \alpha$$

For this specification, we substitute rP or rp for d in formulas (4.1) and (4.2). From (4.2) we get

$$n_0 = \frac{t^2 pq}{r^2 p^2} = \frac{t^2}{r^2} \frac{q}{p} \tag{4.2'}$$

Formula (4.3) is unchanged.

4.5   RARE ITEMS—INVERSE SAMPLING

In estimating n from formulas (4.1), (4.2) and (4.2)', the sampler inserts his best advance estimate of the population proportion P. If P is known to be between 30 and 70%, as in the example in Section 4.1, accurate estimation of P is not crucial. But with a rare item (e.g., $P \leq 10\%$), the necessary n for a specified relative error r is 11 times as large when $P = 1\%$ as when $P = 10\%$. In this situation (P small but not well known in advance), Haldane's (1945) method of continuing sampling until m of the rare items have been found in the sample has one important advantage. The method is usually called *inverse sampling*.

If n is the sample size at which the mth rare item appears, ($m > 1$), an unbiased estimate of P is $p = (m-1)/(n-1)$. For N very large, P small, and $m \geq 10$, a good approximation to $V(p)$ may be shown to be $mP^2 Q/(m-1)^2$. Hence, $\text{cv}(p) \doteq (mQ)^{1/2}/(m-1) < \sqrt{m}/(m-1)$, which will be a close upper limit if P is small. Thus, by fixing m in advance, we can control the value of $\text{cv}(p)$ without advance knowledge of P. The value $m = 27$ gives $\text{cv}(p) < 20\%$, but $m = 102$ is needed for $\text{cv}(p) < 10\%$. The value of n with this method is a random variable, but will be large if P is small.

4.6 THE FORMULA FOR n WITH CONTINUOUS DATA

Most commonly, we wish to control the *relative* error r in the estimated population total or mean. With a simple random sample having mean $\bar{y}$, we want

$$\Pr\left(\left|\frac{\bar{y} - \bar{Y}}{\bar{Y}}\right| \geq r\right) = \Pr\left(\left|\frac{N\bar{y} - N\bar{Y}}{N\bar{Y}}\right| \geq r\right) = \Pr\left(|\bar{y} - \bar{Y}| \geq r\bar{Y}\right) = \alpha$$

where α is a small probability. We assume that $\bar{y}$ is normally distributed: from theorem 2.2, corollary 1, its standard error is

$$\sigma_{\bar{y}} = \sqrt{\frac{N-n}{N}} \frac{S}{\sqrt{n}}$$

Hence

$$r\bar{Y} = t\sigma_{\bar{y}} = t\sqrt{\frac{N-n}{N}} \frac{S}{\sqrt{n}} \tag{4.4}$$

Solving for n gives

$$n = \left(\frac{tS}{r\bar{Y}}\right)^2 \Big/ \left[1 + \frac{1}{N}\left(\frac{tS}{r\bar{Y}}\right)^2\right]$$

Note that the population characteristic on which n depends is its coefficient of variation $S/\bar{Y}$. This is often more stable and easier to guess in advance than S itself.

As a first approximation we take

$$n_0 = \left(\frac{tS}{r\bar{Y}}\right)^2 = \frac{1}{C}\left(\frac{S}{\bar{Y}}\right)^2 \tag{4.5}$$

substituting an advance estimate of $(S/\bar{Y})$. The quantity C is the desired $(\text{cv})^2$ of the sample estimate.

If n_0/N is appreciable we compute n as in (4.3)

$$n = \frac{n_0}{1 + (n_0/N)} \tag{4.3}$$

If instead of the relative error r we wish to control the absolute error d in $\bar{y}$, we take $n_0 = t^2 S^2/d^2 = S^2/V$, where V is the desired variance of $\bar{Y}$.

Example. In nurseries that produce young trees for sale it is advisable to estimate, in late winter or early spring, how many healthy young trees are likely to be on hand, since this determines policy toward the solicitation and acceptance of orders. A study of sampling methods for the estimation of the total numbers of seedlings was undertaken by Johnson (1943). The data that follow were obtained from a bed of silver maple seedlings 1 ft wide and 430 ft long. The sampling unit was 1 ft of the length of the bed, so that $N = 430$. By complete enumeration of the bed if was found that $\bar{Y} = 19$, $S^2 = 85.6$, these being the true population values.

With simple random sampling, how many units must be taken to estimate $\bar{Y}$ within 10%, apart from a chance of 1 in 20? From (4.5) we obtain

$$n_0 = \frac{t^2 S^2}{r^2 \bar{Y}^2} = \frac{(4)(85.6)}{(1.9)^2} = 95$$

Since n_0/N is not negligible, we take

$$n = \frac{95}{1 + \frac{95}{430}} = 78$$

Almost 20% of the bed has to be counted in order to attain the precision desired.

The formulas for n given here apply only to simple random sampling in which the sample mean is used as the estimate of $\bar{Y}$. The appropriate formulas for other methods of sampling and estimation are presented with the discussion of these techniques.

4.7   ADVANCE ESTIMATES OF POPULATION VARIANCES

The nursery example is atypical in that the population variance S^2 was known. In practice, there are four ways of estimating population variances for sample size determinations: (1) by taking the sample in two steps, the first being a simple random sample of size n_1 from which estimates s_1^2 or p_1 of S^2 or P and the required n will be obtained; (2) by the results of a pilot survey; (3) by previous sampling of the same or a similar population; and (4) by guesswork about the structure of the population, assisted by some mathematical results.

Method 1 gives the most reliable estimates of S^2 or P, but it is not often used, since it slows up the completion of the survey. When the method is feasible, Cox (1952), following work by Stein (1945), shows how to compute n from A_1^2 or p_1 so that the final estimate $\bar{y}$ or p will have a preassigned variance V, a preassigned limit of error d, or a preassigned coefficient of variation. The first sample is assumed large enough to neglect terms of order $1/n_1^2$. A few results are quoted.

The results given here assume $n_1 \le n$, the size of the final sample. When this is not so, see Cox (1952).

Estimation of $\bar{Y}$ with given cv $= \sqrt{C}$

The results assume y_i normally distributed. If s_1^2 is the estimated variance from the first sample, take additional units to make the final sample size

$$n = \frac{s_1^2}{C\bar{y}_1^2}\left(1 + 8C + \frac{s_1^2}{n_1 \bar{y}_1^2} + \frac{2}{n_1}\right) \tag{4.6}$$

The mean $\bar{y}$ of the final sample is slightly biased. Take $\hat{\bar{Y}} = \bar{y}(1 - 2C)$.

Estimation of $\bar{Y}$ with Variance V

Take additional units to make the total sample size

$$n = \frac{s_1^2}{V}\left(1 + \frac{2}{n_1}\right) \tag{4.7}$$

If S were known exactly, the required sample size would be S^2/V. The effect of not knowing S is to increase the average size by the factor $(1 + 2/n_1)$.

Estimation of P with Variance V

Let p_1 be the estimate of P from the first sample. The combined size of the first two samples should be

$$n = \frac{p_1 q_1}{V} + \frac{3 - 8p_1 q_1}{p_1 q_1} + \frac{1 - 3p_1 q_1}{Vn_1} \tag{4.8}$$

The first term on the right is the size required if P is known to be equal to p_1. With this method, the ordinary binomial estimate p made from the complete sample of size n is slightly biased. To correct for bias, take

$$\hat{P} = p + \frac{V(1 - 2p)}{pq}$$

Estimation of P with given cv $= \sqrt{C}$

Take

$$n = \frac{q_1}{Cp_1} + \frac{3}{p_1 q_1} + \frac{1}{Cp_1 n_1} \tag{4.9}$$

The estimate is $\hat{P} = p - Cp/q$. In all results given above the fpc is ignored.

Example. A sampler wishes to estimate P with a coefficient of variation of 0.1(10%). He guesses that P will lie somewhere between 5 and 20%. This range is too wide to give a good initial estimate of the required n. Since the cv of P is $\sqrt{Q/nP}$, it is easily verified that $n = 400$ is adequate for $P = 20\%$, but $n = 1900$ will be needed if P is only 5%.

Accordingly, he takes an initial sample with $n_1 = 396$ and finds $p_1 = 0.101$. Since $\sqrt{C} = 0.1$, $C = 0.01$. Equation 4.9 gives

$$n = \frac{(0.899)}{(0.01)(0.101)} + \frac{3}{(0.0908)} + \frac{1}{(0.01)(40)} = 926$$

The combined sample gives $np = 88$; $p = 88/926 = 0.0950$. The correction for bias, Cp/q, amounts to 0.0011, giving a final estimate of 0.094 or 9.4%.

The second method, a small pilot survey, serves many purposes, especially if the feasibility of the main survey is in doubt. If the pilot survey it itself a simple random sample, the preceding methods apply. But often the pilot work is restricted to a part of the population that is convenient to handle or that will reveal the magnitude of certain problems. Allowance must be made for the selective nature of the pilot when using its results to estimate S^2 or P. For instance, a common practice is to confine the pilot work to a few clusters of units. Thus the computed s^2 measures mostly the variation within a cluster and may be an underestimate of the relevant S^2. The relation between intra- and intercluster variation is discussed in Chapter 9. The same problem arises in cluster sampling for proportions, in which the formula pq/n may underestimate the effect of variation among clusters. Cornfield (1951) gives a good illustration of the estimation of sample size in cluster sampling for proportions.

Method 3—the use of results from previous surveys—points to the value of making available, or at least keeping accessible, any data on standard deviations obtained in previous surveys. Unfortunately, the cost of computing standard deviations in complex surveys is high, even with electronic machines, and frequently only those s.d.'s needed to give a rough idea of the precision of the principal estimates are computed and recorded. If suitable past data are found, the value of S^2 may require adjustment for time changes. With skew data in which $\bar{Y}$ is changing with time, S^2 is often found to change at a rate lying somewhere between $k\bar{Y}$ and $k\bar{Y}^2$, where k is a constant. Thus, if $\bar{Y}$ is thought to have increased by 10% in the time interval since the previous survey, we might increase our initial estimate of S^2 by 10 to 20%.

Finally, a serviceable estimate of S^2 can sometimes be made from relatively little information about the nature of the population. In early studies of the numbers of wireworms in soils, a tool was used to take a sample ($9 \times 9 \times 5$ in.) of the topsoil. For estimating n, the sampler needed to know the standard deviation of the number of wireworms found in a boring with the tool. If wireworms were distributed at random over the topsoil, the number found in a small volume would follow the Poisson distribution, for which $S^2 = \bar{Y}$. Since there might be some tendency for wireworms to congregate, it was decided to assume $S^2 = 1.2\bar{Y}$, the factor 1.2 being an arbitrary safety factor. Although $\bar{Y}$ was not known, the values of $\bar{Y}$ that are of economic importance with respect to crop damage could be delineated. These two pieces of information made it possible to determine sample sizes that proved satisfactory.

Deming (1960) shows how some simple mathematical distributions may be used to estimate S^2 from a knowledge of the range and a general idea of the shape of the distribution. If the distribution is like a binomial, with a proportion p of the observations at one end of the range and a proportion q at the other end, $S^2 = pqh^2$, where h is the range. When $p = q = \frac{1}{2}$, the value of $S^2 = 0.25h^2$ is the maximum possible for a given range h. Other useful relations are that $S^2 = 0.083h^2$ for a rectangular distribution, $S^2 = 0.056h^2$ for a distribution shaped like a right triangle, and $S^2 = 0.042h^2$ for an isosceles triangle.

These relations do not help much if h is large or poorly known. However, if h is large, good sampling practice is to stratify the population (Chapter 5) so that within any stratum the range is much reduced. Usually the shape also becomes simpler (closer to rectangular) within a stratum. Consequently, these relations are effective in predicting S^2, hence n, within individual strata.

4.8 SAMPLE SIZE WITH MORE THAN ONE ITEM

In most surveys information is collected on more than one item. One method of determining sample size is to specify margins of error for the items that are regarded as most vital to the survey. An estimation of the sample size needed is first made separately for each of these important items.

When the single item estimations of n have been completed, it is time to take stock of the situation. It may happen that the n's required are all reasonably close. If the largest of the n's falls within the limits of the budget, this n is selected. More commonly, there is a sufficient variation among the n's so that we are reluctant to choose the largest, either from budgetary considerations or because this will give an over-all standard of precision substantially higher than originally contemplated. In this event the desired standard of precision may be relaxed for certain of the items, in order to permit the use of a smaller value of n.

In some cases the n's required for different items are so discordant that certain of them must be dropped from the inquiry; with the resources available the precision expected for these items is totally inadequate. The difficulty may not be merely one of sample size. Some items call for a different type of sampling from others. With populations that are sampled repeatedly, it is useful to amass information about those items that can be combined economically in a general survey and those that necessitate special methods. As an example, a classification of items into four types, suggested by experience in regional agricultural surveys, is shown in Table 4.1. In this classification, a general survey means one in which the units are fairly evenly distributed over some region as, for example, by a simple random sample.

TABLE 4.1

An Example of Different Types of Item in
Regional Surveys

Type	Characteristics of Item	Type of Sampling Needed
1	Widespread throughout the region, occurring with reasonable frequency in all parts.	A general survey with low sampling ratio.
2	Widespread throughout the region but with low frequency.	A general survey, but with a higher sampling ratio.
3	Occurring with reasonable frequency in most parts of the region, but with more sporadic distribution, being absent in some parts and highly concentrated in others.	For best results, a stratified sample with different intensities in different parts of the region (Chapter 5). Can sometimes be included in a general survey with supplementary sampling.
4	Distribution very sporadic or concentrated in a small part of the region.	Not suitable for a general survey. Requires a sample geared to its distribution.

4.9 SAMPLE SIZE WHEN ESTIMATES ARE WANTED FOR SUBDIVISIONS OF THE POPULATION

It is often planned to present estimates not only for the population as a whole but for certain subdivisions. If these can be identified in advance, as with different geographical regions, a separate calculation of n is made for each region. Suppose that the mean of each subdivision is to be estimated with a specified variance V. For the ith subdivision, we have $n_i = S_i^2 / V$, so that the total sample size $n = \sum S_i^2 / V$. The individual S_i^2 will, on the average, be smaller than S^2, the population variance, but often they are only slightly smaller. Thus, if there are k subdivisions, $n \doteq kS^2 / V$, whereas if only the estimate for the population as a whole were wanted we would take $n = S^2 / V$.

Thus if estimates with variance V are wanted for each of k subdivisions the sample size may approach k times the n needed for an over-all estimate of the same precision. This point tends to be overlooked in calculations of sample size by persons inexperienced in survey methods.

If the subdivisions represent classifications by variables such as age, sex, income, and years of schooling, the subdivision to which a person belongs is not known until the sample has been taken. Advance sample size estimates can still be made if the proportions π_i of the units that belong to the various subdivisions are known. If a simple random sample of size n is selected, the expected size of sample from the ith subdivision is $n\pi_i$. The average variance of the mean from this subdivision is

$$V(\bar{y}_i) = E\left(\frac{S_i^2}{n_i}\right) \doteq \frac{S_i^2}{n\pi_i} \tag{4.10}$$

if $n\pi_i$ is large. Hence we require $n \doteq S_i^2/\pi_i V$ in order to make $V(\bar{y}_i) = V$. If this is to hold for every subdivision,

$$n \doteq \max\left(\frac{S_i^2}{\pi_i V}\right) \tag{4.11}$$

If the subdivisions are into classes like age, income, S_i^2/π_i may be less than S^2 for central classes, but may be large for an extreme class with small π_i. In this event, we may either have to increase the value of V in this subdivision or find some way of identifying units in this subdivision in advance so that they can be sampled at a higher rate. The method of double sampling (Chapter 12) is sometimes useful for this purpose.

The demands on sample size are still greater in analytical studies in which the specifications are

$$V(\bar{y}_i - \bar{y}_j) \le V \tag{4.12}$$

for every pair of subdivisions (domains). In this case

$$n \doteq \max_{i,j} \frac{1}{V}\left(\frac{S_i^2}{\pi_i} + \frac{S_j^2}{\pi_j}\right) \tag{4.13}$$

If the S_i^2 are not very different from S^2, n will be $2kS^2/V$ when the k domains are of equal size, and still greater otherwise. The effect of fpc terms, neglected in this discussion, is to reduce the required n's to some extent.

4.10 SAMPLE SIZE IN DECISION PROBLEMS

A more logical approach to the determination of sample size can sometimes be developed when a practical decision is to be made from the results of the sample. The decision will presumably be more soundly based if the sample estimate has a low error than if it has a high error. We may be able to calculate, in monetary terms, the loss $l(z)$ that will be incurred in a decision through an error of amount z in the estimate. Although the actual value of z is not predictable in advance, sampling theory enables us to find the frequency distribution $f(z, n)$ of z which, for a specified sampling method will depend on the sample size n. Hence the *expected loss* for a given size of sample is

$$L(n) = \int l(z) f(z, n)\, dz \tag{4.14}$$

The purpose in taking the sample is to diminish this loss. If $C(n)$ is the cost of a sample of size n, a reasonable procedure is to choose n to minimize

$$C(n) + L(n) \tag{4.15}$$

since this is the total cost involved in taking the sample and in making decisions from its results. The choice of n determines both the optimum size of sample and the most advantageous degree of precision.

Alternatively, the same approach can be presented in terms of the monetary *gain* that accrues from having the sample information, instead of in terms of the loss that arises from errors in the sample information. If monetary gain is used, we construct an expected gain $G(n)$ from a sample of size n, where $G(n)$ is zero if no sample is taken. We *maximize*

$$G(n) - C(n)$$

In this form the principle is equivalent to the rule in classical economics that profit is to be maximized.

The simplest application occurs when the loss function, $l(z)$, is λz^2, where λ is a constant. It follows that

$$L(n) = \lambda E(z^2) \tag{4.16}$$

For instance, if $\hat{\bar{Y}}$ is the sample estimate of $\bar{Y}$, and $z = \hat{\bar{Y}} - \bar{Y}$,

$$L(n) = \lambda V(\hat{\bar{Y}}) = \frac{\lambda S^2}{n} - \frac{\lambda S^2}{N} \tag{4.17}$$

if simple random sampling is used.

The simplest type of cost function for the sample is

$$C(n) = c_0 + c_1 n \tag{4.18}$$

where c_0 is the overhead cost. By differentiation, the value of n that minimizes cost plus loss is

$$n = \sqrt{\lambda S^2 / c_1} \tag{4.19}$$

A more general form of this result is given by Yates (1960). The same analysis applies to any method of sampling and estimation in which the variance of the estimate is inversely proportional to n and the cost is a linear function of n.

Blythe (1945) describes the application of this principle to the estimation of the volume of timber in a lot for selling purposes (see exercise 4.11). Nordin (1944) discusses the optimum size of sample for estimating potential sales in a market that a manufacturer intends to enter. If the sales can be forecast accurately, the amount of fixed equipment and the production per unit period can be allocated to maximize the manufacturer's expected profit. Grundy et al. (1954, 1956) consider the optimum size of a second sample when the results of a first sample are already known.

This approach has received substantial further development from workers on statistical decision theory. Generalizations include the substitution of utility for money value as a scale on which to measure costs and losses, the explicit use of subjective prior information about unknown parameters by expressing this information as "prior" probability distributions of the unknown parameters, and the investigation of different types of cost and loss functions and of qualitative as well as quantitative data. For a comprehensive account of the method, see Raiffa and Schlaifer (1961). Although it is still not evident how frequently decision problems will be amenable to complete solution by this approach, the method has value in stimulating clear thinking about the important factors in a good decision. One area that appears suitable for applications is the sampling of lots of articles in a mass-production process in order to decide whether to accept or reject the lot on the basis of its estimated quality. Sittig (1951) considers the economics of sample-size determination, taking account of costs of inspection and the costs incurred through defective articles in accepted lots and good articles in rejected lots.

4.11   THE DESIGN EFFECT (*Deff*)

With the more complex sampling plans described later in this book, a useful quantity is the *design effect* (*deff*) of the plan (Kish, 1965). He describes this as the ratio of the variance of the estimate obtained from the (more complex) sample to the variance of the estimate obtained from a simple random sample of the same number of units. The design effect has two primary uses—in sample-size estimation and in appraising the efficiency of more complex plans. For instance, in estimating the proportion of people who possess some attribute, it is often convenient to use the household instead of the person as a sampling unit. As noted in Chapter 3, the formula PQ/n cannot be used with these plans. For estimating the proportion who had seen a doctor (section 3.12), a simple random sample of households gave $v(p) = 0.00520$ as against $pq/n = 0.00197$ for an equal-sized simple random sample of persons. An estimate of the *deff* for this cluster sample and this variate is $520/197 = 2.6$. When the sampling fractions are small, we can therefore estimate sample size by calculating the n (number of persons) needed with a simple random sample of persons and multiplying by 2.6. By noting *deff* ratios in this way for the important variates with a complex plan, we can use the simple formulas in this chapter for estimating the sample size with the complex plan and also judge whether the complex plan is advantageous in efficiency relative to its cost and complexity.

Estimating the *deff* from the results of a complex sample may require some algebra. We need to show how these results provide, if possible, unbiased estimates of S^2 or of PQ. Examples of these calculations are given for stratified random sampling in section 5A.11 and for cluster sampling with clusters of equal sizes in section 9.3.

EXERCISES

4.1 In a district containing 4000 houses the percentage of owned houses is to be estimated with a s.e. of not more than 2% and the percentage of two-car households with a s.e. of not more than 1%. (The figures 2 and 1% are the absolute values, not the cv's.) The true percentage of owners is thought to lie between 45 and 65% and the percentage of two-car households between 5 and 10%. How large a sample is necessary to satisfy both aims?

4.2 In the population of 676 petition sheets (Table 2.2, page 28) how large must the sample be if the total number of signatures is to be estimated with a margin of error of 1000, apart from a 1 in 20 chance? Assume that the value of s^2 given on page 28 is the population S^2.

4.3 A survey is to be made of the prevalence of the common diseases in a large population. For any disease that affects at least 1% of the individuals in the population, it is desired to estimate the total number of cases, with a coefficient of variaton of not more than 20%. (a) What size of simple random sample is needed, assuming that the presence of the disease can be recognized without mistakes? (b) What size is needed if total cases are wanted separately for males and females, with the same precision?

4.4 In a wireworm survey the number of wireworms per acre is to be estimated with a limit of error of 30%, at the 95% probability level, in any field in which wireworm density exceeds 200,000 per acre in the top 5 in. of soil. The sampling tool measures $9 \times 9 \times 5$ in. deep. Assuming that the number of wireworms in a single sample follows a distribution slightly more variable than the Poisson, we take $S^2 = 1.2\bar{Y}$. What size of simple random sample is needed? (1 acre = 43,560 sq ft.)

4.5 The following coefficients of variation per unit were obtained in a farm survey in Iowa, the unit being an area 1 mile square (data of R. J. Jessen):

Item	Estimated cv (%)
Acres in farms	38
Acres in corn	39
Acres in oats	44
Number of family workers	100
Number of hired workers	110
Number of unemployed	317

A survey is planned to estimate acreage items with a cv of $2\frac{1}{2}\%$ and numbers of workers (excluding unemployed) with a cv of 5%. With simple random sampling, how many units are needed? How well would this sample be expected to estimate the number of unemployed?

4.6 By experimental sampling, the mean value of a random variate is to be estimated with variance $V = 0.0005$. The values of the random variate for the first 20 samples drawn are shown on p. 87. How many more samples are needed? (Use equation 4.7.)

4.7 A household survey is designed to estimate the proportion of families possessing certain attributes. For the principal items of interest, the value of P is expected to lie between 30 and 70%. With simple random sampling, how large are the values of n necessary to estimate the following means with a standard error not exceeding 3%? (a) The

Sample Number	Value of Random Variate	Sample Number	Value of Random Variate
1	0.0725	11	0.0712
2	0.0755	12	0.0748
3	0.0759	13	0.0878
4	0.0739	14	0.0710
5	0.0732	15	0.0754
6	0.0843	16	0.0712
7	0.0727	17	0.0757
8	0.0769	18	0.0737
9	0.0730	19	0.0704
10	0.0727	20	0.0723

over-all mean P? (*b*) The *individual* means P_i for the income classes—under \$5000; \$5000 to \$10,000; over \$10,000. ($i = 1, 2, 3$)? (*c*) The differences between the means $(P_i - P_j)$ for every pair of the classes in (b)? Give a separate answer for (*a*), (*b*), and (*c*). Income statistics indicate that the proportions of families with incomes in the three classes above are 50, 38, and 12%.

4.8   The 4-year colleges in the United States were divided into classes of four different sizes according to their 1952–1953 enrollments. The standard deviations within each class are shown below.

	Class			
	1	2	3	4
Number of students	<1000	1000–3000	3000–10,000	over 10,000
S_i	236	625	2008	10,023

If you know the class boundaries but not the values of S_i, how well can you guess the S_i values by using simple mathematical figures (section 4.7)? No college has less than 200 students and the largest has about 50,000 students.

4.9   With a quadratic loss function and a linear cost function, as in section 4.10, S^2 is reduced to S'^2 by a superior sampling plan, c_0, c_1, and λ remaining unchanged. If n', V' denote the new optimum sample size and the accompanying $V(\hat{Y})$, show that $n' < n$ and that $V' < V$.

4.10   If the loss function due to an error in $\bar{y}$ is $\lambda|\bar{y} - \bar{Y}|$ and if the cost $C = c_0 + c_1 n$, show that with simple random sampling, ignoring the fpc, the most economical value of n is

$$\left(\frac{\lambda S}{c_1\sqrt{2\pi}}\right)^{2/3}$$

4.11   (Adapted from Blythe, 1945). The selling price of a lot of standing timber is UW, where U is the price per unit volume and W is the volume of timber on the lot. The number N of logs on the lot is counted, and the average volume per log is estimated from a simple random sample of n logs. The estimate is made and paid for by the seller and is provisionally accepted by the buyer. Later, the buyer finds out the exact volume purchased, and the seller reimburses him if he has paid for more than was delivered. If he has paid for less than was delivered, the buyer does not mention the fact.

Construct the seller's loss function. Assuming that the cost of measuring n logs is cn, find the optimum value of n. The standard deviation of the volume per log may be denoted by S and the fpc ignored.

4.12   (a) The presence or absence of each of two characteristics is to be measured on each unit in a simple random sample from a large population. If P_1, P_2 are the percentages of units in the population that possess characteristics 1 and 2, a client wishes to estimate $(P_1 - P_2)$ with a standard error not exceeding two percentage points. What sample size do you suggest if the client thinks that P_1 and P_2 both lie between 40 and 60% and that the characteristics are independently distributed on the units?

(b) Suppose that in (a) the client thinks that the characteristics are positively correlated, but does not know the correlation. You suggest an initial sample of 200, with the following results.

Characteristics		Number of units
1	2	
Yes	Yes	72
Yes	No	44
No	Yes	14
No	No	70
		200

What sample size do you now recommend to estimate $(P_1 - P_2)$ with a standard error $\leq 2\%$?

4.13   (a) Suppose you are estimating the sex ratio, which is close to equality, and could sample households of four persons father, mother, two children. Ignoring the small proportion of families with identical twins, find the *deff* factor for a simple random sample of n households versus one of the $4n$ persons.

(b) Would identical twin families lower or raise the *deff* factor?

CHAPTER 5

Stratified Random Sampling

5.1 DESCRIPTION

In stratified sampling the population of N units is first divided into subpopulations of $N_1, N_2, \ldots, N_L$ units, respectively. These subpopulations are nonoverlapping, and together they comprise the whole of the population, so that

$$N_1 + N_2 + \cdots + N_L = N$$

The subpopulations are called *strata*. To obtain the full benefit from stratification, the values of the N_h must be known. When the strata have been determined, a sample is drawn from each, the drawings being made independently in different strata. The sample sizes within the strata are denoted by $n_1, n_2, \ldots, n_L$, respectively.

If a simple random sample is taken in each stratum, the whole procedure is described as *stratified random sampling.*

Stratification is a common technique. There are many reasons for this; the principal ones are the following.

1. If data of known precision are wanted for certain subdivisions of the population, it is advisable to treat each subdivision as a "population" in its own right.

2. Administrative convenience may dictate the use of stratification; for example, the agency conducting the survey may have field offices, each of which can supervise the survey for a part of the population.

3. Sampling problems may differ markedly in different parts of the population. With human populations, people living in institutions (e.g., hotels, hospitals, prisons) are often placed in a different stratum from people living in ordinary homes because a different approach to the sampling is appropriate for the two situations. In sampling businesses we may possess a list of the large firms, which are placed in a separate stratum. Some type of area sampling may have to be used for the smaller firms.

4. Stratification may produce a gain in precision in the estimates of characteristics of the whole population. It may possible to divide a heterogeneous population

into subpopulations, each of which is internally homogeneous. This is suggested by the name *strata*, with its implication of a division into layers. If each stratum is homogeneous, in that the measurements vary little from one unit to another, a precise estimate of any stratum mean can be obtained from a small sample in that stratum. These estimates can then be combined into a precise estimate for the whole population.

The theory of stratified sampling deals with the properties of the estimates from a stratified sample and with the best choice of the sample sizes n_h to obtain maximum precision. In this development it is taken for granted that the strata have already been constructed. The problems of how to construct strata and of how many strata there should be are postponed to a later stage (section 5A.7).

5.2 NOTATION

The suffix h denotes the stratum and i the unit within the stratum. The notation is a natural extension of that previously used. The following symbols all refer to stratum h.

$$N_h \qquad\qquad\qquad \text{total number of units}$$

$$n_h \qquad\qquad\qquad \text{number of units in sample}$$

$$y_{hi} \qquad\qquad\qquad \text{value obtained for the } i\text{th unit}$$

$$W_h = \frac{N_h}{N} \qquad\qquad\qquad \text{stratum weight}$$

$$f_h = \frac{n_h}{N_h} \qquad\qquad\qquad \text{sampling fraction in the stratum}$$

$$\bar{Y}_h = \frac{\sum_{i=1}^{N_h} y_{hi}}{N_h} \qquad\qquad\qquad \text{true mean}$$

$$\bar{y}_h = \frac{\sum_{i=1}^{n_h} y_{hi}}{n_h} \qquad\qquad\qquad \text{sample mean}$$

$$S_h^2 = \frac{\sum_{i=1}^{N_h} (y_{hi} - \bar{Y}_h)^2}{N_h - 1} \qquad\qquad\qquad \text{true variance}$$

Note that the divisor for the variance is $(N_h - 1)$.

5.3 PROPERTIES OF THE ESTIMATES

For the population mean per unit, the estimate used in stratified sampling is $\bar{y}_{st}$ (*st* for *stratified*), where

$$\bar{y}_{st} = \frac{\sum\limits_{h=1}^{L} N_h \bar{y}_h}{N} = \sum\limits_{h=1}^{L} W_h \bar{y}_h \tag{5.1}$$

where $N = N_1 + N_2 + \cdots + N_L$.

The estimate $\bar{y}_{st}$ is not in general the same as the sample mean. The sample mean, $\bar{y}$, can be written as

$$\bar{y} = \frac{\sum\limits_{h=1}^{L} n_h \bar{y}_h}{n} \tag{5.2}$$

The difference is that in $\bar{y}_{st}$ the estimates from the individual strata receive their correct weights N_h/N. It is evident that $\bar{y}$ coincides with $\bar{y}_{st}$ provided that in every stratum

$$\frac{n_h}{n} = \frac{N_h}{N} \qquad \text{or} \qquad \frac{n_h}{N_h} = \frac{n}{N} \qquad \text{or} \qquad f_h = f$$

This means that the sampling fraction is the same in all strata. This stratification is described as stratification with *proportional* allocation of the n_h. It gives a *self-weighting* sample. If numerous estimates have to be made, a self-weighting sample is time-saving.

The principal properties of the estimate $\bar{y}_{st}$ are outlined in the following theorems. The first two theorems apply to stratified sampling in general and are not restricted to stratified random sampling; that is, the sample from any stratum need not be a simple random sample.

Theorem 5.1. If in every stratum the sample estimate $\bar{y}_h$ is unbiased, then $\bar{y}_{st}$ is an unbiased estimate of the population mean $\bar{Y}$.

Proof.

$$E(\bar{y}_{st}) = E \sum\limits_{h=1}^{L} W_h \bar{y}_h = \sum\limits_{h=1}^{L} W_h \bar{Y}_h$$

since the estimates are unbiased in the individual strata. But the population mean $\bar{Y}$ may be written

$$\bar{Y} = \frac{\sum\limits_{h=1}^{L} \sum\limits_{i=1}^{N_h} y_{hi}}{N} = \frac{\sum\limits_{h=1}^{L} N_h \bar{Y}_h}{N} = \sum\limits_{h=1}^{L} W_h \bar{Y}_h$$

This completes the proof.

Theorem 5.2. If the samples are drawn independently in different strata,

$$V(\bar{y}_{st}) = \sum_{h=1}^{L} W_h^2 V(\bar{y}_h) \tag{5.3}$$

where $V(\bar{y}_h)$ is the variance of $\bar{y}_h$ over repeated samples from stratum h.

Proof. Since

$$\bar{y}_{st} = \sum_{h=1}^{L} W_h \bar{y}_h \tag{5.4}$$

$\bar{y}_{st}$ is a linear function of the $\bar{y}_h$ with fixed weights W_h. Hence we may quote the result in statistics for the variance of a linear function.

$$V(\bar{y}_{st}) = \sum_{h=1}^{L} W_h^2 V(\bar{y}_h) + 2 \sum_{h=1}^{L} \sum_{j>h}^{L} W_h W_j \operatorname{Cov}(\bar{y}_h \bar{y}_j) \tag{5.5}$$

But since samples are drawn independently in different strata, all covariance terms vanish. This gives the result (5.3).

To summarize theorems 5.1 and 5.2: if $\bar{y}_h$ is an unbiased estimate of $\bar{Y}_h$ in every stratum, and sample selection is independent in different strata, then $\bar{y}_{st}$ is an unbiased estimate of $\bar{Y}$ with variance $\sum W_h^2 V(\bar{y}_h)$.

The important point about this result is that the variance of $\bar{y}_{st}$ depends only on the variances of the estimates of the individual stratum means $\bar{Y}_h$. If it were possible to divide a highly variable population into strata such that all items had the same value within a stratum, we could estimate $\bar{Y}$ without any error. Equation (5.4) shows that it is the use of the correct stratum weights N_h/N in making the estimate $\bar{y}_{st}$ that leads to this result.

Theorem 5.3. For stratified random sampling, the variance of the estimate $\bar{y}_{st}$ is

$$V(\bar{y}_{st}) = \frac{1}{N^2} \sum_{h=1}^{L} N_h(N_h - n_h) \frac{S_h^2}{n_h} = \sum_{h=1}^{L} W_h^2 \frac{S_h^2}{n_h}(1 - f_h) \tag{5.6}$$

Proof. Since $\bar{y}_h$ is an unbiased estimate of $\bar{Y}_h$, theorem 5.2 can be applied. Furthermore, by theorem 2.2, applied to an individual stratum,

$$V(\bar{y}_h) = \frac{S_h^2}{n_h} \frac{N_n - n_h}{N_h}$$

By substitution into the result of theorem 5.2, we obtain

$$V(\bar{y}_{st}) = \frac{1}{N^2} \sum_{h=1}^{L} N_h^2 V(\bar{y}_h) = \frac{1}{N^2} \sum_{h=1}^{L} N_h(N_h - n_h) \frac{S_h^2}{n_h} = \sum W_h^2 \frac{S_h^2}{n_h}(1 - f_h)$$

Some particular cases of this formula are given in the following corollaries.

Corollary 1. If the sampling fractions n_h/N_h are negligible in all strata,

$$V(\bar{y}_{st}) = \frac{1}{N^2} \sum \frac{N_h{}^2 S_h{}^2}{n_h} = \sum \frac{W_h{}^2 S_h{}^2}{n_h} \tag{5.7}$$

This is the appropriate formula when finite population corrections can be ignored.

Corollary 2. With proportional allocation, we substitute

$$n_h = \frac{n N_h}{N}$$

in (5.6). The variance reduces to

$$V(\bar{y}_{st}) = \sum \frac{N_h}{N} \frac{S_h{}^2}{n} \left(\frac{N-n}{N}\right) = \frac{1-f}{n} \sum W_h S_h{}^2 \tag{5.8}$$

Corollary 3. If sampling is proportional and the variances in all strata have the same value, $S_w{}^2$, we obtain the simple result

$$V(\bar{y}_{st}) = \frac{S_w{}^2}{n} \left(\frac{N-n}{N}\right) \tag{5.9}$$

Theorem 5.4. If $\hat{Y}_{st} = N\bar{y}_{st}$ is the estimate of the population total Y, then

$$V(\hat{Y}_{st}) = \sum N_h (N_h - n_h) \frac{S_h{}^2}{n_h} \tag{5.10}$$

This follows at once from theorem 5.3.

Example. Table 5.1 shows the 1920 and 1930 number of inhabitants, in thousands, of 64 large cities in the United States. The data were obtained by taking the cities which ranked fifth to sixty-eighth in the United States in total number of inhabitants in 1920. The cities are arranged in two strata, the first containing the 16 largest cities and the second the remaining 48 cities.

The total number of inhabitants in all 64 cities in 1930 is to be estimated from a sample of size 24. Find the standard error of the estimated total for (1) a simple random sample, (2) a stratified random sample with proportional allocation, (3) a stratified random sample with 12 units drawn from each stratum.

This population resembles the populations of many types of business enterprise in that some units—the large cities—contribute very substantially to the total and display much greater variability than the remainder.

The stratum totals and sums of squares are given under Table 5.1. Only the 1930 data are used in this example: the 1920 data appear in a later example.

For the complete population in 1930, we find

$$Y = 19{,}568, \qquad S^2 = 52{,}448$$

The three estimates of Y are denoted by $\hat{Y}_{ran}$, $\hat{Y}_{prop}$, and $\hat{Y}_{equal}$.

1. For simple random sampling

$$V(\hat{Y}_{ran}) = \frac{N^2 S^2}{n} \frac{N-n}{N} = \frac{(64)^2 (52{,}448)}{24} \left(\frac{40}{64}\right) = 5{,}594{,}453$$

TABLE 5.1

SIZES OF 64 CITIES (IN 1000's) IN 1920 AND 1930

	1920 Size (x_{hi})				1930 Size (y_{hi})		
	Stratum				Stratum		
$h = 1$	2			1	2		
797	314	172	121	900	364	209	113
773	298	172	120	822	317	183	115
748	296	163	119	781	328	163	123
734	258	162	118	805	302	253	154
588	256	161	118	670	288	232	140
577	243	159	116	1238	291	260	119
507	238	153	116	573	253	201	130
507	237	144	113	634	291	147	127
457	235	138	113	578	308	292	100
438	235	138	110	487	272	164	107
415	216	138	110	442	284	143	114
401	208	138	108	451	255	169	111
387	201	136	106	459	270	139	163
381	192	132	104	464	214	170	116
324	180	130	101	400	195	150	122
315	179	126	100	366	260	143	134

Note. Cities are arranged in the same order in both years.

Totals and sums of squares

	1920		1930	
Stratum	$\sum (x_{hi})$	$\sum (x_{hi}^2)$	$\sum (y_{hi})$	$\sum (y_{hi}^2)$
1	8,349	4,756,619	10,070	7,145,450
2	7,941	1,474,871	9,498	2,141,720

from theorem 2.2, corollary 2. The standard error is

$$\sigma(\hat{Y}_{ran}) = 2365$$

2. For the individual strata the variances are

$$S_1^2 = 53,843, \qquad S_2^2 = 5581$$

Note that the stratum with the largest cities has a variance nearly 10 times that of the other stratum.

In proportional allocation, we have $n_1 = 6$, $n_2 = 18$. From (5.7), multiplying by N^2, we have

$$V(\hat{Y}_{prop}) = \frac{N-n}{n} \sum N_h S_h^2$$

$$= \tfrac{40}{24}[(16)(53,843) + (48)(5581)] = 1,882,293$$

$$\sigma(\hat{Y}_{prop}) = 1372$$

3. For $n_1 = n_2 = 12$ we use the general formula (5.9):

$$V(\hat{Y}_{equal}) = \sum N_h(N_h - n_h)\frac{S_h^2}{n_h}$$

$$= \frac{(16)(4)(53,843)}{12} + \frac{(48)(36)(5581)}{12} = 1,090,827$$

$$\sigma(\hat{Y}_{equal}) = 1044$$

In this example equal sample sizes in the two strata are more precise than proportional allocation. Both are greatly superior to simple random sampling.

5.4 THE ESTIMATED VARIANCE AND CONFIDENCE LIMITS

If a simple random sample is taken within each stratum, an unbiased estimate of S_h^2 (from theorem 2.4) is

$$s_h^2 = \frac{1}{n_h - 1} \sum_{i=1}^{n_h} (y_{hi} - \bar{y}_h)^2 \tag{5.11}$$

Hence we obtain the following.

Theorem 5.5. With stratified random sampling, an unbiased estimate of the variance of $\bar{y}_{st}$ is

$$v(\bar{y}_{st}) = s^2(\bar{y}_{st}) = \frac{1}{N^2} \sum_{h=1}^{L} N_h(N_h - n_h)\frac{s_h^2}{n_h} \tag{5.12}$$

An alternative form for computing purposes is

$$s^2(\bar{y}_{st}) = \sum_{h=1}^{L} \frac{W_h^2 s_h^2}{n_h} - \sum_{h=1}^{L} \frac{W_h s_h^2}{N} \tag{5.13}$$

The second term on the right represents the reduction due to the fpc.

In order to compute this estimate, there must be at least two units drawn from every stratum. Estimation of the variance when stratification is carried to the point at which only one unit is chosen per stratum is discussed in section 5A.12.

The formulas for confidence limits are as follows.

Population mean:	$\bar{y}_{st} \pm ts(\bar{y}_{st})$	(5.14)
Population total:	$N\bar{y}_{st} \pm tNs(\bar{y}_{st})$	(5.15)

These formulas assume that $\bar{y}_{st}$ is normally distributed and that $s(\bar{y}_{st})$ is well determined, so that the multiplier t can be read from tables of the normal distribution.

If only a few degrees of freedom are provided by each stratum, the usual procedure for taking account of the sampling error attached to a quantity like $s(\bar{y}_{st})$ is to read the t-value from the tables of Student's t instead of from the normal table. The distribution of $s(\bar{y}_{st})$ is in general too complex to allow a strict application of this method. An approximate method of assigning an effective number of degrees of freedom to $s(\bar{y}_{st})$ is as follows (Satterthwaite, 1946).

We may write

$$s^2(\bar{y}_{st}) = \frac{1}{N^2} \sum_{h=1}^{L} g_h s_h^2, \quad \text{where } g_h = \frac{N_h(N_h - n_h)}{n_h}$$

The effective number of degrees of freedom n_e is

$$n_e = \frac{\left(\sum g_h s_h^2 \right)^2}{\sum \dfrac{g_h^2 s_h^4}{n_h - 1}} \tag{5.16}$$

The value of n_e always lies between the smallest of the values $(n_h - 1)$ and their sum. The approximation takes account of the fact that S_h^2 may vary from stratum to stratum. It requires the assumption that the y_{hi} are normal, since it depends on the result that the variance of s_h^2 is $2\sigma_h^4/(n_h - 1)$. If the distribution of y_{hi} has positive kurtosis, the variance of s_h^2 will be larger than this and formula 5.16 overestimates the effective degrees of freedom.

5.5 OPTIMUM ALLOCATION

In stratified sampling the values of the sample sizes n_h in the respective strata are chosen by the sampler. They may be selected to minimize $V(\bar{y}_{st})$ for a specified cost of taking the sample or to minimize the cost for a specified value of $V(\bar{y}_{st})$.

The simplest cost function is of the form

$$\text{cost} = C = c_0 + \sum c_h n_h \tag{5.17}$$

Within any stratum the cost is proportional to the size of sample, but the cost per unit c_h may vary from stratum to stratum. The term c_0 represents an overhead cost. This cost function is appropriate when the major item of cost is that of taking the measurements on each unit. If travel costs between units are substantial, empirical and mathematical studies suggest that travel costs are better represented by the expression $\sum t_h \sqrt{n_h}$ where t_h is the travel cost per unit (Beardwood et al., 1959). Only the linear cost function (5.17) is considered here.

Theorem 5.6. In stratified random sampling with a linear cost function of the form (5.17), the variance of the estimated mean $\bar{y}_{st}$ is a minimum for a specified cost C, and the cost is a minimum for a specified variance $V(\bar{y}_{st})$, when n_h is proportional to $W_h S_h / \sqrt{c_h}$.

Proof. We have

$$C = c_0 + \sum_{h=1}^{L} c_h n_h \tag{5.17}$$

$$V = V(\bar{y}_{st}) = \sum_{h=1}^{L} \frac{W_h^2 S_h^2}{n_h}(1 - f_h) = \sum_{h=1}^{L} \frac{W_h^2 S_h^2}{n_h} - \sum_{h=1}^{L} \frac{W_h^2 S_h^2}{N_h} \tag{5.18}$$

Our problems are either (1) to choose the n_h so as to minimize V for specified C, or (2) to choose the n_h so as to minimize C for specified V. It happens that apart from their final steps, the problems have the same solution. Choosing the n_h to minimize V for fixed C or C for fixed V are both equivalent to minimizing the product

$$V'C' = \left(V + \sum \frac{W_h^2 S_h^2}{N_h} \right)(C - c_0)$$

$$= \left(\sum \frac{W_h^2 S_h^2}{n_h} \right)\left(\sum c_h n_h \right) \tag{5.19}$$

Stuart (1954) has noted that (5.19) may be minimized neatly by use of the Cauchy–Schwarz inequality. If a_h, b_h are two sets of L positive numbers, this inequality comes from the identity

$$\left(\sum a_h^2 \right)\left(\sum b_h^2 \right) - \left(\sum a_h b_h \right)^2 = \sum_i \sum_{j>i} (a_i b_j - a_j b_i)^2 \tag{5.20}$$

It follows from (5.20) that

$$\left(\sum a_h^2 \right)\left(\sum b_h^2 \right) \ge \left(\sum a_h b_h \right)^2 \tag{5.21}$$

equality occurring if and only if b_h / a_h is constant for all h. In (5.19) take

$$a_h = \frac{W_h S_h}{\sqrt{n_h}}, \qquad b_h = \sqrt{c_h n_h}, \qquad a_h b_h = W_h S_h \sqrt{c_h}$$

The inequality (5.21) gives

$$V'C' = \left(\sum \frac{W_h^2 S_h^2}{n_h} \right)\left(\sum c_h n_h \right) = \left(\sum a_h^2 \right)\left(\sum b_h^2 \right) \ge \left(\sum W_h S_h \sqrt{c_h} \right)^2$$

Thus, no choice of the n_h can make $V'C'$ smaller than $\left(\sum W_h S_h \sqrt{c_h} \right)^2$. The minimum value occurs when

$$\frac{b_h}{a_h} = \frac{n_h\sqrt{c_h}}{W_h S_h} = \text{constant} \tag{5.22}$$

as stated in the theorem.

In terms of the total sample size n_h in a stratum, we have

$$\frac{n_h}{n} \equiv \frac{W_h S_h/\sqrt{c_h}}{\sum (W_h S_h/\sqrt{c_h})} = \frac{N_h S_h/\sqrt{c_h}}{\sum (N_h S_h/\sqrt{c_h})} \tag{5.23}$$

This theorem leads to the following rules of conduct. In a given stratum, take a larger sample if

1. The stratum is larger.
2. The stratum is more variable internally.
3. Sampling is cheaper in the stratum.

One further step is needed to complete the allocation. Equation (5.23) gives the n_h in terms of n, but we do not yet know what value n has. The solution depends on whether the sample is chosen to meet a specified total cost C or to give a specified variance V for $\bar{y}_{st}$. If *cost* is fixed, substitute the optimum values of n_h in the cost function (5.17) and solve for n. This gives

$$n = \frac{(C - c_0)\sum (N_h S_h/\sqrt{c_h})}{\sum (N_h S_h\sqrt{c_h})} \tag{5.24}$$

If V is fixed, substitute the optimum n_h in the formula for $V(\bar{y}_{st})$. We find

$$n = \frac{\left(\sum W_h S_h \sqrt{c_h}\right)\sum W_h S_h/\sqrt{c_h}}{V + (1/N)\sum W_h S_h^2} \tag{5.25}$$

where $W_h = N_h/N$.

An important special case arises if $c_h = c$, that is, if the cost per unit is the same in all strata. The cost becomes $C = c_0 + cn$, and optimum allocation for fixed cost reduces to optimum allocation for fixed sample size. The result in this special case is as follows.

Theorem 5.7. In stratified random sampling $V(\bar{y}_{st})$ is minimized for a fixed total size of sample n if

$$n_h = n\frac{W_h S_h}{\sum W_h S_h} = n\frac{N_h S_h}{\sum N_h S_h} \tag{5.26}$$

This allocation is sometimes called *Neyman allocation*, after Neyman (1934), whose proof gave the result prominence. An earlier proof by Tschuprow (1923) was later discovered.

A formula for the minimum variance with fixed n is obtained by substituting the value of n_h in (5.26) into the general formula for $V(\bar{y}_{st})$. The result is

$$V_{min}(\bar{y}_{st}) = \frac{\left(\sum W_h S_h\right)^2}{n} - \frac{\sum W_h S_h^2}{N} \tag{5.27}$$

The second term on the right represents the fpc.

5.6 RELATIVE PRECISION OF STRATIFIED RANDOM AND SIMPLE RANDOM SAMPLING

If intelligently used, stratification nearly always results in a smaller variance for the estimated mean or total than is given by a comparable simple random sample. It is not true, however, that *any* stratified random sample gives a smaller variance than a simple random sample. If the values of the n_h are far from optimum, stratified sampling may have a higher variance. In fact, even stratification with optimum allocation for fixed total sample size may give a higher variance, although this result is an academic curiosity rather than something likely to happen in practice.

In this section a comparison is made between simple random sampling and stratified random sampling with proportional and optimum allocation. This comparison shows how the gain due to stratification is achieved.

The variances of the estimated *means* are denoted by V_{ran}, V_{prop}, and V_{opt}, respectively.

Theorem 5.8. If terms in $1/N_h$ are ignored relative to unity,

$$V_{opt} \leq V_{prop} \leq V_{ran} \tag{5.28}$$

where the optimum allocation is for fixed n, that is, with $n_h \propto N_h S_h$.

Proof.

$$V_{ran} = (1-f)\frac{S^2}{n} \tag{5.29}$$

$$V_{prop} = \frac{(1-f)}{n}\sum W_h S_h^2 = \frac{\sum W_h S_h^2}{n} - \frac{\sum W_h S_h^2}{N} \tag{5.30}$$

[from equation (5.8), section 5.3]

$$V_{opt} = \frac{\left(\sum W_h S_h\right)^2}{n} - \frac{\sum W_h S_h^2}{N} \tag{5.31}$$

[from equation (5.27), section 5.5]

From the standard algebraic identity for the analysis of variance of a stratified population, we have

$$(N-1)S^2 = \sum_h \sum_i (y_{hi} - \bar{Y})^2$$

$$= \sum_h \sum_i (y_{hi} - \bar{Y}_h)^2 + \sum_h N_h(\bar{Y}_h - \bar{Y})^2$$

$$= \sum_h (N_h - 1)S_h^2 + \sum_h N_h(\bar{Y}_h - \bar{Y})^2 \qquad (5.32)$$

If terms in $1/N_h$ are negligible and hence also in $1/N$, (5.32) gives

$$S^2 = \sum W_h S_h^2 + \sum W_h(\bar{Y}_h - \bar{Y})^2 \qquad (5.33)$$

Hence

$$V_{ran} = (1-f)\frac{S^2}{n} = \frac{(1-f)}{n}\sum W_h S_h^2 + \frac{(1-f)}{n}\sum W_h(\bar{Y}_h - \bar{Y})^2 \qquad (5.34)$$

$$= V_{prop} + \frac{(1-f)}{n}\sum W_h(\bar{Y}_h - \bar{Y})^2 \qquad (5.35)$$

By the definition of V_{opt}, we must have $V_{prop} \geq V_{opt}$. By (5.30) and (5.31) their difference is

$$V_{prop} - V_{opt} = \frac{1}{n}\left[\sum W_h S_h^2 - \left(\sum W_h S_h\right)^2\right]$$

$$= \frac{1}{n}\left[\sum W_h(S_h - \bar{S})^2\right] \qquad (5.36)$$

where $\bar{S} = \sum W_h S_h$ is a weighted mean of the S_h.

From (5.35) and (5.36), with terms in $1/N_h$ negligible,

$$V_{ran} = V_{opt} + \frac{1}{n}\sum W_h(S_h - \bar{S})^2 + \frac{(1-f)}{n}\sum W_h(\bar{Y}_h - \bar{Y})^2 \qquad (5.37)$$

To summarize, in equation (5.37) there are two components of the decrease in variance as we change from simple random sampling to optimum allocation. The first component (term on the extreme right) comes from the elimination of differences among the stratum means; the second (middle term on the right) comes from elimination of the effect of differences among the stratum standard deviations. The second component represents the difference in variance between optimum and proportional allocation.

If terms in $1/N_h$ are not negligible, substitution for S^2 from (5.32) leads to

$$V_{ran} = V_{prop} + \frac{(1-f)}{n(N-1)}\left[\sum N_h(\bar{Y}_h - \bar{Y})^2 - \frac{1}{N}\sum (N-N_h)S_h^2\right] \qquad (5.38)$$

instead of to (5.35).

It follows that proportional stratification gives a higher variance than simple random sampling if

$$\sum N_h(\bar{Y}_h - \bar{Y})^2 < \frac{1}{N} \sum (N - N_h)S_h^2 \tag{5.39}$$

Mathematically, this can happen. Suppose that the S_h^2 are all equal to S_w^2, so that proportional allocation is optimum in the sense of Neyman. Then (5.39) becomes

$$\sum N_h(\bar{Y}_h - \bar{Y})^2 < (L-1)S_w^2$$

or

$$\frac{\sum N_h(\bar{Y}_h - \bar{Y})^2}{L-1} < S_w^2 \tag{5.40}$$

Those familiar with the analysis of variance will recognize this relation as implying that the mean square among strata is smaller than the mean square within strata, that is, that the F-ratio is less than 1.

5.7 WHEN DOES STRATIFICATION PRODUCE LARGE GAINS IN PRECISION?

The ideal variate for stratification is the value of y itself—the quantity to be measured in the survey. If we could stratify by the values of y, there would be no overlap between strata, and the variance within strata would be much smaller than the over-all variance, particularly if there were many strata. This situation is illustrated by the example in section 5.3, page 94. The population consisted of the sizes (numbers of inhabitants) of 64 cities in 1930, stratified by size. Although there were only two strata, proportional stratification reduced the s.e. ($\hat{Y}$) from 2365 to 1372. Stratification with $n_1 = n_2 = 12$, which is optimum under Neyman allocation, produced a further reduction to 1044.

In practice, of course, we cannot stratify by the values of y. But some important applications come close to this situation, and therefore give large gains in precision, by satisfying the following three conditions.

1. The population is composed of institutions varying widely in size.
2. The principal variables to be measured are closely related to the sizes of the institutions.
3. A good measure of size is available for setting up the strata.

Examples are businesses of a specific kind, for example, groceries (in surveys dealing with the volume of business or number of employees), schools (in surveys related to numbers of pupils), hospitals (in studies of patient load), and income tax returns (for items highly correlated with taxable income). In the United States farms also vary greatly in size as measured by total acreage or gross income, but common farm items, such as the production of particular crops or types of livestock, often exhibit only a moderate correlation with farm size, so that the gains from stratification by farm size are not huge.

If the size of the institution remains stable through time, at least for short periods, then its best practical measure is usually the size of the institution on some recent occasion when a census was taken. The example in section 5.3 illustrates the situation in which good previous data are available. Table 5.2 shows the S_h and the resulting optimum $n_h \propto N_h S_h$, when the allocation is made from 1920 and 1930 data, respectively.

The 1920 data indicate an n_1 of 11.56, as against a "true" optimum of 12.21 for the 1930 data. When rounded to integers, both sets of data give the same allocation—a sample size of 12 from each stratum.

TABLE 5.2

CALCULATION OF THE OPTIMUM ALLOCATION

Stratum	N_h	1920 Data			1930 Data		
		S_h	$N_h S_h$	n_h	S_h	$N_h S_h$	n_h
1	16	163.30	2612.80	11.56	232.04	3712.64	12.21
2	48	58.55	2810.40	12.44	74.71	3586.08	11.79
Totals	64		5423.20	24.00		7298.72	24.00

Note that the optimum sampling fraction is 75% in stratum 1 but only 25% in stratum 2. It is often found that because of the high variability of the stratum consisting of the largest institutions, the formula calls for 100% sampling in this stratum. Indeed, the allocation may call for more than 100% sampling (see section 5.8). Note also that the S_h are smaller in 1920 than in 1930. The 1920 data give an overoptimistic impression of the precision to be obtained in a 1930 survey. As mentioned in section 4.7, the possibility of a change in the levels of the S_h should always be considered when using past data, even though an allowance for change may have to be something of a guess.

Geographic stratification, in which the strata are compact areas such as counties or neighbourhoods in a city, is common—often for administrative convenience or because separate data are wanted for each stratum. It is usually accompanied by some increase in precision because many factors operate to make people living or crops growing in the same area show similarities in their principal characteristics. The gains from geographic stratification, however, are generally modest. For example, Table 5.3 shows data published by Jessen (1942) and Jessen and Houseman (1944) on the effectiveness of geographic stratification for a number of typical farm economic items.

Four sizes of stratum are represented—the township, the county, the "type of farming" area, and the state. To give some idea of the relative sizes of the strata, there are about 1600 townships, 100 counties, and 5 areas in Iowa.

In the table the precision of a method of stratification is taken as inversely proportional to the value of $V(\bar{y}_{st})$ given by the method. Thus the relative precision of method 1 to method 2 is the ratio $V_2(\bar{y}_{st})/V_1(\bar{y}_{st})$, expressed as a percentage. The data shown are averages over the numbers of items given in the second column. The county is taken as a standard in each case. As indicated, the gains in precision are moderate. In Iowa the use of 1600 strata (townships) compared with no stratification (state) increases the precision by about 30%; that is, it reduces the variance by about 25%.

TABLE 5.3

RELATIVE PRECISION OF DIFFERENT KINDS OF GEOGRAPHIC
STRATIFICATION (IN PER CENT)

		Stratum			
State	No. of Items	Township	County	Type of Farming Area	State
Iowa, 1938	18	115	100	96	91
Iowa, 1939	19	121	100	97	91
Florida, 1942					
Citrus fruit area	14	144	100	..	
Truck farming area	15	111	100	..	
California, 1942	17	113	100	97	

As regards proportional versus optimum stratification, there are two situations in which optimum stratification wins handsomely. The first is the case, already discussed, in which the population consists of large and small institutions, stratified by some measure of size. The variances $S_h{}^2$ are usually much greater for the large institutions than for the small, making proportional stratification inefficient. The second situation is found in surveys in which some strata are much more expensive to sample than others. The influence of the factors $\sqrt{c_h}$ may make proportional allocation poor.

When planning an allocation in which the estimated n_h do not differ greatly from proportionality, it is worthwhile to estimate how much larger $V(\bar{y}_{st})$ or $V(\hat{Y}_{st})$ become if proportional allocation is used. The optima in the allocation problem are rather flat (see section 5A.2) and the increase in variance may turn out surprisingly small. Moreover, the superiority of the optimum, as computed from estimated values of the S_h, is always exaggerated because of the errors in the estimated S_h. The simplicity and the self-weighting feature of proportional allocation are probably worth a 10 to 20% increase in variance.

5.8 ALLOCATION REQUIRING MORE THAN 100 PER CENT SAMPLING

As mentioned in section 5.7, the formula for the optimum may produce an n_h in some stratum that is larger than the corresponding N_h. Consider the example on city sizes in section 5.3. A sample of 24 cities, distributed between two strata, called for 12 cities out of 16 in the first stratum and 12 out of 48 in the second. Had the sample size been 48, the allocation would demand 24 cities out of 16 in the first stratum. The best that can be done is to take all cities in the stratum, leaving 32 cities for the second stratum instead of the 24 postulated by the formula. This problem arises only when the over-all sampling fraction is substantial and some strata are much more variable than others. It has occurred in practice on several occasions.

If the original allocation gives $n_1 > N_1$ when there are more than two strata, the optimum revised allocation is:

$$\tilde{n}_1 = N_1; \qquad \tilde{n}_h = (n - N_1) \frac{W_h S_h}{\sum\limits_{2}^{L} W_h S_h}, \qquad (h \geq 2) \qquad (5.41)$$

provided that $\tilde{n}_h \leq N_h$ for $h \geq 2$. If it should happen that $\tilde{n}_2 > N_2$, we change the allocation to

$$\tilde{n}_1 = N_1; \qquad \tilde{n}_2 = N_2; \qquad \tilde{n}_h = (n - N_1 - N_2) \frac{W_h S_h}{\sum\limits_{3}^{L} W_h S_h}, \qquad (h \geq 3) \quad (5.41)'$$

provided that $\tilde{n}_h \leq N_h$ for $h \geq 3$. We continue this process until every $\tilde{n}_h \leq N_h$. The resulting allocation may be shown to be optimum for given n, as would be expected.

Care must be taken to use the correct formula for $V(\bar{y}_{st})$. The general formula (5.6) in section 5.3 is correct if the $\tilde{n}_h$ given by the *revised* optimum allocation are substituted. Formula (5.27) for $V_{min}(\bar{y}_{st})$,

$$V_{min}(\bar{y}_{st}) = \frac{(\sum W_h S_h)^2}{n} - \frac{\sum W_h S_h^2}{N} \qquad (5.27)$$

no longer holds. If $\sum'$ denotes summation over the strata in which $\tilde{n}_h < N_h$, an alternative correct formula is

$$V_{min}(\bar{y}_{st}) = \frac{(\sum' W_h S_h)^2}{n'} - \frac{\sum' W_h S_h^2}{N} \qquad (5.42)$$

where n' is the revised total sample size in these strata.

5.9 ESTIMATION OF SAMPLE SIZE WITH CONTINUOUS DATA

Formulas for the determination of n under an estimated optimum allocation were given in section 5.5. The present section presents formulas for any allocation, with some useful special cases. It is assumed that the estimate has a specified variance V. If, instead, the margin of error d (section 4.4) has been specified, $V = (d/t)^2$, where t is the normal deviate corresponding to the allowable probability that the error will exceed the desired margin.

Estimation of the Population Mean $\bar{Y}$

Let s_h be the estimate of S_h and let $n_h = w_h n$ where the w_h have been chosen. In these terms the anticipated $V(\bar{y}_{st})$ (from theorem 5.3, section 5.3) is

$$V = \frac{1}{n} \sum \frac{W_h^2 s_h^2}{w_h} - \frac{1}{N} \sum W_h s_h^2 \tag{5.43}$$

with $W_h = N_h/N$. This gives, as a general formula for n,

$$n = \frac{\sum \dfrac{W_h^2 s_h^2}{w_h}}{V + \dfrac{1}{N} \sum W_h s_h^2} \tag{5.44}$$

If the fpc is ignored, we have, as a first approximation,

$$n_0 = \frac{1}{V} \sum \frac{W_h^2 s_h^2}{w_h} \tag{5.45}$$

If n_0/N is not negligible, we may calculate n as

$$n = \frac{n_0}{1 + \dfrac{1}{NV} \sum W_h s_h^2} \tag{5.46}$$

In particular cases the formulas take various forms that may be more convenient for computation. A few are given.

Presumed optimum allocation (for fixed n): $w_h \propto W_h s_h$.

$$n = \frac{(\sum W_h s_h)^2}{V + \dfrac{1}{N} \sum W_h s_h^2} \tag{5.47}$$

Proportional allocation: $w_h = W_h = N_h/N$.

$$n_o = \frac{\sum W_h s_h^2}{V}, \qquad n = \frac{n_0}{1 + \dfrac{n_0}{N}} \tag{5.48}$$

Estimation of the Population Total

If V is the desired $V(\hat{Y}_{st})$, the principal formulas are as follows
General:

$$n = \frac{\sum \dfrac{N_h^{\,2} s_h^{\,2}}{w_h}}{V + \sum N_h s_h^{\,2}} \tag{5.49}$$

Presumed optimum (for fixed n):

$$n = \frac{\left(\sum N_h s_h\right)^2}{V + \sum N_h s_h^{\,2}} \tag{5.50}$$

Proportional:

$$n_0 = \frac{N}{V} \sum N_h s_h^{\,2}, \qquad n = \frac{n_0}{1 + \dfrac{n_0}{N}} \tag{5.51}$$

Example. This example comes from a paper by Cornell (1947), which describes a sample of United States colleges and universities drawn in 1946 by the U.S. Office of Education in order to estimate enrollments for the 1946–1947 academic year. The illustration is for the population of 196 teachers' colleges and normal schools. These were arranged in seven strata, of which one small stratum will be ignored. The first five strata were constructed by size of institution: the sixth contained colleges for women only. Estimates s_h of the S_h were computed from results for the 1943–1944 academic year. An "optimum," stratification based on these s_h was employed.

The objective was a coefficient of variation of 5% in the estimated total enrollment. In 1943 the total enrollment for this group of colleges was 56,472. Thus the desired standard error is

$$(0.05)(56,472) = 2824$$

so that the desired variance is

$$V = (2824)^2 = 7,974,976$$

It may be objected that enrollments will be greater in 1946 than in 1943 and that allowance should be made for this increase. Actually, the calculation assumes only that the cv per college remains the same in 1943 and 1946—an assumption that may not be unreasonable.

Table 5.4 shows the values of N_h, s_h, and $N_h s_h$, which were known before determining n.

The appropriate formula for n is (5.50), which applies to an "optimum" allocation for estimating a total. With only 196 units in this population, it is improbable that the fpc will be negligible. However, for purposes of illustration, a first approximation ignoring the fpc will be sought. This is

$$n_0 = \frac{\left(\sum N_h s_h\right)^2}{V} = \frac{(26,841)^2}{7,974,976} = 90.34$$

TABLE 5.4

DATA FOR ESTIMATING SAMPLE SIZE

Stratum	N_h	s_h	$N_h s_h$	n_h
1	13	325	4,225	9
2	18	190	3,420	7
3	26	189	4,914	10
4	42	82	3,444	7
5	73	86	6,278	13
6	24	190	4,560	10
Totals	196		26,841	56

Adjustment is obviously needed. For the correct n in (5.50), we have

$$n = \frac{n_0}{1 + \dfrac{1}{V}\sum N_h s_h^2} = \frac{90.34}{1 + \dfrac{4{,}640{,}387}{7{,}974{,}976}} = 57.1$$

A sample size of 56 was chosen.* The n_h for individual strata appear in the right-hand column of Table 5.4.

5.10 STRATIFIED SAMPLING FOR PROPORTIONS

If we wish to estimate the proportion of units in the population that fall into some defined class C, the ideal stratification is attained if we can place in the first stratum every unit that falls in C, and in the second every unit that does not. Failing this, we try to construct strata such that the proportion in class C varies as much as possible from stratum to stratum.

Let

$$P_h = \frac{A_h}{N_h}, \qquad p_h = \frac{a_h}{n_h}$$

be the proportions of units in C in the hth stratum and in the sample from that stratum, respectively. For the proportion in the whole population, the estimate appropriate to stratified random sampling is

$$p_{st} = \sum \frac{N_h p_h}{N} \tag{5.52}$$

Theorem 5.9. With stratified random sampling, the variance of p_{st} is

$$V(p_{st}) = \frac{1}{N^2} \sum \frac{N_h^2 (N_h - n_h)}{N_h - 1} \frac{P_h Q_h}{n_h} \tag{5.53}$$

* The arithmetical results differ slightly from those given by Cornell (1947).

Proof.   This is a particular case of the general theorem for the variance of the estimated mean. From theorem 5.3

$$V(\bar{y}_{st}) = \frac{1}{N^2} \Sigma N_h(N_h - n_h)\frac{S_h^2}{n_h} \qquad (5.54)$$

Let y_{hi} be a variate which has the value 1 when the unit is in C, and zero otherwise. In section 3.2, equation 3.4, it was shown that for this variate

$$S_h^2 = \frac{N_h}{N_h - 1}P_hQ_h \qquad (5.55)$$

This gives the result.

Note.   In nearly all applications, even if the fpc is not negligible, terms in $1/N_h$ will be negligible, and the slightly simpler formula

$$V(p_{st}) = \frac{1}{N^2} \Sigma N_h(N_h - n_h)\frac{P_hQ_h}{n_h} = \Sigma \frac{W_h^2 P_hQ_h}{n_h}(1 - f_h) \qquad (5.56)$$

can be used.

Corollary 1.   When the fpc can be ignored,

$$V(p_{st}) = \Sigma W_h^2 \frac{P_hQ_h}{n_h} \qquad (5.57)$$

Corollary 2.   With proportional allocation,

$$V(p_{st}) = \frac{N-n}{N} \frac{1}{nN} \Sigma \frac{N_h^2 P_hQ_h}{N_h - 1} \qquad (5.58)$$

$$\doteq \frac{1-f}{n} \Sigma W_h P_hQ_h \qquad (5.59)$$

For the sample estimate of the variance, substitute $p_hq_h/(n_h - 1)$ for the unknown P_hQ_h/n_h in any of the formulas above.

The best choice of the n_h in order to minimize $V(p_{st})$ follows from the general theory in section 5.5.

Minimum Variance for Fixed Total Sample Size.

$$n_h \propto N_h\sqrt{N_h/(N_h - 1)}\sqrt{P_hQ_h} \doteq N_h\sqrt{P_hQ_h}$$

Thus

$$n_h \doteq n\frac{N_h\sqrt{P_hQ_h}}{\Sigma N_h\sqrt{P_hQ_h}} \qquad (5.60)$$

Minimum Variance for Fixed Cost, where Cost $= c_0 + \sum c_h n_h$.

$$n_h \doteq n \frac{N_h \sqrt{P_h Q_h / c_h}}{\sum N_h \sqrt{P_h Q_h / c_h}} \tag{5.61}$$

The value of n is found as in section 5.5.

5.11 GAINS IN PRECISION IN STRATIFIED SAMPLING FOR PROPORTIONS

If the costs per unit are the same in all strata, two useful working rules are that (*a*) the gain in precision from stratified random over simple random sampling is small or modest unless the P_h vary greatly from stratum to stratum, and (*b*) optimum allocation for fixed n gains little over proportional allocation if all P_h lie between 0.1 and 0.9.

TABLE 5.5

RELATIVE PRECISION OF STRATIFIED AND SIMPLE RANDOM SAMPLING

P_h	Simple $nV(p)/(1-f)$ $= PQ$	Stratified $nV(p_{st})/(1-f)$ $= \frac{1}{3}\sum P_h Q_h$	Relative Precision (%)
0.4, 0.5, 0.6	2500	2433	103
0.3, 0.5, 0.7	2500	2233	112
0.2, 0.5, 0.8	2500	1900	132
0.1, 0.5, 0.9	2500	1433	174

To illustrate the first result, Table 5.5 compares stratified random sampling (proportional allocation) with simple random sampling for three strata of equal sizes ($W_h = \frac{1}{3}$). Four cases are included, the first having $P_h = 0.4, 0.5$, and 0.6 in the three strata and the last (and most extreme) having $P_h = 0.1, 0.5$, and 0.9. The next two columns show the variances of the estimated proportion, multiplied by $n/(1-f)$, and the last gives the relative precisions of stratified to simple random sampling. The gain in precision is large only in the last two cases.

To compare proportional with optimum allocation for fixed n, it will be found that apart from the multiplier $(1-f)$,

$$V_{opt} = \frac{(\sum W_h \sqrt{P_h Q_h})^2}{n}, \qquad V_{prop} = \frac{\sum W_h P_h Q_h}{n} \tag{5.62}$$

The relative precision of proportional to optimum allocation is therefore

$$\frac{V_{opt}}{V_{prop}} = \frac{(\sum W_h \sqrt{P_h Q_h})^2}{\sum W_h P_h Q_h} \tag{5.63}$$

If all P_h lie between the two values P_0 and $(1-P_0)$, we are interested in the smallest value the relative precision will take. For simplicity, we consider two strata of equal size ($W_1 = W_2$). The minimum relative precision is attained when $P_1 = \frac{1}{2}$ and $P_2 = P_0$. The relative precision then becomes

$$\frac{V_{opt}}{V_{prop}} = \frac{(0.5 + \sqrt{P_0 Q_0})^2}{2(0.25 + P_0 Q_0)} \tag{5.64}$$

Some values of this function are given in table 5.6. Even with P_0 equal to 0.1, or as high as 0.9, the relative precision is 94%. In most cases the simplicity and the self-weighting feature of proportional stratification more than compensate for this slight loss in precision.

The limitations of the example should be noted. It does not take account of differential costs of sampling in different strata. In some surveys the P_h are very small, but they range from, say, 0.001 to 0.05 in different strata. Here there would be a more substantial gain from optimum stratification.

TABLE 5.6

RELATIVE PRECISION OF PROPORTIONAL TO OPTIMUM ALLOCATION

P_0	0.4 or 0.6	0.3 or 0.7	0.2 or 0.8	0.1 or 0.9	0.05 or 0.95
RP(%)	100.0	99.8	98.8	94.1	86.6

5.12   ESTIMATION OF SAMPLE SIZE WITH PROPORTIONS

Formulas can be deduced from the more general formulas in section 5.9. Let V be the desired variance in the estimate of the proportion P for the whole population. The formulas for the two principal types of allocation are as follows:

Proportional:

$$n_0 = \frac{\sum W_h p_h q_h}{V}, \qquad n = \frac{n_0}{1 + \dfrac{n_0}{N}} \tag{5.65}$$

Presumed optimum:

$$n_0 = \frac{(\sum W_n \sqrt{p_h q_h})^2}{V}, \qquad n = \frac{n_0}{1 + \dfrac{1}{NV}\sum W_h p_h q_h} \tag{5.66}$$

where n_0 is the first approximation, which ignores the fpc, and n is the corrected value taking account of the fpc. In the development of these formulas, the factors $N_h/(N_h - 1)$ have been taken as unity.

These results apply to the estimate of a *proportion*. If it is preferable to think in terms of percentages, the same formulas apply if p_h, q_h, V, and so forth, are expressed as percentages. For the estimation of the total number in the population in class C, that is, of NP, all variances are multiplied by N^2.

EXERCISES

5.1   In a population with $N = 6$ and $L = 2$ the values of y_{hi} are 0, 1, 2 in stratum 1 and 4, 6, 11 in stratum 2. A sample with $n = 4$ is to be taken. (*a*) Show that the optimum n_h under Neyman allocation, when rounded to integers, are $n_h = 1$ in stratum 1 and $n_h = 3$ in stratum 2. (*b*) Compute the estimate $\bar{y}_{st}$ for every possible sample that can be drawn under optimum allocation and under proportional allocation. Verify that the estimates are unbiased. Hence find $V_{opt}(\bar{y}_{st})$ and $V_{prop}(\bar{y}_{st})$ directly. (*c*) Verify that $V_{opt}(\bar{y}_{st})$ agrees with the formula given in equation (5.6) and that $V_{prop}(\bar{y}_{st})$ agrees with the formula given in equation (5.8), page 93. (*d*) Use of formula (5.27), page 99, to compute $V_{opt}(\bar{y}_{st})$ is slightly incorrect because it does not allow for the fact that the n_h were rounded to integers. How well does it agree with the corrected value?

5.2   The households in a town are to be sampled in order to estimate the average amount of assets per household that are readily convertible into cash. The households are stratified into a high-rent and a low-rent stratum. A house in the high-rent stratum is thought to have about nine times as much assets as one in the low-rent stratum, and S_h is expected to be proportional to the square root of the stratum mean.

There are 4000 households in the high-rent stratum and 20,000 in the low-rent stratum. (*a*) How would you distribute a sample of 1000 households between the two strata? (*b*) If the object is to estimate the difference between assets per household in the two strata, how should the sample be distributed?

5.3   The following data show the stratification of all the farms in a county by farm size and the average acres of corn (maize) per farm in each stratum. For a sample of 100 farms, compute the same sizes in each stratum under (*a*) proportional allocation, (*b*) optimum allocation. Compare the precisions of these methods with that of simple random sampling.

Farm Size (acres)	Number of Farms N_h	Average Corn Acres $\bar{Y}_h$	Standard Deviation S_h
0–40	394	5.4	8.3
41–80	461	16.3	13.3
81–120	391	24.3	15.1
121–160	334	34.5	19.8
161–200	169	42.1	24.5
201–240	113	50.1	26.0
241–	148	63.8	35.2
Total or mean	2010	26.3	

5.4   Prove the result stated in formula (5.38), section 5.6:

$$V_{ran} = V_{prop} + \frac{(1-f)}{n(N-1)}\left[\sum N_h(\bar{Y}_h - \bar{Y})^2 - \frac{1}{N}\sum(N-N_h)S_h^2\right]$$

5.5   A sampler has two strata with relative sizes W_1, W_2. He believes that S_1, S_2 can be taken as equal but thinks that c_2 may be between $2c_1$ and $4c_1$. He would prefer to use proportional allocation but does not wish to incur a substantial increase in variance compared with optimum allocation. For a given cost $C = c_1n_1 + c_2n_2$, ignoring the fpc, show that

$$\frac{V_{prop}(\bar{y}_{st})}{V_{opt}(\bar{y}_{st})} = \frac{W_1c_1 + W_2c_2}{(W_1\sqrt{c_1} + W_2\sqrt{c_2})^2}$$

If $W_1 = W_2$, compute the relative increases in variance from using proportional allocation when $c_2/c_1 = 2, 4$.

5.6   A sampler proposes to take a stratified random sample. He expects that his field costs will be of the form $\sum c_h n_h$. His advance estimates of relevant quantities for the two strata are as follows.

Stratum	W_h	S_h	C_h
1	0.4	10	\$4
2	0.6	20	\$9

(a) Find the values of n_1/n and n_2/n that minimize the total field cost for a given value of $V(\bar{y}_{st})$. (b) Find the sample size required, under this optimum allocation, to make $V(\bar{y}_{st}) = 1$. Ignore the fpc. (c) How much will the total field cost be?

5.7   After the sample in exercise 5.6 is taken, the sampler finds that his field costs were actually \$2 per unit in stratum 1 and \$12 in stratum 2. (a) How much greater is the field cost than anticipated? (b) If he had known the correct field costs in advance, could he have attained $V(\bar{y}_{st}) = 1$ for the original estimated field cost in exercise 5.6? (Hint. The Cauchy–Schwarz inequality, page 97, with $V' = 1$, gives the answer to this question without finding the new allocation.)

5.8   In a stratification with two strata, the values of the W_h and S_h are as follows.

Stratum	W_h	S_h
1	0.8	2
2	0.2	4

Compute the sample sizes n_1, n_2 in the two strata needed to satisfy the following conditions. Each case requires a separate computation. (Ignore the fpc.) (a) The standard error of the estimated population mean $\bar{y}_{st}$ is to be 0.1 and the total sample size $n = n_1 + n_2$ is to be minimized. (b) The standard error of the estimated mean of each stratum is to be 0.1. (c) The standard error of the difference between the two estimated stratum means is to be 0.1, again minimizing the total size of sample.

5.9   With two strata, a sampler would like to have $n_1 = n_2$ for administrative convenience, instead of using the values given by the Neyman allocation. If $V(\bar{y}_{st})$, $V_{opt}(\bar{y}_{st})$ denote the variances given by the $n_1 = n_2$ and the Neyman allocations, respectively, show that the fractional increase in variance

$$\frac{V(\bar{y}_{st}) - V_{opt}(\bar{y}_{st})}{V_{opt}(\bar{y}_{st})} = \left(\frac{r-1}{r+1}\right)^2$$

where $r = n_1/n_2$ as given by Neyman allocation. For the strata in exercise 5.8, case a, what would the fractional increase in variance be by using $n_1 = n_2$ instead of the optimum?

5.10   If the cost function is of the form $C = c_0 + \sum t_h \sqrt{n_h}$, where c_0 and the t_h are known numbers, show that in order to minimize $V(\bar{y}_{st})$ for fixed total cost n_h must be proportional to

$$\left(\frac{W_h^2 S_h^2}{t_h}\right)^{2/3}$$

Find the n_h for a sample of size 1000 under the following conditions.

Stratum	W_h	S_h	t_h
1	0.4	4	1
2	0.3	5	2
3	0.2	6	4

5.11   If $V_{prop}(\bar{y}_{st})$ is the variance of the estimated mean from a stratified random sample of size n with proportional allocation and $V(\bar{y})$ is the variance of the mean of a simple random sample of size n, show that the ratio

$$\frac{V_{prop}(\bar{y}_{st})}{V(\bar{y})}$$

does not depend on the size of sample but that the ratio

$$\frac{V_{min}(\bar{y}_{st})}{V_{prop}(\bar{y}_{st})}$$

decreases as n increases. (This implies that optimum allocation for fixed n becomes more effective in relation to proportional allocation as n increases.) [Use formulas (5.8 and 5.27).]

5.12   Compare the values obtained for $V(p_{st})$ under proportional allocation and optimum allocation for fixed sample size in the following two populations. Each stratum is of equal size. The fpc may be ignored.

Population 1		Population 2	
Stratum	P_h	Stratum	P_h
1	0.1	1	0.01
2	0.5	2	0.05
3	0.9	3	0.10

What general result is illustrated by these two populations?

5.13   Show that in the estimation of proportions the results corresponding to theorem 5.8 are as follows.

$$V_{ran} = V_{prop} + \frac{(1-f)}{n} \sum W_h (P_h - P)^2$$

$$V_{prop} = V_{opt} + \frac{\sum W_h (\sqrt{P_h Q_h} - \sqrt{\overline{P_h Q_h}})^2}{n}$$

where

$$\sqrt{\overline{P_h Q_h}} = \sum W_h \sqrt{P_h Q_h}.$$

5.14   In a firm, 62% of the employees are skilled or unskilled males, 31% are clerical females, and 7% are supervisory. From a sample of 400 employees the firm wishes to estimate the proportion that uses certain recreational facilities. Rough guesses are that the facilities are used by 40 to 50% of the males, 20 to 30% of the females, and 5 to 10% of the supervisors. (*a*) How would you allocate the sample among the three groups? (*b*) If the true proportions of users were 48, 21, and 4%, respectively, what would the s.e. of the estimated proportion p_{st} be? (*c*) What would the s.e. of p be from a simple random sample with $n = 400$?

5.15   Formula (5.27) for the minimum variance of $\bar{y}_{st}$ under Neyman allocation reads as follows.

$$V_{min}(\bar{y}_{st}) = \frac{(\sum W_h S_h)^2}{n} - \frac{\sum W_h S_h^2}{N}$$

A student comments: "Since $\sum W_h S_h^2 > (\sum W_h S_h)^2$ unless all the S_h are equal, the formula must be wrong because as n approaches N it will give a negative value for $V(\bar{y}_{st})$." Is the formula or the student wrong?

5.16   By formula (5.26) for Neyman allocation, the sampling fraction in stratum h is $f_h = n_h/N_h = nS_h/N \sum W_j S_j$. The situations in which this formula calls for more than 100% sampling in a stratum ($f_h > 1$) are therefore likely to be those in which the overall sampling fraction n/N is fairly substantial and one stratum has unusually high variability. The following is an example for a small population, with $N = 100$, $n = 40$.

Stratum	N_h	S_h	Optimum n_h
1	60	2	15
2	30	4	15
3	10	15	10
	100		40

(*a*) Verify that the optimum n_h are as shown in the right column. (*b*) Calculate $V(\bar{y}_{st})$ by formula (5.6) and by formula (5.42) and show that both give $V(\bar{y}_{st}) = 0.12$.

Further Aspects of Stratified Sampling

5A.1 EFFECTS OF DEVIATIONS FROM THE OPTIMUM ALLOCATION

This chapter discusses a number of special topics in the practical use of stratified sampling. Sections 5A.1 to 5A.8, 5A.10, and 5A.15 deal with problems that may come up in the planning of the sample; the remaining sections deal with techniques of analysis of results. The present section considers the loss in precision by failure to achieve an optimum allocation of the sample.

Suppose that it is intended to use optimum allocation for given n. The sample size n_h' in stratum h should be

$$n_h' = \frac{n(W_h S_h)}{\sum W_h S_h} \tag{5A.1}$$

From equation (5.27), page 99, the resulting minimum variance is

$$V_{min}(\bar{y}_{st}) = \frac{1}{n}(\sum W_h S_h)^2 - \frac{1}{N}\sum W_h S_h^2 \tag{5A.2}$$

In practice, since the S_h are not known, we can only approximate this allocation. If $\hat{n}_h$ is the sample size used in stratum h, the variance actually attained, from equation (5.6), page 92, is

$$V(\bar{y}_{st}) = \sum \frac{W_h^2 S_h^2}{\hat{n}_h} - \frac{1}{N}\sum W_h S_h^2 \tag{5A.3}$$

The increase in variance caused by the imperfect allocation is

$$V(\bar{y}_{st}) - V_{min}(\bar{y}_{st}) = \sum \frac{W_h^2 S_j^2}{\hat{n}_h} - \frac{1}{n}(\sum W_h S_h)^2 \tag{5A.4}$$

In the first term on the right substitute for $W_h S_h$ in terms of n_h' from (5A.1). This gives the interesting result

$$V(\bar{y}_{st}) - V_{min}(\bar{y}_{st}) = \frac{(\sum W_h S_h)^2}{n^2}\left(\sum \frac{n_h'^2}{\hat{n}_h} - n\right)$$

$$= \frac{(\sum W_h S_h)^2}{n^2}\sum \frac{(\hat{n}_h - n_h')^2}{\hat{n}_h} \tag{5A.5}$$

Reverting to equation (5A.2), if the fpc (last term on the right) is negligible, we see that

$$\frac{V_{min}(\bar{y}_{st})}{n} = \frac{(\sum W_h S_h)^2}{n^2} \tag{5A.6}$$

Hence the proportional increase in variance resulting from deviations from the optimum allocation is

$$\frac{V(\bar{y}_{st}) - V_{min}(\bar{y}_{st})}{V_{min}(\bar{y}_{st})} = \frac{1}{n}\sum_{h=1}^{L}\frac{(\hat{n}_h - n_h')^2}{\hat{n}_h} \tag{5A.7}$$

where $\hat{n}_h$ is the actual and n_h' the optimum sample size in stratum h. If the fpc is not negligible, the = sign in (5A.7) becomes $\geq$.

Let $g_h = |\hat{n}_h - n_h'|/\hat{n}_h$ be the absolute difference in the sample sizes in stratum h, expressed as a fraction of the actual sample size n_h. Then (5A.7) becomes

$$\frac{V - V_{min}}{V_{min}} = \sum_{h=1}^{L}\frac{\hat{n}_h}{n}g_h^2 \tag{5A.8}$$

a weighted mean of the g_h^2. A conservative upper limit to $(V - V_{min})/V_{min}$ is therefore g^2, where g is the largest proportional difference in any stratum. Thus, if $g = 0.2$ or 20%, the proportional increase in variance cannot exceed $(0.2)^2$ or 4%. If $g = 30\%$ the proportional increase in variance is at most 9%. In this sense the optimum can be described as flat.

TABLE 5A.1

EFFECTS OF DEVIATIONS FROM OPTIMUM ALLOCATION

Stratum	n_h' (opt)	$\hat{n}_h$ (act)	$\dfrac{|\hat{n}_h - n_h'|}{\hat{n}_h}$	$\dfrac{(\hat{n}_h - n_h')^2}{\hat{n}_h}$
1	200	150	0.33	16.7
2	100	120	0.17	3.3
3	40	70	0.43	12.9
Total	340	340	—	32.9

Furthermore, (5A.8) suggests that the upper limit g^2 will often overestimate the proportional increase in variance by a substantial amount. Table 5A.1 gives an example with three strata for $n = 340$. Optimum allocation requires sample sizes of 200, 100, and 40, whereas the sizes actually used are 150, 120, and 70.

Since the value of g is 0.43 (stratum 3), the rough rule gives 18% as the proportional increase in variance. From the column on the right, the actual increase is seen to be $32.9/340 = 9.7\%$.

Evans (1951) examined the same question in terms of the effects of errors in the estimated S_h and developed an approximate rule showing whether an estimated optimum is likely to be more precise than proportional allocation. He supposes that the coefficient of variation of the estimated S_h is the same in all strata. This assumption is appropriate when the S_h have been estimated from a preliminary sample of the same size in each stratum. He shows how to compute the size of a preliminary sample needed to make an "optimum" allocation better, on the average, than proportional allocation. Previously, Sukhatme (1953) showed that a small initial sample usually gives a high probability that "optimum" allocation will be superior to proportional stratification. See Sukhatme and Sukhatme (1970), p. 88.

5A.2   EFFECTS OF ERRORS IN THE STRATUM SIZES

For a desirable type of stratification, the stratum totals N_h may not be known exactly, being derived from census data that are out of date. Instead of the true stratum proportions W_h, we have estimates w_h. The sample estimate of $\bar{Y}$ is $\sum w_h \bar{y}_h$.

In general terms, the consequences of using weights that are in error are as follows.

1. The sample estimate is *biased*. Because of the bias, we measure the accuracy of the estimate by its mean square error about $\bar{Y}$ rather than by its variance about its own mean (see section 1.9).

2. The bias remains constant as the sample size increases. Consequently, a size of sample is always reached for which the estimate is less accurate than simple random sampling, and all the gain in precision from stratification is lost.

3. The usual estimate $s(\bar{y}_{st})$ underestimates the true error of $\bar{y}_{st}$, since it does not contain the contribution of the bias to the error.

To justify these statements, note that in repeated sampling the mean value of the estimate is $\sum w_h \bar{Y}_h$. The bias therefore amounts to

$$\sum (w_h - W_h) \bar{Y}_h$$

It is independent of the size of the sample. In finding the mean square error (MSE) of the estimate, it is easy to verify that the variance term is given by the usual formula, with w_h in place of W_h. Hence

$$\text{MSE}(\bar{y}_{st}) = \sum \frac{w_h^2 S_h^2}{n_h}(1 - f_h) + \left[\sum (w_h - W_h)\bar{Y}_h\right]^2 \qquad (5A.9)$$

This expression was given by Stephan (1941). Finally, the usual formula for $s^2(\bar{y}_{st})$ is clearly an unbiased estimate of the first term in (5A.9) but takes no account of the second term.

Example. This illustrates the loss of precision from incorrect weights when stratification is (*a*) slightly effective, (*b*) highly effective. Consider a large population with $S^2 = 1$, divisible into two strata with $W_1 = 0.9$, $W_2 = 0.1$. We will assume $S_1 = S_2 = S_h$. Then, neglecting terms in $1/N_h$,

$$S^2 \doteq \sum W_h S_h^2 + \sum W_h(\bar{Y}_h - \bar{Y})^2 \qquad (5A.10)$$
$$= S_h^2 + W_1 W_2(\bar{Y}_1 - \bar{Y}_2)^2$$

that is,

$$1 = S_h^2 + 0.09(\bar{Y}_1 - \bar{Y}_2)^2$$

In (*a*) take $\bar{Y}_1 - \bar{Y}_2 = 1$. Then $S_h^2 = 0.91$, and proportional stratification with correct weights reduces the variance by 9%, compared with simple random sampling.

In (*b*) take $\bar{Y}_1 - \bar{Y}_2 = 3$, giving $S_h^2 = 0.19$, a reduction in variance of more than 80%.

With two strata and incorrect weights, the bias may be written

$$(w_1 - W_1)(\bar{Y}_1 - \bar{Y}_2)$$

since $(w_1 - W_1) = -(w_2 - W_2)$. Suppose that the estimated weights are $w_1 = 0.92$ and $w_2 = 0.08$. The bias amounts to $(0.02)(1) = 0.02$ in (*a*) and to 0.06 in (*b*). Hence we have the following comparable MSE's for a sample of size n.

Simple random sampling: $V(\bar{y}) = \dfrac{1}{n}$

Stratified random sampling:

$$(a): \quad \text{MSE}(\bar{y}_{st}) = \frac{0.91}{n} + 0.0004$$

$$(b): \quad \text{MSE}(\bar{y}_{st}) = \frac{0.19}{n} + 0.0036$$

As Table 5A.2 shows, simple random sampling begins to win relative to (*a*) at $n = 300$. There is little to choose between the two methods, however, up to $n = 1000$.

In (*b*), with more at stake, stratification is superior up to $n = 200$, although most of the potential gain has already been lost at this sample size. Beyond $n = 300$ stratification becomes markedly inferior to simple random sampling. Accurate estimation of the W_h is particularly important when stratification is highly effective or when the sample size is large.

In some surveys a large preliminary sample of size n' can be taken in order to estimate the W_h. This technique, known as *double sampling* or *two-phase sampling*, has numerous applications and is discussed in Chapter 12. It will be shown

TABLE 5A.2

COMPARABLE VALUES OF MSE($\bar{y}$)

| | Simple | Stratified Random | |
n	Random	(a)	(b)
50	0.0200	0.0186	0.0074
100	0.0100	0.0095	0.0055
200	0.0050	0.0049	0.0045
300	0.0033	0.0034	0.0042
400	0.0025	0.0027	0.0041
1000	0.0010	0.0013	0.0038

that with double sampling the mean square error of $\bar{y}_{st}$ is approximately

$$\frac{\sum W_h S_h^2}{n} + \frac{\sum W_h(\bar{Y}_h - \bar{Y})^2}{n'} \qquad (5A.11)$$

By comparing this MSE with S^2/n, as given by equation (5A.10), we see that most of the gain from stratification is retained provided that n' is much greater than n. To put it more generally, a set of estimated weights preserves most of the potential gain from stratification if the weights are much more accurately estimated than they would be from a simple random sample of size n.

5A.3 THE PROBLEM OF ALLOCATION WITH MORE THAN ONE ITEM

Since the best allocation for one item will not in general be best for another, some compromise must be reached in a survey with numerous items. The first step is to reduce the items considered in the allocation to a relatively small number thought to be most important. If good previous data are available, we can then compute the optimum allocation for each item separately and see to what extent there is disagreement. In a survey of a specialized type the correlations among the items may be high and the allocations may differ relatively little.

Example. Data given by Jessen (1942) illustrate a farm survey of this kind. The state of Iowa was divided into five geographic regions, each denoted by its major agricultural enterprise. Suppose that these regions are to be used as strata in a survey on dairy farming. The three items of most interest are the number of cows milked per day, the number of gallons of milk per day, and the total annual cash receipts from dairy products. From a survey made in 1938, the estimated standard deviations s_h within strata are shown in table 5A.3. In Table 5A.4 the optimum Neyman allocations based on these s_h are given for the individual items in a sample of 1000 farms.

TABLE 5A.3
STANDARD DEVIATIONS WITHIN STRATA

Stratum	$W_h = \dfrac{N_h}{N}$	s_h Cows Milked	s_h Gallons of Milk	s_h Receipts for Dairy Products ($)
Northeast dairy	0.197	4.6	11.7	332
Cash grain	0.191	3.4	9.8	357
Western livestock	0.219	3.3	7.0	246
Southern pasture	0.184	2.8	6.5	173
Eastern livestock	0.208	3.7	9.8	279

TABLE 5A.4
SAMPLE SIZES WITHIN STRATA ($n = 1000$)
Allocation

Stratum	Proportional	Optimum for Cows	Optimum for Gallons	Optimum for Receipts	Average m_h
Northeast dairy	197	254	258	236	250
Cash grain	191	182	209	246	212
Western livestock	219	203	171	194	189
Southern pasture	184	145	134	115	131
Eastern livestock	208	216	228	209	218

TABLE 5A.5
EXPECTED VARIANCES OF THE ESTIMATED MEAN

Type of allocation	Cows	Gallons	Receipts
Optimum	0.0127	0.0800	76.9
Compromise	0.0128	0.0802	77.6
Proportional	0.0131	0.0837	80.9

The individual optimum allocations differ only moderately from each other. With one exception, all three deviate in the same direction from a proportional allocation. Thus, in the first stratum, proportional allocation suggests 197 farms, and the individual allocations lead to numbers between 236 and 258. The average of the optimum sample sizes for the three items, shown in the right-hand column, provides a satisfactory compromise allocation.

Table 5A.5 shows the expected sampling variances of $\bar{y}_{st}$, as given by the individual optima, the compromise, and the proportional allocations. The formulas are as follows.

$$v_{opt} = \frac{(\sum W_h s_h)^2}{n}, \qquad v_{comp} = \sum \frac{(W_h s_h)^2}{m_h}, \qquad v_{prop} = \frac{\sum W_h s_h^2}{n}$$

The compromise allocation gives results almost as precise as if it were possible to use separate optimum allocations for each item. What is more noteworthy is that proportional allocation is only slightly less precise than the compromise or the individual optima. Furthermore, Table 5A.5 overestimates the precision of the optima and of the compromise, since these allocations were made from estimated variances. This result is another illustration of the flatness of the optimum mentioned in section 5A.1.

5A.4 OTHER METHODS OF ALLOCATION WITH MORE THAN ONE ITEM

An alternative compromise allocation suggested by Chatterjee (1967) is to choose the n_h that minimize the *average* of the proportional increases in variance from (5A.7), taken over the variables. If j denotes a variable this amounts to choosing

$$n_h = n \sqrt{\sum_j n_{jh}'^2} \Big/ \sum_h \sqrt{\sum_j n_{jh}'^2} \tag{5A.12}$$

where n_{jh}' is the optimum sample size in stratum h for variable j. For the data in Table 5A.4, where the individual optima differ only slightly, Chatterjee's n_h vary from the average m_h in Table 5A.4 by, at most, one unit in any stratum.

In some surveys the optimum allocations for individual variates differ so much that there is no obvious compromise. Some principle is needed to determine the allocation to be used. Two useful ones suggested by Yates (1960) are presented.

The first applies to surveys with a specialized objective, in which the loss due to an error of given size in an estimate can be measured in terms of money or utility, as discussed in section 4.10. With k variates and quadratic loss functions, it may be reasonable to express the total expected loss as a linear function of the variances of the estimated population means or totals. For the means,

$$L = \sum_j^k a_j V(\bar{y}_{jst}) = \sum_j^k a_j \sum_h^L W_h^2 S_{jh}^2 \left(\frac{1}{n_h} - \frac{1}{N_h} \right) \tag{5A.13}$$

where S_{jh}^2 is the variance of the jth variate in stratum h. Interchange of the order of summation gives

$$L = \sum_h \frac{W_h^2}{n_h} \left(\sum_j a_j S_{jh}^2 \right) - \frac{1}{N} \sum_h W_h \left(\sum_j a_j S_{jh}^2 \right) \tag{5A.14}$$

With a linear function for the costs of sampling, we have

$$C = c_0 + \sum c_h n_h \tag{5A.15}$$

Minimizing the product of $(C - c_0)$ and the first term in L (the term depending on the n_h) gives, by the Cauchy–Schwarz inequality,

$$n_h \propto \frac{W_h}{\sqrt{c_h}} \sqrt{\sum_j a_j S_{jh}^2} \tag{5A.16}$$

The constant of proportionality is found by satisfying the constraint given for L or C. For instance, suppose that the value of L is specified and that the fpc term may be ignored. We have

$$n_h = \frac{n(W_h A_h / \sqrt{c_h})}{\sum (W_h A_h / \sqrt{c_h})} \tag{5A.17}$$

where $A_h = \sqrt{\sum_j a_j S_{jh}^2}$. The required total sample size is, from (5A.14),

$$n = \frac{1}{L} \left(\sum_h \frac{W_h A_h}{\sqrt{c_h}} \right) \left(\sum_h W_h A_h \sqrt{c_h} \right) \tag{5A.18}$$

In the second approach we specify the desired variance V_j for *each* variate. For population means this implies that

$$\sum_{h=1}^{L} \frac{W_h^2 S_{jh}^2}{n_h} - \sum_{h=1}^{L} \frac{W_h S_{jh}^2}{N} \leq V_j \qquad (j = 1, 2, \ldots, k) \tag{5A.19}$$

Inequality signs are used because the most economical allocation may supply variances smaller than the desired V_j for some items.

In this approach the cost C [equation (5A.15)] is minimized subject to the tolerances V_j and the conditions $0 \leq n_h \leq N_h$. The problem is one in nonlinear programming. Algorithms for its solution have been given by Hartley and Hocking (1963), Chatterjee (1966), Zukhovitsky and Avdeyeva (1966), and Huddleston et al. (1970). Earlier, Dalenius (1957) gave an ingenious graphical solution, while Yates (1960) and Kokan (1963) developed methods of successive approximation, illustrated in the second edition of this book.

A useful first step is, of course, to work out the optimum allocation for each variate separately and find the cost of satisfying its tolerance. Take the variate, say y_1, for which the cost C_1 is highest and examine whether the optimum n_h values for y_1 satisfy all the other $(k - 1)$ tolerances. If so, we use this allocation and the problem is solved, because no other allocation will satisfy the tolerance V_1 for y_1 at a cost as low as C_1.

By working a series of examples in a related problem, Booth and Sedransk (1969) have pointed out that in default of a computer program a good approximation to the solution of Yates' second problem can often be obtained by solving the easier first problem. Specify that L in (5A.13) shall have the value $V^* = \sum a_j V_j$, where the V_j are the desired individual tolerances and the a_j are made inversely proportional to the V_j. Thus with two variates, $a_1 = V_2/(V_1 + V_2)$, $a_2 = V_1/(V_1 + V_2)$, and

$$V^* = \frac{2V_1 V_2}{(V_1 + V_2)} \qquad (5A.20)$$

Example. (Four strata, two variates.)   The data and the application of the approximate method are shown in columns (1) to (6) of Table 5A.6. The problem is to find the smallest n for which

$$V(\bar{y}_{1st}) \leqq 0.04, \qquad V(\bar{y}_{2st}) \leq 0.01$$

TABLE 5A.6
ARTIFICIAL DATA FOR FOUR STRATA, TWO VARIATES

Column	(1)	(2)	(3)	(4)	(5)	(6)	(7)
Stratum	W_h	S_{1h}^2	S_{2h}^2	$A_h^2 = \sum_j a_j S_{jh}^2$	$W_h A_h$	n_h	$\tilde{n}_h$
1	0.4	25	1	5.8	0.963	206	194
2	0.3	25	4	8.2	0.859	183	180
3	0.2	25	16	17.8	0.844	180	187
4	0.1	25	64	56.2	0.750	160	171
				Totals	3.416	729	732

By working out the optimum allocation for each variate separately it is easily verified that $n = 625$ is needed to satisfy the first constraint and $n = 676$ is needed to satisfy the second. However, $n = 676$ with its allocation does not satisfy the first constraint giving 0.0589 instead of 0.04 for V_1. An iterative solution to satisfy both constraints (presented in the second edition) gave $\tilde{n} = 732$, with the $\tilde{n}_h$ shown in column (7) of Table 5A.6.

To use the Booth and Sedransk approach with $V_1 = 0.04$, $V_2 = 0.01$, we specify

$$L = 0.2\,V(\bar{y}_{1st}) + 0.8\,V(\bar{y}_{2st}) = \frac{2(0.04)(0.01)}{(0.05)} = 0.016$$

From (5A.18), with $c_h = 1$, we have

$$n = \frac{\left(\sum W_h A_h\right)^2}{L} = \frac{(3.416)^2}{0.016} = 729$$

using column (5) of Table 5A.6. From (5A.17), column (5) also leads to the n_h values, shown in column (6) of Table 5A.6. As columns (6) and (7) show, the two solutions n_h and $\tilde{n}_h$ agree well.

As Booth and Sedransk note, $n \leq \tilde{n}$ in all problems of this type, since n satisfies the single constraint $L = V^*$, but it need not satisfy the constraint on every variate, whereas the $\tilde{n}$ allocation satisfies L as well as the individual constraints.

5A.5　TWO-WAY STRATIFICATION WITH SMALL SAMPLES

Suppose that there are two criteria of stratification, say by R rows and C columns, making RC cells. If $n \geq RC$, every cell can be represented in the sample. A problem arises when $n < RC$, and we would like the sample to give proportional representation to each criterion of stratification. In a simple method developed by Bryant, Hartley, and Jessen (1960) the technique requires only that n exceed the greater of R and C.

To illustrate this method, suppose that a small population of 165 schools has been stratified by size of city into five classes and by average expenditure per pupil into four classes. The numbers of schools m_{ij} and the proportions of schools $P_{ij} = m_{ij}/165$ in each of the 20 cells are shown in Table 5A.7.

TABLE 5A.7

NUMBER AND PROPORTION OF SCHOOLS IN EACH CELL

Size of City		Expenditure per Pupil				Totals		$n_{i.}$
		A	B	C	D			
I	m_{1j}	15	21	17	9	$m_{1.}$	62	
	P_{1j}	0.091	0.127	0.103	0.055	$P_{1.}$	0.376	4
II	m_{2j}	10	8	13	7	$m_{2.}$	38	
	P_{2j}	0.061	0.049	0.079	0.042	$P_{2.}$	0.231	2
III	m_{3j}	6	9	5	8	$m_{3.}$	28	
	P_{3j}	0.036	0.055	0.030	0.049	$P_{3.}$	0.170	2
IV	m_{4j}	4	3	6	6	$m_{4.}$	19	
	P_{4j}	0.024	0.018	0.036	0.036	$P_{4.}$	0.114	1
V	m_{5j}	3	2	5	8	$m_{5.}$	18	
	P_{5j}	0.018	0.012	0.030	0.049	$P_{5.}$	0.109	1
Totals	$m_{.j}$	38	43	46	38		165	
	$P_{.j}$	0.230	0.261	0.278	0.231		1.000	
	$n_{.j}$	2	3	3	2			

The objective is to give each school an approximately equal chance of selection while giving each marginal class its proportional representation. In this illustration $n = 10$. Compute the numbers $n_{i.} = nP_{i.}$ and $n_{.j} = nP_{.j}$, where these products are rounded to the nearest integers (with a further minor adjustment, if needed, so that the $n_{i.}$ and the $n_{.j}$ both add to n). These numbers are shown in Table 5A.7.

The next step is to draw $n = 10$ cells with probability $n_{i.}n_{.j}/n^2$ for the ijth cell. This is done by constructing an $n \times n$ square (Table 5A.8). In row 1 one column is drawn at random. In row 2 one of the remaining columns is drawn at random, and so on. At the end, each row and column contains one unit. (This draw is most quickly made by a random permutation of the numbers 1 to 10.) The results of one draw are indicated by ×'s in Table 5A.8.

TABLE 5A.8

10 × 10 SQUARE FOR DRAWING THE SAMPLE

Row		1 (A)	2 (A)	3 (B)	4 (B)	5 (B)	6 (C)	7 (C)	8 (C)	9 (D)	10 (D)
1	I	×									
2	I				×						
3	I		×								
4	I							×			
5	II						×				
6	II								×		
7	III			×							
8	III									×	
9	IV					×					
10	V										×

Note that columns 1 and 2 are assigned to marginal stratum A, since $n_{.1} = 2$. Similarly, rows 1 through 4 are assigned to marginal stratum I, since $n_{1.} = 4$, and so on. This completes the allocation of the sample to the 20 cells. The allocation appears in more compact form in Table 5A.9. Two schools are drawn at random from the 15 schools in cell IA, and so on. The probability that a school in row i, column j is drawn is proportional to $n_{i.}n_{.j}/P_{ij}$. Thus the probabilities are not equal, although they will be approximately so if $P_{ij} \doteq n_{i.}n_{.j}/n^2$.

An unbiased estimate of the mean per school is

$$\bar{y}_U = \frac{1}{n}\sum \frac{n^2 P_{ij}}{n_{i.}n_{.j}} y_{ij}$$

where y_{ij} is the sample *total* in the ijth cell. If, however, $P_{ij} \doteq n_{i.}n_{.j}/n^2$, the sample mean $\bar{y}$ is probably preferable, since its bias should be negligible. A sample estimate of variance is available for both the unbiased and biased estimates,

TABLE 5A.9

ALLOCATION OF THE SAMPLE TO THE 20 CELLS

	A	B	C	D	Total
I	2	1	1	0	4
II	0	0	2	0	2
III	0	1	0	1	2
IV	0	1	0	0	1
V	0	0	0	1	1
Total	2	3	3	2	10

provided that n is at least twice the greater of R and C and that at least two units are drawn in every row and column.

If P_{ij} differs markedly from $n_{i.}n_{.j}/n^2$ in some cells, an extra step keeps the probabilities of selection of schools more nearly constant. After computing the $n_{i.}$ and $n_{.j}$, examine the quantities $D_{ij} = nP_{ij} - n_{i.}n_{.j}/n$, after rounding them to integers. If, in any cell, D_{ij} is a positive integer, automatically assign D_{ij} units to this cell. Reduce n, the $n_{i.}$, and the $n_{.j}$ by the amounts required by this fixed allocation and carry out the remaining allocation as before.

5A.6 CONTROLLED SELECTION

Another technique for this problem with small samples was named *controlled selection* by Goodman and Kish (1950). A simple illustration given by Hess, Riedel, and Fitzpatrick (1976), who applied the method for sampling hospitals, shows the basic idea. The principal stratification is by size of hospital (two strata). Representation of each of two types of ownership of the hospitals is also desirable, but only one unit (hospital) is to be drawn from each principal stratum. In

TABLE 5A.10

ORDERING OF UNITS WITHIN STRATA FOR CONTROLLED SELECTION

Original Order Stratum		Revised Order Stratum	
I-Large Hospital	II-Small Hospital	Large Hospital	Small Hospital
1	1	1	3'
2	2	2	4'
3'	3'	3'	5'
4'	4'	4'	1
	5'		2

numbering the units within strata, a prime (') indicates one ownership type, and absence of a prime indicates the other type. Table 5A.10 (left side) shows the units in the two strata.

If unit 1 or 2 is drawn from stratum I, we would like to draw units 3', 4', or 5' from stratum II, so that both types of stratification are present with $n = 2$. Similarly, 3' or 4' (stratum I) is desired with 1 or 2 (stratum II). Controlled selection makes the probability of these desired combinations as high as is mathematically possible, while retaining equal probability selection within strata and therefore unbiased estimates by the usual formulas for stratified sampling. The purpose is either increased accuracy for given n or a saving in field costs.

With stratified random sampling the probability of a desired combination is $(.5)(.6) + (.5)(.4) = .5$. This probability can be increased to .9 by two simple changes in sample selection. Rearrange the units in stratum II so that the desired combinations (3', 4', 5') with 1 and 2 in stratum I come first, as on the right in Table 5A.10. Then draw a random number r between 1 and 100 and use it to select the units from *both* strata. In stratum I, $1 \le r \le 25$ selects unit 1, $26 \le r \le 50$ selects unit 2, and so on, so as to give each unit the desired one fourth probability of being chosen. Similarly, in stratum II, $1 \le r \le 20$ selects 3', $21 \le r \le 40$ selects 4', and so on. Hence, if $1 \le r \le 20$, we select (1, 3'), if $20 \le r \le 25$, we select (1, 4') and so on. The joint selections and their probabilities are as follows.

Pair:	(1, 3')	(1, 4')	(2, 4')	(2, 5')	(3', 5')	(3', 1)	(4', 1)	(4', 2)
Probability	.20	.05	.15	.10	.10	.15	.05	.20

The only nondesired combination is (3', 5'). Thus the total probability of the desired combinations is .90.

Since sampling is not independent in the two strata, the formulas for $V(\bar{y}_{st})$ and $v(\bar{y}_{st})$ do not apply. Hess, Riedel, and Fitzpatrick (1976), give approximate formulas. This monograph also gives an algorithm for the application of controlled selection in problems with more strata, larger n, and more complex controls.

For another approach using balanced incomplete block designs, see Avadhani and Sukhatme (1973).

5A.7   THE CONSTRUCTION OF STRATA

This topic raises several questions. What is the best characteristic for the construction of strata? How should the boundaries between the strata be determined? How many strata should there be? For a single item or variable y the best characteristic is clearly the frequency distribution of y itself. The next best is presumably the frequency distribution of some other quantity highly correlated with y. Given the number of strata, the equations for determining the best stratum boundaries under proportional and Neyman allocation have been worked out by Dalenius (1957), and quicker approximate methods by several workers. We will consider Neyman allocation, since it is usually superior to proportional allocation in populations in which gains from stratification are greatest. It is assumed at first that the strata are set up by using the value of y itself.

Let y_0, y_L be the smallest and largest values of y in the population. The problem is to find intermediate stratum boundaries y_1, y_2, $\cdots$, y_{L-1} such that

$$V(\bar{y}_{st}) = \frac{1}{n} \left(\sum_{h=1}^{L} W_h S_h \right)^2 - \frac{1}{N} \sum_{h=1}^{L} W_h S_h^2 \tag{5A.21}$$

is a minimum. If the fpc is ignored, it is sufficient to minimize $\sum W_h S_h$. Since y_h appears in this sum only in the terms $W_h S_h$ and $W_{h+1} S_{h+1}$, we have

$$\frac{\partial}{\partial y_h} (\sum W_h S_h) = \frac{\partial}{\partial y_h} (W_h S_h) + \frac{\partial}{\partial y_h} (W_{h+1} S_{h+1})$$

Now if $f(y)$ is the frequency function of y,

$$W_h = \int_{y_{h-1}}^{y_h} f(t) \, dt, \qquad \frac{\partial W_h}{\partial y_h} = f(y_h) \tag{5A.22}$$

Further,

$$W_h S_h^2 = \int_{y_{h-1}}^{y_h} t^2 f(t) \, dt - \frac{\left[\int_{y_{h-1}}^{y_h} t f(t) \, dt \right]^2}{\int_{y_{h-1}}^{y_h} f(t) \, dt} \tag{5A.23}$$

Differentiation of (5A.23) gives

$$S_h^2 \frac{\partial W_h}{\partial y_h} + 2 W_h S_h \frac{\partial S_h}{\partial y_h} = y_h^2 f(y_h) - 2 y_h \mu_h f(y_h) + \mu_h^2 f(y_h) \tag{5A.24}$$

where μ_h is the mean of y in stratum h. Add $S_h^2 \, \partial W_h / \partial y_h$ to the left side, and the equal quantity $S_h^2 f(y_h)$ to the right side. This gives, on dividing by $2 S_h$,

$$\frac{\partial (W_h S_h)}{\partial y_h} = S_h \frac{\partial W_h}{\partial y_h} + W_h \frac{\partial S_h}{\partial y_h} = \frac{1}{2} f(y_h) \frac{(y_h - \mu_h)^2 + S_h^2}{S_h} \tag{5A.25}$$

Similarly we find

$$\frac{\partial (W_{h+1} S_{h+1})}{\partial y_h} = -\frac{1}{2} f(y_h) \frac{(y_h - \mu_{h+1})^2 + S_{h+1}^2}{S_{h+1}} \tag{5A.26}$$

Hence the calculus equations for y_h are

$$\frac{(y_h - \mu_h)^2 + S_h^2}{S_h} = \frac{(y_h - \mu_{h+1})^2 + S_{h+1}^2}{S_{h+1}} \quad (h = 1, 2, \ldots, L-1) \tag{5A.27}$$

Unfortunately, these equations are ill adapted to practical computation, since both μ_h and S_h depend on y_h. A quick approximate method, due to Dalenius and

Hodges (1959), is presented for minimizing $\sum W_h S_h$. Let

$$Z(y) = \int_{y_0}^{y} \sqrt{f(t)}\, dt \tag{5A.28}$$

If the strata are numerous and narrow, $f(y)$ should be approximately constant (rectangular) within a given stratum. Hence,

$$W_h = \int_{y_{h-1}}^{y_h} f(t)\, dt \doteq f_h(y_h - y_{h-1}) \tag{5A.29}$$

$$S_h \doteq \frac{1}{\sqrt{12}}(y_h - y_{h-1}) \tag{5A.30}$$

$$Z_h - Z_{h-1} = \int_{y_{h-1}}^{y_h} \sqrt{f(t)}\, dt \doteq \sqrt{f_h}(y_h - y_{h-1}) \tag{5A.31}$$

where f_h is the "constant" value of $f(y)$ in stratum h. By substituting these approximations, we find

$$\sqrt{12} \sum_{h=1}^{L} W_h S_h \doteq \sum_{h=1}^{L} f_h(y_h - y_{h-1})^2 \doteq \sum_{h=1}^{L} (Z_h - Z_{h-1})^2 \tag{5A.32}$$

Since $(Z_L - Z_0)$ is fixed, it is easy to verify that the sum on the right is minimized by making $(Z_h - Z_{h-1})$ constant.

Given $f(y)$, the rule is to form the cumulative of $\sqrt{f(y)}$ and choose the y_h so that they create equal intervals on the cum $\sqrt{f(y)}$ scale. Table 5A.11 illustrates the use of the rule.

TABLE 5A.11

CALCULATION OF STRATUM BOUNDARIES BY THE CUM $\sqrt{f(y)}$ RULE

$\dfrac{\text{Industrial Loans}}{\text{Total Loans}}\%$	$f(y)$	Cum $\sqrt{f(y)}$	$\dfrac{\text{Industrial Loans}}{\text{Total Loans}}\%$	$f(y)$	Cum $\sqrt{f(y)}$
0–5	3464	58.9	50–55	126	340.3
5–10	2516	109.1	55–60	107	350.6
10–15	2157	155.5	60–65	82	359.7
15–20	1581	195.3	65–70	50	366.8
20–25	1142	229.1	70–75	39	373.0
25–30	746	256.4	75–80	25	378.0
30–35	512	279.0	80–85	16	382.0
35–40	376	298.4	85–90	19	386.4
40–45	265	314.7	90–95	2	387.8
45–50	207	329.1	95–100	3	389.5

Example. The data show the frequency distribution of the percentage of bank loans devoted to industrial loans in a population of 13,435 banks of the United States (McEvoy, 1956). The distribution is skew, with its mode at the lower end. In the cum $\sqrt{f}$ column, $58.9 = \sqrt{3436}$, $109.1 = \sqrt{3464} + \sqrt{2516}$, and so on.

Suppose that we want five strata. Since the total of cum $\sqrt{f}$ is 389.5, the division points should be at 77.9, 155.8, 233.7, and 311.6 on this scale. The nearest available points are as follows:

	Stratum				
	1	2	3	4	5
Boundaries	0–5%	5–15%	15–25%	25–45%	45–100%
Interval on cum $\sqrt{f}$	58.9	96.6	73.6	85.6	74.8

The first two intervals, 58.9 and 96.6, are rather unequal, but cannot be improved on without a finer subdivision of the original classes.

If the class intervals in the original distribution of y are of unequal length, a slight change is needed. When the interval changes from one of length d to one of length ud, the value of $\sqrt{f}$ for the second interval is multiplied by $\sqrt{u}$ when forming cum $\sqrt{f}$.

Another method, proposed by Sethi (1963), is to work out the boundaries given by the calculus equations (5A.27) for a standard continuous distribution resembling the study population. For the normal and various χ^2 distributions, Sethi has tabulated the optimum boundaries for Neyman, equal, and proportional allocation for $L \le 6$. If one of these distributions seems to approximate that in the study population, the boundaries can be read from Sethi's tables.

Two further approximate methods require some trial and error. From relations (5A.32), the Dalenius-Hodges rule is roughly equivalent to making $W_h S_h$ constant, as conjectured earlier by Dalenius and Gurney (1951). A similar rule is that of Ekman (1959), who makes $W_h(y_h - y_{h-1})$ constant.

In comparisons on some theoretical and eight study populations, Cochran (1961) found that the cum. $\sqrt{f}$ rule and the Ekman rule worked consistently well (the Sethi method was not tried). In a study of United States hospital bed capacity, whose distribution resembles χ^2 with 1 degree of freedom, Hess, Sethi, and Balakrishnan (1966) found the Ekman method slightly superior to cum. $\sqrt{f}$ and Sethi's for $L > 2$, while Murthy (1967) also reports good performance by Ekman's method.

The relations (5A.32) have an interesting consequence. If $W_h S_h$ is constant, Neyman allocation gives a *constant* sample size $n_h = n/L$ in all strata. For the approximate methods, the comparisons that have been made suggest that the simple rule $n_h = n/L$ is satisfactory.

Thus far we have made the unrealistic assumption that stratification can be based on the values of y itself. In practice, some other variable x is used (perhaps the value of y at a recent census). Dalenius (1957) develops equations for the boundaries of x that minimize $\sum W_h S_{yh}$, given a knowledge of the regression of y on x. If this regression is nonlinear, these boundaries may differ considerably from those that are optimum when x itself is the variable to be measured. The equations indicate, however, that if the regression of y on x is linear and the correlation between y and x is high within all strata the two sets of boundaries should be nearly the same. Let

$$y = \alpha + \beta x + e$$

where $E(e) = 0$ for all x and e, x are uncorrelated. The variance of e within stratum h is S_{eh}^2. Then the x-boundaries that make $V(\bar{y}_{st})$ a minimum satisfy the equations (Dalenius, 1957).

$$\frac{\beta^2[(x_h - \mu_{xh})^2 + S_{xh}^2] + 2S_{eh}^2}{\beta S_{xh}\sqrt{1 + S_{eh}^2/\beta^2 S_{xh}^2}} = \frac{\beta^2[(x_h - \mu_{x,h+1})^2 + S_{x,h+1}^2] + 2S_{e,h+1}^2}{\beta S_{x,h+1}\sqrt{1 + S_{e,h+1}^2/\beta^2 S_{x,h+1}^2}}$$

If $S_{eh}^2/\beta^2 S_{xh}^2$ is small for all h, these equations reduce to the form (5A.27) that gives optimum boundaries for x. But $S_{eh}^2/\beta^2 S_{xh}^2 = (1 - \rho_h^2)/\rho_h^2$ where ρ_h is the correlation between y and x within stratum h.

Although more investigation is needed, this result suggests that the cum $\sqrt{f}$ rule applied to x should give an efficient stratification for another variable y that has a linear regression on x with high correlation. Some numerical results by Cochran (1961) support this conjecture. Moreover, if the ρ_h are only moderate, as will happen when the number of strata is increased, failure to use the optimum x-boundaries should have a less deleterious effect on y.

The preceding discussion is, of course, mainly relevant to the sampling of institutions stratified by some measure of size. The situation is different when one set of variables is closely related to y_1 and another set, with a markedly different frequency distribution, is closely related to y_2. One possibility is to seek compromise stratum boundaries that meet the desired tolerances on $V(\bar{y}_{1st})$ and $V(\bar{y}_{2st})$, following a general approach given in section 5A.4, but computational methods have not been worked out.

In geographical stratification the problem is less amenable to a mathematical approach, since there are so many different ways in which stratum boundaries may be formed. The usual procedure is to select a few variables that have high correlations with the principal items in the survey and to use a combination of judgment and trial and error to construct boundaries that are good for these selected variables. Since the gains in precision from stratification are likely to be modest, it is not worthwhile to expend a great deal of effort in improving boundaries. Bases of stratification for economic items have been discussed by Stephan (1941) and Hagood and Bernert (1945) and for farm items by King and McCarty (1941).

5A.8 NUMBER OF STRATA

The two questions relevant to a decision about the number of strata L are (a) at what rate does the variance of $\bar{y}_{st}$ decrease as L is increased? (b) How is the cost of the survey affected by an increase in L?

As regards (a), suppose first that strata are constructed by the values of y. To take the simplest case, let the distribution of y be rectangular in the interval $(a, a+d)$. Then S_y^2, before stratification, is $d^2/12$, so that with a simple random sample of size n, $V(\bar{y}) = d^2/12n$. If L strata of equal size are created, the variance within any stratum is $S_{yh}^2 = d^2/12L^2$. Hence, for a stratified sample, with $W_h = 1/L$ and $n_h = n/L$,

$$V(\bar{y}_{st}) = \frac{1}{n}\left(\sum_{h=1}^{L} W_h S_{yh}\right)^2 = \frac{1}{n}\left(\sum_{h=1}^{L}\frac{1}{L}\cdot\frac{d}{\sqrt{12L}}\right)^2 = \frac{d^2}{12nL^2} = \frac{V(\bar{y})}{L^2} \quad (5A.33)$$

Thus with a rectangular distribution the variance of $\bar{y}_{st}$ decreases inversely as the *square* of the number of strata. Rather remarkably, this relation continues to hold, roughly, when actual skew distributions with finite range are stratified with the optimum choice of boundaries for Neyman allocation. In eight distributions of data of the type likely to occur in practice, Cochran (1961) found that the average values of $V(\bar{y}_{st})/V(\bar{y})$ were 0.232, 0.098, and 0.053 for $L = 2, 3, 4$, as compared with 0.250, 0.111, and 0.062 for the rectangular distribution.

These results, which suggest that multiplicaton of strata is profitable, give a misleading picture of what happens when some other variable x is used to construct the strata. If $\phi(x) = E(y|x)$ is the regression of y on x, we may write

$$y = \phi(x) + e \quad (5A.34)$$

where ϕ and e are uncorrelated. Hence

$$S_y^2 = S_\phi^2 + S_e^2 \quad (5A.35)$$

By the preceding results, creation of L optimal strata for x may reduce S_ϕ^2 to S_ϕ^2/L^2 if $\phi(x)$ is linear or at a smaller rate if $\phi(x)$ is nonlinear. But the term S_e^2 is not reduced by stratification on x. As L increases, a value is reached sooner or later at which the term S_e^2 dominates. Further increases in L will produce only a trivial proportional reduction in $V(\bar{y}_{st})$.

How quickly the point of diminishing returns is reached depends on a number of factors—particularly the relative sizes of S_e^2 and S_ϕ^2 and the nature of $\phi(x)$. Only a few examples from actual data are available in the literature. To supplement them, a simple theoretical approach is used. Suppose that the optimum choice of stratum boundaries by means of x, with samples of equal size n/L in each stratum, reduces $V(\bar{x}_{st})$ at a rate proportional to $1/L^2$. Thus

$$V(\bar{x}_{st}) = \frac{L}{n}\sum_{h=1}^{L} W_h^2 S_{xh}^2 = \frac{S_x^2}{nL^2} \quad (5A.36)$$

Suppose also that the regression of y on x is linear, that is,

$$y = \alpha + \beta x + e \tag{5A.37}$$

where S_e^2 is constant. Then,

$$V(\bar{y}_{st}) = \frac{L}{n} \sum_{h=1}^{L} W_h^2 S_{yh}^2 = \frac{L\beta^2}{n} \sum_{h=1}^{L} W_h^2 S_{xh}^2 + \frac{LS_e^2}{n} \sum_{h=1}^{L} W_h^2 \tag{5A.38}$$

For any set of L strata, $\sum W_h^2 \geq \frac{1}{L}$. Using (5A.36), we have

$$V(\bar{y}_{st}) \geq \frac{1}{n}\left(\frac{\beta^2 S_x^2}{L^2} + S_e^2\right) = \frac{S_y^2}{n}\left[\frac{\rho^2}{L^2} + (1-\rho^2)\right] \tag{5A.39}$$

where ρ is the correlation between y and x in the unstratified population.

With this model, Table 5A.12 shows $V(\bar{y}_{st})/V(\bar{y})$ for $\rho = 0.99$, 0.95, 0.90, and 0.85 and $L = 2$ to 6, assuming that relation (5A.39) is an equality. The right-hand columns of the table give $V(\bar{y}_{st})/V(\bar{y})$ for three sets of actual data, described under the table, in which x is the value of y at some earlier time.

The results for the regression model indicate that unless ρ exceeds 0.95, little reduction in variance is to be expected byond $L = 6$. Data sets 2 and 3 support this conclusion, although some further increase in L might be profitable with the

TABLE 5A.12

$V(\bar{y}_{st})/V(\bar{y})$ as a Function of L for the Linear Regression Model and for Some Actual Data

	Linear Regression Model				Data, Set		
	$\rho =$						
L	0.99	0.95	0.90	0.85	1	2	3
2	0.265	0.323	0.392	0.458	0.197	0.295	0.500
3	0.129	0.198	0.280	0.358	0.108	0.178	0.375
4	0.081	0.154	0.241	0.323	0.075	0.142	0.244
5	0.059	0.134	0.222	0.306	0.065	0.105	0.241
6	0.047	0.123	0.212	0.298	0.050	0.104	0.212
∞	0.020	0.098	0.190	0.277	—	—	—

		Type of Data			
Set	Data	x	y	Source	
1	College enrollments	1952	1958	Cochran (1961)	
2	City sizes	1940	1950	Cochran (1961)	
3	Family incomes	1929	1933	Dalenius and Gurney (1951)	

college enrollment data (set 1). In two comparisons on survey data, Hess, Sethi, and Balakrishnan (1966) found that $V(\bar{y}_{st})$ decreased faster with L than (5A.39) predicts, which suggests that model (5A.37) is oversimplified.

To complete this analysis, we require a cost function that shows how the cost depends on L. Dalenius (1957) suggests the relation $C = LC_s + nC_n$. The cost ratio C_s/C_n will vary with the type of survey. An increase in the number of strata involves extra work in planning and drawing the sample and increases the number of weights used in computing the estimates, unless they are self-weighting. In some surveys almost no change is required in the organization of the field work; in others a separate field unit is set up in each stratum. Whatever the form of the cost function, the results in Table 5A.12 suggest that if an increase in L beyond 6 necessitates any substantial decrease in n in order to keep the cost constant the increase will seldom be profitable.

The discussion in this section is confined to surveys in which only over-all estimates are to be made. If estimates are wanted also for geographic subdivisions of the population, the argument for a larger number of strata is stronger.

5A.9 STRATIFICATION AFTER SELECTION OF THE SAMPLE (POSTSTRATIFICATION)

With some variables that are suitable for stratification, the stratum to which a unit belongs is not known until the data have been collected. Personal characteristics such as age, sex, race, and educational level are common examples. The stratum sizes N_h may be obtainable fairly accurately from official statistics, but the units can be classified into the strata only after the sample data are known. We assume here that W_h, N_h are known.

One procedure is to take a simple random sample of size n and classify the units. Instead of the sample mean $\bar{y}$, we use the estimate $\bar{y}_W = \sum W_h \bar{y}_h$, where $\bar{y}_h$ is the mean of the sample units that fall in stratum h, and $W_h = N_h/N$. This method is almost as precise as *proportional* stratified sampling, provided that (a) the sample is reasonably large, say >20, in every stratum, and (b) the effects of errors in the weights W_h can be ignored (see section 5A.2).

To show this, let m_h be the number of units in the sample that fall in stratum h, where m_h will vary from sample to sample. For samples in which the m_h are fixed and all m_h exceed zero,

$$V(\bar{y}_W) = \sum \frac{W_h^2 S_h^2}{m_h} - \frac{1}{N} \sum W_h S_h^2 \qquad (5A.40)$$

The average value of $V(\bar{y}_W)$ in repeated samples of size n must now be calculated. This requires a little care, since one or more of the m_h could be zero. If this happened, two or more strata would have to be combined before making the estimate, and a less precise estimate would be produced. With increasing n, the probability that any m_h is zero becomes so small that the contribution to the variance from this source is negligible.

If the case in which m_h is zero is ignored, Stephan (1945) has shown that to terms of order n^{-2}

$$E\left(\frac{1}{m_h}\right) = \frac{1}{nW_h} + \frac{1-W_h}{n^2W_h^2}$$ (5A.41)

Hence

$$E[V(\bar{y}_W)] = \frac{1-f}{n} \sum W_h S_h^2 + \frac{1}{n^2} \sum (1-W_h)S_h^2$$ (5A.42)

The first term is the value of $V(\bar{y}_{st})$ for proportional stratification. The second represents the increase in variance that arises because the m_h do not distribute themselves proportionally. But

$$\frac{1}{n^2} \sum (1-W_h)S_h^2 = \frac{1}{n}\left(\frac{L}{n}\right)\bar{S}_h^2 - \frac{1}{n^2} \sum W_h S_h^2 = \frac{1}{n\bar{n}_h}\bar{S}_h^2 - \frac{1}{n^2} \sum W_h S_h^2$$ (5A.43)

where $\bar{S}_h^2$ is the average of the S_h^2 and $\bar{n}_h = n/L$ is the average number of units per stratum. Thus, if the S_h^2 do not differ greatly, the increase is about $(L-1)/L\bar{n}_h$ times the variance for proportional stratification, ignoring the fpc. The increase will be small if $\bar{n}_h$ is reasonably large.

This method can also be applied to a sample that is already stratified by another factor, for example, into five geographic regions, provided that the W_h are known separately within each region. This twofold stratification is widely employed in U.S. National Surveys: see Bean (1970) for a description of the estimation formulas in the Health Interview Survey of the National Center for Health Statistics.

5A.10 QUOTA SAMPLING

In another method that has been used in opinion and market research surveys the n_h required in each stratum is computed in advance so that stratification is proportional. The enumerator is instructed to continue sampling until the necessary "quota" has been obtained in each stratum. The most common variables for stratification are geographic area, age, sex, race, and some measure of economic level. If the enumerator were to choose persons at random within the geographic areas and assign each to his appropriate stratum, the method would be identical with stratified random sampling. A considerable amount of field work would be required to fill all quotas, however, since in the later stages most of the persons approached would fall in quotas already filled.

To expedite the filling of quotas, some latitude is allowed to the enumerator regarding the persons or households to be included. The amount of latitude varies with the agency but, in general, quota sampling may be described as stratified sampling with a more or less nonrandom selection of units within strata. For this reason, sampling-error formulas cannot be applied with confidence to the results of quota samples. A number of comparisons between the results of quota and probability samples are summarized by Stephan and McCarthy (1958), who give an excellent critique of the performance of both types of survey. The quota method seems likely to produce samples that are biased on characteristics such as income, education, and occupation, although it often agrees well with the probability samples on questions of opinion and attitude.

5A.11   ESTIMATION FROM A SAMPLE OF THE GAIN DUE TO STRATIFICATION

When a stratified random sample has been taken, it may be of interest, as a guide to the conduct of future surveys, to appraise the gain in precision relative to simple random sampling.

The data available from the sample are the values of N_h, n_h, $\bar{y}_h$, and s_h^2. From section 5.4, the estimated variance of the weighted mean from the stratified sample is, by formula (5.13),

$$v(\bar{y}_{st}) = \sum \frac{W_h^2 s_h^2}{n_h} - \sum \frac{W_h s_h^2}{N}$$

The problem is to compare this variance with an estimate of the variance of the mean that would have been obtained from a simple random sample. One procedure sometimes used calculates the familiar mean square deviation from the sample mean,

$$s^2 = \frac{\sum (y_{hi} - \bar{y})^2}{n-1}$$

where the strata are ignored. This is taken as an estimate of s^2, so that $\hat{V}_{ran} = (N-n)s^2/Nn$ for the mean of a simple random sample. This method works well enough if the allocation is proportional, since a simple random sample distributes itself approximately proportionally among strata. But if an allocation far from proportional has been adopted, the sample actually taken does not resemble a simple random sample, and this s^2 may be a poor estimator. A general procedure is given, the proof being due to J. N. K. Rao (1962).

Theorem 5A.1.   Given the results of a stratified random sample, an unbiased estimator of V_{ran}, the variance of the mean of a simple random sample from the same population is

$$V_{ran} = \frac{(N-n)}{n(N-1)} \left[\frac{1}{N} \sum_h^L \frac{N_h}{n_h} \sum_j^{n_h} y_{hj}^2 - \bar{y}_{st}^2 + v(\bar{y}_{st}) \right] \qquad (5A.44)$$

where $v(\bar{y}_{st})$ is the usual unbiased estimator of $V(\bar{y}_{st})$.

Proof.

$$V_{ran} = \frac{(N-n)}{nN}S^2 = \frac{(N-n)}{n(N-1)}\left[\frac{1}{N}\sum_h^L\sum_j^{N_h} y_{hj}^2 - \bar{Y}^2\right] \qquad (5A.45)$$

Now

$$\frac{1}{N}E\left(\sum_h^L\frac{N_h}{n_h}\sum_j^{n_h} y_{hj}^2\right) = \frac{1}{N}\sum_h^L\sum_j^{N_h} y_{hj}^2 \qquad (5A.46)$$

Also, since $v(\bar{y}_{st})$ and $\bar{y}_{st}$ are unbiased estimators of $V(\bar{y}_{st})$ and $\bar{Y}$, respectively,

$$Ev(\bar{y}_{st}) = V(\bar{y}_{st}) = E(\bar{y}_{st}^2) - \bar{Y}^2 \qquad (5A.47)$$

and hence an unbiased estimator of $\bar{Y}^2$ in (5A.45) is

$$\bar{y}_{st}^2 - v(\bar{y}_{st}) \qquad (5A.48)$$

From (5A.46) and (5A.48) it follows that an unbiased sample estimator of V_{ran} in (5A.45) is

$$V_{ran} = \frac{(N-n)}{n(N-1)}\left[\frac{1}{N}\sum_h^L\frac{N_h}{n_h}\sum_j^{n_h} y_{hj}^2 - \bar{y}_{st}^2 + v(\bar{y}_{st})\right] \qquad (5A.44)$$

This proves Theorem 5A.1.

With proportional allocation, $(N_h/n_h = N/n)$, the first two terms inside the square brackets become $(1/n)$ times the within-sample sum of squares $= (n-1)s^2/n$. Formula (5A.44) then reduces to

$$v_{ran} = \frac{(N-n)}{n(N-1)}\left[\frac{(n-1)}{n}s^2 + v(\bar{y}_{st})\right] \qquad (5A.49)$$

If n is large, $(n-1) \doteq n$, $(N-1) \doteq N$, and the term in $v(\bar{y}_{st})$ is of order $1/n$ relative to the term in s^2 in (5A.49). Hence

$$v_{ran} \doteq \frac{(N-n)}{nN}s^2 \qquad (5A.50)$$

for proportional allocation. In the general case the corresponding simplification (n large) is

$$v_{ran} \doteq \frac{(N-n)}{nN}\left[\frac{1}{N}\sum\frac{N_h}{n_h}\sum^{n_h} y_{hj}^2 - \bar{y}_{st}^2\right] \qquad (5A.51)$$

Example. The calculations are illustrated from the first three strata in the sample of teachers' colleges (section 5.9). The data in Table 5A.13 are for the later 1946 sample. The means represent enrollment per college in thousands. The s_h^2 values are slightly higher than in the second edition, owing to a correction.

TABLE 5A.13

BASIC DATA FROM A STRATIFIED SAMPLE OF TEACHERS'
COLLEGES

Stratum	N_h	n_h	$\bar{y}_h$	s_h^2	$\dfrac{N_h}{n_h}\left(\sum^{n_h} y_{hj}^2\right)$
1	13	9	2.200	1.8173	83.920
2	18	7	1.638	0.0735	49.429
3	26	10	0.992	0.0859	27.596
	57	26			160.945

With n small we use formula (5A.44). We find $\bar{y}_{st} = 1.4715$. The values of the y_{hj} for the sample were not reported, but the figures in the right-hand column can be obtained from preceding columns of Table 5A.13. The formulas work out as follows.

$$v(\bar{y}_{st}) = \frac{1}{N^2} \sum \frac{N_h(N_h - n_h)}{n_h} s_h^2 = 0.00497$$

$$v_{ran} = \frac{31}{(26)(56)}\left[\frac{160.945}{57} - (1.4715)^2 + 0.00497\right] = 0.01412$$

Stratification appears to have reduced the variance to about one third of the value for a simple random sample, the estimated *deff* factor (section 4.11) being $0.00497/0.01412 = 0.35$.

5A.12 ESTIMATION OF VARIANCE WITH ONE UNIT PER STRATUM

If the population is highly variable and many effective criteria for stratification are known, stratification may be carried to the point at which the sample contains only one unit in each stratum. In this event the formulas previously given for estimating $V(\hat{Y}_{st})$ and $V(\bar{y}_{st})$ cannot be used. With L even, an estimate may be attempted by grouping the strata in pairs thought beforehand to have roughly equal true stratum totals. The allocation into pairs should be made before seeing the sample results, for reasons that will become evident.

Let the sample observations in a typical pair be y_{j1}, y_{j2}, where j goes from 1 to $L/2$. Let $\hat{Y}_{j1} = N_{j1}y_{j1}$, $\hat{Y}_{j2} = N_{j2}y_{j2}$ be the estimated stratum totals. Now

$$\hat{Y}_{j1} - \hat{Y}_{j2} = (Y_{j1} - Y_{j2}) + (\hat{Y}_{j1} - Y_{j1}) - (\hat{Y}_{j2} - Y_{j2}) \qquad (5A.52)$$

Hence, averaging over all samples from this pair,

$$E(\hat{Y}_{j1} - \hat{Y}_{j2})^2 = (Y_{j1} - Y_{j2})^2 + N_{j1}(N_{j1} - 1)S_{j1}^2 + N_{j2}(N_{j2} - 1)S_{j2}^2 \qquad (5A.53)$$

For $V(\hat{Y}_{st})$ consider the estimate

$$v_1(\hat{Y}_{st}) = \sum_{j=1}^{L/2} (\hat{Y}_{j1} - \hat{Y}_{j2})^2 \tag{5A.54}$$

By (5A.53) the expected value of this quantity is

$$Ev_1(\hat{Y}_{st}) = \sum_{h=1}^{L} N_h(N_h - 1)S_h^2 + \sum_{j=1}^{L/2} (Y_{j1} - Y_{j2})^2 \tag{5A.55}$$

The first term on the right is the correct variance (by theorem 5.4 with $n_h = 1$). The second term represents a positive bias, whose size depends on the success attained in selecting pairs of strata whose true totals differ little. The form of the estimate (5A.54) warns that construction of pairs by making the *sample* estimated totals differ as little as possible can give a serious underestimate. The technique is called the method of "collapsed strata."

With L odd, at least one group must of course, be of size different from 2. The extension of the estimate (5A.54) to G groups of any chosen sizes $L_j \geq 2$ is

$$v_1(\hat{Y}_{st}) = \sum_{j=1}^{G} \frac{L_j}{L_j - 1} \sum_{k=1}^{L_j} (\hat{Y}_{jk} - \hat{Y}_j/L_j)^2 \tag{5A.56}$$

where $\hat{Y}_j$ is the estimated total for group j. For $L_j = 2$, when $\hat{Y}_j = \hat{Y}_{j1} + \hat{Y}_{j2}$, this form agrees with (5A.54). As with (5A.54), the expectation of this $v_1(\hat{Y}_{st})$ gives the correct variance $V(\hat{Y}_{st})$, plus a positive bias found by substituting Y_{jk} and Y_j for $\hat{Y}_{jk}$ and $\hat{Y}_j$ in (5A.56).

When an auxiliary variate A_h is known for each stratum that predicts the stratum total Y_h, Hansen, Hurwitz, and Madow (1953) suggested the alternative variance estimator

$$v_2(\hat{Y}_{st}) = \sum_{j=1}^{G} \frac{L_j}{L_j - 1} \sum_{k=1}^{L_j} (\hat{Y}_{jk} - A_{jk}\hat{Y}_j/A_j)^2 \tag{5A.57}$$

If A_h is a good predictor, the positive bias term in v_2, coming from the deviations $(\hat{Y}_{jk} - A_{jk}\hat{Y}_j/A_j)^2$, is likely to be smaller than the corresponding term in v_1, although unlike v_1, v_2 also gives a biased estimate of the term in the S_h^2 in $V(\hat{Y}_{st})$. Hartley, Rao, and G. Kiefer (1969) found v_2 less biased than v_1 in two of three populations, with little difference in the third.

These authors developed a method that does not involve the collapsing of strata. This method uses one or more auxiliary variates x_{1h}, x_{2h}, and so forth, on which the true stratum means $\bar{Y}_h$ are thought to have a linear regression. If y_h is the sample value in stratum h, the method uses the deviations

$$d_h = y_h - \bar{y} - \sum_i b_i(x_{ih} - \bar{x}_i) \tag{5A.58}$$

The variance-covariance matrix of the d_h can be expressed as a linear function of the σ_h^2, plus certain bias terms. By inverting this relation, estimates $\hat{\sigma}_h^2$ are obtained, where $\sigma_h^2 = (N_h - 1)S_h^2/N_h$, giving

$$v_3(\hat{Y}_{st}) = \sum_{h=1}^{L} N_h^2 \hat{\sigma}_h^2 \tag{5A.59}$$

The method appears promising and extends to ratio estimates, but the authors warn that additional comparisons with "collapsed strata" methods are needed.

Using a different approach, Fuller (1970) developed a method of stratum construction that provides an unbiased sample estimate of $V(\hat{Y}_{st})$ with one unit per stratum. For simplicity suppose that $N/n = N/L = k$ (an integer). Select a random number r between 1 and k. The first stratum consists of the units numbered from $(r+1)$ up to $(r+k)$, the second those numbered from $(r+k+1)$ up to $(r+2k)$, and so on, the last (Lth and nth stratum) those numbered from $r+(n-1)k+1$ to $N = nk$ and those from 1 to r. At first sight, this last stratum may look a poor choice. As Fuller notes, however, this method can work well in geographic stratification with areal units. Here, stratification usually leans on the notion that units near one another tend to be similar. By numbering units in serpentine fashion, one can have y_N near y_1, so that the stratum that includes both y_N and y_1 an also be internally homogeneous. The estimate $v(\hat{Y}_{st})$ is a weighted sum of the differences $(y_h - y_{h+1})^2$.

The circular method would be less effective for a population showing a rising trend from y_1 to y_N, in which the stratum including both y_1 and y_N would have large internal variability. For this situation Fuller gives a second plan, slightly more complex, which should give good precision with a rising trend and also furnishes an unbiased estimate of $V(\hat{Y}_{st})$.

5A.13   STRATA AS DOMAINS OF STUDY

This section deals with surveys in which the primary purpose is to make comparisons between different strata, assumed to be identifiable in advance. The rules for allocating the sample sizes to the strata are different from those that apply when the objective is to make over-all population estimates. If there are only two strata, we might choose n_1, n_2 to minimize the variance of the difference $(\bar{y}_1 - \bar{y}_2)$ between the estimated strata means. Omitting the fpc's for reasons given in section 2.14, we have

$$V(\bar{y}_1 - \bar{y}_2) = \frac{S_1^2}{n_1} + \frac{S_2^2}{n_2} \tag{5A.60}$$

With a linear cost function

$$C = c_0 + c_1 n_1 + c_2 n_2 \tag{5A.61}$$

V is minimized when

$$n_1 = \frac{\dfrac{nS_1}{\sqrt{c_1}}}{S_1/\sqrt{c_1} + S_2/\sqrt{c_2}}, \qquad n_2 = \frac{\dfrac{nS_2}{\sqrt{c_2}}}{S_1/\sqrt{c_1} + S_2/\sqrt{c_2}} \tag{5A.62}$$

With L strata, $L > 2$, the optimum allocation depends on the amounts of precision desired for different comparisons. For instance, the cost might be minimized subject to the set of $L(L-1)/2$ conditions that $V(\bar{y}_h - \bar{y}_i) \leq V_{hi}$, where the values of V_{hi} are chosen according to the precision considered necessary for a satisfactory comparison of strata h and i.

Frequently a simpler method of allocation is adequate, especially if the S_h and c_h do not differ greatly. One approach is to minimize the *average* variance of the difference between all $L(L-1)/2$ pairs of strata, that is, to minimize

$$\bar{V} = \frac{2}{L} \left(\frac{S_1^2}{n_1} + \frac{S_2^2}{n_2} + \cdots + \frac{S_L^2}{n_L} \right) \tag{5A.63}$$

$\bar{V}$ is minimized, for fixed C, by the rule in (5A.62),

$$n_h \propto \frac{S_h}{\sqrt{c_h}} \tag{5A.64}$$

This rule may result in certain pairs of strata being more precisely compared and others less precisely than is felt appropriate. An alternative is to select the n_h so that the s.e. of the difference is the same, say $\sqrt{V}$, for every pair of strata. This amounts to making $S_h^2/n_h = V/2$ for every stratum. For a fixed cost this method gives less over-all precision than the first method. The reader may verify that the two optimum allocations give

$$\bar{V} = \frac{2(\sum S_h \sqrt{c_h})^2}{L(C-c_0)}, \qquad V = \frac{2(\sum S_h^2 c_h)}{(C-c_0)} \tag{5A.65}$$

It follows from the Cauchy–Schwarz inequality that V is always greater than $\bar{V}$ unless $S_h \sqrt{c_h} = $ constant. If V is substantially greater than $\bar{V}$, a compromise allocation can sometimes be found, after a little trial and error, that will give an average variance close to $\bar{V}$ and also keep $V(\bar{y}_h - \bar{y}_i)$ reasonably constant.

Sometimes the objective is to obtain estimates for each stratum as well as over-all estimates for the whole population. In planning the survey, we might specify the following conditions.

$$V(\bar{y}_h) = \frac{S_h^2}{n_h}(1-f_h) \leq V_h, \qquad V(\bar{y}_{st}) = \sum \frac{N_h^2 S_h^2}{n_h}(1-f_h) \leq V$$

The fpc terms are now included, since the purpose is to specify the precision with which the means in the finite population are to be estimated. The conditions on the $V(\bar{y}_h)$ determine lower limits to the values of the n_h. If these lower limits are found to satisfy the condition on $V(\bar{y}_{st})$, the allocation problem is solved. When the condition on $V(\bar{y}_{st})$ is not satisfied, Dalenius (1957) has indicated a graphical approach.

More complex problems arise when the $L = 2^k$ strata represent all combinations of k factors each at two levels, and the objective is to estimate the average effects of the factors. If the stratum or cell to which any member of the population belongs is known in advance of sampling, a sample of desired size n_h can be drawn from stratum h. For 2, 3, or 4 factors Sedransk (1967) has given methods for finding the n_h that minimize the cost under different specifications about the variances of the estimated average effects of the factors and the desired power of a test for interactions.

5A.14 ESTIMATING TOTALS AND MEANS OVER SUBPOPULATIONS

Frequently the subpopulations or domains of study are represented in all strata. If stratification is geographic, for example, separate estimates may be wanted, over the whole population, for males and females, for different age groups, for users and nonusers of Blank's toothpaste, and the like. The problem presents some complications. The basic formulas were given by Yates (1953) with further discussion and proofs by Durbin (1958) and Hartley (1959). Methods applicable to a single stratum are discussed in sections 2.10 and 2.11.

The following notation applies to the units in stratum h that lie in domain j.

Notation.

Number of units: N_{hj}, $\sum_j N_{hj} = N_h$

Number in sample: n_{hj}, $\sum_j n_{hj} = n_h$

Measurement on individual unit: y_{hij}

Sample mean: $\bar{y}_{hj} = \sum_{i=1}^{n_{hj}} \frac{y_{hij}}{n_{hj}}$

Domain mean: $\bar{Y}_{hj} = \sum_{i=1}^{N_{hj}} \frac{y_{hij}}{N_{hj}}$

The population total and mean for domain j over all strata are, respectively,

$$Y_j = \sum_h N_{hj} \bar{Y}_{hj}, \qquad \bar{Y}_j = \frac{Y_j}{N_j}$$

where $N_j = \sum_h N_{hj}$.

The complication arises because the n_{hj} are random variables. If the N_{hj} were known, the problem would be simple. As estimates of Y_j and $\bar{Y}_j$, we could use

$$\hat{Y}_h' = \sum_h N_{hj} \bar{y}_{hj}, \qquad \hat{\bar{Y}}_j' = \frac{\hat{Y}_j'}{N_j}$$

By the method in section 2.12, the ordinary formula for $V(\bar{y}_{hj})$ is still valid, provided all $n_{hj} > 0$. Thus

$$V(\hat{Y}_j') = \sum_h \frac{N_{hj}^2 S_{hj}^2}{n_{hj}} \left(1 - \frac{n_{hj}}{N_{hj}}\right) \tag{5A.66}$$

where S_{hj}^2 is the variance among units in domain j within stratum h. In applications, however, the N_{hj} are rarely known.

Estimating Domain Totals

In default of the N_{hj}, each stratum total of the domain is estimated as in section 2.13. These totals are added to obtain an estimated domain total, that is,

$$\hat{Y}_j = \sum_h \frac{N_h}{n_h} \sum_i^{n_{hj}} y_{hij} \tag{5A.67}$$

The true and estimated variance of $\hat{Y}_j$ are found by the device used in section 2.13. A variate y'_{hi} is introduced that equals y_{hij} for all units in domain j and equals zero for all other units in the population. As shown in section 2.13 this gives for the estimated variance

$$v(\hat{Y}_j) = \sum_h \frac{N_h^2}{n_h(n_h - 1)} (1 - f_h) \left[\sum_i^{n_{hj}} y_{hij}^2 - \frac{(\sum y_{hij})^2}{n_h} \right] \tag{5A.68}$$

Estimating Domain Means

In order to estimate the domain mean Y_j/N_j, a sample estimate of N_j is required. An unbiased estimate is

$$\hat{N}_j = \sum_h \frac{N_h}{n_h} n_{hj} \tag{5A.69}$$

Hence we take

$$\hat{\bar{Y}}_j = \frac{\hat{Y}_j}{\hat{N}_j} = \frac{\sum_h (N_h/n_h) \sum_i y_{hij}}{\sum_h (N_h/n_h) n_{hj}} \tag{5A.70}$$

With proportional stratification, $\hat{\bar{Y}}_j$ reduces to the ordinary sample mean of the units that fall in domain j. In the general case, this estimate is known as a *combined ratio estimate*, discussed later in section 6.11. To show it, introduce another dummy variate x'_{hi} which equals 1 for every unit in domain j and 0 for all other units, where i now goes from 1 to N_h. Clearly,

$$\bar{x}_h' = \frac{\sum_i^{n_h} x_{hi}'}{n_h} = \frac{n_{hj}}{n_h}, \qquad \bar{y}_h' = \frac{\sum_i^{n_h} y_{hi}}{n_h} = \frac{\sum_i^{n_{hj}} y_{hij}}{n_h} = \frac{n_{hj}}{n_h} \bar{y}_{hj} \tag{5A.71}$$

so that the estimated domain mean may be written

$$\hat{\bar{Y}}_j = \frac{\sum_h (N_h/n_h) \sum_i y_{hij}}{\sum_h (N_h/n_h) n_{hj}} = \frac{\sum_h N_h \bar{y}_h'}{\sum_h N_h \bar{x}_h'} = \frac{\bar{y}_{st}}{\bar{x}_{st}'} \tag{5A.72}$$

This is the formula for the combined ratio estimate for the two variables y_{hi}' and x_{hi}'. From section 6.11, the estimated variance may be expressed approximately as

$$v(\hat{\bar{Y}}_j) \doteq \frac{1}{\hat{N}_j^2} \sum_h \frac{N_h^2 (1-f_h)}{n_h(n_h-1)} \sum_i^{n_h} [y_{hi}' - \hat{\bar{Y}}_j x_{hi}' - (\bar{y}_h' - \hat{\bar{Y}}_j \bar{x}_h')]^2 \tag{5A.73}$$

The second summation may be written

$$\sum_i^{n_h} (y_{hi}' - \hat{\bar{Y}}_j x_{hi}')^2 - n_h (\bar{y}_h' - \hat{\bar{Y}}_j \bar{x}_h')^2 = \sum_i^{n_{hj}} (y_{hij} - \hat{\bar{Y}}_j)^2 - \frac{n_{hj}^2}{n_h}(\bar{y}_{hj} - \hat{\bar{Y}}_j)^2 \tag{5A.74}$$

using (5A.71). Furthermore, the first term in (5A.74) can be expressed alternatively as

$$\sum_i^{n_{hj}} (y_{hij} - \bar{y}_{hj})^2 + n_{hj}(\bar{y}_{hj} - \hat{\bar{Y}}_j)^2$$

Inserting these results in (5A.73) gives, finally, for the estimated variance,

$$v(\hat{\bar{Y}}_j) \doteq \frac{1}{\hat{N}_j^2} \sum_h \frac{N_h^2 (1-f_h)}{n_h(n_h-1)} \left[\sum_i (y_{hij} - \bar{y}_{hj})^2 + n_{hj}\left(1 - \frac{n_{hj}}{n_h}\right)(\bar{y}_{hj} - \hat{\bar{Y}}_j)^2 \right] \tag{5A.75}$$

The term on the right represents a between-stratum contribution to the variance. Differences among strata means are not entirely eliminated from the variance of the estimated mean of any subpopulation. The between-stratum contribution is small if the terms $1 - n_{hj}/n_h$ are small, that is, if the subpopulation is almost as large as the complete population.

As Durbin (1958) has pointed out, (5A.75) applies also to means estimated for the whole population, if the sample is incomplete for any reason such as non-response, provided, of course, that $\hat{\bar{Y}}_j$ is the estimate used. In this event $\hat{\bar{Y}}_j$ is interpreted as the estimated mean for the part of the population that would give a response under the methods of data collection employed. There is, however, an additional complication, in that the "nonreponse" part of the population often has a different mean from the "response" part. Thus $\hat{\bar{Y}}_j$ is a biased estimate of the mean of the whole population, and this bias contribution is not included in (5A.75).

5A.15 SAMPLING FROM TWO FRAMES

An early example of the use of a sample from a list B of large businesses in combination with a sample from an areal frame A that covers the complete population is the 1949 sample survey of retail stores taken by the Census Bureau and described in Hansen, Hurwitz, and Madow (1953, p. 516 ff.). The objectives in this combined use of an incomplete list frame and a complete areal frame were to gain increased accuracy and save money. Large businesses can sometimes be sampled cheaply by a combination of mail and telephone; moreover, being large, they are often the businesses that have the largest variance for the y variables being measured. Placing them in a separate stratum with optimum allocation (or 100 per cent sampling if this seems close to optimum) can produce substantial increases in accuracy. In the retail stores survey, businesses in the list frame that were present in the area sample were identified and removed from the area sample, so that the population being sampled fell into two distinct strata. This process, called *unduplication*, was performed by the field supervisor in advance of sampling. Unduplication is sometimes complicated and subject to errors. Some of the practical difficulties are discussed by Hansen, Hurwitz and Jabine (1963), as well as various ways in which incomplete lists can be helpful.

Identification of the sample members from the frame A sample that belong to the list frame B sometimes requires measurement of the y variables for these members. The sampler then has three samples at his disposal in which y has been measured—a sample from stratum a (the part of A that does not belong to B) and two independent samples from stratum B. One, denoted by the suffix ab, is the sample obtained from A and identified as belonging to B, and one is obtained by direct sampling of frame B. With frame A complete and simple random sampling from both frames, Hartley (1962) proposed that both B samples be used in the poststratified estimate

$$\hat{Y} = N_a \bar{y}_a + N_{ab}(p\bar{y}_{ab} + q\bar{y}_B) \tag{5A.76}$$

where $\bar{y}_a$, $\bar{y}_{ab}$, $\bar{y}_B$ denote the respective sample means. The weighting factors p and q for the two samples that belong to frame B, with $p + q = 1$, are chosen to minimize $V(\hat{Y})$ under a cost function of the form

$$C = c_A n_A + c_B n_B \tag{5A.77}$$

In (5A.76) the stratum sizes $N_{ab} = N_B$ and $N_a = (N_A - N_B)$ will, of course, be known.

With $S_B^2 > S_a^2$ and $c_B < c_A$, Hartley showed that this method can give large reductions in $V(\hat{Y})$ as compared with sampling from frame A only, even if the frame A sample is poststratified into the two strata a and $B = ab$.

The problem becomes more difficult if frame A is also incomplete, two frames A and B, with some duplication, being required to obtain complete coverage of the population. For poststratification, there are three distinct strata: a (units in A alone); ab (units in both A and B); and b (units in B alone). The three strata cannot be sampled directly, samples of sizes n_A, n_B having to be drawn from frames A and B. Furthermore, the strata sizes N_a, N_{ab}, N_b will not usually be known. For simple random sampling from frames A and B, Hartley (1962) suggested the estimate

$$\hat{Y} = \frac{N_A}{n_A} y_a + p\frac{N_A}{n_A} y_{ab} + q\frac{N_B}{n_B} y_{ba} + \frac{N_B}{n_B} y_b \qquad (5A.78)$$

where, as before, $p+q = 1$ and the y's are sample *totals* in the strata, the suffixes ab and ba denoting the samples in the duplicate stratum found in n_A and n_B. Hartley determined p and q to minimize $V(\hat{Y})$ for fixed cost. Improvements in Hartley's estimate (5A.78) have been given by Lund (1968) and Fuller and Burmeister (1972), essentially by using better estimates of N_a, N_{ab}, N_b than are implied in Hartley's estimate. Fuller and Burmeister (1972) also dealt with the case in which frame A is areal, with subsampling of the areal units. Hartley (1974) gives a general approach to two-frame sampling, applicable to any sample design in the two frames.

EXERCISES

5A.1   In planning a survey of sales in a certain type of store, with $n = 550$, good estimates of S_h are available from a previous survey in two of the three strata. The third stratum consists of new stores and stores that had no sales in the previous survey, so that a value for S_3 has to be guessed. If S_3 is actually 10, compute $V(\bar{y}_{st})$ as given by an estimated Neyman allocation when S_3 is guessed as (a) 5, (b) 20. Show that in both cases the proportional increase in variance over the true optimum is slightly over 2%.

Stratum	W_h	True S_h	Estimated S_h (a)	(b)
1	0.3	30	30	30
2	0.6	20	20	20
3	0.1	10	5	20

5A.2   Show that if all S_h, except S_L, are correctly estimated and S_L is estimated as $\hat{S}_L = S_L(1+\lambda)$, the proportional increase in $V_{opt}(\bar{y}_{st})$, using $\hat{S}_L$ instead of the true S_L for Neyman allocation, is

$$\frac{\lambda^2 n_L'(n - n_L')}{(1+\lambda)n^2}$$

where n_L' is the sample size in stratum L under true Neyman allocation. Verify that this formula agrees with the results in exercise 5A.1. (The agreement is not exact because of the rounding of the n_h to integers.) Hence show that a 50% underestimation of S_L has the same effect as 100% overestimation.

5A.3   If there are two strata and if ϕ is the ratio of the actual n_1/n_2 to the Neyman optimum n_1/n_2, show that whatever the values of N_1, N_2, S_1, and S_2, the ratio $V_{min}(\bar{y}_{st})/V(\bar{y}_{st})$ is never less than $4\phi(1+\phi)^2$ when the fpc's are negligible.

5A.4   The results of a simple random sample with $n = 1000$ can be classified into three "strata," with $\bar{y}_h = 10.2$, 12.6, and 17.1, $s_h^2 = 10.82$ (the same in each stratum), and $s^2 = 17.66$. The estimated stratum weights are $w_h = 0.5$, 0.3, 0.2, respectively. These weights are known to be inexact, but it is thought that all are correct within 5%, so that the worst cases are either $W_h = 0.525, 0.285$, and 0.190 or $W_h = 0.475, 0.315$, and 0.210. By the methods of section 5A.2, would you recommend stratification? (Where needed, assume that $\bar{y}_h = \bar{Y}_h$ and $s_h{}^2 = S_h{}^2$.)

5A.5   In a stratified random sample with two variates the objective is to satisfy the specifications

$$V(\bar{y}_{1st}) \le V_1; \qquad V(\bar{y}_{2st}) \le V_2$$

for minimum cost $C = \sum c_h n_h$. The fpc's can be ignored. (a) Prove the result by Chatterjee (1972) that a compromise allocation is necessary if

$$\frac{\sum W_h S_{2h} \sqrt{c_h}}{\sum W_h (S_{1h}^2/S_{2h}) \sqrt{c_h}} \le \frac{V_2}{V_1} \le \frac{\sum W_h (S_{2h}^2/S_{1h}) \sqrt{c_h}}{\sum W_h S_{1h} \sqrt{c_h}}.$$

(b) If V_2/V_1 equals or exceeds the upper limit, the optimum allocation for y_1 satisfies both tolerances, with a corresponding result about the lower limit.

5A.6   A survey with three strata is planned to estimate the percentage of families who have accounts in savings banks and the average amount invested per family. Advance estimates of the percentages P_h and the within-stratum S_h for the amount invested are as follows.

Stratum	W_h	$P_h(\%)$	$S_h(\$)$
1	0.6	20	90
2	0.3	40	180
3	0.1	70	520

Compute the smallest sample sizes n and the n_h that satisfy the following requirements: (a) The percentage of families is to be estimated with s.e. $= 2$ and the average amount invested with s.e. $= \$5$. (b) The percentage of families is to be estimated with s.e. $= 1.5$ and the average amount invested with s.e. $= \$5$.

Part (b) requires a compromise allocation, either by a computer program or the method in the second edition, p. 123. The allocation $n_h = 371, 344, 315$, with $n = 1030$ satisfies both tolerances. Show that the Booth–Sedransk method (section 5A.4) gives $n_h/n = 0.431$, $0.326, 0.243$. This allocation would require $n = 1073$ to meet both tolerances.

5A.7   The table at top of p. 148 shows the frequency distribution of a population of 911 city sizes for cities from 10,000 to 60,000, arranged in classes of 2000. To shorten the calculations, a coded y' and values of $\sqrt{f}$, cum. $\sqrt{f}$, cum. f, fy', and $\sum fy'^2$ are given. Apply the Dalenius-Hodges rule to create two strata for optimum allocation in the sense of Neyman. Find the values of W_h and S_h for each of your strata. Verify (a) that the optimum sample sizes are almost the same in the two strata and (b) by finding S^2 for the whole population, that

$$\frac{V(\bar{y})}{V_{opt}(\bar{y}_{st})} \doteq 4.8$$

5A.8   The right triangular distribution $f(y) = 2(1-y)$, $0 < y < 1$, is divided into two strata at the point a. (a) Show that

$$W_1 = a(2-a), \qquad W_2 = (1-a)^2$$

$$S_1{}^2 = \frac{a^2(6-6a+a^2)}{18(2-a)^2}, \qquad S_2{}^2 = \frac{(1-a)^2}{18}$$

(b) Show that under the cum $\sqrt{f}$ rule the best choice of a is $1-1/\sqrt[3]{4}=0.37$ and that with this boundary the optimum n_1/n_2 is about $\frac{27}{25}$ and $V(\bar{y}_{st})$ is about 27% of the value given by simple random sampling.

5A.9 In both exercises 5A.7 and 5A.8, show that the Ekman rule $W_h(y_h - y_{h-1}) =$ constant agrees very closely with the cum. $\sqrt{f}$ rule in determining the stratum boundaries.

	f	y'	$\sqrt{f}$	Cum f	Cum $\sqrt{f}$	fy'
10 —	205	0	14.3	205	14.3	0
12 —	135	1	11.6	340	25.9	135
14 —	106	2	10.3	446	36.2	212
16 —	82	3	9.1	528	45.3	246
18 —	61	4	7.8	589	53.1	244
20 —	42	5	6.5	631	59.6	210
22 —	32	6	5.7	663	65.3	192
24 —	30	7	5.5	693	70.8	210
26 —	27	8	5.2	720	76.0	216
28 —	18	9	4.2	738	80.2	162
30 —	22	10	4.7	760	84.9	220
32 —	21	11	4.6	781	89.5	231
34 —	19	12	4.4	800	93.9	228
36 —	16	13	4.0	816	97.9	208
38 —	14	14	3.7	830	101.6	196
40 —	17	15	4.1	847	105.7	255
42 —	9	16	3.0	856	108.7	144
44 —	8	17	2.8	864	111.5	136
46 —	11	18	3.3	875	114.8	198
48 —	9	19	3.0	884	117.8	171
50 —	7	20	2.6	891	120.4	140
52 —	4	21	2.0	895	122.4	84
54 —	5	22	2.2	900	124.6	110
56 —	5	23	2.2	905	126.8	115
58 —	6	24	2.4	911	129.2	144
Totals	911		129.2			4407

$$\sum fy'^2 = 50,395$$

5A.10 A sum of $5000 is available for a stratified sample. In the notation of section 5A.8 the cost function is thought to be, roughly, $C = 200L + 10n$ and

$$V(\bar{y}_{st}) \doteq \frac{S^2}{n}\left[\frac{\rho^2}{L^2}+(1-\rho^2)\right]$$

where ρ is the correlation between the variate used to construct the strata and the variate to be measured in the survey. Compute the optimum L for $\rho = 0.95, 0.9,$ and 0.8. What is a good compromise number of strata to use for all three values of ρ?

5A.11 The following data are derived from a stratified sample of tire dealers taken in March 1945 (Deming and Simmons, 1946). The dealers were assigned to strata according to the number of new tires held at a previous census. The sample means $\bar{y}_h$ are the mean numbers of new tires per dealer. (a) Estimate the gain in precision due to the stratification. (b) Compare this result with the gain that would have been attained from proportional allocation.

Stratum Boundaries	N_h	W_h	$\bar{y}_h$	$s_h{}^2$	n_h
1–9	19,850	0.8032	4.1	34.8	3000
10–19	3,250	0.1315	13.0	92.2	600
20–29	1,007	0.0407	25.0	174.2	340
30–39	606	0.0245	38.2	320.4	230
Totals	24,713	0.9999			4170

5A.12 A population has two strata of relative sizes $W_1 = 0.8$, $W_2 = 0.2$ and within-stratum variances $S_1{}^2 = 100$, $S_2{}^2 = 400$. A stratified random sample is to be taken to satisfy the following requirements: (i) the means of each stratum are to be estimated with variance ≤ 1; (ii) $V(\bar{y}_{st}) \leq 0.5$. Ignoring the fpc, find the values of n_1, n_2 that satisfy all three requirements for minimum $n = n_1 + n_2$.

Hint. Note that $-\partial V(\bar{y}_{st})/\partial n_1 > -\partial V(\bar{y}_{st})/\partial n_2$ if $n_1 < 2n_2$. Fuller (1966) has discussed various methods of handling this problem.

5A.13 In an example due to Nordbotten (1956) and worked by Kokan (1963), a survey is planned to estimate total employment Y_1 and the value of production Y_2 in establishments manufacturing furniture. When establishments are stratified by size, the N_h and rough estimates of the S_{ih}^2 are as follows.

Stratum	N_h	S_{1h}^2	S_{2h}^2
Large	600	200	500,000
Small	1,000	10	4,000
	1,600		

The requirement that estimates of Y_1 and Y_2 not be in error by more than 6% ($P = 0.95$) amounts to tolerances

$$V_1 = V(\bar{y}_{1st}) \leq 0.0351: V_2 = V(\bar{y}_{2st}) \leq 56.25$$

Show that the optimum allocation for y_1 with $n_1 = 450$, $n_2 = 167$, $n = 617$ satisfies both tolerances. Note that in this problem the fpc cannot be ignored.

5A.14 In stratified random sampling with one unit per stratum, assume that the strata can be grouped into pairs with $N_{j1} = N_{j2} = N_j$ ($j = 1, 2, \cdots, L/2$). An alternative sampling method draws two units at random from each pair of strata. Show that for this method

$$V(\hat{Y}_{st2}) = \frac{(N_j - 1)}{(2N_j - 1)} \sum_{j=1}^{L/2} \left[2N_j(N_j - 1) \sum_{i=1}^{2} S_{ji}^2 + (Y_{j1} - Y_{j2})^2 \right]$$

Hence show that the expected value of the "collapsed strata" estimate $v_1(\hat{Y}_{st})$ in formula (5A.54), section 5A.12, overestimates $V(\hat{Y}_{st2})$, the variance that would apply if strata twice as large were used.

CHAPTER 6

Ratio Estimators

6.1 METHODS OF ESTIMATION

One feature of theoretical statistics is the creation of a large body of theory that discusses how to make good estimates from data. In the development of theory specifically for sample surveys, relatively little use has been made of this knowledge. I think there are two principal reasons. First, in surveys that contain a large number of items, there is a great advantage, even with computers, in estimation procedures that require little more than simple addition, whereas the superior methods of estimation in statistical theory, such as maximum likelihood, may necessitate a series of successive approximations before the estimate can be found. Second, as noted in section 1.4, there has been a difference in attitude in the two lines of research. Most of the estimation methods in theoretical statistics assume that we know the functional form of the frequency distribution followed by the data in the sample, and the method of estimation is carefully geared to this type of distribution. The preference in sample survey theory has been to make, at most, limited assumptions about this frequency distribution (that it is very skew or rather symmetrical) and to leave its specific functional form out of the discussion. This preference leads to the use of simple methods of estimation that work well under a range of types of frequency distributions. This attitude is a reasonable one for handling surveys in which the type of distribution may change from one item to another and when we do not wish to stop and examine all of them before deciding how to make each estimate.

Consequently, estimation techniques for sample survey work are at present restricted in scope. Two techniques will now be considered—the ratio method in this chapter and the linear regression method in Chapter 7.

6.2 THE RATIO ESTIMATOR

In the ratio method an auxiliary variate x_i, correlated with y_i, is obtained for each unit in the sample. The population total X of the x_i must be known. In practice, x_i is often the value of y_i at some previous time when a complete census was taken. The aim in this method is to obtain increased precision by taking

150

advantage of the correlation between y_i and x_i. At present we assume simple random sampling.

The ratio estimate of Y, the population total of the y_i, is

$$\hat{Y}_R = \frac{y}{x}X = \frac{\bar{y}}{\bar{x}}X \tag{6.1}$$

where y, x are the sample totals of the y_i and x_i, respectively.

If x_i is the value of y_i at some previous time the ratio method uses the sample to estimate the relative change Y/X that has occurred since that time. The estimated relative change y/x is multiplied by the known population total X on the previous occasion to provide an estimate of the current population total. If the ratio y_i/x_i is nearly the same on all sampling units, the values of y/x vary little from one sample to another, and the ratio estimate is of high precision. In another application x_i may be the total acreage of a farm and y_i the number of acres sown to some crop. The ratio estimate will be successful in this case if all farmers devote about the same percentage of their total acreage to this crop.

If the quantity to be estimated is $\bar{Y}$, the population mean value of y_i, the ratio estimate is

$$\hat{\bar{Y}}_R = \frac{y}{x}\bar{X}$$

Frequently we wish to estimate a ratio rather than a total or mean, for example, the ratio of corn acres to wheat acres, the ratio of expenditures on labor to total expenditures, or the ratio of liquid assets to total assets. The sample estimate is $\hat{R} = y/x$. In this case X need not be known. The use of ratio estimates for this purpose has already been discussed in sections 2.11 and (with cluster sampling for proportions) 3.12.

Example. Table 6.1 shows the number of inhabitants (in 1000's) in each of a simple random sample of 49 cities drawn from the population of 196 large cities discussed in section 2.15. The problem is to estimate the total number of inhabitants in the 196 cities in 1930. The true 1920 total, X, is assumed to be known. Its value is 22,919.

The example is a suitable one for the ratio estimate. The majority of the cities in the sample show an increase in size from 1920 to 1930 of the order of 20%. From the sample data we have

$$y = \sum y_i = 6262, \qquad x = \sum x_i = 5054$$

Consequently the ratio estimate of the 1930 total for all 196 cities is

$$\hat{Y}_R = \frac{y}{x}X = \frac{6262}{5054}(22,919) = 28,397$$

The corresponding estimate based on the sample mean per city is

$$\hat{Y} = N\bar{y} = \frac{(196)(6262)}{49} = 25,048$$

The correct total in 1930 is 29,351.

TABLE 6.1

Sizes of 49 Large United States Cities (in 1000's) in 1920 (x_i) and 1930 (y_i)

x_i	y_i	x_i	y_i	x_i	y_i
76	80	2	50	243	291
138	143	507	634	87	105
67	67	179	260	30	111
29	50	121	113	71	79
381	464	50	64	256	288
23	48	44	58	43	61
37	63	77	89	25	57
120	115	64	63	94	85
61	69	64	77	43	50
387	459	56	142	298	317
93	104	40	60	36	46
172	183	40	64	161	232
78	106	38	52	74	93
66	86	136	139	45	53
60	57	116	130	36	54
46	65	46	53	50	58
				48	75

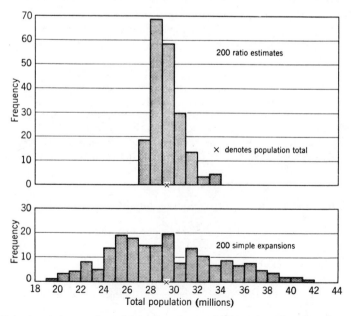

Fig. 6.1 Experimental comparison of the ratio estimate with the estimate based on the sample mean.

Figure 6.1 shows the ratio estimate and the estimate based on the sample mean per city for each of 200 simple random samples of size 49 drawn from this population. A substantial improvement in precision from the ratio method is apparent.

6.3 APPROXIMATE VARIANCE OF THE RATIO ESTIMATE

The distribution of the ratio estimate has proved annoyingly intractable because both y and x vary from sample to sample. The known theoretical results fall short of what we would like to know for practical applications. The principal results are stated first without proof.

The ratio estimate is consistent (this is obvious). It is biased, except for some special types of population, although the bias is neglible in large samples. The limiting distribution of the ratio estimate, as n becomes very large, is normal, subject to some mild restrictions of the type of population from which we are sampling. In samples of moderate size the distribution shows a tendency to positive skewness in the kinds of populations for which the method is most often used. We have an exact formula for the bias but for the sampling variance of the estimate only an approximation valid in large samples.

These results amount to saying that there is no difficulty if the sample is large enough so that (a) the ratio is nearly normally distributed and (b) the large-sample formula for its variance is valid. As working rule, the large-sample results may be used if the sample size exceeds 30 and is also large enough so that the coefficients of variation of $\bar{x}$ and $\bar{y}$ are both less than 10%.

Theorem 6.1. The ratio estimates of the population total Y, the population mean, $\bar{Y}$, and the population ratio Y/X are, respectively,

$$\hat{Y}_R = \frac{\bar{y}}{\bar{x}}X, \qquad \hat{\bar{Y}}_R = \frac{\bar{y}}{\bar{x}}\bar{X}, \qquad \hat{R} = \frac{\bar{y}}{\bar{x}}$$

In a simple random sample of size n (n large)

$$V(\hat{Y}_R) \doteq \frac{N^2(1-f)}{n}\left[\frac{\sum\limits_{i=1}^{N}(y_i - Rx_i)^2}{N-1}\right] \tag{6.2}$$

$$V(\hat{\bar{Y}}_R) \doteq \frac{1-f}{n}\left[\frac{\sum\limits_{i=1}^{N}(y_i - Rx_i)^2}{N-1}\right] \tag{6.3}$$

$$V(\hat{R}) \doteq \frac{1-f}{n\bar{X}^2}\left[\frac{\sum\limits_{i=1}^{N}(y_i - Rx_i)^2}{N-1}\right] \tag{6.4}$$

where $f = n/N$ is the sampling fraction. The method used in theorem 2.5 shows that (6.2), (6.3), and (6.4) are also approximations to the mean square errors of the estimator in these formulas.

The argument leading to the approximate result (6.4) was given in theorem 2.5. Since $\hat{Y}_R = \bar{X}\hat{R}$, $\hat{Y}_R = N\bar{X}\hat{R}$, the other two results follow immediately.

Corollary 1. There are various alternative forms of the result. Since $\bar{Y} = R\bar{X}$, we may write

$$V(\hat{Y}_R = \frac{N^2(1-f)}{n(N-1)} \sum_{i=1}^{N} [(y_i - \bar{Y}) - R(x_i - \bar{X})]^2$$

$$= \frac{N^2(1-f)}{n(N-1)} [\sum (y_i - \bar{Y})^2 + R^2 \sum (x_i - \bar{X})^2$$

$$- 2R \sum (y_i - \bar{Y})(x_i - \bar{X})]$$

The correlation coefficient ρ between y_i and x_i in the finite population is defined by the equation

$$\rho = \frac{E(y_i - \bar{Y})(x_i - \bar{X})}{\sqrt{E(y_i - \bar{Y})^2 E(x_i - \bar{X})^2}} = \frac{\sum_{i}^{N} (y_i - \bar{Y})(x_i - \bar{X})}{(N-1)S_y S_x}$$

This leads to the result

$$V(\hat{Y}_R) = \frac{N^2(1-f)}{n}(S_y^2 + R^2 S_x^2 - 2R\rho S_y S_x) \tag{6.5}$$

An equivalent form is

$$V(\hat{Y}_R) = (1-f)\frac{Y^2}{n}\left(\frac{S_y^2}{\bar{Y}^2} + \frac{S_x^2}{\bar{X}^2} - \frac{2S_{yx}}{\bar{Y}\bar{X}}\right) \tag{6.6}$$

where $S_{yx} = \rho S_y S_x$ is the covariance between y_i and x_i. This relation may also be written as

$$V(\hat{Y}_R) = (1-f)\frac{Y^2}{n}(C_{yy} + C_{xx} - 2C_{yx}) \tag{6.7}$$

where C_{yy}, C_{xx} are the squares of the coefficients of variation (cv) of y_i and x_i respectively, and C_{yx} is the relative covariance.

Corollary 2. Since $\hat{Y}_R$, $\hat{\bar{Y}}_R$, and $\hat{R}$ differ only by known multipliers, the coefficient of variation (i.e., the standard error divided by the quantity being estimated) is the same for all three estimates. From (6.7) the square of this cv is

$$(cv)^2 = \frac{V(\hat{Y}_R)}{Y^2} = \frac{1-f}{n}(C_{yy} + C_{xx} - 2C_{yx}) \tag{6.8}$$

The quantity $(cv)^2$ has been called the *relative variance* by Hansen et al. (1953). Its use avoids repetition of variance formulas for related quantities like the estimated population total and mean.

6.4 ESTIMATION OF THE VARIANCE FROM A SAMPLE

From equation (6.2),

$$V(\hat{Y}_R) \doteq \frac{N^2(1-f)}{n} \frac{\sum\limits_{}^{N}(y_i - Rx_i)^2}{N-1}$$

As already mentioned in section 2.11, we take

$$\frac{\sum\limits_{}^{n}(y_i - \hat{R}x_i)^2}{(n-1)}$$

as a sample estimate of the population variance. This estimate has a bias of order $1/n$.

For the estimated variance, $v(\hat{Y}_R)$, this gives

$$v(\hat{Y}_R) = \frac{N^2(1-f)}{n(n-1)} \sum_{i=1}^{n}(y_i - \hat{R}x_i)^2 \tag{6.9}$$

This result may be expressed in several different ways. For example,

$$v(\hat{Y}_R) = \frac{N^2(1-f)}{n(n-1)}(\sum y_i^2 + \hat{R}^2 \sum x_i^2 - 2\hat{R} \sum y_i x_i) \tag{6.10}$$

$$= \frac{N^2(1-f)}{n}(s_y^2 + \hat{R}^2 s_x^2 - 2\hat{R}s_{yx}) \tag{6.11}$$

where $s_{yx} = \sum(y_i - \bar{y})(x_i - \bar{x})/(n-l)$ is the sample covariance between y_i and x_i.

There are two alternative formulas for the sample estimate of the variance. Since $\hat{Y}_R = N\bar{X}\hat{R}$, one form for $\hat{R}$ is

$$v_1(\hat{R}) = \frac{(1-f)}{n\bar{X}^2}(s_y^2 + \hat{R}^2 s_x^2 - 2\hat{R}s_{yx}) \tag{6.12}$$

Since, however, $\hat{R} = \bar{y}/\bar{x}$, the quantity $\bar{X}$ need not be known and is sometimes not known when estimating R. This suggests the alternative form

$$v_2(\hat{R}) = \frac{(1-f)}{n\bar{x}^2}(s_y^2 + \hat{R}^2 s_x^2 - 2\hat{R}s_{yx}) \tag{6.13}$$

This form could also be used for $v(\hat{Y}_R)$, taking $v_2(\hat{Y}_R) = X^2 v_2(\hat{R})$.

This raises the question: If $\bar{X}$ is known, is v_1 preferable to v_2? The answer is not at present clear. P. S. R. S. Rao and J. N. K. Rao (1971) studied the biases in v_1 and

v_2 analytically in an infinite population with the linear regression model

$$y_i = \alpha + \beta x_i + e_i$$

with

$$E(e_i|x_i) = 0, \ V(e_i|x_i) = \delta x_i^t, \ E(e_i e_j|x_i x_j) = 0,$$

where $0 \le t \le 2$ and x_i has a gamma distribution $ax^{h-1}e^{-x}$. The range $0 \le t \le 2$ was studied because in applications the residual variance of y_i is thought to increase with x_i at different rates in different populations. They found v_2 less biased for $0 \le t \le 3/2$, but also less stable for $t = 0$ or $t = 1$.

6.5 CONFIDENCE LIMITS

If the sample is large enough so that the normal approximation applies, confidence limits for Y and R may be obtained.

$$Y: \hat{Y}_R \pm z\sqrt{v(\hat{Y}_R)} \tag{6.14}$$

$$R: \hat{R} \pm z\sqrt{v(\hat{R})} \tag{6.15}$$

where z is the normal deviate corresponding to the chosen confidence probability.

In section 6.3 it was suggested that the normal approximation holds reasonably well if the sample size is at least 30 and is large enough so that the cv's of $\bar{y}$ and $\bar{x}$ are both less than 0.1. When these conditions do not apply, the formula for $v(\hat{R})$ tends to give values that are too low and the positive skewness in the distribution of $\hat{R}$ may become noticeable.

An alternative method of computing confidence limits, which takes some account of the skewness of the distribution of $\hat{R}$, has been used in biological assay (Fieller, 1932; Paulson, 1942). The method requires that $\bar{y}$ and $\bar{x}$ follow a bivariate normal distribution, so that $(\bar{y} - R\bar{x})$ is normally distributed. It follows that in simple random samples the quantity

$$\frac{\bar{y} - R\bar{x}}{\sqrt{[(N-n)/Nn]}\sqrt{s_y^2 + R^2 s_x^2 - 2Rs_{yx}}} \tag{6.16}$$

is approximately normally distributed with mean zero and unit standard deviation.

The value of R is unknown, but any contemplated value of R which makes this normal deviate large enough may be regarded as rejected by the sample data. Consequently, confidence limits for R are found by setting (6.16) equal to $\pm z$ and solving the resulting quadratic equation for R. The confidence limits are approximate since the two roots of the quadratic are imaginary with some samples. Such cases become rare if the cv's of $\bar{y}$ and $\bar{x}$ are less than 0.3.

After some manipulation, the two roots may be expressed as

$$R = \hat{R}\frac{(1-z^2 c_{\bar{y}\bar{x}}) \pm z\sqrt{(c_{\bar{y}\bar{y}}+c_{\bar{x}\bar{x}}-2c_{\bar{y}\bar{x}})-z^2(c_{\bar{y}\bar{y}}c_{\bar{x}\bar{x}}-c_{\bar{y}\bar{x}}^2)}}{1-z^2 c_{\bar{x}\bar{x}}} \qquad (6.17)$$

where

$$c_{\bar{y}\bar{y}} = \frac{N-n}{Nn}\frac{s_y^2}{\bar{y}^2}$$

is the square of the estimated cv of $\bar{y}$, with analogous definitions of $c_{\bar{y}\bar{x}}$ and $c_{\bar{x}\bar{x}}$. If $z^2 c_{\bar{y}\bar{y}}$, $z^2 c_{\bar{x}\bar{x}}$, and $z^2 c_{\bar{y}\bar{x}}$ are all small relative to 1, the limits reduce to

$$R = \hat{R} \pm z\sqrt{c_{\bar{y}\bar{y}}+c_{\bar{x}\bar{x}}-2c_{\bar{y}\bar{x}}}$$

This expression is the same as the normal approximation (6.15).

Evan with bivariate normality, the Fieller limits have been criticized as not conservative enough. James, Wilkinson, and Venables (1975) explain the nature of the difficulty and present an alternative method.

6.6 COMPARISON OF THE RATIO ESTIMATE WITH THE MEAN PER UNIT

The type of estimate of Y that was studied in preceding chapters is $N\bar{y}$, where $\bar{y}$ is the mean per unit for the sample (in simple random sampling) or a weighted mean per unit (in stratified random sampling). Estimates of this kind are called estimates based on the *mean per unit* or estimates obtained by *simple expansion*.

Theorem 6.2 In large samples, with simple random sampling, the ratio estimate $\hat{Y}_R$ has a smaller variance than the estimate $\hat{Y} = N\bar{y}$ obtained by simple expansion, if

$$\rho > \frac{1}{2}\left(\frac{S_x}{\bar{X}}\right)\Big/\left(\frac{S_y}{\bar{Y}}\right) = \frac{\text{coefficient of variation of } x_i}{2(\text{coefficient of variation of } y_i)}$$

Proof. For $\hat{Y}$ we have

$$V(\hat{Y}) = \frac{N^2(1-f)}{n}S_y^2$$

For the ratio estimate we have from (6.5)

$$V(\hat{Y}_R) = \frac{N^2(1-f)}{n}(S_y^2 + R^2 S_x^2 - 2R\rho S_y S_x)$$

Hence the ratio estimate has the smaller variance if

$$S_y^2 + R^2 S_x^2 - 2R\rho S_y S_x < S_y^2$$

If $R = \bar{Y}/\bar{X}$ is positive, this condition becomes

$$\rho > \frac{RS_x}{2S_y} = \frac{1}{2}\left(\frac{S_x}{\bar{X}}\right) \Big/ \left(\frac{S_y}{\bar{Y}}\right) = \frac{1}{2}\frac{\text{cv}(x)}{\text{cv}(y)} \tag{6.18}$$

6.7 CONDITIONS UNDER WHICH THE RATIO ESTIMATOR IS A BEST LINEAR UNBIASED ESTIMATOR

A well-known result in regression theory indicates the type of population under which the ratio estimate may be called the best among a wide class of estimates. The result was first proved for infinite populations. Brewer (1963b) and Royall (1970a) extended the result to finite populations. The result holds if two conditions are satisfied.

1. The relation between y_i and x_i is a straight line through the origin.
2. The variance of y_i about this line is proportional to x_i.

A "best linear unbiased estimator" is defined as follows. Consider all estimators $\hat{Y}$ of Y that are linear functions of the sample values y_i, that is, that are of the form

$$l_1 y_1 + l_2 y_2 + \cdots + l_n y_n$$

where the l's do not depend on the y_i, although they may be functions of the x_i. The choice of l's is restricted to those that give unbiased estimation of Y. The estimator with the smallest variance is called the best linear unbiased estimator (BLUE).

Formally, Brewer and Royall assume that the N population values (y_i, x_i) are a random sample from a superpopulation in which

$$y_i = \beta x_i + \varepsilon_i \tag{6.19}$$

where the ε_i are independent of the x_i and $x_i > 0$. In arrays in which x_i is fixed, ε_i has mean 0 and variance λx_i. The $x_i (i = 1, 2, \ldots, N)$ are known.

In the randomization theory used thus far in this book, the finite population total Y has been regarded as a fixed quantity. Under model (6.19), on the other hand, $Y = \beta X + \sum\limits^{N} \varepsilon_i$ is a random variable. In defining an unbiased estimator under this model, Brewer and Royall use a concept of unbiasedness which differs from that in randomization theory. They regard an estimator $\hat{Y}$ as unbiased if $E(\hat{Y}) = E(Y)$ in repeated selections of the finite population and sample under the model. Such an estimator might be called *model-unbiased*.

Theorem 6.3. Under model (6.19) the ratio estimator $\hat{Y}_R = X\bar{y}/\bar{x}$ is a best linear unbiased estimator for any sample, random or not, selected solely according to the values of the x_i.

Proof. Since $E(\varepsilon_i|x_i) = 0$ in repeated sampling, it follows from (6.19) that

$$Y = \beta X + \sum_{}^{N} \varepsilon_i : E(Y) = \beta X \qquad (6.20)$$

Furthermore, with the model (6.19) any linear estimator $\hat{Y}$ is of the form

$$\hat{Y} = \sum_{}^{n} l_i y_i = \beta \sum_{}^{n} l_i x_i + \sum_{}^{n} l_i \varepsilon_i \qquad (6.21)$$

If we keep the n sample values x_i fixed in repeated sampling under the model (6.19),

$$E(\hat{Y}) = \beta \sum_{}^{n} l_i x_i : V(\hat{Y}) = \lambda \sum_{}^{n} l_i^2 x_i \qquad (6.22)$$

From (6.20) and (6.22), $\hat{Y}$ is clearly model-unbiased if $\sum_{}^{n} l_i x_i \equiv X$. Minimizing $V(\hat{Y})$ under this condition by a Lagrange multiplier gives

$$2 l_i x_i = c x_i : l_i = \text{constant} = X/n\bar{x} \qquad (6.23)$$

The constant must have the value $X/n\bar{x}$ in order to satisfy the model-unbiased condition $l \sum_{}^{n} x_i = X$. Hence the BLUE estimator $\hat{Y}$ is $n\bar{y}X/n\bar{x} = X\bar{y}/\bar{x} = \hat{Y}_R$, the usual ratio estimator. This completes the proof.

Furthermore, from (6.20) and (6.21), with $l = X/n\bar{x}$,

$$\hat{Y}_R - Y = \sum_{}^{n} l_i \varepsilon_i - \sum_{}^{N} \varepsilon_i = (X/n\bar{x})(\sum_{}^{n} \varepsilon_i) - \sum_{}^{N} \varepsilon_i \qquad (6.24)$$

$$= \frac{(X - n\bar{x})}{n\bar{x}} \sum_{}^{n} \varepsilon_i - \sum_{}^{N-n} \varepsilon_i \qquad (6.25)$$

where $\sum^{N-n}$ denotes the sum over the $(N-n)$ population values that are *not* in the sample. Hence

$$V(\hat{Y}_R) = \frac{\lambda(X - n\bar{x})^2(n\bar{x})}{(n\bar{x})^2} + \lambda(X - n\bar{x}) = \frac{\lambda(X - n\bar{x})X}{n\bar{x}} \qquad (6.26)$$

A model-unbiased estimator of λ from this sample is easily shown to be

$$\hat{\lambda} = \sum_{}^{n} \frac{1}{x_i}(y_i - \hat{R}x_i)^2/(n-1) \qquad (6.27)$$

where $\hat{R} = \bar{y}/\bar{x}$, as usual. This value may be substituted in (6.26) to give a model-unbiased sample estimate of $V(\hat{Y}_R)$.

The practical relevance of these results is that they suggest the conditions under which the ratio estimator is superior not only to $\bar{y}$ but is the best of a whole class of estimators. When we are trying to decide what kind of estimate to use, a graph in which the sample values of y_i are plotted against those of x_i is helpful. If this graph shows a straight line relation passing through the origin and if the variance of the points y_i about the line seems roughly proportional to x_i, the ratio estimator will be hard to beat.

Sometimes the variance of the y_i in arrays in which x_i is fixed is *not* proportional to x_i. If this residual variance is of the form $\lambda v(x_i)$, where $v(x_i)$ is known, Brewer and Royall showed that the BLUE estimator becomes

$$\hat{Y} = X \frac{\sum\limits_{i}^{n} w_i y_i x_i}{\sum\limits^{n} w_i x_i^2} \tag{6.28}$$

where $w_i = 1/v(x_i)$. In a population sample of Greece, Jessen et al. (1947) judged that the residual variance increased roughly as x_i^2. This suggests a weighted regression with $w_i = 1/x_i^2$, which gives

$$\hat{Y} = \frac{X\left(\sum\limits^{n} w_i y_i x_i\right)}{\sum\limits^{n} (w_i x_i^2)} = \frac{X}{n} \sum\limits^{n} \left(\frac{y_i}{x_i}\right) \tag{6.29}$$

For a given population and given n, $V(\hat{Y}_R)$ in (6.26) is clearly minimized, given every $x_i > 0$, when the sample consists of the n *largest* x_i in the population. In 16 small natural populations of the type to which ratio estimates have been applied, Royall (1970) found for samples having $n = 2$ to 12 that selection of the n largest x_i usually increased the accuracy of $\hat{Y}_R$.

In summary, the Brewer-Royall results show that the assumption of a certain type of model leads to an unbiased ratio estimator and formulas for $V(\hat{Y}_R)$ and $v(\hat{Y}_R)$ that are simple and exact for any $n > 1$. The results might be used in practice in cases where examination of the y, x pairs from the available data suggests that the model is reasonably correct. The variance formulas (6.26) and (6.27) appear to be sensitive to inaccuracy in the model, although this issue needs further study.

Further work by Royall and Herson (1973) discusses the type of sample distribution needed with respect to the x_i in order that $\hat{Y}_R$ remains unbiased when there is a polynomial regression of y_i on x_i.

6.8 BIAS OF THE RATIO ESTIMATE

In general, the ratio estimate has a bias of order $1/n$. Since the s.e. of the estimate is of order $1/\sqrt{n}$, the quantity (bias/s.e.) is also of order $1/\sqrt{n}$ and becomes negligible as n becomes large. In practice, this quantity is usually unimportant in samples of moderate size. Its value in small samples is of interest, however, in stratified sampling with many strata, where we may wish to compute and examine ratio estimates in individual strata with small samples in the strata. Two useful results about the bias are presented.

The first gives the leading term in the bias when it is expanded in a Taylor's series.

$$\hat{R} - R = \frac{\bar{y}}{\bar{x}} - R = \frac{\bar{y} - R\bar{x}}{\bar{x}}$$

Write

$$\frac{1}{\bar{x}} = \frac{1}{\bar{X} + (\bar{x} - \bar{X})} = \frac{1}{\bar{X}}\left(1 + \frac{\bar{x} - \bar{X}}{\bar{X}}\right)^{-1} \doteq \frac{1}{\bar{X}}\left(1 - \frac{\bar{x} - \bar{X}}{\bar{X}}\right) \qquad (6.30)$$

Hence,

$$\hat{R} - R \doteq \frac{\bar{y} - R\bar{x}}{\bar{X}}\left(1 - \frac{\bar{x} - \bar{X}}{\bar{X}}\right) \qquad (6.31)$$

Now

$$E(\bar{y} - R\bar{x}) = \bar{Y} - R\bar{X} = 0$$

so that the leading term in the bias comes from the second term inside the brackets. Furthermore,

$$E\bar{y}(\bar{x} - \bar{X}) = E(\bar{y} - \bar{Y})(\bar{x} - \bar{X}) = \frac{1-f}{n}\rho S_y S_x \qquad (6.32)$$

by theorem 2.3 (p. 25) and the definition of ρ. Also,

$$E\bar{x}(\bar{x} - \bar{X}) = E(\bar{x} - \bar{X})^2 = \frac{1-f}{n}S_x^2$$

Hence the leading term in the bias is

$$E(\hat{R} - R) \doteq \frac{1-f}{n\bar{X}^2}(RS_x^2 - \rho S_y S_x) \qquad (6.33)$$

$$\doteq \frac{1-f}{n}(C_{xx} - C_{yx})R \qquad (6.34)$$

For a rigorous justification of (6.34), see David and Sukhatme (1974). Now the leading term in $V(\hat{R})$ is

$$V(\hat{R}) = \frac{(1-f)}{n\bar{X}^2}(S_y^2 + R^2 S_x^2 - 2R\rho S_y S_x) \qquad (6.35)$$

from (6.5), substituting $\hat{R} = \hat{Y}_R/N\bar{X}$.

From (6.33) and (6.35) the leading term in the quantity (bias/s.e.), which is the same for $\hat{R}$, $\hat{Y}_R$, and $\hat{\bar{Y}}_R$, may be expressed as

$$\frac{\text{(bias)}}{\text{s.e.}} = \text{cv}(\bar{x}) \frac{(RS_x - \rho S_y)}{(R^2 S_x^2 - 2R\rho S_y S_x + S_y^2)^{1/2}} \tag{6.36}$$

where $\text{cv}(\bar{x}) = \sqrt{1-f} S_x / \sqrt{n} \bar{X}$. By substituting sample estimates of the terms in (6.36), Kish, Namboodiri, and Pillai (1962) computed the (bias/s.e.) values for numerous items in various national and more localized studies. In the national studies nearly all the (bias/s.e.) values were <0.03 and almost the only values >0.10 in their studies occurred for a single stratum with $n_h = 6$ small hospitals.

The second result, due to Hartley and Ross (1954), gives an exact result for the bias and an upper bound to the ratio of the bias to the standard error. Consider the covariance, in simple random samples of size n, of the quantities $\hat{R}$ and $\bar{x}$. We have

$$\text{cov}(\hat{R}, \bar{x}) = E\left(\frac{\bar{y}}{\bar{x}} \cdot \bar{x}\right) - E(\hat{R}) E(\bar{x}) \tag{6.37}$$

$$= \bar{Y} - \bar{X} E(\hat{R}) \tag{6.38}$$

Hence

$$E(\hat{R}) = \frac{\bar{Y}}{\bar{X}} - \frac{1}{\bar{X}} \text{cov}(\hat{R}, \bar{x}) = R - \frac{1}{\bar{X}} \text{cov}(\hat{R}, \bar{x}) \tag{6.39}$$

Thus the bias in $\hat{R}$ is $-\text{cov}(\hat{R}, \bar{x})/\bar{X}$. Unlike the Taylor approximation (6.33) to the bias, this expression is exact.
Furthermore,

$$|\text{bias in } \hat{R}| = \frac{|\rho_{\hat{R}, \bar{x}} \sigma_{\hat{R}} \sigma_{\bar{x}}|}{\bar{X}}$$

$$\leq \frac{\sigma_{\hat{R}} \sigma_{\bar{x}}}{\bar{X}}$$

since $\hat{R}$ and $\bar{x}$ cannot have a correlation >1. Hence

$$\frac{|\text{bias in } \hat{R}|}{\sigma_{\hat{R}}} \leq \frac{\sigma_{\bar{x}}}{\bar{X}} = \text{cv of } \bar{x} \tag{6.40}$$

The same bound applies, of course, to the bias in $\hat{Y}_R$ and $\hat{\bar{Y}}_R$. Thus, if the cv of $\bar{x}$ is less than 0.1, the bias may safely be regarded as negligible in relation to the s.e.

6.9 ACCURACY OF THE FORMULAS FOR THE VARIANCE AND ESTIMATED VARIANCE

With small samples, say $n < 30$ and C_{xx} large, it has long been suspected that the large-sample formulas given for $V(\hat{R})$ and $v(\hat{R})$ are underestimates. By a Taylor series expansion, Sukhatme (1954) expressed the error in $V(\hat{R})$ in terms of the bivariate moments of y and x. Unfortunately, the result is too complicated to lead to a useful guide for practical applications.

If y and x follow a bivariate normal distribution, Sukhatme's result simplifies considerably. Let

$$V_1 = (C_{yy} + C_{xx} - 2C_{yx})/n$$

denote the first approximation to the relative variance of $\hat{R}$, ignoring the fpc. To terms of order $1/n^2$,

$$E\left(\frac{\hat{R}-R}{R}\right)^2 \doteq V_1\left(1 + \frac{3C_{xx}}{n} + \frac{6C_{xx}}{n} \cdot \frac{\rho^2 C_{yy} + C_{xx} - 2C_{yx}}{C_{yy} + C_{xx} - 2C_{yx}}\right) \qquad (6.41)$$

Since the right-hand term inside the parentheses is less than $6C_{xx}/n$, this gives

$$E\left(\frac{\hat{R}-R}{R}\right)^2 < V_1\left(1 + \frac{9C_{xx}}{n}\right) \qquad (6.42)$$

to terms of order $1/n^2$. Now C_{xx}/n is the square of the coefficient of variation of $\bar{x}$. Thus, if n is large enough so that the cv of $\bar{x}$ is less, than 0.1, use of V_1 should not underestimate by more than 9%. In practice, the multiplier 9 in (6.42) appears to be unduly high as compared with (6.41). For instance, if $C_{xx} = C_{yy}$, (6.41) reduces to

$$E\left(\frac{\hat{R}-R}{R}\right)^2 \doteq V_1\left[1 + \frac{C_{xx}}{n}(6 - 3\rho)\right] \qquad (6.43)$$

Since ρ is almost always positive in applications of the ratio method, a multiplier between 3 and 6 is more representative. However, the effects of nonnormality in y and x also enter into the term of order $1/n^2$.

From a Monte Carlo study by Rao (1968) of small natural populations, some illustrative results on the biases of the large-sample formulas for $V(\hat{R})$ and $v_1(\hat{R})$ will be quoted for simple random samples from eight populations each with $N > 30$. The populations are described in Rao (1969). The formulas for $V(\hat{R})$ and $v_1(\hat{R})$, given in (6.4) and (6.12), will be appraised as estimators of the true $\text{MSE}(\hat{R})$.

For a given population the quantities $100[\text{MSE}(\hat{R}) - V(\hat{R})]/\text{MSE}(\hat{R})$ and $100[\text{MSE}(\hat{R}) - Ev_1(\hat{R})]/\text{MSE}(\hat{R})$ are the percent underestimates of the true $\text{MSE}(\hat{R})$. The averages of these percents in the eight populations were as shown in Table 6.2.

In these data the percent underestimation in $V(\hat{R})$ scarcely declines at all with increasing n. This is explained in part by the circumstance that in one population $V(\hat{R})$ gave an *overestimate* which declined with n. The indications from these results are that the biases in $v_1(\hat{R})$ are much more serious in small samples than those in $\hat{R}$ itself, and are unsatisfactory at least up to $n = 12$. For $n = 4$, Koop

TABLE 6.2
Average percent underestimation of MSE($\hat{R}$)

Estimator	n = 4	6	8	12
$V(\hat{R})$ in (6.4)	14	14	14	12
$v_1(\hat{R})$ in (6.12)	31	23	21	18

(1968) found underestimations in $v_1(\hat{R})$ averaging 25% in three populations with $N=20$.

An alternative estimator to $v_1(\hat{R})$ that looks promising as regards bias reduction is presented in section 6.17.

6.10 RATIO ESTIMATES IN STRATIFIED RANDOM SAMPLING

There are two ways in which a ratio estimate of the population total Y can be made. One is to make a *separate* ratio estimate of the total of each stratum and add these totals. If y_h, x_h are the sample totals in the hth stratum and X_h is the stratum total of the x_{hi}, this estimate $\hat{Y}_{Rs}$ (s for *separate*) is

$$\hat{Y}_{Rs} = \sum_h \frac{y_h}{x_h} X_h = \sum_h \frac{\bar{y}_h}{\bar{x}_h} X_h \tag{6.44}$$

No assumption is made that the true ratio remains constant from stratum to stratum. The estimate requires a knowledge of the separate totals X_h.

Theorem 6.4. If an independent simple random sample is drawn in each stratum and sample sizes are large in all strata,

$$V(\hat{Y}_{Rs}) = \sum_h \frac{N_h^2(1-f_h)}{n_h}(S_{yh}^2 + R_h^2 S_{xh}^2 - 2R_h \rho_h S_{yh} S_{xh}) \tag{6.45}$$

where $R_h = Y_h/X_h$ is the true ratio in stratum h, and ρ_h is defined as before in each stratum.

Proof. Apply formula (6.2), section 6.3, for a simple random sample to give in stratum h,

$$V(\hat{Y}_{Rh}) = \frac{N_h^2(1-f_h)}{n_h}(S_{yh}^2 + R_h^2 S_{xh}^2 - 2R_h \rho_h S_{yh} S_{xh}) \tag{6.46}$$

Since $\hat{Y}_{Rs} = \sum_h \hat{Y}_{Rh}$ and sampling is independent in each stratum, $V(\hat{Y}_{Rs}) = \sum V(\hat{Y}_{Rh})$ and the result (6.45) follows.

This formula is valid only if the sample in each stratum is large enough so that the approximate variance formula applies to each stratum. This limitation should be noted in practical applications.

Moreover, when the n_h are small and the number of strata L is large, the bias in $\hat{Y}_{Rs}$ may not be negligible in relation to its standard error, as the following crude argument suggests.

In a single stratum we have seen (section 6.8) that

$$\frac{|\text{bias in } \hat{Y}_{Rh}|}{\sigma(\hat{Y}_{Rh})} \leq \text{cv of } \bar{x}_h$$

If the bias has the same sign in all strata, as may happen, the bias in $\hat{Y}_{Rs}$ will be roughly L times that in $\hat{Y}_{Rh}$. But the standard error of $\hat{Y}_{Rs}$ is only of the order of $\sqrt{L}$ times that of $\hat{Y}_{Rh}$. Hence the ratio

$$\frac{|\text{bias in } \hat{Y}_{Rs}|}{\sigma(\hat{Y}_{Rs})}$$

is of order

$$\sqrt{L}(\text{cv of } \bar{x}_h)$$

For example, with 50 strata and the cv of $\bar{x}_h$ about 0.1 in each stratum, the bias in $\hat{Y}_{Rs}$ might be as large as 0.7 times its standard error. The contribution of the bias to the mean square error of $\hat{Y}_{Rs}$ would then be about one third.

Although in practice the bias is usually much smaller than its upper bound, the danger of bias with the separate ratio estimate should be kept in mind if $\sqrt{L}(\text{cv of } \bar{x}_h)$ exceeds, say, 0.3.

6.11 THE COMBINED RATIO ESTIMATE

An alternative estimate is derived from a single *combined* ratio (Hansen, Hurwitz, and Gurney, 1946). From the sample data we compute

$$\hat{Y}_{st} = \sum_h N_h \bar{y}_h, \qquad \hat{X}_{st} = \sum_h N_h \bar{x}_h \tag{6.47}$$

These are the standard estimates of the population totals Y and X, respectively, made from a stratified sample. The combined ratio estimate, $\hat{Y}_{Rc}$ (c for *combined*) is

$$\hat{Y}_{Rc} = \frac{\hat{Y}_{st}}{\hat{X}_{st}} X = \frac{\bar{y}_{st}}{\bar{x}_{st}} X \tag{6.48}$$

where $\bar{y}_{st} = \hat{Y}_{st}/N$, $\bar{x}_{st} = \hat{X}_{st}/N$ are the estimated population means from a stratified sample.

The estimate $\hat{Y}_{Rc}$ does not require a knowledge of the X_h, but only of X.

The combined estimate is much less subject to the risk of bias than the separate estimate. Using the approach of Hartley and Ross in section 6.8, we have, writing $\hat{R}_c = \bar{y}_{st}/\bar{x}_{st}$,

$$\text{cov}(\hat{R}_c, \bar{x}_{st}) = E\left(\frac{\bar{y}_{st}}{\bar{x}_{st}} \cdot \bar{x}_{st}\right) - E(\hat{R}_c)E(\bar{x}_{st})$$

$$= \bar{Y} - \bar{X}E(\hat{R}_c) \tag{6.49}$$

Hence

$$E(\hat{R}_c) = R - \frac{1}{\bar{X}}\text{cov}(\hat{R}_c, \bar{x}_{st})$$

and

$$\frac{|\text{bias in } \hat{R}_c|}{\sigma_{\hat{R}_c}} = \frac{|\rho_{\hat{R}_c, \bar{x}_{st}} \cdot \sigma_{\bar{x}_{st}}|}{\bar{X}} \le \text{cv of } \bar{x}_{st}. \tag{6.50}$$

Thus the biases in $\hat{R}_c$, $\hat{Y}_{Rc}$ are negligible relative to their standard errors, provided only that the cv of $\bar{x}_{st}$ is less than 0.1.

Theorem 6.5. If the total sample size n is large,

$$V(\hat{Y}_{Rc}) = \sum_h \frac{N_h^2(1-f_h)}{n_h}(S_{yh}^2 + R^2 S_{xh}^2 - 2R\rho_h S_{yh} S_{xh}) \tag{6.51}$$

Proof. This follows the same argument as theorem 2.5. In the present case the key equation is

$$(\hat{Y}_{Rc} - Y) = \frac{N\bar{X}}{\bar{x}_{st}}(\bar{y}_{st} - R\bar{x}_{st}) \doteq N(\bar{y}_{st} - R\bar{x}_{st}) \tag{6.52}$$

Now consider the variate $u_{hi} = y_{hi} - Rx_{hi}$. The right side of (6.52) is $N\bar{u}_{st}$, where $\bar{u}_{st}$ is the weighted mean of the variate u_{hi} in a stratified sample. Furthermore, the population mean $\bar{U}$ of u_{hi} is zero, since $R = \bar{Y}/\bar{X}$.

Hence we may apply to $\bar{u}_{st}$ theorem 5.3 for the variance of the estimated mean from a stratified random sample. This gives

$$V(\hat{Y}_{Rc}) = N^2 V(\bar{u}_{st}) = \sum_h \frac{N_h(N_h - n_h)}{n_h} S_{uh}^2 \tag{6.53}$$

where

$$S_{uh}^2 = \frac{1}{N_h - 1}\sum_{i=1}^{N_h}(u_{hi} - \bar{U}_h)^2$$

$$= \frac{1}{N_h - 1}\sum_{i=1}^{N_h}[(y_{hi} - \bar{Y}_h) - R(x_{hi} - \bar{X}_h)]^2$$

When the quadratic is expanded, result (6.51) is obtained.

From equations (6.45) and (6.51) it is interesting to note that the approximate variances of $\hat{Y}_{Rs}$ and $\hat{Y}_{Rc}$ assume the same general form, the difference being that the population ratios R_h in the individual strata in (6.45) are all replaced by R in (6.51).

6.12 COMPARISON OF THE COMBINED AND SEPARATE ESTIMATES

We may write

$$V(\hat{Y}_{Rc}) - V(\hat{Y}_{Rs})$$

$$= \sum_h \frac{N_h^2(1-f_h)}{n_h}[(R^2 - R_h^2)S_{xh}^2 - 2(R - R_h)\rho_h S_{yh}S_{xh}]$$

$$= \sum \frac{N_h^2(1-f_h)}{n_h}[(R - R_h)^2 S_{xh}^2 + 2(R_h - R)(\rho_h S_{yh}S_{xh} - R_h S_{xh}^2)]$$

In situations in which the ratio estimate is appropriate the last term on the right is usually small. (It vanishes if within each stratum the relation between y_{hi} and x_{hi} is a straight line through the origin.) Thus, unless R_h is constant from stratum to stratum, the use of a separate ratio estimate in each stratum is likely to be more precise if the sample in each stratum is large enough so that the approximate formula for $V(\hat{Y}_{Rs})$ is valid, and the cumulative bias that can affect $\hat{Y}_{Rs}$ (section 6.10) is negligible. With only a small sample in each stratum, the combined estimate is to be recommended unless there is good empirical evidence to the contrary.

For sample estimates of these variances we substitute sample estimates of R_h and R in the appropriate places. The sample mean squares s_{yh}^2 and s_{xh}^2 are substituted for the corresponding variances and the sample covariance for the term $\rho_h S_{yh}S_{xh}$. The sample mean square and covariance must be calculated separately for each stratum.

Example. The data come from a census of all farms in Jefferson County, Iowa. In this example y_{hi} represents acres in corn and x_{hi} acres in the farm. The population is divided into *two* strata, the first stratum containing farms of as many as 160 acres. We assume a sample of 100 farms. When stratified sampling is used, we will suppose that 70 farms are taken from stratum 1 and 30 from stratum 2, this being roughly the optimum allocation. The data are given in Table 6.3. The last three quantities, Q_h, V_h', and V_h'', are auxiliary quantities to be used in the computations, the last two being defined later.

We consider five methods of estimating the population mean corn acres per farm. The fpc are ignored.

1. Simple random sample: mean per farm estimate.

$$V_1 = \frac{S_y^2}{n} = \frac{620}{100} = 6.20$$

TABLE 6.3

DATA FROM JEFFERSON COUNTY, IOWA

Strata	Size (farm acres)	N_h	S_{yh}^2	S_{yxh}	S_{xh}^2	R_h
1	0–160	1580	312	494	2055	0.2350
2	More than 160	430	922	858	7357	0.2109
For complete pop.		2010	620	1453	7619	0.2242

Strata	$\bar{Y}_h$	$\bar{X}_h$	n_h	$Q_h = W_h^2/n_h$	V_h'	V_h''
1	19.40	82.56	70	0.008828	193	194
2	51.63	244.85	30	0.001525	887	907
For complete pop.	26.30	117.28	100			

2. Simple random sample: ratio estimate.

$$V_2 - \frac{1}{n}(S_y^2 + R^2 S_x^2 - 2RS_{yx})$$

$$= \frac{1}{100}[620 + (0.2242)^2(7619) - 2(0.2242)(1453)]$$

$$= 3.51$$

3. Stratified random sample: mean per farm estimate.

$$V_3 = \sum \frac{W_h^2}{n_h} S_{yh}^2 = \sum Q_h S_{yh}^2 = 4.16$$

4. Stratified random sample: ratio estimate using a separate ratio in each stratum.

$$V_4 = \sum Q_h (S_{yh}^2 + R_h^2 S_{xh}^2 - 2R_h S_{yxh}) = \sum Q_h V_h' = 3.06$$

5. Stratified random sampling: ratio estimate using a combined ratio.

$$V_5 = \sum Q_h (S_{yh}^2 + R^2 S_{xh}^2 - 2RS_{yxh}) = \sum Q_h V_h'' = 3.10$$

The relative precisions of the various methods can be summarized as follows.

Sampling method	Method of Estimation	Relative Precision
1. Simple random	Mean per farm	100
2. Simple random	Ratio	177
3. Stratified random	Mean per farm	149
4. Stratified random	Separate ratio	203
5. Stratified random	Combined ratio	200

The results bring out an interesting point of wide application. Stratification by size of farm accomplishes the same general purpose as a ratio estimate in which the denominator is farm size. Both devices diminish the effect of variations in farm size on the sampling error of the estimated mean corn acres per farm. For instance, the gain in precision from a ratio estimate is 77% when simple random sampling is used, but it is only 36% (203 against 149) when stratified sampling is used.

In the design of surveys there may be a choice between introducing a factor into the stratification or utilizing it in the method of estimation. The best decision depends on the circumstances. Relevant points are: (a) some factors, for example, geographical location, are more easily introduced into the stratification than into the method of estimation; (b) the issue depends on the nature of the relation between y_i and x_i. All simple methods of estimation work most effectively with a linear relation. With a complex or discontinuous relation, stratification may be more effective, since, if there are enough strata, stratification will eliminate the effects of almost any kind of relation between y_i and x_i. (c) If some important variates are roughly proportional to x_i, but others are roughly proportional to another variate z_i, it is better to use x_i and z_i as denominators in ratio estimates than to stratify by one of them.

6.13 SHORT-CUT COMPUTATION OF THE ESTIMATED VARIANCE

If $n_h = 2$ in all strata, Keyfitz (1957) has given simple methods for computing the approximations to the estimated variances of $\hat{Y}_{RC}$ or $\hat{R}_C$ or, more generally, of functions of one or more variables of the form $\hat{Y}_{st}$. For $\hat{R}_c$ we have

$$\hat{R}_c = \frac{\hat{Y}_{st}}{\hat{X}_{st}} = \frac{\sum_h \hat{Y}_h}{\sum_h \hat{X}_h} = \frac{\sum_h \frac{N_h}{2}(y_{h1} + y_{h2})}{\sum_h \frac{N_h}{2}(x_{h1} + x_{h2})} \tag{6.54}$$

The Keyfitz method uses the identity that for $n_h = 2$,

$$2s_{yh}^2 = 2\sum_{i=1}^{2} (y_{hi} - \bar{y}_h)^2 \equiv (y_{h1} - y_{h2})^2 = (dy_h)^2 \tag{6.55}$$

where $dy_h = (y_{h1} - y_{h2})$. Hence

$$v(\hat{Y}_h) = \left(\frac{N_h}{2}\right)^2 2(1-f_h)s_{yh}^2 = (1-f_h)(y_{h1}' - y_{h2}')^2 = (1-f_h)(dy_h')^2 \tag{6.56}$$

where $y_{hi}' = N_h y_{hi}/2$. Similarly, for the sample estimate of the covariance,

$$\text{cov}(\hat{Y}_h \hat{X}_h) = (1-f_h)(dy_h')(dx_h') \tag{6.57}$$

Now

$$\hat{R}_c - R \doteq \frac{\hat{Y}_{st}}{\hat{X}_{st}} - R \doteq \frac{\hat{Y}_{st} - R\hat{X}_{st}}{X} = \frac{Y}{X}\left(\frac{\hat{Y}_{st}}{Y} - \frac{\hat{X}_{st}}{X}\right) \tag{6.58}$$

Since sampling is independent in different strata,

$$v(\hat{R}_c) = \left(\frac{Y}{X}\right)^2 \sum_h (1-f_h) \left[\left(\frac{dy_h'}{Y}\right)^2 + \left(\frac{dx_h'}{X}\right)^2 - 2\left(\frac{dy_h'}{Y}\right)\left(\frac{dx_h'}{X}\right)\right] \qquad (6.59)$$

$$= \left(\frac{Y}{X}\right)^2 \sum_h (1-f_h) \left(\frac{dy_h'}{Y} - \frac{dx_h'}{X}\right)^2 \qquad (6.60)$$

Keyfitz (1957) has extended this method to cover poststratified estimators and multistage sampling, and to give variances of differences of estimates from successive surveys in periodic samples. Woodruff (1971) gives a general approach that handles nonlinear estimators, unequal probabilities of selection, and samples of size n_h in the strata. As an illustration of Woodruff's approach, consider a function $f(\hat{\mathbf{Y}})$ where $\hat{\mathbf{Y}}$ represents the vector or set of m variables $\hat{Y}_j = \sum_h \hat{Y}_{jh}$. With simple random sampling in stratum h, the $\hat{Y}_{jh}$ are of the form $(N_h/n_h)\sum y_{jhi}$, the sum extending over the n_h sample units in stratum h. By Taylor's approximation,

$$f(\hat{\mathbf{Y}}) - f(\mathbf{Y}) \doteq \sum_j^m \frac{\partial f}{\partial Y_j}(\hat{Y}_j - Y_j) = \sum_j \sum_h \frac{\partial f}{\partial Y_j}(\hat{Y}_{jh} - Y_{jh}) \qquad (6.61)$$

The trick is to reverse the order of summation, writing

$$f(\hat{\mathbf{Y}}) - f(\mathbf{Y}) \doteq \sum_h \sum_j \frac{\partial f}{\partial Y_j}(\hat{Y}_{jh} - Y_{jh}) = \sum_h (\hat{U}_h - U_h) \qquad (6.62)$$

where

$$\hat{U}_h = \sum_j \frac{\partial f}{\partial Y_j} \hat{Y}_{jh} = \frac{N_h}{n_h} \sum_i^{n_h} \left(\sum_j \frac{\partial f}{\partial Y_j} y_{jhi}\right) = \sum_i^{n_h} u_{hi} \qquad (6.63)$$

By evaluating the $(\partial f/\partial Y_j)$ at the estimates $\hat{Y}_j$, the u_{hi} can be calculated from (6.63) for each sample unit in stratum h. With simple random sampling in the strata, the usual formula for the estimated variance of the sum $\hat{U}_h$ applies. Hence, from (6.62), an approximate sample estimate of the variance of $f(\hat{\mathbf{Y}})$ is

$$v[f(\hat{\mathbf{Y}})] \doteq \sum_h \frac{(N_h - n_h)n_h}{N_h} \frac{\sum_i^{n_h} (u_{hi} - \bar{u}_h)^2}{(n_h - 1)} \qquad (6.64)$$

The advantage of this approach is that the covariances of the $\hat{Y}_{jh}$ need not be calculated.

To illustrate the u_{hi} for the estimator $\hat{R}_c = \hat{Y}/\hat{X}$, it helps to write $\hat{R}_c = \hat{Y}_1/\hat{Y}_2$, so that

$$f(\mathbf{Y}) = \frac{Y_1}{Y_2}: \quad \frac{\partial f}{\partial Y_2} = \frac{1}{Y_2}: \quad \frac{\partial f}{\partial Y_2} = -\frac{Y_1}{Y_2^2} \qquad (6.65)$$

Hence, from (6.63), with $n_h = 2$,

$$u_{hi} = \frac{N_h}{n_h}\left(\frac{y_{1hi}}{\hat{Y}_2} - \frac{\hat{Y}_1}{\hat{Y}_2^2}y_{2hi}\right) = \frac{N_h}{n_h}\frac{(y_{1hi} - \hat{R}_c y_{2hi})}{\hat{Y}_2} \tag{6.66}$$

In terms of the original y_{hi}, x_{hi},

$$u_{hi} = \frac{N_h}{n_h}\frac{(y_{hi} - \hat{R}_c x_{hi})}{\hat{X}} \tag{6.67}$$

Its estimated variance is approximated as in (6.64).

The Keyfitz method is used, for instance, in the Health Interview Survey of the National Center for Health Statistics. The sampling unit is a cluster unit—a county or group of contiguous counties. Each unit is divided into segments of about 9 households, an average of 13 segments being chosen from a sampled unit. The variables y_{h1}, y_{h2} might be the numbers of persons with a specific illness and x_{h1}, x_{h2} the total numbers of persons in the samples from the two units in stratum h.

In addition to the initial geographic stratification of the counties, the persons in the sample are poststratified by age, sex, and color. Thus, instead of the estimator $\hat{Y}_{Rc} = X\hat{R}_c$ of the total number ill, the estimator is

$$\hat{Y}_{PS} = \sum_a X_a \hat{R}_a = \sum_a X_a \frac{\hat{Y}_a}{\hat{X}_a} \tag{6.68}$$

where a represents an age-sex-color class and X_a is the known total population in this class. Here we have a function of *two* sets of random variables $- \hat{Y}_a$ and $\hat{X}_a$. Furthermore, for many reasons, y_{ahi} and $y_{a'hi}$ for two different classes a and a' in a cluster unit may be correlated; for example with an infectious disease the number of cases may be high for all classes in the unit. Applying (6.63) and ignoring the fpc's, we have with $n_h = 2$,

$$v(\hat{Y}_{PS}) = \sum_h \left[\sum_a X_a\left(\frac{dy_{ah}'}{\hat{X}_a} - \frac{\hat{Y}_a dx_{ah}'}{\hat{X}_a^2}\right)\right]^2 \tag{6.69}$$

$$\doteq \sum_h \left[\sum_a X_a\left(\frac{\hat{Y}_a}{\hat{X}_a}\right)\left(\frac{dy_{ah}'}{\hat{Y}_a} - \frac{dx_{ah}'}{\hat{X}_a}\right)\right]^2 \tag{6.70}$$

where $dy_{ah}' = N_h(y_{ah1} - y_{ah2})/2$.

In this application the Keyfitz method also handles further complexities of the survey—selection of primary units with unequal probabilities, adjustments for nonresponse, and use of the method of collapsed strata. Furthermore, since computer time even with this simple method permits variance calculations for only a limited number of items, charts of the relation between s.d. $(\hat{Y})/\hat{Y}$ and $\hat{Y}$ are given for different types of item to help in predicting s.d. $(\hat{Y})$ for items for which the s.d. has not been computed. Bean (1970) gives a clear presentation of these methods and results.

6.14 OPTIMUM ALLOCATION WITH A RATIO ESTIMATE

The optimum allocation of the n_h may be different with a ratio estimate than with a mean per unit. Consider first the variate $\hat{Y}_{Rs}$. From theorem 6.4 its variance is

$$V(\hat{Y}_{Rs}) = \sum_h \frac{N_h(N_h - n_h)}{n_h}(S_{yh}^2 + R_h^2 S_{xh}^2 - 2R_h\rho_h S_{yh}S_{xh})$$

$$= \sum_h \frac{N_h(N_h - n_h)}{n_h} S_{dh}^2, \quad \text{with } S_{dh}^2 = \frac{1}{N_h - 1}\sum_{i=1}^{N_h} d_{hi}^2 \qquad (6.71)$$

where $d_{hi} = y_{hi} - R_h x_{hi}$ is the deviation of y_{hi} from $R_h x_{hi}$. By the methods given in Chapter 5 for finding optimum allocation, it follows that (6.71) is minimized subject to a total cost of the form $\sum c_h n_h$, when

$$n_h \propto \frac{N_h S_{dh}}{\sqrt{c_h}}$$

With a mean per unit it will be recalled that for minimum variance n_h is chosen proportional to $N_h S_{yh}/\sqrt{c_h}$.

In the planning of a sample, the allocation with a ratio estimate may appear a little perplexing, because it seems difficult to speculate about likely values of S_{dh}. Two rules are helpful. With a population in which the ratio estimate is a best linear unbiased estimate, S_{dh} will be roughly proportional to $\sqrt{\bar{X}_h}$ (by theorem 6.3). In this case the n_h should be proportional to $N_h\sqrt{\bar{X}_h}/\sqrt{c_h}$. Sometimes the variance of d_{hi} may be more nearly proportional to $\bar{X}_h^2$. This leads to the allocation of n_h proportional to $N_h\bar{X}_h/\sqrt{c_h}$, that is, to the stratum total of x_{hi}, divided by the square root of the cost per unit. An example of this type is discussed by Hansen, Hurwitz, and Gurney (1946) for a sample designed to estimate sales of retail stores.

If the estimate $\hat{Y}_{Rc}$ is to be used, the same general argument applies.

Example. The different methods of allocation can be compared from data collected in a complete enumeration of 257 commercial peach orchards in North Carolina in June 1946 (Finkner, 1950). The purpose was to determine the most efficient sampling procedure for estimating commercial peach production in this area. Information was obtained on the number of peach trees and the estimated total peach production in each orchard. The high correlation between these two variables suggested the use of a ratio estimate. One very large orchard was omitted.

For this illustration, the area is divided geographically into three strata. The number of peach trees in an orchard is denoted by x_{hi} and the estimated production in bushels of peaches by y_{hi}. Only the first ratio estimate $\hat{Y}_{Rs}$ (based on a separate ratio in each stratum) will be considered, since the principle is the same for both types of stratified ratio estimate.

Four methods of allocation are compared: (a) n_h proportional to N_h, (b) n_h proportional to $N_h S_{yh}$, (c) n_h proportional to $N_h\sqrt{\bar{X}_h}$, and (d) n_h proportional to $N_h\bar{X}_h = X_h$. The sample size is 100. The data for these comparisons are summarized in Table 6.4.

TABLE 6.4

DATA FROM THE NORTH CAROLINA PEACH SURVEY

Strata	S_{xh}^2	S_{yxh}	S_{yh}^2	S_{xh}	S_{yh}	$\bar{X}_h$	$\bar{Y}_h$	R_h	S_{dh}^2
1	5186	6462	8699	72.01	93.27	53.80	69.48	1.29133	658
2	2367	3100	4614	48.65	67.93	31.07	43.64	1.40475	573
3	4877	4817	7311	69.83	85.51	56.97	66.39	1.16547	2706
Pop.	3898	4434	6409	62.43	80.06	44.45	56.47	1.27053	1433

Strata	N_h	(a)	$N_h S_{yh}$	(b)	$\sqrt{\bar{X}_h}$	$N_h \sqrt{\bar{X}_h}$	(c)	$N_h \bar{X}_h$	(d)
1	47	18	4384	22	7.33	344.5	20	2529	22
2	118	46	8016	40	5.57	657.3	39	3666	32
3	91	36	7781	38	7.55	687.1	41	5184	46
Pop.	256	100	20181	100	20.45	1688.9	100	11379	100

The upper part of the table shows the basic data. The method employed to calculate the four variances was first to find the n_h for each type of allocation. These values are shown in the columns headed (a) through (d) in the lower part of the table. Thus, with allocation (a), $n_h = n N_h/N$, so that in the first stratum

$$n_1 = \frac{(100)(47)}{256} = 18$$

When the n_h have been obtained, the corresponding $V(\hat{Y}_{Rs})$ is found by substituting in the formula

$$V(\hat{Y}_{Rs}) = \sum_h \frac{N_h(N_h - n_h)}{n_h} S_{dh}^2$$

where

$$S_{dh}^2 = S_{yh}^2 + R_h^2 S_{xh}^2 - 2R_h S_{yxh}$$

The quantities S_{dh}^2 are given on the extreme right of the top half of Table 6.4.

TABLE 6.5

COMPARISON OF FOUR METHODS OF ALLOCATION

Method of Allocation: n_h Proportional to	Variance				Relative Precision
	Strata			Total	
	1	2	3		
1. N_h	49,824	105,833	376,215	531,872	100
2. $N_h S_{yh}$	35,144	131,847	343,446	510,437	104
3. $N_h \sqrt{\bar{X}_h}$	41,750	136,964	300,312	479,026	111
4. $N_h \bar{X}_h$	35,144	181,710	240,888	457,742	116

The variances and relative precisions are shown in Table 6.5.

There is not much to choose among the different allocations, as would be expected, since the n_h do not differ greatly in the four methods. Method 4, in which allocation is proportional to the total number of peach trees in the stratum, appears a trifle superior to the others.

6.15 UNBIASED RATIO-TYPE ESTIMATES

As we have noted, estimates of the ratio type that are unbiased or subject to a smaller bias than $\hat{R}$ or $\hat{Y}_R$ may be useful in surveys with many strata and small samples in each stratum if the separate ratio estimate seems appropriate. Three methods that give unbiased estimates and three methods that remove the term of order $1/n$ in the bias [see (6.33)] will be discussed briefly.

In comparing these methods, relevant questions are: (a) Does the MSE of the method compare favorably with that of the ordinary ratio estimate? (b) Does the method provide a satisfactory sample estimate of variance? This is a difficulty with $\hat{R}$, as we have seen. We first describe the methods. The unbiased methods require knowledge of $\bar{X}$.

Unbiased Methods

One estimate, due to Hartley and Ross (1954), can be derived by starting with the mean $\bar{r}$ of the ratios y_i/x_i and correcting it for bias.

$$\bar{r} = \frac{1}{n} \sum_{i=1}^{n} r_i = \frac{1}{n} \sum_{i=1}^{n} \frac{y_i}{x_i}$$

Now

$$\frac{1}{N} \sum_{i=1}^{N} r_i(x_i - \bar{X}) = \frac{1}{N} \sum_{i=1}^{N} \frac{y_i}{x_i} \cdot x_i - \left(\frac{1}{N} \sum_{i=1}^{N} r_i\right)\bar{X}$$

$$= \bar{Y} - \bar{X}E(r_i) = \bar{X}[R - E(r_i)] \qquad (6.72)$$

But in simple random sampling $E(\bar{r}) = E(r_i)$. Hence

$$\text{bias in } \bar{r} = E(\bar{r}) - R = -\frac{1}{\bar{X}N} \sum_{i=1}^{N} r_i(x_i - \bar{X}) \qquad (6.73)$$

By theorem 2.3, an unbiased sample estimate of

$$\frac{1}{N-1} \sum_{i=1}^{N} r_i(x_i - \bar{X})$$

is

$$\frac{1}{n-1} \sum_{i=1}^{n} r_i(x_i - \bar{x}) = \frac{n}{n-1}(\bar{y} - \bar{r}\bar{x})$$

On substituting into (6.73), the estimate $\bar{r}$, corrected for bias, becomes

$$\hat{R}_{HR} = \bar{r} + \frac{n(N-1)}{(n-1)N\bar{X}}(\bar{y} - \bar{r}\bar{x}) \tag{6.74}$$

The corresponding unbiased estimate of the population total $\hat{Y}$ is

$$\hat{R}_{HR}X = \bar{r}X + \frac{n(N-1)}{n-1}(\bar{y} - \bar{r}\bar{x}) \tag{6.75}$$

By similar arguments, another unbiased estimate (Mickey, 1959) is derived from the n ratios $\hat{R}_i$ obtained by *removing* each unit in turn from the sample, so that $\hat{R}_i = \sum y / \sum x$ over the remaining $(n-1)$ members. If $\hat{R}_-$ denotes the mean of the $\hat{R}_i$, Mickey's estimate is

$$\hat{R}_M = \hat{\bar{R}}_- + \frac{n(N-n+1)}{N\bar{X}}(\bar{y} - \hat{\bar{R}}_-\bar{x}). \tag{6.76}$$

As a third method, Lahiri (1951) showed that the ordinary ratio estimate $\hat{R}$ is unbiased if the sample is drawn with probability proportional to $\sum x_i$. Perhaps the simplest method of doing this (Midzuno, 1951) is to draw the first member of the sample with probability proportional to x_i. The remaining $(n-1)$ members of the sample are drawn with equal probability. It is easy to prove (exercise 6.10) that with this method the probability that a specific sample is drawn is proportional to $\sum\limits^{n} x_i$, and that $\hat{R} = \sum\limits^{n} y_i / \sum\limits^{n} x_i$ is unbiased for this method of sample selection.

Methods with bias of order $1/n^2$

These methods consist of an adjustment to $\hat{R}$. The first, due to Quenouille (1956), is applicable to a broad class of statistical problems in which the proposed estimate has a bias of order $1/n$. It has been given the name of the *jackknife* method, to denote a tool with many uses. The utility of this method for ratio estimates was pointed out by Durbin (1959).

Ignoring the fpc for the moment, the bias of estimates like $\hat{R}$ may be expanded in a series of the form

$$E(\hat{R}) = R + \frac{b_1}{n} + \frac{b_2}{n^2} + \cdots \tag{6.77}$$

If $n = mg$, let the sample be divided at random into g groups of size m. From (6.77)

$$E(g\hat{R}) = gR + \frac{b_1}{m} + \frac{b_2}{gm^2} + \cdots \tag{6.78}$$

Now let $\hat{R}_j$ be the ordinary ratio $\sum y / \sum x$, computed from the sample after *omitting* the jth group. Since $\hat{R}_j$ is obtained from a simple random sample of size

$m(g-1)$, we have

$$E(\hat{R}_j) = R + \frac{b_1}{(g-1)m} + \frac{b_2}{(g-1)^2 m^2} + \cdots \tag{6.79}$$

Hence

$$E[(g-1)\hat{R}_j] = (g-1)R + \frac{b_1}{m} + \frac{b_2}{(g-1)m^2} + \cdots \tag{6.80}$$

Subtraction from (6.78) gives, to order n^{-2},

$$E[g\hat{R} - (g-1)\hat{R}_j] = R - \frac{b_2}{g(g-1)m^2} = R - \frac{b_2}{n^2}\frac{g}{(g-1)}$$

The bias is now of order $1/n^2$. We can construct g estimates of this type, one for each group. Quenouille's estimator (the jackknife) is the average of these g estimates, that is,

$$\hat{R}_Q = g\hat{R} - (g-1)\hat{R}_- \tag{6.81}$$

where $\hat{R}_-$ is the average of the g quantities $\hat{R}_j$. As Quenouille showed, the variance of $\hat{R}_Q$ differs from that of $\hat{R}$ by terms of order $1/n^2$. Any increase in variance due to this adjustment for bias should therefore be negligible in moderately large samples. The choice $m = 1$, $g = n$ seems best with the jackknife in small samples.

If the fpc cannot be ignored, the leading term in the bias of $\hat{R}$, as in (6.33), is of the form $b_1(1-f)/n$. It can be shown (exercise 6.10) that in order to remove both the terms in $1/n$ and $1/N$, we need

$$\hat{R}_Q = w\hat{R} - (w-1)\hat{R}_- \tag{6.82}$$

where $w = g[1 - (n-m)/N]$, or with $m = 1$, $g = n$, $w = n[1 - (n-1)/N]$.

Beale's (1962) estimator is

$$\hat{R}_B = \frac{\bar{y} + [(1-f)/n](s_{yx}/\bar{x})}{\bar{x} + [(1-f)/n](s_x^2/\bar{x})} = \hat{R} \cdot \frac{1 + [(1-f)/n]c_{yx}}{1 + [(1-f)/n]c_{xx}} \tag{6.83}$$

while Tin's (1965) is the closely related quantity

$$\hat{R}_T = \hat{R}\left[1 - \frac{(1-f)}{n}\left(\frac{s_x^2}{\bar{x}^2} - \frac{s_{yx}}{\bar{y}\bar{x}}\right)\right] = \hat{R}\left[1 - \frac{(1-f)}{n}(c_{xx} - c_{yx})\right] \tag{6.84}$$

where $s_{yx} = \sum_{}^{n}(y_i - \bar{y})(x_i - \bar{x})/(n-1)$, $s_x^2 = \sum_{}^{n}(x_i - \bar{x})^2/(n-1)$, so that c_{yx}, c_{xx} are the sample relative covariance and relative variance of x.

The structure of $\hat{R}_T$ may be seen by noting that from (6.34) the leading term in the expected value of $\hat{R}$ may be written

$$R\left[1 + \frac{(1-f)}{n}(C_{xx} - C_{yx})\right]$$

Evidently, $\hat{R}_T$ is adjusting $\hat{R}$ by a sample estimate of the needed bias correction. $\hat{R}_B$ and $\hat{R}_T$ are identical to terms of order $1/n$ and in general perform very similarly.

6.16 COMPARISON OF THE METHODS

Example.   The artificial population in Table 6.6 contains three strata with $N_h = 4$, $n_h = 2$ in each stratum. The population was deliberately constructed so that (*a*) R_h varies markedly from stratum to stratum, thus favoring a *separate* ratio estimate $\hat{Y}_{Rs}$, and (*b*) $\hat{R}_h$ overestimates in each stratum, with the threat of a serious cumulative bias in $\hat{Y}_{Rs}$. The

TABLE 6.6

A SMALL ARTIFICIAL POPULATION

Stratum

	I		II		III	
	y	x	y	x	y	x
	2	2	2	1	3	1
	3	4	5	4	7	3
	4	6	9	8	9	4
	11	20	24	23	25	12
Totals	20	32	40	36	44	20
R_h	0.625		1.111		2.200	

following methods of estimating the population *total* Y were compared.

Simple expansion: $\sum N_h \bar{y}_h$

Combined ratio: $(\bar{y}_{st}/\bar{x}_{st})X$

Separate ratio: $\sum (\bar{y}_h/\bar{x}_x)X_h = \sum \hat{R}_h X_h$

The remaining estimates, the *separate* Hartley-Ross, Lahiri, and Quenouille, Beale, and Tin methods, have the same form as the separate ratio estimate, except that $\hat{R}_{(HR)h}$, $\hat{R}_{(L)h}$, and so on, replace $\hat{R}_h$. (For $n_h = 2$, the Hartley-Ross and Mickey methods are identical.) There are $6^3 = 216$ possible samples. Results are exact apart from rounding errors. For help in some computations I am indebted to Dr. Joseph Sedransk.

The results in Table 6.7 show some interesting features. For the combined ratio estimate, the contribution of the (bias)^2 to the mean square error is trivial, despite the extreme conditions, but this estimate does poorly as regards variance because of the wide variation in the R_h. As judged by the MSE, the separate ratio estimate is much more accurate than the combined estimate, but it is badly biased. Of the unbiased methods, Hartley-Ross shows relatively high variance, as it has been found to do for $n_h = 2$ in some studies on natural populations. The Lahiri method does particularly well. This population suits the Lahiri method because one unit in each stratum has unusually high values of both y_i and x_i,

TABLE 6.7

RESULTS FOR DIFFERENT ESTIMATES OF Y

Method	Variance	(Bias)^2	MSE
Simple expansion	820.3	0.0	820.3
Combined ratio	262.8	6.5	269.3
Separate ratio	35.9	24.1	60.0
Separate Hartley-Ross	153.6	0.0	153.6
Separate Lahiri	19.6	0.0	19.6
Separate Quenouille	42.9	1.1	44.0
Separate Beale	28.9	8.0	36.9
Separate Tin	28.6	5.7	34.3

with the consequences that this unit has a high probability of being drawn and that samples containing this unit give good estimates of R_h.

The Quenouille, Beale, and Tin methods all produced substantial decreases in bias as compared with the separate ratio estimate, and all had smaller MSE's, so that in this example they achieved their principal purposes.

The study by J. N. K. Rao (1969) of natural populations cited in section 6.9 compared the Quenouille, Beale, and Tin methods for $n = 2, 4, 6, 8$, on 15 such populations. For $n = 2$, the most severe test, the medians and the upper quartiles of the quantities | bias|/$\sqrt{\text{MSE}}$ were as follows: $\hat{R}_Q$, 3%, 7%; $\hat{R}_B$, 8%, 12%; $\hat{R}_T$, 8%, 19%; as against 15%, 20% for $\hat{R}$. The more complex methods appear to help materially as regards bias in these tiny samples.

The same study compared the MSE's of five of the estimates in this section with that of $\hat{R}$. (Lahiri's method was omitted, since the study was confined to simple random samples.) For each method the ratio 100 $\text{MSE}(\hat{R}_Q)/\text{MSE}(\hat{R})$, and so forth, was calculated for each population.

For $n = 4$, Quenouille's and Mickey's estimates were slightly inferior to $\hat{R}$ in these populations but, for $n \geq 6$, all methods had average MSE's very close to those of $\hat{R}$. For a very small sample from a *single* population, this study suggests that these more complex methods have no material advantage in accuracy over $\hat{R}$. But the fact that they reduce bias with little or no increase in MSE in a single stratum should give them an advantage in a separate ratio estimate with numerous strata having small samples.

Under a linear regression model, comparisons of the MSE's of these methods for small n by P.S.R.S. Rao and J. N. K. Rao (1971), Hutchinson (1971), and J. N. K. Rao and Kuzik (1974) gave results in general agreeing with those from the natural populations.

6.17 IMPROVED ESTIMATION OF VARIANCE

One estimator worth consideration for moderate or small samples was suggested by Tukey (1958) for the jackknife (Quenouille's) estimate $\hat{R}_Q$. With g

groups, $n = mg$, and f negligible, $\hat{R}_Q$ is the average of the g quantities

$$\hat{R}'_j = g\hat{R} - (g-1)\hat{R}_j$$

where $\hat{R}_j = \sum y / \sum x$, computed after omitting group j. If the $\hat{R}'_j$ could be regarded as g independent estimates of R then, with simple random sampling, an unbiased estimate of $V(\hat{R}_Q)$ would be

$$v(\hat{R}_Q) = \frac{(1-f)}{g} \frac{\sum_{}^{g} (\hat{R}'_j - \hat{R}_Q)^2}{(g-1)} \tag{6.85}$$

Since $\hat{R}'_j - \hat{R}_Q = -(g-1)(\hat{R}_j - \hat{R}_-)$, (6.85) is more easily calculated as

$$v(\hat{R}_Q) = \frac{(1-f)(g-1)}{g} \sum_{}^{g} (\hat{R}_j - \hat{R}_-)^2 \tag{6.86}$$

where $\hat{R}_-$ is the mean of the g quantities $\hat{R}_j$.

The $\hat{R}'_j$ or $\hat{R}_j$ for different j are, of course, not independent, and formula (6.86) is an approximation. So far, the analytical properties of $v(\hat{R}_Q)$ have been established only for large samples. Arvesen (1969) showed that for a broad class of estimates including $\hat{R}_Q$ that are symmetrical in the elements of the sample [estimates known as Hoeffding's U-Statistics (1948)], the formula $v(\hat{R}_Q)$ becomes unbiased either for fixed g or for $g = n$ as n becomes large.

From Rao's (1969) study of eight natural populations, small-sample average percent underestimations in the standard $v_1(\hat{R})$ as an estimate of the true $\text{MSE}(\hat{R})$ were reported in section 6.9 for $n = 4, 6, 8, 12$. The corresponding average percent biases in $v(\hat{R}_Q)$ are shown for comparison in Table 6.8; these are the averages of the eight numbers: $100[v(\hat{R}_Q) - \text{MSE}(\hat{R}_Q)]/\text{MSE}(\hat{R}_Q)$.

Table 6.8 also gives the averages of the quantities $100 [v(\hat{R}_Q) - \text{MSE}(\hat{R})]/\text{MSE}(\hat{R})$. These averages are of interest to the investigator who uses $\hat{R}$ but is willing to replace $v_1(\hat{R})$ by $v(\hat{R}_Q)$ as an estimator of $V(\hat{R})$ if it seems less biased. In view of the biases in $\hat{R}$ and $\hat{R}_Q$, the comparisons of $V(\hat{R}_Q)$ are made with the MSE's as more appropriate.

TABLE 6.8
AVERAGE PERCENT BIAS IN ESTIMATORS OF VARIANCE

	$n =$			
Average of	4	6	8	12
$100[v(\hat{R}_Q) - \text{MSE}(\hat{R}_Q)]/\text{MSE}(\hat{R}_Q)$	+11%	+10%	+6%	+1%
$100[v(\hat{R}_Q) - \text{MSE}(\hat{R})]/\text{MSE}(\hat{R})$	+11%	+10%	+6%	+1%
$100[v_1(\hat{R}) - \text{MSE}(\hat{R})]/\text{MSE}(\hat{R})$	-31%	-23%	-21%	-18%

In these populations $v(\hat{R}_Q)$ is a slight overestimate of both $\text{MSE}(\hat{R}_Q)$ and $\text{MSE}(\hat{R})$, while $v_1(\hat{R})$ has substantial negative biases in these small samples.

The stability of $v(\hat{R}_Q)$ relative to that of $v_1(\hat{R})$, as judged by the squares of the coefficients of variation of these variance estimates, was poor in these samples. In studies by Rao and Beegle (1968) of $v(\hat{R}_Q)$ and $v_1(\hat{R})$ under a linear regression model of y on x in an infinite population with x normal, however, $v(\hat{R}_Q)$ and $v_1(\hat{R})$ appeared about equally stable for $n = 4$ to $n = 12$.

With a separate ratio estimate and numerous strata these results suggest that $\sum X_h^2 v(\hat{R}_{Qh})$ is superior to $\sum X_h^2 v_1(\hat{R}_h)$ as an estimator of $V(\hat{Y}_{Rs})$. The former is likely to be freer from bias and both should have adequate stability. But with only a few strata the issue is questionable until further comparisons appear.

6.18   COMPARISON OF TWO RATIOS

In analytical surveys it is frequently necessary to estimate the difference $\hat{R} - \hat{R}'$ between two ratios and to compute the standard error of $\hat{R} - \hat{R}'$. The formulas given here are for the *estimated* variance of $\hat{R} - \hat{R}'$, since these are the ones most commonly required. The fpc terms are omitted for reasons presented in section 2.14.

Simple random sampling is assumed at first. Three cases can be distinguished.

The Two Ratios Are Independent

This occurs when the units are classified into two distinct classes and we wish to compare ratios estimated separately in the two classes. For instance, in a study of household expenditures, a simple random sample of households might be subdivided into owned and rented houses in order to compare the proportions of income spent on upkeep of the house in the two classes. If the estimated ratios are denoted by $\hat{R} = \bar{y}/\bar{x}$, $\hat{R}' = \bar{y}'/\bar{x}'$, then

$$v(\hat{R} - \hat{R}) = v(\hat{R}) + v(\hat{R}') \tag{6.87}$$

The Two Ratios Have the Same Denominator

When the unit is a cluster of families, we might wish to compare the proportion of adult males who use electric shavers with the proportion who use razors. In any unit, y = number of adult males using electric shavers, y' = number of adult males using razors, and x = total number of adult males.

$$\hat{R} - \hat{R}' = \frac{\bar{y} - \bar{y}'}{\bar{x}}$$

If $d_i = y_i - y_i'$, the estimated variance of $\hat{R} - \hat{R}'$ may be computed as

$$v(\hat{R} - \hat{R}') \doteq \frac{1}{n(n-1)\bar{x}^2} \sum_{i=1}^{n} [d_i - (\hat{R} - \hat{R}')x_i]^2 \tag{6.88}$$

The Two Ratios Have Different Denominators But May Be Correlated

An example is the comparison of the proportion of men who smoke with the proportion of women who smoke, in a survey in which the unit is a cluster of houses. Mathematically, this is the most general case.

$$v(\hat{R} - \hat{R}') = v(\hat{R}) + v(\hat{R}') - 2 \operatorname{cov} (\hat{R}\hat{R}') \tag{6.89}$$

The only unfamiliar term is cov $(\hat{R}\hat{R}')$. Writing, in the usual way,

$$\hat{R} - R \doteq \frac{\bar{y} - R\bar{x}}{\bar{X}} \qquad \hat{R}' - R' \doteq \frac{\bar{y}' - R'\bar{x}'}{\bar{X}'}$$

we have

$$\operatorname{cov} (\hat{R}\hat{R}') \doteq \frac{1}{n\bar{X}\bar{X}'} \operatorname{cov} (y_i - Rx_i)(y_i' - R'x_i')$$

A sample estimate may be computed as follows:

$$\operatorname{cov} (\hat{R}\hat{R}') \doteq \frac{1}{n(n-1)\bar{x}\bar{x}'} \sum_{}^{n} (y_i y_i' - \hat{R}y_i'x_i - \hat{R}'y_i x_i' + \hat{R}\hat{R}'x_i x_i') \tag{6.90}$$

Example. The 1954 field trial of the Salk polio vaccine was conducted among children in the first three grades in all schools in a number of counties. The counties were not randomly selected, since those with a history of previous polio attacks were favored, but for this illustration, it will be assumed that they are a random sample from some population.

Children whose parents did not give permission to participate in the trial were called the "not inoculated" group and, of course, received no shots. Half of the children who received permission were given three shots of an inert liquid and were called the "placebo" group. From the data in Table 6.9, compare the frequencies $\hat{R}$, $\hat{R}'$ of paralytic polio in the "not inoculated" and "placebo" groups. To reduce the amount of data, the comparison is restricted to 34 counties, each having more than 4000 children in the two groups combined.

In these data any variation in the polio attack rate from county to county would produce a positive correlation between $\hat{R}$ and $\hat{R}'$.

The following quantities are derived from the totals.

$$\text{Placebo:} \quad \hat{R} = \frac{88}{167.4} = 0.525687, \qquad \bar{x} = \frac{167.4}{34} = 4.9235$$

$$\text{Not inoculated:} \quad \hat{R}' = \frac{99}{284.6} = 0.347857, \qquad \bar{x}' = \frac{284.6}{34} = 8.3706$$

For $v(\hat{R})$, $v(\hat{R}')$ and cov $(\hat{R}\hat{R}')$, all uncorrected sums of squares and products among the four variates are required.

$$v(\hat{R}) = \frac{1}{n(n-1)\bar{x}^2} (\sum y^2 - 2\hat{R} \sum yx + \hat{R}^2 \sum x^2)$$

$$= \frac{1}{(34)(33)(4.9235)^2}[(564) - (1.05137)(822.2) + (0.27635)(1661.92)]$$

$$= 0.00584$$

TABLE 6.9

Number of Children (x, x') and of Paralytic Cases (y, y') per County

x*	x'	y†	y'	x	x'	y	y'
4.1	2.4	0	0	13.8	25.6	3	3
3.5	8.0	1	6	10.5	8.1	2	0
4.1	6.1	7	2	21.6	25.9	10	7
2.6	4.6	2	1	3.5	6.7	2	2
2.4	1.5	2	1	6.8	7.3	3	8
2.2	1.9	0	0	2.3	3.7	0	1
1.1	4.0	1	1	2.6	2.9	2	0
1.6	4.0	1	2	6.0	11.1	3	1
5.7	7.8	1	4	11.0	14.8	7	11
3.3	11.0	3	7	19.4	42.5	11	14
1.0	3.8	0	1	6.8	13.7	6	2
2.0	5.2	1	0	1.2	4.0	3	1
8.3	19.0	4	4	5.4	9.3	11	6
1.0	3.7	1	5	1.7	2.6	0	2
1.1	4.2	0	1	2.1	2.3	0	0
2.3	6.8	1	2	1.5	2.6	0	0
1.9	3.5	0	2	3.0	4.0	0	2
				Totals			
				167.4	284.6	88	99

* x, x' = numbers of "placebo" and "not inoculated" children (in 1000's)

† y, y' = numbers of paralytic polio cases in the placebo and not inoculated groups

Similarly, we find $v(\hat{R}') = 0.00240$.

$$\text{cov}(\hat{R}\hat{R}') = \frac{1}{n(n-1)\bar{x}\bar{x}'}(\sum yy' - \hat{R}\sum y'x - \hat{R}'\sum yx' + \hat{R}\hat{R}'\sum xx')$$

$$= \frac{(497) - (0.52569)(844.6) - (0.34786)(1397.4) + (0.52569)(0.34786)(2690.8)}{(34)(33)(4.9235)(8.3706)}$$

$$= 0.00127$$

Hence

$$\text{s.e.}(\hat{R} - \hat{R}') = \sqrt{0.00584 + 0.00240 - 0.00254} = 0.0754$$

Since $\hat{R} - \hat{R}' = 0.1778$, the difference approaches significance at the 5% level (the distribution of $\hat{R} - \hat{R}'$ may be somewhat skew for this size of sample). A possible explanation is that the not-inoculated children may have had more natural protection against polio than the placebo children.

The same problem may arise in stratified samples in which the domains of study cut across strata. If $\hat{R}_h$, $\hat{R}_h{}'$ appear to vary from stratum to stratum, the comparison will probably be based on an examination of the values of $\hat{R}_h - \hat{R}_h{}'$ in individual strata. By finding the standard errors of $\hat{R}_h - \hat{R}_h{}'$ it is possible to determine whether these differences vary from stratum to stratum and, if not, to compute an efficient over-all difference.

If the $\hat{R}_h$, $\hat{R}_h{}'$ exhibit no real variation from stratum to stratum, it may be sufficient to compare the combined estimates $\hat{R}_c$ and $\hat{R}_c{}'$. As before,

$$v(\hat{R}_c - \hat{R}_c{}') = v(\hat{R}_c) + v(\hat{R}_c{}') - 2\,\mathrm{cov}\,(\hat{R}_c\hat{R}_c{}') \qquad (6.91)$$

where, putting $d_{hi} = (y_{hi} - \bar{y}_h) - \hat{R}_c(x_{hi} - \bar{x}_h)$,

$$v(\hat{R}_c) = \frac{1}{\bar{x}_{st}^2} \sum_h \frac{N_h^2}{n_h(n_h - 1)} \sum_i d_{hi}^2 \qquad (6.92)$$

$$\mathrm{cov}\,(\hat{R}_c\hat{R}_c{}') = \frac{1}{\bar{x}_{st}\bar{x}_{st}{}'} \sum_h \frac{N_h^2}{n_h(n_h - 1)} \sum_i d_{hi}d_{hi}{}' \qquad (6.93)$$

A more thorough discussion of the comparison of ratios, including shortcut computing formulas when the sample permits them, has been given by Kish and Hess (1959b).

6.19 RATIO OF TWO RATIOS

In some applications we want to estimate the ratio R/R' of two ratios. Thus, in the preceding example (section 6.18), we might be interested in the ratio R/R' of the paralytic polio case rates for "placebo" and "not inoculated" children, or the ratio of the proportions of males and females in the labor force from a cluster sample. If data on (y,x) are available in the same sample for two time periods, the quantity might be the ratio of the weekly expenditures on food per household at the two times.

With a simple random sample (e.g., of clusters) the sample estimate of R/R' is $\hat{R}/\hat{R}' = (\bar{y}/\bar{x})/\bar{y}'/\bar{x}')$, sometimes called a *double ratio estimate*. As with the single ratio the leading term in the bias of $\hat{R}/\hat{R}'$ is of order $1/n$, but is more complex than for $\hat{R}$ or $\hat{R}'$. We may write

$$\hat{R}/\hat{R}' = (R/R')(1 + \delta\bar{y})(1 + \delta\bar{x}')/(1 + \delta\bar{x})(1 + \delta\bar{y}') \qquad (6.94)$$

where $\delta\bar{y}$ denotes $(\bar{y} - \bar{Y})/\bar{Y}$, and so forth. When this expression is expanded, there are six quadratic terms of order $1/n$ that enter into the bias of $\hat{R}/\hat{R}'$. Rao and Pereira (1968) give an exact expression for the bias.

Formula (6.8) for the relative variance $V(\hat{R})/R^2 = C_{\hat{R}\hat{R}}$ of a ratio can be written

$$C_{\hat{R}\hat{R}} = C_{\bar{y}\bar{y}} + C_{\bar{x}\bar{x}} + 2C_{\bar{y}\bar{x}}$$

From this, the leading term in $V(\hat{R}/\hat{R}')$ is

$$V(\hat{R}/\hat{R}') \doteq \left(\frac{R}{R'}\right)^2 (C_{\hat{R}\hat{R}} + C_{\hat{R}'\hat{R}'} - 2C_{\hat{R}\hat{R}'}) \tag{6.95}$$

where

$$C_{\hat{R}\hat{R}'} = C_{\bar{y}\bar{y}'} + C_{\bar{x}\bar{x}'} - C_{\bar{y}\bar{x}'} - C_{\bar{y}'\bar{x}} \tag{6.96}$$

For the corresponding sample estimate $v(\hat{R}/\hat{R}')$ we substitute sample estimates of the terms in (6.95).

Example. For the ratio of placebo and not inoculated case rates, $\hat{R}/\hat{R} = 0.52569/0.34786 = 1.511$. Estimate the s.e. of this ratio. The computations in the preceding example give

$$\hat{C}_{\hat{R}\hat{R}} = \frac{0.00584}{(0.5257)^2} = 0.0211; \qquad \hat{C}_{\hat{R}'\hat{R}'} = \frac{0.00240}{(0.3479)^2} = 0.0198;$$

$$\hat{C}_{\hat{R}\hat{R}'} = \frac{0.00127}{(0.5257)(0.3479)} = 0.00694$$

$$v(\hat{R}/\hat{R}') = (1.511)^2 (0.0211 + 0.0198) - 0.0139) = 0.0617$$

$$\text{s.e. } (\hat{R}/\hat{R}') = 0.248$$

The double ratio estimate has occasionally been used in place of $\hat{Y}_R = \hat{R}X$ to estimate a population *total* Y, as suggested by Keyfitz (Yates, 1960). Suppose that $\hat{R}' = (\bar{y}'/\bar{x}')$ is known for the *same* sample from a previous period and that $R' = \bar{Y}'/\bar{X}'$ is also known. If $\hat{R}'/R'$ has been found, say, to be slightly > 1, we might argue intuitively that $\hat{R}$ is also likely to give an overestimate of R that should be adjusted downward by dividing it by the ratio $\hat{R}'/R'$. This leads to the double ratio estimate $\hat{Y}_{DR}$.

$$\hat{Y}_{DR} = \frac{R'}{\hat{R}'}(\hat{R}X) = \frac{\hat{R}}{\hat{R}'}(R'X) \tag{6.97}$$

Since the relative variance of $\hat{Y}_R$ is $C_{\hat{R}\hat{R}}$, while that of $\hat{Y}_{DR}$ is

$$C_{\hat{R}\hat{R}} + C_{\hat{R}'\hat{R}'} - 2C_{\hat{R}\hat{R}'} \tag{6.98}$$

the double ratio will give a more precise estimate in large samples if the correlation between $\hat{R}$ and $\hat{R}'$ is high enough.

6.20 MULTIVARIATE RATIO ESTIMATES

Olkin (1958) has extended the ratio estimate to the situation in which p auxiliary x-variables $(x_1, x_2, \ldots x_p)$ are available. For the population total, the proposed estimate, say $\hat{Y}_{MR}$ for *multivariate ratio*, is

$$\hat{Y}_{MR} = W_1 \frac{\bar{y}}{\bar{x}_1} X_1 + W_2 \frac{\bar{y}}{\bar{x}_2} X_2 + \cdots + w_p \frac{\bar{y}}{\bar{x}_p} X_p$$

$$= W_1 \hat{Y}_{R_1} + W_2 \hat{Y}_{R_2} + \cdots + W_p \hat{Y}_{R_p}$$

where the W_i are weights to be determined to maximize the precision of $\hat{Y}_{MR}$, subject to $\sum W_i = 1$. This type of estimate appears appropriate when the regression of y on $x_i, x_2, \ldots x_p$ is linear and passes through the origin. The population totals X_i must be known.

The method is described for two x-variates, since this should be the most frequent application. We have

$$\hat{Y}_{MR} - Y = W_1(\hat{Y}_{R_1} - Y) + W_2(\hat{Y}_{R_2} - Y)$$

Hence, assuming negligible bias,

$$V(\hat{Y}_{MR}) = W_1^2 V(\hat{Y}_{R_1}) + 2W_1 W_2 \text{ cov } (\hat{Y}_{R_1} \hat{Y}_{R_2}) + W_2^2 V(\hat{Y}_{R_2})$$
$$= W_1^2 V_{11} + 2W_1 W_2 V_{12} + W_2^2 V_{22} \tag{6.99}$$

where $V_{11} = V(\hat{Y}_{R_1})$, etc. The values of W_1, W_2, that minimize the variance, subject to $W_1 + W_2 = 1$, are found to be

$$W_1 = \frac{V_{22} - V_{12}}{V_{11} + V_{22} - 2V_{12}}, \qquad W_2 = \frac{V_{11} - V_{12}}{V_{11} + V_{22} - 2V_{12}}$$

and the minimum variance is

$$V_{min}(\hat{Y}_{MR}) = \frac{V_{11} V_{22} - V_{12}^2}{V_{11} + V_{22} - 2V_{12}} \tag{6.100}$$

With p variates, it is necessary to compute the inverse V^{ij} of the matrix V_{ij}. Then the optimum $W_i = \sum_i / \sum$, where $\sum_i$ is the sum of the elements in the ith column of V^{ij} and $\sum$ is the sum of all the p^2 elements of V^{ij}. The minimum variance is $1/\sum$.

In practice, the weights are determined from estimated variances and covariances v_{ij}. From (6.7) in section 6.3,

$$v_{11} = \frac{(1-f)\hat{Y}^2}{n}(c_{yy} + c_{11} - 2c_{y1})$$

$$v_{22} = \frac{(1-f)\hat{Y}^2}{n}(c_{yy} + c_{22} - 2c_{y2})$$

where $c_{yy} = s_y^2/\bar{y}^2$, etc. The covariance can be expressed as

$$v_{12} = \frac{(1-f)\hat{Y}^2}{n}(c_{yy} + c_{12} - c_{y1} - c_{y2})$$

A convenient method of computation is first to obtain the matrix

$$C = \begin{vmatrix} c_{yy} & c_{y1} & c_{y2} \\ c_{y1} & c_{11} & c_{12} \\ c_{y2} & c_{12} & c_{22} \end{vmatrix}$$

If $v_{ij}' = n v_{ij}/(1-f)\hat{Y}^2$, the matrix v_{ij}' is easily obtained by taking diagonal contrasts in C, that is,

$$v_{11}' = c_{yy} + c_{11} - c_{y1} - c_{y1}$$

$$v_{12}' = c_{yy} + c_{12} - c_{y1} - c_{y2} \qquad \text{etc.}$$

The factor $(1-f)\,\hat{Y}^2/n$ is not needed when computing the w_i, but it must be inserted when computing the minimum variance. Thus

$$v_{min}(\hat{Y}_{MR}) = \frac{(1-f)\,\hat{Y}^2}{n} \frac{(v_{11}'v_{22}' - v_{12}'^2)}{(v_{11}' + v_{22}' - 2v_{12}')} \tag{6.101}$$

In view of the amount of computation involved, this estimate will probably be restricted to smaller surveys of specialized scope. The method is capable of giving a marked increase in precision over $\hat{Y}_{R_1}$ or $\hat{Y}_{R_2}$ alone.

6.21 PRODUCT ESTIMATORS

If an auxiliary variate x has a negative correlation with y where x and y are variates that take only positive values, a natural analogue of the ratio estimator is the product estimator, for which

$$\hat{\bar{Y}}_p = \frac{\bar{x}}{\bar{X}}\bar{y}; \qquad \hat{Y}_p = N\frac{\bar{x}}{\bar{X}}\bar{y} \tag{6.102}$$

By the usual Taylor series expansion, the analogue of (6.8) for the product estimator in a large simple random sample is

$$(cv)^2 = \frac{(1-f)}{n}(C_{yy} + C_{xx} + 2C_{yx}) \tag{6.103}$$

where $(cv)^2$ is the square of the coefficient of variation of either $\hat{\bar{Y}}_p$ of $\hat{Y}_p$. P.S.R.S. Rao and Mudholkar (1967) have extended Olkin's multivariate ratio estimator to a weighted combination of ratio estimators (for x_i positively correlated with y) and product estimators (for x_i negatively correlated with y).

EXERCISES

6.1 A pilot survey of 21 households gave the following data for numbers of members (x), children (y_1), cars (y_2), and TV sets (y_3).

x	y_1	y_2	y_3	x	y_1	y_2	y_3	x	y_1	y_2	y_3
5	3	1	3	2	0	0	1	6	3	2	0
2	0	1	1	3	1	1	1	4	2	1	1
4	1	2	0	2	0	2	0	4	2	1	1
4	2	1	1	6	4	2	1	3	1	0	1
6	4	1	1	3	1	0	0	2	0	2	1
3	1	1	2	4	2	1	1	4	2	1	1
5	3	1	1	5	3	1	1	3	1	1	1

Assuming that the total population X is known, would you recommend that ratio estimates be used instead of simple expansions for estimating total numbers of children, cars, and TV sets?

6.2   In a field of barley the grain, y_i, and the grain plus straw, x_i, were weighted for each of a large number of sampling units located at random over the field. The total produce (grain plus straw) of the whole field was also weighed. The following data were obtained: $c_{yy} = 1.13$, $c_{yx} = 0.78$, $c_{xx} = 1.11$. Compute the gain in precision obtained by estimating the grain yield of the field from the ratio of grain to total produce instead of from the mean yield of grain per unit.

It requires 20 min to cut, thresh, and weigh the grain on each unit, 2 min to weigh the straw on each unit, and 2 hr to collect and weigh the total produce of the field. How many units must be taken per field in order that the ratio estimate may be more economical than the mean per unit?

6.3   For the data in Table 6.1, $\hat{Y}_R = 28{,}367$ and $c_{\bar{y}\bar{y}} = 0.0142068$, $c_{\bar{y}\bar{x}} = 0.0146541$, $c_{\bar{x}\bar{x}} = 0.0156830$. Compute the 95% quadratic confidence limits for Y and compare them with the limits found by the normal approximation.

6.4   The values of y and x are measured for each unit in a simple random sample from a population. If $\bar{X}$, the population mean of x, is known, which of the following procedures do you recommend for estimating $\bar{Y}/\bar{X}$? (a) Always use $\bar{y}/\bar{X}$. (b) Sometimes use $\bar{y}/\bar{X}$ and sometimes $\bar{y}/\bar{x}$. (c) Always use $\bar{y}/\bar{x}$. Give reasons for your answer.

6.5   The following data are for a small artificial population with $N = 8$ and two strata of equal size.

	Stratum 1		Stratum 2
x_{1i}	y_{1i}	x_{2i}	y_{2i}
2	0	10	7
5	3	18	15
9	7	21	10
15	10	25	16

For a stratified random sample in which $n_1 = n_2 = 2$, compare the MSE's of $\hat{Y}_{Rs}$ and $\hat{Y}_{Rc}$ by working out the results for all possible samples. To what extent is the difference in MSE's due to biases in the estimates?

6.6.   In exercise 6.5 compute the variance given by using Lahiri's method of sample selection within each stratum and a separate ratio estimate.

6.7   Forty-five states of the United States (excluding the five largest) were arranged in nine strata with five states each, states in the same stratum having roughly the same ratio of 1950 to 1940 population. A stratified random sample with $n_h = 2$ gave the following results for 1960 population (y) and 1950 population (x), in millions.

				Stratum					
	1	2	3	4	5	6	7	8	9
y_{h1}	0.23	0.63	0.97	2.54	4.67	4.32	4.56	1.79	2.18
x_{h1}	0.13	0.50	0.91	2.01	3.93	3.96	4.06	1.91	1.90
y_{h2}	4.95	2.85	0.61	6.07	3.96	1.41	3.57	1.86	1.75
x_{h2}	2.78	2.38	0.53	4.84	3.44	1.33	3.29	2.01	1.32

Given that the 1950 population total X is 97.94, estimate the 1960 population by the combined ratio estimate. Find the standard error of your estimate by Keyfitz' short-cut method (section 6.13). The correct 1960 total was 114.99. Does your estimate agree with this figure within sampling errors?

6.8   In the example of a bivariate ratio estimate given by Olkin, a sample of 50 cities was drawn from a population of 200 large cities. The variates y, x_1, x_2 are the numbers of inhabitants per city in 1950, 1940, and 1930, respectively. For the population, $\bar{Y} = 1699$, $\bar{X}_1 = 1482$, $\bar{X}_2 = 1420$ (in 100's) and, for the sample, $\bar{y} = 1896$, $\bar{x}_1 = 1693$, $\bar{x}_2 = 1643$. The C matrix as defined in section 6.20 is

	y	x_1	x_2
y	1.213	1.241	1.256
x_1	1.241	1.302	1.335
x_2	1.256	1.335	1.381

Estimate $\bar{Y}$ by (a) the sample mean, (b) the ratio of 1950 to 1940 numbers of inhabitants, and (c) the bivariate ratio estimate. Compute the estimated standard error of each estimate.

6.9   Prove that with Midzuno's method of sample selection (section 6.15) the probability that any specific sample will be drawn is

$$\frac{(n-1)!(N-n)!}{(N-1)!} \frac{\sum\limits_{}^{n}(x_i)}{X}$$

6.10   In small populations the leading term in the bias of $\hat{R}$ in simple random samples of size n is of the form

$$E(\hat{R} - R) = \frac{b_1(1-f)}{n} = \frac{b_1}{n} - \frac{b_1}{N}$$

where b_1 does not depend on n, N. If $n = mg$ and the sample is divided at random into g groups of size m, let $\hat{R}_j = \sum y / \sum x$ taken over the remaining $(n-m)$ sample members when group j is omitted from the sample. Show that in the bias of the estimate

$$w\hat{R} - (w-1)\hat{R}_j$$

both terms in b_1 vanish if $w = g[1-(n-m)/N]$.

C H A P T E R 7

Regression Estimators

7.1 THE LINEAR REGRESSION ESTIMATE

Like the ratio estimate, the linear regression estimate is designed to increase precision by the use of an auxiliary variate x_i that is correlated with y_i. When the relation between y_i and x_i is examined, it may be found that although the relation is approximately linear, the line does not go through the origin. This suggests an estimate based on the linear regression of y_i on x_i rather than on the ratio of the two variables.

We suppose that y_i and x_i are each obtained for every unit in the sample and that the population mean $\bar{X}$ of the x_i is known. The linear regression estimate of $\bar{Y}$, the population mean of the y_i, is

$$\bar{y}_{lr} = \bar{y} + b(\bar{X} - \bar{x}) \tag{7.1}$$

where the subscript lr denotes *linear regression* and b is an estimate of the change in y when x is increased by unity. The rationale of this estimate is that if $\bar{x}$ is below average we should expect $\bar{y}$ also to be below average by an amount $b(\bar{X} - \bar{x})$ because of the regression of y_i on x_i. For an estimate of the population total Y, we take $\hat{Y}_{lr} = N\bar{y}_{lr}$.

Watson (1937) used a regression of leaf area on leaf weight to estimate the average area of the leaves on a plant. The procedure was to weigh all the leaves on the plant. For a small sample of leaves, the area and the weight of each leaf were determined. The sample mean leaf area was then adjusted by means of the regression on leaf weight. The point of the application is, of course, that the weight of a leaf can be found quickly but determination of its area is more time consuming.

This example illustrates a general situation in which regression estimates are helpful. Suppose that we can make a rapid estimate x_i of some characteristic for every unit and can also, by some more costly method, determine the correct value y_i of the characteristic for a simple random sample of the units. A rat expert might make a quick eye estimate of the number of rats in each block in a city area and then determine, by trapping, the actual number of rats in each of a simple random sample of the blocks. In another application described by Yates (1960), an eye

189

estimate of the volume of timber was made on each of a population of $\frac{1}{10}$-acre plots, and the actual timber volume was measured for a sample of the plots. The regression estimate

$$\bar{y}+b(\bar{X}-\bar{x})$$

adjusts the sample mean of the actual measurements by the regression of the actual measurements on the rapid estimates. The rapid estimates need not be free from bias. If $x_i - y_i = D$, so that the rapid estimate is perfect except for a constant bias D, then with $b = 1$ the regression estimate becomes

$$\bar{y}+(\bar{X}-\bar{x})=\bar{X}+(\bar{y}-\bar{x})$$

$$= (\text{pop. mean of rapid estimate})+(\text{adjustment for bias})$$

If no linear regression model is assumed, our knowledge of the properties of the regression estimate is of the same scope as our knowledge for the ratio estimate. The regression estimate is consistent, in the trivial sense that when the sample comprises the whole population, $\bar{x} = \bar{X}$, and the regression estimate reduces to $\bar{Y}$. As will be shown, the estimate is in general biased, but the ratio of the bias to the standard error becomes small when the sample is large. We possess a large-sample formula for the variance of the estimate, but more information is needed about the distribution of the estimate in small samples and about the value of n required for the practical use of large-sample results.

By a suitable choice of b, the regression estimate includes as particular cases both the mean per unit and the ratio estimate. Obviously if b is taken as zero, $\bar{y}_{lr}$ reduces to $\bar{y}$. If $b = \bar{y}/\bar{x}$,

$$\bar{y}_{lr} = \bar{y}+\frac{\bar{y}}{\bar{x}}(\bar{X}-\bar{x})=\frac{\bar{y}}{\bar{x}}\bar{X}=\hat{\bar{Y}}_R \qquad (7.2)$$

7.2 REGRESSION ESTIMATES WITH PREASSIGNED b

Although, in most applications, b is estimated from the results of the sample, it is sometimes reasonable to choose the value of b in advance. In repeated surveys, previous calculations may have been shown that the sample values of b remain fairly constant; or, if x is the value of y at a recent census, general knowledge of the population may suggest that b is not far from unity, so that $b = 1$ is chosen. Since the sampling theory of regression estimates when b is preassigned is both simple and informative, this case is considered first.

Theorem 7.1. In simple random sampling, in which b_0 is a preassigned constant, the linear regression estimate

$$\bar{y}_{lr} = \bar{y}+b_0(\bar{X}-\bar{x})$$

is unbiased, with variance

$$V(\bar{y}_{lr}) = \frac{1-f}{n} \frac{\sum\limits_{i=1}^{N} [(y_i - \bar{Y}) - b_0(x_i - \bar{X})]^2}{N-1} \tag{7.3}$$

$$= \frac{1-f}{n} (S_y^2 - 2b_0 S_{yx} + b_0^2 S_x^2) \tag{7.4}$$

Note that no assumption is required about the relation between y and x in the finite population.

Proof. Since b_0 is constant in repeated sampling,

$$E(\bar{y}_{lr}) = E(\bar{y}) + b_0 E(\bar{x} - \bar{X}) = \bar{Y} \tag{7.5}$$

by theorem 2.1. Furthermore, $\bar{y}_{lr}$ is the sample mean of the quantities $y_i - b_0(x_i - \bar{X})$, whose population mean is $\bar{Y}$. Hence, by theorem 2.2,

$$V(\bar{y}_{lr}) = \frac{1-f}{n} \cdot \frac{\sum\limits_{i=1}^{N} [(y_i - \bar{Y}) - b_0(x_i - \bar{X})]^2}{N-1} \tag{7.6}$$

$$= \frac{1-f}{n} (S_y^2 - 2b_0 S_{yx} + b_0^2 S_x^2) \tag{7.7}$$

Corollary. *An unbiased sample estimate of $V(\bar{y}_{lr})$ is*

$$v(\bar{y}_{lr}) = \frac{1-f}{n} \frac{\sum\limits_{i=1}^{n} [(y_i - \bar{y}) - b_0(x_i - \bar{x})]^2}{n-1} \tag{7.8}$$

$$= \frac{1-f}{n} (s_y^2 - 2b_0 s_{yx} + b_0^2 s_x^2) \tag{7.9}$$

This follows at once by applying theorem 2.4 to the variate $y_i - b_0(x_i - \bar{X})$.

A natural question at this point is: What is the best value of b_0? The answer is given in theorem 7.2.

Theorem 7.2. The value of b_0 that minimizes $V(\bar{y}_{lr})$ is

$$b_0 = B = \frac{S_{yx}}{S_x^2} = \frac{\sum\limits_{i=1}^{N} (y_i - \bar{Y})(x_i - \bar{X})}{\sum\limits_{i=1}^{N} (x_i - \bar{X})^2} \tag{7.10}$$

which may be called the linear regression coefficient of y on x in the finite population. Note that B does not depend on the properties of any sample that is drawn, and therefore could theoretically be preassigned. The resulting minimum variance is

$$V_{min}(\bar{y}_{lr}) = \frac{1-f}{n} S_y^2 (1-\rho^2) \qquad (7.11)$$

where ρ is the population correlation coefficient between y and x.

Proof. In expression (7.4), for $V(\bar{y}_{lr})$, put

$$b_0 = B + d = \frac{S_{yx}}{S_x^2} + d \qquad (7.12)$$

This gives

$$V(\bar{y}_{lr}) = \frac{1-f}{n} \left[S_y^2 - 2S_{yx}\left(\frac{S_{yx}}{S_x^2} + d\right) + S_x^2\left(\frac{S_{yx}^2}{S_x^4} + 2d\frac{S_{yx}}{S_x^2} + d^2\right)\right]$$

$$= \frac{1-f}{n} \left[\left(S_y^2 - \frac{S_{yx}^2}{S_x^2}\right) + d^2 S_x^2 \right] \qquad (7.13)$$

Clearly, this is minimized when $d = 0$. Since $\rho^2 = S_{yx}^2/S_y^2 S_x^2$,

$$V_{min}(\bar{y}_{lr}) = \frac{1-f}{n} S_y^2 (1-\rho^2) \qquad (7.14)$$

The same analysis may be used to show how far b_0 can depart from B without incurring a substantial loss of precision. From (7.13) and (7.14),

$$V(\bar{y}_{lr}) = \frac{1-f}{n}[S_y^2(1-\rho^2) + (b_0 - B)^2 S_x^2] \qquad (7.15)$$

$$= V_{min}(\bar{y}_{lr})\left[1 + \frac{(b_0 - B)^2 S_x^2}{S_y^2(1-\rho^2)}\right] \qquad (7.16)$$

Since $BS_x = \rho S_y$, this may be written

$$V((\bar{y}_{lr}) = V_{min}(\bar{y}_{lr})\left[1 + \left(\frac{b_0}{B} - 1\right)^2 \frac{\rho^2}{(1-\rho^2)}\right] \qquad (7.17)$$

Thus, if the proportional increase in variance is to be less than α, we must have

$$\left|\frac{b_0}{B} - 1\right| < \sqrt{\alpha(1-\rho^2)/\rho^2} \qquad (7.18)$$

For example, if $\rho = 0.7$, the increase in variance is less than 10%, ($\alpha = 0.1$), provided that

$$\left|\frac{b_0}{B} - 1\right| < \sqrt{(0.1)(0.51)/(0.49)} = 0.32$$

Expression (7.18) makes it clear that in order to ensure a *small* proportional increase in variance b_0/B must be close to 1 if ρ is very high but can depart substantially from 1 if ρ is only moderate.

7.3 REGRESSION ESTIMATES WHEN b IS COMPUTED FROM THE SAMPLE

Theorem 7.2 suggests that if b must be computed from the sample an effective estimate is likely to be the familiar least squares estimate of B, that is,

$$b = \frac{\sum_{i=1}^{n} (y_i - \bar{y})(x_i - \bar{x})}{\sum_{i=1}^{n} (x_i - \bar{x})^2} \tag{7.19}$$

The theory of linear regression plays a prominent part in statistical methodology. The standard results of this theory are not entirely suitable for sample surveys because they require the assumptions that the population regression of y on x is linear, that the residual variance of y about the regression line is constant, and that the population is infinite. If the first two assumptions are violently wrong, a linear regression estimate will probably not be used. However, in surveys in which the regression of y on x is thought to be approximately linear, it is helpful to be able to use $\bar{y}_{lr}$ without having to assume exact linearity or constant residual variance.

Consequently we present an approach that makes no assumption of any specific relation between y_i and x_i. As in the analogous theory for the ratio estimate, only large-sample results are obtained.

With b as in (7.19), the linear regression estimator of $\bar{Y}$ in simple random samples is

$$\bar{y}_{lr} = \bar{y} + b(\bar{X} - \bar{x}) = \bar{y} - b(\bar{x} - \bar{X}) \tag{7.20}$$

The estimator $\bar{y}_{lr}$, like $\bar{y}_R$, will be shown in section 7.7 to have a bias of order $1/n$. In finding the sampling error of $\bar{y}_{lr}$, replace the sample b in (7.20) by the population regression coefficient B in (7.10). In Theorem 7.3 the error committed in this approximation will be shown to be of order $1/\sqrt{n}$ relative to the terms retained. We first examine the relation between b and B.

Introduce the variate e_i defined by the relation

$$e_i = y_i - \bar{Y} - B(x_i - \bar{X}) \tag{7.21}$$

Two properties of the e_i are that $\sum_{}^{N} e_i = 0$ and

$$\sum^{N} e_i(x_i - \bar{X}) = \sum^{N} (y_i - \bar{Y})(x_i - \bar{X}) - B \sum^{N} (x_i - \bar{X})^2 = 0 \tag{7.22}$$

by definition of B. Now

$$b = \sum_{i}^{n} y_i(x_i - \bar{x}) / \sum_{i}^{n} (x_i - \bar{x})^2$$

$$= \sum_{i}^{n} [\bar{Y} + B(x_i - \bar{X}) + e_i](x_i - \bar{x}) / \sum (x_i - \bar{x})^2$$

$$= B + \sum_{i}^{n} e_i(x_i - \bar{x}) / \sum_{i}^{n} (x_i - \bar{x})^2 \tag{7.23}$$

A result needed in Theorem 7.3 is that $(b - B)$ is of order $1/\sqrt{n}$. By theorem 2.3, $\sum_{i}^{n} e_i(x_i - \bar{x})/(n-1)$ is an unbiased estimate of $\sum^{N} e_i(x_i - \bar{X})/(N-1)$ which, by (7.22), is zero. Thus $\sum e_i(x_i - \bar{x})/(n-1)$ is distributed about a zero mean in repeated samples. Since the standard error of a sample covariance is known to be of order $1/\sqrt{n}$, $\sum_{i}^{n} e_i(x_i - \bar{x})/(n-1)$ is of order $1/\sqrt{n}$. But $\sum (x_i - \bar{x})^2/(n-1) = s_x^2$ is of order unity. Hence $(b - B)$, which from (7.23) is the ratio of these two quantities, is of order $1/\sqrt{n}$.

Theorem 7.3. If b is the least squares estimate of B and

$$\bar{y}_{lr} = \bar{y} + b(\bar{X} - \bar{x}) \tag{7.24}$$

then in simple random samples of size n, with n large,

$$V(\bar{y}_{lr}) \doteq \frac{(1-f)}{n} S_y^2(1 - \rho^2) \tag{7.25}$$

where $\rho = S_{yx}/S_y S_x$ is the population correlation between y and x.

Proof. The sampling error of $\bar{y}_{lr}$ arises from the quantity

$$\bar{y}_{lr} - \bar{Y} = \bar{y} - \bar{Y} + b(\bar{X} - \bar{x}) \tag{7.26}$$

As an approximation, replace $\bar{y}_{lr}$ by

$$\tilde{y}_{lr} = \bar{y} + B(\bar{X} - \bar{x}) \tag{7.27}$$

where B is the population linear regression coefficient of y on x. The error committed in this approximation is $(B - b)(\bar{X} - \bar{x})$. This quantity is of order $1/n$ in a simple random sample of size n, since $(b - B)$ and $(\bar{x} - \bar{X})$ are both of order $1/\sqrt{n}$. But the sampling error in $\bar{y}_{lr}$ is of order $1/\sqrt{n}$, since it is the error in the sample mean of the variate $(y_i - Bx_i)$. Hence the leading term in $E(\bar{y}_{lr} - \bar{Y})^2$ is $V(\tilde{y}_{lr})$. By (7.11), in large samples,

$$E(\bar{y}_{lr} - \bar{Y})^2 \doteq V(\tilde{y}_{lr}) = \frac{(1-f)}{n} S_y^2(1 - \rho^2) \tag{7.28}$$

7.4 SAMPLE ESTIMATE OF VARIANCE

As a sample estimate of $V(\bar{y}_{lr})$, valid in large samples, we may use

$$v(\bar{y}_{lr}) = \frac{1-f}{n(n-2)} \sum_{i=1}^{n} [(y_i - \bar{y}) - b(x_i - \bar{x})]^2 \tag{7.29}$$

$$= \frac{1-f}{n(n-2)} \left\{ \sum (y_i - \bar{y})^2 - \frac{[\sum (y_i - \bar{y})(x_i - \bar{x})]^2}{\sum (x_i - \bar{x})^2} \right\} \tag{7.30}$$

the latter being the usual short-cut computing formula. The derivation is as follows.

In theorem 7.3, equation (7.28), we had, since $S_y^2(1 - \rho^2) = S_e^2$,

$$V(\bar{y}_{lr}) \doteq \frac{(1-f)}{n} S_e^2$$

From theorem 2.4, an unbiased estimate of S_e^2 is

$$s_e^2 = \frac{1}{n-1} \sum_{i=1}^{n} (e_i - \bar{e})^2$$

Now, from equation (7.21), it follows that

$$e_i - \bar{e} = (y_i - \bar{y}) - B(x_i - \bar{x}) = [(y_i - \bar{y}) - b(x_i - \bar{x})] + (b - B)(x_i - \bar{x}) \tag{7.31}$$

The second term on the right, of order $1/\sqrt{n}$, may be neglected in relation to the first term, which is of order unity. Hence in large samples we may use

$$\frac{1}{n-1} \sum_{i=1}^{n} [(y_i - \bar{y}) - b(x_i - \bar{x})]^2 \tag{7.32}$$

as an estimate of S_e^2. The divisor $(n-2)$ instead of $(n-1)$ is suggested in (7.29) and (7.30) because it is used in standard regression theory and is known to give an unbiased estimate of S_e^2 if the population is infinite and the regression is linear.

7.5 LARGE-SAMPLE COMPARISON WITH THE RATIO ESTIMATE AND THE MEAN PER UNIT

For these comparisons the sample size n must be large enough so that the approximate formulas for the variances of the ratio and regression estimates are valid. The three comparable variances for the estimated population mean $\bar{Y}$ are as follows.

$$V(\bar{y}_{lr}) = \frac{N-n}{Nn} S_y^2 (1-\rho^2) \qquad \text{(regression)}$$

$$V(\bar{y}_R) = \frac{N-n}{Nn} (S_y^2 + R^2 S_x^2 - 2R\rho S_y S_x) \qquad \text{(ratio)}$$

$$V(\bar{y}) = \frac{N-n}{Nn} S_y^2 \qquad \text{(mean per unit)}$$

It is apparent that the variance of the regression estimate is smaller than that of the mean per unit unless $\rho = 0$, in which case the two variances are equal.

The variance of the regression estimate is less than that of the ratio estimate if

$$-\rho^2 S_y^2 < R^2 S_x^2 - 2R\rho S_y S_x \qquad (7.33)$$

This is equivalent to the inequalities

$$(\rho S_y - RS_x)^2 > 0 \qquad \text{or} \qquad (B-R)^2 > 0 \qquad (7.34)$$

Thus the regression estimate is more precise than the ratio estimate unless $B = R$. This occurs when the relation between y_i and x_i is a straight line through the origin.

Example. The precision of the regression, ratio, and mean per unit estimates from a simple random sample can be compared by using data collected in the complete enumeration of peach orchards described on p. 172. In this example, y_i is the estimated peach production in an orchard and x_i the number of peach trees in the orchard. We will compare the estimates of the total production of the 256 orchards, made from a sample of 100 orchards. It is doubtful whether the sample is large enough to make the variance formulas fully valid, since the cv's of $\bar{y}$ and $\bar{x}$ are both somewhat higher than 10%, but the example will serve to illustrate the computations. The basic data are as follows.

$$S_y^2 = 6409 \qquad S_{yx} = 4434 \qquad S_x^2 = 3898$$

$$R = 1.270 \qquad \rho = 0.887 \qquad n = 100 \qquad N = 256$$

$$V(\hat{Y}_{lr}) = \frac{N(N-n)}{n} S_y^2 (1-\rho^2)$$

$$= \frac{(256)(156)}{100} (6409)(1-0.787) = 545{,}000$$

$$V(\hat{Y}_R) = \frac{N(N-n)}{n} (S_y^2 + R^2 S_x^2 - 2RS_{yx})$$

$$= \frac{(256)(156)}{100} [6409 + (1.613)(3898) - 2(1.270)(4434)]$$

$$= 573{,}000$$

$$V(\hat{Y}) = \frac{N(N-n)}{n} S_y^2 = 2{,}559{,}000$$

There is little to choose between the regression and the ratio estimates, as might be expected from the nature of the variables. Both techniques are greatly superior to the mean per unit.

The preceding result on the superiority of the linear regression estimate is strictly a large-sample result. In small samples on natural populations the regression estimate appears disappointing in performance. In eight natural populations of the type in which the ratio estimate has been used. Rao (1969) found in a Monte Carlo study that the average of the ratios $MSE(\hat{Y}_{lr})/MSE(\hat{Y}_R)$ was 1.15 for $n = 12$, 1.36 for $n = 8$, and 1.51 for $n = 6$. These lower efficiencies of $\hat{Y}_{lr}$ were not due to greater biases in the regression estimates, the corresponding variance ratios being almost the same.

7.6 ACCURACY OF THE LARGE-SAMPLE FORMULAS FOR $V(\bar{y}_{lr})$ AND $v(\bar{y}_{lr})$

No general analytical results are available on the accuracy of the approximate formulas (7.25) for $V(\bar{y}_{lr})$ and (7.29) for $v(\bar{y}_{lr})$ in moderate or small samples. The approximate estimators in (7.25) and (7.29) are

$$V(\bar{y}_{lr}) = \frac{(1-f)}{n} S_y^2 (1 - \rho^2) \tag{7.25}$$

$$v(\bar{y}_{lr}) = \frac{1-f}{n(n-2)} \sum_i^n [(y_i - \bar{y}) - b(x_i - \bar{x})]^2 \tag{7.29}$$

Suppose that the y_i for $i = 1, 2, \ldots N$ are a random sample from an infinite population under the model

$$y_i = \alpha + \beta x_i + \varepsilon_i \tag{7.35}$$

where for fixed x_i, the ε_i are independently distributed with mean 0, variance $\sigma_\varepsilon^2 = \sigma_y^2 (1 - \rho^2)$. With this model, Cochran (1942) gave the result that to terms of order $1/n^2$.

$$EV(\bar{y}_{lr}) = \frac{(1-f)}{n} \sigma_y^2 (1 - \rho^2) \left(1 + \frac{1}{n-3} + \frac{2G_1^2}{n^2} \right) \tag{7.36}$$

where $G_1 = k_{3x}/\sigma_x^3$ is Fisher's measure of relative skewness of the distribution of x. Since $S_y^2(1 - \rho^2)$ in (7.25) is an unbiased estimate of $\sigma_y^2(1 - \rho^2)$ under this model, (7.36) suggests that with x symmetrically distributed the percent underestimation by $V(\bar{y}_{lr})$ is $100/(n-2)$ with this model.

From Monte Carlo studies on eight small natural populations, (Rao, 1968), Table 7.1 shows for $n = 6, 8, 12$, the average percent underestimation of the variance of $\bar{y}_{lr}$ by the approximations $V(\bar{y}_{lr})$ and $v(\bar{y}_{lr})$.

TABLE 7.1

AVERAGE PERCENT UNDERESTIMATION OF THE VARIANCE OF $\bar{y}_{lr}$

Estimator	n		
	6	8	12
$V(\bar{y}_{lr})$ in (7.25)	38	34	28
$v(\bar{y}_{lr})$ in (7.29)	48	42	33

For x_i symmetrical, formula (7.36) suggests percent underestimations by $V(\bar{y}_{lr})$ of 25, 17, and 10% for $n = 6$, 8, 12. The percents for $V(\bar{y}_{lr})$ in Table 7.1 are substantially higher, by amounts judged unlikely to be accounted for by skewness in x in these populations—more likely by deficiencies in the linear model. The underestimations are still greater for the sample estimates of variance $v(\bar{y}_{lr})$. Furthermore, comparison with Table 6.2, page 164, which applies to the ratio estimate in the same eight populations, shows that the percent underestimations in V and v for $\bar{y}_{lr}$ are at least twice those for $\bar{y}_R$ in samples of the same size.

7.7 BIAS OF THE LINEAR REGRESSION ESTIMATE

The estimator $\bar{y}_{lr}$ has a bias of order $1/n$ in simple random sampling. We have

$$E(\bar{y}_{lr}) = \bar{Y} - Eb(\bar{x} - \bar{X}) \tag{7.37}$$

Thus one expression for the bias is $-Eb(\bar{x} - \bar{X}) = -\text{cov}(b, \bar{x})$. The leading term in the bias turns out to be

$$\frac{-(1-f)}{n} \frac{Ee_i(x_i - \bar{X})^2}{S_x^2} \tag{7.38}$$

This term represents a contribution from the *quadratic* component of the regression of y on $\bar{x}$. Thus, if a sample plot of y_i against x_i appears approximately linear, there should be little risk of major bias in $\bar{y}_{lr}$.

To show (7.38) requires some algebraic development. By (7.23), page 194,

$$b = B + \frac{\sum_{i}^{n} e_i(x_i - \bar{x})}{\sum_{i}^{n}(x_i - \bar{x})^2} \tag{7.39}$$

Replace $\sum_{i}^{n}(x_i - \bar{x})^2$ by its leading term, nS_x^2. Also write

$$\sum_{i}^{n} e_i(x_i - \bar{x}) = \sum_{i}^{n} e_i(x_i - \bar{X}) + n\bar{e}(\bar{X} - \bar{x}) \tag{7.40}$$

Hence the leading term in the bias $-Eb(\bar{x}-\bar{X})$ of $\bar{y}_{lr}$ is the average of

$$\frac{-\sum\limits_{}^{n} e_i(x_i-\bar{X})(\bar{x}-\bar{X})}{nS_x^2}+\frac{\bar{e}(\bar{x}-\bar{X})^2}{S_x^2} \qquad (7.41)$$

Let $u_i = e_i(x_i-\bar{X})$. By (7.22) its population mean $\bar{U}=0$. The average value of the *first* term in (7.41) may therefore be written

$$\frac{-E(\bar{u}-\bar{U})(\bar{x}-\bar{X})}{S_x^2}=-\frac{(1-f)}{n}\frac{E(u_i-\bar{U})(x_i-\bar{X})}{S_x^2} \qquad (7.42)$$

by theorem (2.3) (p. 25) for the average value of a sample covariance in simple random sampling. This in turn equals (7.38), namely

$$-\frac{(1-f)}{n}\frac{Ee_i(x_i-\bar{X})^2}{S_x^2}$$

In the second term in (7.41), $\bar{e}$ is $O(1/\sqrt{n})$ and $(\bar{x}-\bar{X})^2$ is $O(1/n)$, so that this term is of smaller order than (7.38). Thus (7.38) is the leading term in the bias of $\bar{y}_{lr}$.

7.8 THE LINEAR REGRESSION ESTIMATOR UNDER A LINEAR REGRESSION MODEL

Suppose that the finite population values y_i $(i = 1, 2, \ldots N)$ are randomly drawn from an infinite superpopulation in which

$$y = \alpha + \beta x + \varepsilon \qquad (7.43)$$

where the ε are independent, with means 0 and variance σ_ε^2 for fixed x. By direct substitution from the model we find that

$$b=\frac{\sum\limits_{}^{n} y_i(x_i-\bar{x})}{\sum\limits_{}^{n}(x_i-\bar{x})^2}=\beta+\frac{\sum\limits_{}^{n}\varepsilon_i(x_i-\bar{x})}{\sum\limits_{}^{n}(x_i-\bar{x})^2} \qquad (7.44)$$

$$\bar{y}_{lr}-\bar{Y}=(\bar{\varepsilon}_n-\bar{\varepsilon}_N)+(\bar{X}-\bar{x})\frac{\sum\limits_{}^{n}\varepsilon_i(x_i-\bar{x})}{\sum\limits_{}^{n}(x_i-\bar{x})^2} \qquad (7.45)$$

where $\bar{\varepsilon}_n$ and $\bar{\varepsilon}_N$ are means over the sample and the finite population. It follows from (7.45) that under this model, $E(\bar{y}_{lr}-\bar{Y})=0$, so that $\bar{y}_{lr}$ is model-unbiased for any size of sample.

As regards variance, it follows from (7.45) that for a given set of x's,

$$V(\bar{y}_{lr})=E(\bar{y}_{lr}-\bar{Y})^2=\sigma_\varepsilon^2\left[\left(\frac{1}{n}-\frac{1}{N}\right)+\frac{(\bar{X}-\bar{x})^2}{\sum\limits_{}^{n}(x_i-\bar{x})^2}\right]. \qquad (7.46)$$

This result holds for any $n > 1$ and any sample selected solely by the values of x. This approach and its generalization to the case of unequal residual variances were given by Royall (1970). Under this model a purposive sampling plan that succeeded in making $\bar{x} = \bar{X}$ would minimize $V(\bar{y}_{lr})$ for given n.

Also, for any sample selected solely according to the values of the x_i, the usual least squares estimator

$$s_\varepsilon^2 = \sum_{}^{n} [(y_i - \bar{y}) - b(x_i - \bar{x})]^2/(n-2) \tag{7.47}$$

is a model-unbiased estimator of σ_ε^2 for $n > 2$.

Thus, in problems in which this model applies, simple exact results about the mean and variance of $\bar{y}_{lr}$ can be established, valid for any sample size > 2 and requiring only sample selection according to the values of the x_i, the random element being supplied gratis by the distribution of the ε's assumed in the model.

7.9   REGRESSION ESTIMATES IN STRATIFIED SAMPLING

As with the ratio estimate, two types of regression estimate can be made in stratified random sampling. In the first estimate $\bar{y}_{lrs}$ (s for separate), a separate regression estimate is computed for each stratum mean, that is,

$$\bar{y}_{lrh} = \bar{y}_h + b_h(\bar{X}_h - \bar{x}_h) \tag{7.48}$$

Then, with $W_h = N_h/N$,

$$\bar{y}_{lrs} = \sum_h W_h \bar{y}_{lrh} \tag{7.49}$$

This estimate is appropriate when it is thought that the true regression coefficients B_h vary from stratum to stratum.

The second regression estimate, $\bar{y}_{lrc}$ (c for combined), is appropriate when the B_h are presumed to be the same in all strata. To compute $\bar{y}_{lrc}$, we first find

$$\bar{y}_{st} = \sum_h W_h \bar{y}_h \qquad \bar{x}_{st} = \sum_h W_h \bar{x}_h$$

Then

$$\bar{y}_{lrc} = \bar{y}_{st} + b(\bar{X} - \bar{x}_{st}) \tag{7.50}$$

The two estimates will be considered first in the case in which the b_h and b are chosen in advance, since their properties are unusually simple in this situation. From section 7.2, $\bar{y}_{lrh}$ is an unbiased estimate of $\bar{Y}_h$, so that $\bar{y}_{lrs}$ is an unbiased estimate of $\bar{Y}$. Since sampling is independent in different strata, it follows from theorem 7.1 that

$$V((\bar{y}_{lrs}) = \sum_h \frac{W_h^2(1-f_h)}{n_h}(S_{yh}^2 - 2b_h S_{yxh} + b_h^2 S_{xh}^2) \tag{7.51}$$

Theorem 7.2 shows that $V(\bar{y}_{lrs})$ is minimized when $b_h = B_h$, the true regression coefficient in stratum h. The minimum value of the variance may be written

$$V_{min}(\bar{y}_{lrs}) = \sum_h \frac{W_h^2(1-f_h)}{n_h}\left(S_{yh}^2 - \frac{S_{yxh}^2}{S_{xh}^2}\right) \qquad (7.52)$$

Turning to the combined estimate with preassigned b, (7.50) shows that $\bar{y}_{lrc}$ is also an unbiased estimate of $\bar{Y}$ in this case. Since $\bar{y}_{lrc}$ is the usual estimate from a stratified sample for the variate $y_{hi} + b(\bar{X} - x_{hi})$, we may apply theorem 5.3 to this variate, giving the result

$$V(\bar{y}_{lrc}) = \sum_h \frac{W_h^2(1-f_h)}{n_h}(S_{yh}^2 - 2bS_{yxh} + b^2S_{xh}^2) \qquad (7.53)$$

The value of b that minimizes this variance is

$$B_c = \sum_h \frac{W_h^2(1-f_h)S_{yxh}}{n_h} \Big/ \sum_h \frac{W_h^2(1-f_h)S_{xh}^2}{n_h} \qquad (7.54)$$

The quantity B_c is a weighted mean of the stratum regression coefficients $B_h = S_{yxh}/S_{xh}^2$. If we write

$$a_h = \frac{W_h^2(1-f_h)}{n_h}S_{xh}^2$$

then $B_c = \sum a_h B_h / \sum a_h$.

From (7.52) and (7.53), with B_c in place of b, we find

$$V_{min}(\bar{y}_{lrc}) - V_{min}(\bar{y}_{lrs}) = \sum a_h B_h^2 - \left(\sum a_h\right) V_c^2$$

$$= \sum a_h (B_h - B_c)^2 \qquad (7.55)$$

This result shows that with the optimum choices the separate estimate has a smaller variance than the combined estimate unless B_h is the same in all strata. These optimum choices would, of course, require advance knowledge of the S_{yxh} and S_{xh}^2 values.

7.10 REGRESSION COEFFICENTS ESTIMATED FROM THE SAMPLE

The preceding analysis is helpful in indicating the type of sample estimates b_h and b that may be efficient when used in regression estimates. With the separate estimate, the analysis suggests that we take

$$b_h = \frac{\sum_i (y_{hi} - \bar{y}_h)(x_{hi} - \bar{x}_h)}{\sum_i (x_{hi} - \bar{x}_h)^2} \qquad (7.56)$$

the within-stratum least squares estimate of B_h.

Applying theorem 7.3 to each stratum, we have

$$V(\bar{y}_{lrs}) = \sum_h \frac{W_h^2(1-f_h)}{n_h} S_{yh}^2(1-\rho_h^2) \tag{7.57}$$

provided that the sample size n_h is large in all strata. To obtain a sample estimate of variance, substitute

$$s_{y\cdot xh}^2 = \frac{1}{n_h-2}\left[\sum_i (y_{hi}-\bar{y}_h)^2 - b_h^2 \sum_i (x_{hi}-\bar{x}_h)^2\right] \tag{7.58}$$

in place of $S_{yh}^2(1-\rho_h^2)$ in (7.57).

The estimate $\bar{y}_{lrs}$ suffers from the same difficulty as the corresponding ratio estimate, in that the ratio of the bias to the standard error may become appreciable. It follows from section 7.7 that the regression estimates $\bar{y}_{lrh}$ in the individual strata may have biases of order $1/n_h$ and the biases may be of the same sign in all strata, so that the over-all bias in $\bar{y}_{lrs}$ may also be of order $1/n_h$. Since the leading term in the bias comes from the quadratic regression of y_{hi} on x_{hi}, as shown in section 7.7, this danger is most acute when the relation between the variates approximates the quadratic rather than the linear type.

With the combined estimate, we saw that the variance is minimized when $b = B_c$, as defined in (7.54). This suggests that we take

$$b_c = \sum_h \frac{W_h^2(1-f_h)}{n_h(n_h-1)} \sum_i (y_{hi}-\bar{y}_h)(x_{hi}-\bar{x}_h) \bigg/ \sum_h \frac{W_h^2(1-f_h)}{n_h(n_h-1)} \sum_i (x_{hi}-\bar{x}_h)^2$$

as a sample estimate of B_c. If the stratification is proportional and if we may replace the (n_h-1) in b_c by n_h, b_c reduces to the familiar pooled least squares estimate

$$b_c' = \sum_h \sum_i (y_{hi}-\bar{y}_h)(x_{hi}-\bar{x}_h) \bigg/ \sum_h \sum_i (x_{hi}-\bar{x}_h)^2$$

In certain circumstances other estimates may be preferable to b_c or b_c'. For instance, if the true regression coefficients B_h are the same in all strata but the residual variances about the regression line differ substantially from one stratum to another, a different weighted mean of the b_h, weighting inversely as the estimated variance, may be more precise. However, the gain in precision as it affects $\bar{y}_{lrc}$ is likely to be small.

Since

$$\begin{aligned}
\bar{y}_{lrc} - \bar{Y} &= \bar{y}_{st} - \bar{Y} + b_c(\bar{X}-\bar{x}_{st}) \\
&= [\bar{y}_{st} - \bar{Y} + B_c(\bar{X}-\bar{x}_{st})] + (b_c - B_c)(\bar{X}-\bar{x}_{st}) \tag{7.59}
\end{aligned}$$

it follows that if sampling errors of b_c are negligible

$$V(\bar{y}_{lrc}) = \sum_h \frac{W_h^2(1-f_h)}{n_h}(S_{yh}^2 - 2B_c S_{yxh} + B_c^2 S_{xh}^2) \tag{7.60}$$

As an estimate of $V(\bar{y}_{lrc})$, we may take

$$v(\bar{y}_{lrc}) = \sum_h \frac{W_h^2(1-f_h)}{n_h(n_h-1)} \sum_i [(y_{hi} - \bar{y}_h) - b_c(x_{hc} - \bar{x}_h)]^2 \qquad (7.61)$$

7.11 COMPARISON OF THE TWO TYPES OF REGRESSION ESTIMATE

Hard and fast rules cannot be given to decide whether the separate or the combined estimate is better in any specific situation. The defects of the separate estimate are that it is more liable to bias when samples are small within the individual strata and that its variance has a larger contribution from sampling errors in the regression coefficients. The defect of the combined estimate is that its variance is inflated if the population regression coefficients differ from stratum to stratum.

If we are confident that the regressions are linear and if B_h appears to be roughly the same in all strata, the combined estimate is to be preferred. If the regressions appear linear (so that the danger of bias seems small) but B_h seems to vary markedly from stratum to stratum, the separate estimate is advisable. If there is some curvilinearity in the regressions when a linear regression estimate is used, the combined estimate is probably safer unless the samples are large in all strata.

Estimators of the regression type that are unbiased have been developed by Mickey (1959) and Williams (1963), but have not yet been extensively tried. Rao (1969) found Mickey's estimator usually inferior to the standard regression and ratio estimators in natural populations. A jackknifed version can also be constructed. With $n = mg$, let $\bar{y}'_{(lr)j}$ be the standard regression estimate, computed from the sample with group j omitted, $(j = 1, 2, \ldots, g)$. Then the jackknifed form is

$$\bar{y}_{(lr)Q} = g\bar{y}_{lr} - (g-1)\left(\sum^g \bar{y}'_{(lr)j}\right)\Big/ g \qquad (7.62)$$

EXERCISES

7.1 An experienced farmer makes an eye estimate of the weight of peaches x_i on each tree in an orchard of $N = 200$ trees. He finds a total weight of $X = 11,600$ lb. The peaches are picked and weighed on a simple random sample of 10 trees, with the following results.

		1	2	3	4	5	6	7	8	9	10	Total
						Tree Number						
Actual wt.	y_i	61	42	50	58	67	45	39	57	71	53	543
Est. wt.	x_i	59	47	52	60	67	48	44	58	76	58	569

As an estimate of the total actual weight Y, we take

$$\hat{Y} = N[\bar{X} + (\bar{y} - \bar{x})]$$

Compute the estimate and find its standard error.

7.2   Does it appear that the linear regression estimate, with the sample least squares b, would give a more precise estimate in 7.1?

7.3   From the sample data in Table 6.1 (p. 152) compute the regression estimate of the 1930 total number of inhabitants in the 196 large cities. Find the approximate standard error of this estimate and compare its precision with that of the ratio estimate.

7.4   In exercise 7.3 find the estimated total number of inhabitants and its standard error if b is taken as 1.

7.5   In the following population with $N = 5$, verify (a) that the regression of y on x is linear and (b) that the linear regression estimate is unbiased in simple random samples with $n = 3$. The (y, x) pairs are $(3, 0)$, $(5, 0)$, $(8, 2)$, $(8, 3)$, $(12, 3)$.

7.6   A rough measurement x, made on each unit, is related to the true measurement y on the unit by the equation

$$x = y + e + d$$

where d is a constant bias and e is an error of measurement, uncorrelated with y, which has mean zero and variance S_e^2 in the population, assumed infinite. In simple random samples of size n compare the variances of (a) the "difference" estimate $[\bar{y} + (X - \bar{x})]$ of the mean $\bar{Y}$ and (b) the linear regression estimate, using the value of b that gives minimum variance. (The variances may depend on S_y^2.)

7.7   By working out all possible cases, compare the MSE's of the separate and combined regression estimates of the total Y of the following population, when simple random samples of size 2 are drawn from each stratum. For each estimate, how much does its bias contribute to the MSE?

Stratum 1		Stratum 2	
x_{1i}	y_{1i}	x_{2i}	y_{2i}
4	0	5	7
6	3	6	12
7	5	8	13

Use the ordinary least squares estimates of the B's, b_h and b_c on pp. 201–2.

7.8   In the population of exercise 7.7, show that if the optimum preassigned B could be used in each case, $V(\hat{Y}_{lrs}) = 4.39$, $V(\hat{Y}_{lrc}) = 4.43$, both estimates being, of course, unbiased.

7.9   By the same method, compare the MSE's of the separate and combined *ratio* estimates in the population in exercise 7.7. Since the ratio Y/X is $8/17 = 0.47$ in stratum 1 and $32/19 = 1.68$ in stratum 2, large sample theory would suggest that $\hat{Y}_{Rs}$ would be superior to $\hat{Y}_{Rc}$. You will find, however, that in these tiny samples, $\hat{Y}_{Rc}$ has the smaller MSE. Its superiority is not due to smaller bias, neither $\hat{Y}_{Rs}$ nor $\hat{Y}_{Rc}$ being materially biased. As another disagreement with large-sample theory, you will find that $\hat{Y}_{Rs}$ and $\hat{Y}_{Rc}$ have smaller MSE's than the corresponding regression estimates.

CHAPTER 8

Systematic Sampling

8.1 DESCRIPTION

This method of sampling is at first sight quite different from simple random sampling. Suppose that the N units in the population are numbered 1 to N in some order. To select a sample of n units, we take a unit at random from the first k units and every kth unit thereafter. For instance, if k is 15 and if the first unit drawn is number 13, the subsequent units are numbers 28, 43, 58, and so on. The selection of the first unit determines the whole sample. This type is called an *every kth* systematic sample.

The apparent advantages of this method over simple random sampling are as follows.

1. It is easier to draw a sample and often easier to execute without mistakes. This is a particular advantage when the drawing is done in the field. Even when drawing is done in an office there may be a substantial saving in time. For instance, if the units are described on cards that are all of the same size and lie in a file drawer, a card can be drawn out every inch along the file as measured by a ruler. This operation is speedy, whereas simple random sampling would be slow. Of course, this method departs slightly from the strict "every kth" rule.

2. Intuitively, systematic sampling seems likely to be more precise than simple random sampling. In effect, it stratifies the population into n strata, which consist of the first k units, the second k units, and so on. We might therefore expect the systematic sample to be about as precise as the corresponding stratified random sample with *one* unit per stratum. The difference is that with the systematic sample the units occur at the same relative position in the stratum, whereas with the stratified random sample the position in the stratum is determined separately by randomization within each stratum (see Fig. 8.1). The systematic sample is spread more evenly over the population, and this fact has sometimes made systematic sampling considerably more precise than stratified random sampling.

One variant of the systematic sample is to choose each unit at or near the center of the stratum; that is, instead of starting the sequence by a random number chosen between 1 and k, we take the starting number as $(k+1)/2$ if k is odd and

SYSTEMATIC SAMPLING

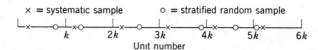

Fig. 8.1 Systematic and stratified random sampling.

either $k/2$ or $(k+2)/2$ if k is even (Madow, 1953). This procedure carries the idea of systematic sampling to its logical conclusion. If y_i can be considered a continuous function of a continuous variable i, there are grounds for expecting that this centrally located sample will be more precise than one randomly located. Limited investigation on some natural populations supports this opinion, although centrally located samples tend to behave erratically. Attention here will be confined to samples with some random element.

TABLE 8.1

THE POSSIBLE SYSTEMATIC SAMPLES FOR $N = 23$, $k = 5$

Systematic sample number

I	II	III	IV	V
1	2	3	4	5
6	7	8	9	10
11	12	13	14	15
16	17	18	19	20
21	22	23		

Since N is not in general an integral multiple of k, different systematic samples from the same finite population may vary by one unit in size. Thus, with $N = 23$, $k = 5$, the numbers of the units in the five systematic samples are shown in Table 8.1. The first three samples have $n = 5$ and the last two have $n = 4$. This fact introduces a disturbance into the theory of systematic sampling. The disturbance is probably negligible if n exceeds 50 and will be ignored, for simplicity, in the presentation of theory. It is unlikely to be large even when n is small.

Another method, suggested by Lahiri in 1952 (see Murthy, 1967, p. 139) provides both a constant sample size and an unbiased sample mean. Regard the N units as arranged round a circle and let k now be the integer nearest to N/n. Select a random number between 1 and N and take every kth unit thereafter, going round the circle until the desired n units have been chosen. Suppose we want $n = 5$ with $N = 23$. Then $k = 5$. If the random number is 19 we take units 19, 1, 6, 11, 16. It is easily verified that every unit has an equal probability of selection with this method. If $n = 4$ units are wanted with $N = 23$ we take $k = 6$.

8.2 RELATION TO CLUSTER SAMPLING

There is another way of looking at systematic sampling. With $N = nk$, the k possible systematic samples are shown in the columns of Table 8.2. It is evident from this table that the population has been divided into k large sampling units, each of which contains n of the original units. The operation of choosing a randomly located systematic sample is just the operation of choosing *one* of these large sampling units at random. Thus systematic sampling amounts to the selection of a *single* complex sampling unit that constitutes the whole sample. A systematic sample is a simple random sample of one cluster unit from a population of k cluster units.

TABLE 8.2

COMPOSITION OF THE k SYSTEMATIC SAMPLES

Sample number

	1	2	$\cdots$	i	$\cdots$	k
	y_1	y_2		y_i		y_k
	y_{k+1}	y_{k+2}		y_{k+i}		y_{2k}
	$\cdots$	$\cdots$		$\cdots$		$\cdots$
	$y_{(n-1)k+1}$	$y_{(n-1)k+2}$		$y_{(n-1)k+i}$		y_{nk}
Means	$\bar{y}_1$	$\bar{y}_2$		$\bar{y}_i$		$\bar{y}_k$

8.3 VARIANCE OF THE ESTIMATED MEAN

Several formulas have been developed for the variance of $\bar{y}_{sy}$, the mean of a systematic sample. The three given below apply to any kind of cluster sampling in which the clusters contain n elements and the sample consists of one cluster. In these formulas we assume $N = nk$.

If $N = nk$, it is easy to verify that $\bar{y}_{sy}$ is an unbiased estimate of $\bar{Y}$ for a randomly located systematic sample.

In the following analysis the symbol y_{ij} denotes the jth member of the ith systematic sample, so that $j = 1, 2, \ldots, n$, $i = 1, 2, \ldots, k$. The mean of the ith sample is denoted by $\bar{y}_i$.

Theorem 8.1. The variance of the mean of a systematic sample is

$$V(\bar{y}_{sy}) = \frac{N-1}{N} S^2 - \frac{k(n-1)}{N} S^2_{wsy} \qquad (8.1)$$

where

$$S^2_{wsy} = \frac{1}{k(n-1)} \sum_{i=1}^{k} \sum_{j=1}^{n} (y_{ij} - \bar{y}_{i.})^2$$

is the variance among units that lie within the same systematic sample. The denominator of this variance, $k(n-1)$, is constructed by the usual rules in the analysis of variance: each of the k samples contributes $(n-1)$ degrees of freedom to the sum of squares in the numerator.

Proof. By the usual identity of the analysis of variance

$$(N-1)S^2 = \sum_i \sum_j (y_{ij} - \bar{Y})^2$$

$$= n \sum_i (\bar{y}_{i.} - \bar{Y})^2 + \sum_i \sum_j (y_{ij} - \bar{y}_{i.})^2$$

But the variance of $\bar{y}_{sy}$ is by definition

$$V(\bar{y}_{sy}) = \frac{1}{k} \sum_{i=1}^{k} (\bar{y}_{i.} - \bar{Y})^2$$

Hence

$$(N-1)S^2 = nk V(\bar{y}_{sy}) + k(n-1)S^2_{wsy} \tag{8.2}$$

The result follows.

Corollary. The mean of a systematic sample is more precise than the mean of a simple random sample if and only if

$$S^2_{wsy} > S^2$$

Proof. If $\bar{y}$ is the mean of a simple random sample of size n,

$$V(\bar{y}) = \frac{N-n}{N} \frac{S^2}{n}$$

From (8.1), $V(\bar{y}_{sy}) < V(\bar{y})$ if and only if

$$\frac{N-1}{N}S^2 - \frac{k(n-1)}{N}S^2_{wsy} < \frac{N-n}{N} \frac{S^2}{n} \tag{8.3}$$

that is, if

$$k(n-1)S^2_{wsy} > \left(N-1-\frac{N-n}{n}\right)S^2 = k(n-1)S^2 \tag{8.4}$$

This important result, which applies to cluster sampling in general, states that systematic sampling is more precise than simple random sampling if the variance within the systematic samples is *larger* than the population variance as a whole. Systematic sampling is precise when units within the same sample are heterogeneous and is imprecise when they are homogeneous. The result is obvious intuitively. If there is little variation within a systematic sample relative to that in the population, the successive units in the sample are repeating more or less the same information.

Another form for the variance is given in theorem 8.2.

Theorem 8.2.

$$V(\bar{y}_{sy}) = \frac{S^2}{n}\left(\frac{N-1}{N}\right)[1+(n-1)\rho_w] \qquad (8.5)$$

where ρ_w is the correlation coefficient between pairs of units that are in the same systematic sample. It is defined as

$$\rho_w = \frac{E(y_{ij}-\bar{Y})(y_{iu}-\bar{Y})}{E(y_{ij}-\bar{Y})^2} \qquad (8.6)$$

where the numerator is averaged over all $kn(n-1)/2$ distinct pairs, and the denominator over all N values of y_{ij}. Since the denominator is $(N-1)S^2/N$, this gives

$$\rho_w = \frac{2}{(n-1)(N-1)S^2}\sum_{i=1}^{k}\sum_{j<u}(y_{ij}-\bar{Y})(y_{iu}-\bar{Y}) \qquad (8.7)$$

Proof.

$$n^2kV(\bar{y}_{sy}) = n^2\sum_{i=1}^{k}(\bar{y}_{i.}-\bar{Y})^2$$

$$= \sum_{i=1}^{k}[(y_{i1}-\bar{Y})+(y_{i2}-\bar{Y})+\cdots+(y_{in}-\bar{Y})]^2$$

The squared terms amount to the total sum of squares of deviations from $\bar{Y}$, that is, to $(N-1)S^2$. This gives

$$n^2kV(\bar{y}_{sy}) = (N-1)S^2+2\sum_{i}\sum_{j<u}(y_{ij}-\bar{Y})(y_{iu}-\bar{Y}) \qquad (8.8)$$

$$= (N-1)S^2+(n-1)(N-1)S^2\rho_w \qquad (8.9)$$

Hence

$$V(\bar{y}_{sy}) = \frac{S^2}{n}\left(\frac{N-1}{N}\right)[1+(n-1)\rho_w] \qquad (8.10)$$

This result shows that positive correlation between units in the same sample inflates the variance of the sample mean. Even a small positive correlation may have a large effect, because of the multiplier $(n-1)$.

The two preceding theorems express $V(\bar{y}_{sy})$ in terms of S^2, hence relate it to the variance for a simple random sample. There is an analogue of theorem 8.2 that expresses $V(\bar{y}_{sy})$ in terms of the variance for a stratified random sample in which the strata are composed of the first k units, the second k units, and so on. In our notation the subscript j in y_{ij} denotes the stratum. The stratum mean is written $\bar{y}_{.j}$.

Theorem 8.3.

$$V(\bar{y}_{sy}) = \frac{S^2_{wst}}{n}\left(\frac{N-n}{N}\right)[1+(n-1)\rho_{wst}] \qquad (8.11)$$

where

$$S^2_{wst} = \frac{1}{n(k-1)} \sum_{j=1}^{n} \sum_{i=1}^{k} (y_{ij} - \bar{y}_{.j})^2 \qquad (8.12)$$

This is the variance among units that lie in the same stratum. The divisor $n(k-1)$ is used because each of the n strata contributes $(k-1)$ degrees of freedom. Furthermore,

$$\rho_{wst} = \frac{E(y_{ij} - \bar{y}_{.j})(y_{iu} - \bar{y}_{.u})}{E(y_{ij} - \bar{y}_{.j})^2} \qquad (8.13)$$

This quantity is the correlation between the deviations from the stratum means of pairs of items that are in the same systematic sample.

$$\rho_{wst} = \frac{2}{n(n-1)(k-1)} \sum_{i=1}^{k} \sum_{j<u} \frac{(y_{ij} - \bar{y}_{.j})(y_{iu} - \bar{y}_{.u})}{S^2_{wst}} \qquad (8.14)$$

The proof is similar to that of theorem 8.2.

Corollary. A systematic sample has the same precision as the corresponding stratified random sample, with one unit per stratum, if $\rho_{wst} = 0$. This follows because for this type of stratified random sample $V(\bar{y}_{st})$ is (theorem 5.3, corollary 3)

$$V(\bar{y}_{st}) = \left(\frac{N-n}{N}\right) \frac{S^2_{wst}}{n} \qquad (8.15)$$

Other formulas for $V(\bar{y}_{sy})$, appropriate to an autocorrelated population, have been given by W. G. and L. H. Madow (1944), who made the first theoretical study of the precision of systematic sampling.

Example. The data in Table 8.3 are for a small artificial population that exhibits a fairly steady rising trend. We have $N = 40$, $k = 10$, $n = 4$. Each column represents a systematic sample, and the rows are the strata. The example illustrates the situation in which the "within-stratum" correlation is positive. For instance, in the first sample each of the four numbers 0, 6, 18, and 26 lies below the mean of the stratum to which it belongs. This is consistently true, with a few exceptions, in the first five systematic samples. In the last five samples, deviations from the strata means are mostly positive. Thus the cross-product terms in ρ_{wst} are predominantly positive. From theorem 8.3 we expect systematic sampling to be less precise than stratified random sampling with one unit per stratum.

The variance $V(\bar{y}_{sy})$ is found directly from the systematic sample totals as

$$V(\bar{y}_{sy}) = V_{sy} = \frac{1}{k} \sum_{i=1}^{k} (\bar{y}_{i.} - \bar{Y})^2 = \frac{1}{n^2 k} \sum_{i=1}^{k} (n\bar{y}_{i.} - n\bar{Y})^2$$

$$= \frac{1}{160} \left[(50)^2 + (58)^2 + \cdots + (88)^2 - \frac{(727)^2}{10} \right] = 11.63$$

For random and stratified random sampling, we need an analysis of variance of the population into "between rows" and "within rows." This is presented in Table 8.4. hence

TABLE 8.3

DATA FOR 10 SYSTEMATIC SAMPLES WITH $n = 4$, $N = kn = 40$

Strata	Systematic sample numbers										Strata means
	1	2	3	4	5	6	7	8	9	10	
I	0	1	1	2	5	4	7	7	8	6	4.1
II	6	8	9	10	13	12	15	16	16	17	12.2
III	18	19	20	20	24	23	25	28	29	27	23.3
IV	26	30	31	31	33	32	35	37	38	38	33.1
Totals	50	58	61	63	75	71	82	88	91	88	72.7

TABLE 8.4

ANALYSIS OF VARIANCE

	df	ss	ms
Between rows (strata)	3	4828.3	
Within strata	36	485.5	$13.49 = S^2_{wst}$
Totals	39	5313.8	$136.25 = S^2$

the variances of the estimated means from simple random and stratified random samples are as follows.

$$V_{ran} = \left(\frac{N-n}{N}\right)\frac{S^2}{n} = \frac{9}{10} \cdot \frac{136.25}{4} = 30.66$$

$$V_{st} = \left(\frac{N-n}{N}\right)\frac{S^2_{wst}}{n} = \frac{9}{10} \cdot \frac{13.49}{4} = 3.04$$

Both stratified random sampling and systematic sampling are much more effective than simple random sampling but, as anticipated, systematic sampling is less precise than stratified random sampling.

Table 8.5 shows the same data, with the order of the observations *reversed* in the second and fourth strata. This has the effect of making ρ_{wst} negative, because it makes the majority of the cross products between deviations from the strata means negative for pairs of observations that lie in the same systematic sample. In the first systematic sample, for instance, the deviations from the strata means are now -4.1, $+4.8$, -5.3, $+4.9$. Of the six products of pairs of deviations, four are negative. Roughly the same situation applies in every systematic sample.

This change does not affect V_{ran} and V_{st}. With systematic sampling, it brings about a dramatic increase in precision, as is seen when the systematic sample totals in Table 8.5 are compared with those in Table 8.3. We now have

$$V_{sy} = \frac{1}{160}\left[(73)^2 + (74)^2 + \cdots + (65)^2 - \frac{(727)^2}{10}\right] = 0.46$$

TABLE 8.5

DATA IN TABLE 8.3, WITH THE ORDER REVERSED IN STRATA II AND IV

Strata	Systematic sample numbers										Strata means
	1	2	3	4	5	6	7	8	9	10	
I	0	1	1	2	5	4	7	7	8	6	4.1
II	17	16	16	15	12	13	10	9	8	6	12.2
III	18	19	20	20	24	23	25	28	29	27	23.3
IV	38	38	37	35	32	33	31	31	30	26	33.1
Totals	73	74	74	72	73	73	73	75	75	65	72.7

It is sometimes possible to exploit this result by numbering the units to create negative correlations within strata. Accurate knowledge of the trends within the population is required. However, as will be seen later, the situation in Table 8.5 is one in which it is difficult to obtain from the sample a good estimate of the standard error of $\bar{y}_{sy}$.

8.4 COMPARISON OF SYSTEMATIC WITH STRATIFIED RANDOM SAMPLING

The performance of systematic sampling in relation to that of stratified or simple random sampling is greatly dependent on the properties of the population. There are populations for which systematic sampling is extremely precise and others for which it it is less precise than simple random sampling. For some populations and some values of n, $V(\bar{y}_{sy})$ may even *increase* when a larger sample is taken—a startling departure from good behavior. Thus it is difficult to give general advice about the situations in which systematic sampling is to be recommended. A knowledge of the structure of the population is necessary for its most effective use.

Two lines of research on this problem have been followed. One is to compare the different types of sampling on artificial populations in which y_i is some simple function of i. The other is to make the comparisons for natural populations. Some of the principal results are presented in the succeeding sections.

8.5 POPULATIONS IN "RANDOM" ORDER

Systematic sampling is sometimes used, for its convenience, in populations in which the numbering of the units is effectively random. This is so in sampling from a file arranged alphabetically by surnames, if the item that is being measured has no relation to the surname of the individual. There will then be no trend or stratification in y_i as we proceed along the file and no correlation between neighboring values.

In this situation we would expect systematic sampling to be essentially equivalent to simple random sampling and to have the same variance. For any single finite population, with given values of n and k, this is not exactly true, because V_{sy}, which is based on only k degrees of freedom, is rather erratic when k is small and may turn out to be either greater or smaller than V_{ran}. There are two results which show that on the average the two variances are equal.

Theorem 8.4. Consider all $N!$ finite populations that are formed by the $N!$ permutations of any set of numbers $y_1, y_2, \ldots, y_N$. Then, on the average over these finite populations,

$$E(V_{sy}) = V_{ran} \tag{8.16}$$

Note that V_{ran} is the same for all permutations.

This result, proved by W. G. and L. H. Madow (1944), shows that if the order of the items in a specific finite population can be regarded as drawn at random from the $N!$ permutations, systematic sampling is on the average equivalent to simple random sampling.

The second approach is to regard the finite population as drawn at random from an infinite superpopulation which has certain properties. The result that is proved does not apply to any single finite population (i.e., to any specific set of values $y_1, y_2, \ldots, y_N$) but to the average of all finite populations that can be drawn from the infinite population.

The symbol $\mathscr{E}$ denotes averages over all finite populations that can be drawn from this superpopulation.

Theorem 8.5. If the variates y_i $(i = 1, 2, \ldots, N)$ are drawn at random from a superpopulation in which

$$\mathscr{E}y_i = \mu, \qquad \mathscr{E}(y_i - \mu)(y_j - \mu) = 0 \quad (i \neq j), \qquad \mathscr{E}(y_i - \mu)^2 = \sigma_i^2$$

Then

$$\mathscr{E}V_{sy} = \mathscr{E}V_{ran}$$

The crucial conditions are that all y_i have the same mean μ, that is, there is no trend, and that no linear correlation exists between the values y_i and y_j at two different points. The variance σ_i^2 may change from point to point in the series.

Proof. For any specific finite population,

$$V_{ran} = \frac{N-n}{Nn} \frac{\sum\limits_{i=1}^{N}(y_i - \bar{Y})^2}{N-1}$$

Now

$$\sum_{i=1}^{N} (y_i - \bar{Y})^2 = \sum_{i=1}^{N} [(y_i - \mu) - (\bar{Y} - \mu)]^2$$

$$= \sum_{i=1}^{N} (y_i - \mu)^2 - N(\bar{Y} - \mu)^2$$

Since y_i and y_j are uncorrelated $(i \neq j)$,

$$\mathscr{E}(\bar{Y} - \mu)^2 = \frac{1}{N^2} \sum_{i=1}^{N} \sigma_i^2 \qquad (8.17)$$

Hence

$$\mathscr{E}V_{ran} = \frac{N-n}{Nn(N-1)} \left(\sum_{i=1}^{N} \sigma_i^2 - N \frac{\sum \sigma_i^2}{N^2} \right) \qquad (8.18)$$

This gives

$$\mathscr{E}V_{ran} = \frac{N-n}{N^2 n} \sum_{i=1}^{N} \sigma_i^2 \qquad (8.19)$$

Turning to V_{sy}, let $\bar{y}_u$ denote the mean of the uth systematic sample. For any specific finite population,

$$V_{sy} = \frac{1}{k} \sum_{u=1}^{k} (\bar{y}_u - \bar{Y})^2 \qquad (8.20)$$

$$= \frac{1}{k} \left[\sum_{u=1}^{k} (\bar{y}_u - \mu)^2 - k(\bar{Y} - \mu)^2 \right] \qquad (8.21)$$

By the theorem for the variance of the mean of an uncorrelated sample from an infinite population,

$$\mathscr{E}V_{sy} = \frac{1}{k} \left(\frac{\sum_{i=1}^{N} \sigma_i^2}{n^2} - \frac{k \sum_{i=1}^{N} \sigma_i^2}{N^2} \right) \qquad (8.22)$$

$$= \frac{N-n}{N^2 n} \sum_{i=1}^{N} \sigma_i^2 = \mathscr{E}V_{ran} \qquad (8.23)$$

8.6 POPULATIONS WITH LINEAR TREND

If the population consists solely of a linear trend, as illustrated in Fig. 8.2, it is fairly easy to guess the nature of the results. From Fig. 8.2, it looks as if V_{sy} and V_{st} (with one unit per stratum) will both be smaller than V_{ran}. Furthermore, V_{sy} will be larger than V_{st}, for if the systematic sample is too low in one stratum it is too low in

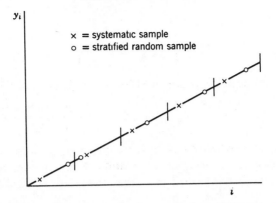

Fig. 8.2 Systematic sampling in a population with linear trend.

all strata, whereas stratified random sampling gives an opportunity for within-stratum errors to cancel.

To examine the effects mathematically, we may assume that $y_i = i$. We have

$$\sum_{i=1}^{N} i = \frac{N(N+1)}{2}, \qquad \sum_{i=1}^{N} i^2 = \frac{N(N+1)(2N+1)}{6}$$

The population variance S^2 is given by

$$S^2 = \frac{1}{N-1}(\sum y_i^2 - N\bar{Y}^2)$$

$$= \frac{1}{N-1}\left[\frac{N(N+1)(2N+1)}{6} - \frac{N(N+1)^2}{4}\right] = \frac{N(N+1)}{12} \qquad (8.24)$$

Hence the variance of the mean of a simple random sample is

$$V_{ran} = \frac{N-n}{N} \cdot \frac{S^2}{n} = \frac{n(k-1)}{N} \cdot \frac{N(N+1)}{12n} = \frac{(k-1)(N+1)}{12} \qquad (8.25)$$

To find the variance within strata, S_w^2, we need only replace N by k in (8.24). This gives

$$V_{st} = \frac{N-n}{N} \cdot \frac{S_w^2}{n} = \frac{n(k-1)}{nk} \cdot \frac{k(k+1)}{12n} = \frac{(k^2-1)}{12n} \qquad (8.26)$$

For systematic sampling, the mean of the second sample exceeds that of the first by 1; the mean of the third exceeds that of the second by 1, and so on. Thus the means $\bar{y}_u$ may be replaced by the numbers $1, 2, \ldots, k$. Hence, by a further application of (8.24),

$$\sum_{u=1}^{k} (\bar{y}_u - \bar{Y})^2 = \frac{k(k^2-1)}{12}$$

This gives

$$V_{sy} = \frac{1}{k} \sum (\bar{y}_u - \bar{Y})^2 = \frac{k^2 - 1}{12} \qquad (8.27)$$

From the formulas (8.25), (8.26), and (8.27) we deduce, as anticipated,

$$V_{st} = \frac{k^2 - 1}{12n} \le V_{sy} = \frac{k^2 - 1}{12} \le V_{ran} = \frac{(k-1)(N+1)}{12} \qquad (8.28)$$

Equality occurs only when $n = 1$. Thus, for removing the effect of a linear trend, suspected or unsuspected, the systematic sample is much more effective than the simple random sample but less effective than the stratified random sample.

8.7   METHODS FOR POPULATIONS WITH LINEAR TRENDS

The performance of systematic sampling in the presence of a linear trend can be improved in several ways. One is to use a centrally located sample. Another is to change the estimate from an unweighted to a weighted mean in which all internal members of the sample have weight unity (before division by n) but different weights are given to the first and last members. If the random number drawn between 1 and k is i, these weights are

$$1 \pm \frac{n(2i - k - 1)}{2(n-1)k} \qquad (8.29)$$

the $+$ sign being used for the first member, the $-$ sign for the last. For any i, the two weights obviously add to 2. The reader may verify that if the population consists of a linear trend and $N = nk$ the weighted sample mean gives the correct population mean. The performance of these *end corrections* has been examined by Yates (1948), to whom they are due.

Bellhouse and Rao (1975) have extended the Yates corrections to the case $N \ne nk$ when the systematic sample is drawn by Lahiri's circular method (section 8.1), which guarantees constant n. As before, the weights different from 1 are applied to the first and last sample numbers *in the original serial order of the population*. For example, if the starting random number in drawing the sample is 19 with $N = 23$, $n = 5$, units 19, 1, 6, 11, 16 constituting the sample, the first and last members are y_1 and y_{19}. Two cases arise.

Case 1.   Small i for which $i + (n-1)k \le N$, so that the n units are obtained without passing over y_N. The weights for the first ($+$) and last ($-$) members are

$$1 \pm \frac{n[2i + (n-1)k - (N+1)]}{2(n-1)k} \qquad (8.30)$$

Case 2. $i + (n-1)k > N$. Let n_2 be the number of sample units obtained after passing over y_N. Thus, with $i = 19$, $n_2 = 4$. The weights for the first ($+$) and last ($-$)

members are

$$1 \pm \frac{n}{2(N-k)} \left[2i + (n-1)k - (N+1) - 2n_2 \frac{N}{n} \right] \qquad (8.31)$$

In both cases the internal sample members receive weight 1 in the sample total. With $N = 23$, $n = k = 5$, $i = 19$, $n_2 = 4$, the first and last weights are $1 \pm (-7/18)$. Hence y_1 receives a weight $11/18$, while y_{19} receives $25/18$.

Two alternative methods attempt to change the method of sample selection so that the sample *mean* is unaffected by a linear trend. With $N = nk$ and n even, a method suggested by Sethi (1965) divides the population into $n/2$ strata of size $2k$, choosing two units equidistant from the end of each stratum. With starting random number i, the $n/2$ pairs of units are those numbered

$$[i + 2jk, 2(j+1)k - i + 1], \qquad j = 0, 1, 2, \ldots \tfrac{1}{2}n - 1 \qquad (8.32)$$

This selection removes the effect of a linear trend in any stratum of $2k$ units, even if the linear slope varies from stratum to stratum. Murthy (1967) has called the method *balanced systematic sampling*.

The *modified* method of Singh et al. (1968) chooses pairs of units equidistant from the ends of the population. With n even, the $n/2$ equidistant pairs that start with unit i $(i = 1, 2, \ldots, k)$ are

$$[i + jk, (N - jk) - i + 1], \qquad j = 0, 1, 2, \ldots \tfrac{1}{2}n - 1 \qquad (8.33)$$

With n odd in these methods, j goes up to $\tfrac{1}{2}(n-1) - 1$ in (8.32) and (8.33). The balanced method (8.32) adds the remaining sample member near the end at $[i + (n-1)k]$; the modified method near the middle at $[i + \tfrac{1}{2}(n-1)k]$. The effect of a linear trend is not completely eliminated in $\bar{y}$ for n odd.

Comparisons of the performances of these two methods with Yates' corrections and with ordinary systematic sampling have been made on superpopulation models representing linear and parabolic trends, periodic and autocorrelated variation (Bellhouse and Rao, 1975), and on a few small natural populations by these authors and by Singh (personal communication). In general the three methods (Yates, balanced, modified) performed similarly, being superior to ordinary systematic sampling in the presence of a linear or parabolic trend.

The population in Table 8.3, p. 211, for example, is one on which these methods should perform very well. Ordinary systematic sampling gave $V_{sy} = 11.63$. Comparable variances for the other methods ($n = 4$, $k = 10$) are: Yates, 1.29; Sethi (balanced), 0.46; Singh (modified), 0.34. The balanced method happens to be that obtained in Table 8.5 by reversal of strata II and IV in Table 8.3.

8.8 POPULATIONS WITH PERIODIC VARIATION

If the population consists of a periodic trend, for example, a simple sine curve, the effectiveness of the systematic sample depends on the value of k. This may be seen pictorially in Fig. 8.3. In this representation the height of the curve is the observation y_i. The sample points A represent the case least favorable to the systematic sample. This case holds whenever k is equal to the period of the sine curve or is an integral multiple of the period. Every observation within the systematic sample is exactly the same, so that the sample is no more precise than a single observation taken at random from the population.

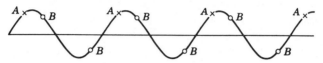

Fig. 8.3   Periodic variation.

The most favorable case (sample B) occurs when k is an odd multiple of the half-period. Every systematic sample has a mean exactly equal to the true population mean, since successive deviations above and below the middle line cancel. The sampling variance of the mean is therefore zero. Between these two cases the sample has various degrees of effectiveness, depending on the relation between k and the wavelength.

Populations that exhibit an exact sine curve are not likely to be encountered in practice. Populations with a more or less definite periodic trend are, however, not uncommon. Examples are the flow of road traffic past a point on a road over 24 hours of the day and store sales over seven days of the week. For estimating an average over a time period, a systematic sample daily at 4 p.m. or every Tuesday would obviously be unwise. Instead, the strategy is to stagger the sample over the periodic curve, for example, by seeing that every weekday is equally represented in the case of store sales.

Some populations have a kind of periodic effect that is less obvious. A series of weekly payrolls in a small sector of a factory may always list the workers in the same order and may contain between 19 and 23 names every week. A systematic sample of 1 in 20 names over a period of weeks might consist mainly of the records of one worker or of the records of two or three workers. Similarly, a systematic sample of names from a city directory might contain too many heads of households, or too many children. If there is time to study the periodic structure, a systematic sample can usually be designed to capitalize on it. Failing this, a simple or stratified random sample is preferable when a periodic effect is suspected but not well known.

In some natural populations quasiperiodic variation may be present that would be difficult to anticipate. L. H. Madow (1946) found evidence pointing this way in a bed of hardwood seedling stock in a rather small population ($N = 420$). Finney (1950) discussed a similar phenomenon in timber volume per strip in the Dehra Dun forest, although in a reexamination of the data Milne (1959) suggested that the apparent periodicity might have been produced by the process of measurement. The effect of quasiperiodicity is that systematic sampling performs poorly at some values of n and particularly well for others. Whether this effect occurs frequently is not known. Matérn (1960) cites examples in which natural forces (e.g., tides) might produce a spatial periodic variation, but he is of the opinion that no clear case has been found in forest surveys.

8.9 AUTOCORRELATED POPULATIONS

With many natural populations, there is reason to expect that two observations y_i, y_j will be more nearly alike when i and j are close together in the series than when they are distant. This happens whenever natural forces induce a slow change as we proceed along the series. In a mathematical model for this effect we may suppose that y_i and y_j are positively correlated, the correlation between them being a function solely of their distance apart, $i - j$, and diminishing as this distance increases. Although this model is oversimplified, it may represent one of the salient features of many natural populations.

In order to investigate whether this model does apply to a population, we can calculate the set of correlations ρ_u for pairs of items that are u units apart and plot this correlation against u. This curve, or the function it represents, is called a *correlogram*. Even if the model is valid, the correlogram will not be a smooth function for any finite population because irregularities are introduced by the finite nature of the population. In a comparison of systematic with stratified random sampling for this model these irregularities make it difficult to derive results for any single finite population. The comparison can be made over the average of a whole series of finite populations, which are drawn at random from an infinite superpopulation to which the model applies. This technique has already been applied in theorem 8.5 and in sections 6.7, 7.8.

Thus we assume that the observations y_i $(i = 1, 2, \ldots, N)$ are drawn from a superpopulation in which

$$\mathscr{E}(y_i) = \mu, \quad \mathscr{E}(y_i - \mu)^2 = \sigma^2, \quad \mathscr{E}(y_i - \mu)(y_{i+u} - \mu) = \rho_u \sigma^2 \tag{8.34}$$

where

$$\rho_u \geq \rho_v \geq 0, \text{ whenever } u < v$$

The drawing of one set of y_i from this superpopulation creates a single finite population of size N.

The average variance for systematic sampling is denoted by

$$\mathscr{E}V_{sy} = \mathscr{E}E(\bar{y}_{sy} - \bar{Y})^2$$

For this class of populations it is easy to show that stratified random sampling is superior to simple random sampling, but no general result can be established about systematic sampling. Within the class there are superpopulations in which systematic sampling is superior to stratified random sampling, but there are also superpopulations in which systematic sampling is inferior to simple random sampling for certain values of k.

A general theorem can be obtained if it is further assumed that the correlogram is concave upwards.

Theorem 8.6. If, in addition to conditions (8.34), we have

$$\delta_u^2 = \rho_{u+1} + \rho_{u-1} - 2\rho_u \geq 0 \quad [u = 2, 3, \ldots, (kn-2)] \tag{8.35}$$

then

$$\mathscr{E}V_{sy} \leq \mathscr{E}V_{st} \leq \mathscr{E}V_{ran} \tag{8.36}$$

for any size of sample. Furthermore, unless $\delta_u^2 = 0$, $u = 2, 3, \ldots, (kn-2)$,

$$\mathscr{E}V_{sy} < \mathscr{E}V_{st} \tag{8.37}$$

A proof has been given by Cochran (1946).

A sketch of the argument for $n = 2$ illustrates the role played by the "concave upwards" condition. In the systematic sample the members of the pair are always k units apart. Hence

$$\mathscr{E}V(\bar{y}_{sy}) = \tfrac{1}{4}(\sigma^2 + \sigma^2 + 2\rho_k\sigma^2) = \tfrac{1}{2}\sigma^2(1+\rho_k) \tag{8.38}$$

With the stratified sample, there are k possible positions for the unit drawn from each stratum, making k^2 combinations of positions. The numbers of combinations $1, 2, \ldots (2k-1)$ units apart are as follows.

Distance	1	$2\ldots(k-1)$	k	$(k+1)\ldots(2k-1)$	Total
Number	1	$2\ldots(k-1)$	k	$(k-1)\ldots$ 1	k^2

Hence the average value of $V(\bar{y}_{st})$, taken over the k^2 combinations, may be written

$$\mathscr{E}V(\bar{y}_{st}) = \frac{\sigma^2}{2k^2}\left[\sum_{u=1}^{k-1} u(2+\rho_u+\rho_{2k-u})+k(1+\rho_k)\right] \tag{8.39}$$

Similarly, $\mathscr{E}V(\bar{y}_{sy})$ may be expressed as

$$\mathscr{E}V(\bar{y}_{sy}) = \frac{\sigma^2}{2k^2}\left[\sum_{u=1}^{k-1} u(2+2\rho_k)+k(1+\rho_k)\right] \tag{8.40}$$

hence

$$\mathscr{E}V(\bar{y}_{st}) - \mathscr{E}V(\bar{y}_{sy}) = \frac{\sigma^2}{2k^2}\left[\sum_{u}^{k-1} u(\rho_u+\rho_{2k-u}-2\rho_k)\right] \tag{8.41}$$

But if

$$\rho_{u+1}+\rho_{u-1} \geq 2\rho_u \quad (u = 2, 3, \ldots)$$

it is easy to show that every term inside the brackets is positive. This completes the proof. In short, the *average* distance apart is k for both the systematic and the stratified sample, but because of the concavity the stratified sample loses more in precision when the distance is less than k than it gains when the distance exceeds k.

Quenouille (1949) has shown that the inequalities in theorem 8.6 remain valid when two of the conditions are relaxed so that

$$\mathscr{E}(y_i) = \mu_i, \qquad \mathscr{E}(y_i - \mu_i)^2 = \sigma_i^2 \qquad (8.42)$$

In this event each of the three average variances is increased by the same amount.

As far as practical applications are concerned, correlograms that are concave upward have been proposed by several writers as models for specific natural populations. The function $\rho_u = \tanh(u^{-3/5})$ was suggested by Fisher and Mackenzie (1922) for the correlation between the weekly rainfall at two weather stations that are a distance u apart; the function $\rho_u = e^{-\lambda u}$ by Osborne (1942) and Matérn (1947) for forestry and land use surveys; and the function $\rho_u = (l-u)/l$ by Wold (1938) for certain types of economic time series.

8.10 NATURAL POPULATIONS

Investigations have been made on a variety of natural populations. The data are described in Table 8.6. The first three studies were made from maps. In the first study the finite population consists of 288 altitudes at successive distances of 0.1 mile in indulating country. In the next two the data are the fractions of the lengths of lines drawn on a cover-type map that lie in a certain type of cover (e.g., grass). These examples might be considered the closest to continuous variation in the mathematical sense.

The next three studies are based on temperatures for 192 consecutive days: (a) 12 in. under the soil, (b) 4 in. under the soil, (c) in air. This trio represents a gradation in the direction of greater influence of erratic day-to-day changes in the weather compared with slow seasonal influences.

The remaining studies deal with plant or tree yields in sequences that lie along a line. In the study on potatoes, which is typical of the group, the finite population consists of the total yields of 96 rows in a field. Since no exhaustive search of the literature has been made, further data may be available.

In some of the studies V_{sy} is compared with the variance V_{st2} for a stratified random sample with strata of size $2k$ and two units per stratum. This comparison is of interest because an unbiased estimate of V_{st2} can be obtained from the sample data. This cannot be done for V_{st1} (with strata of size k and 1 unit per stratum) or for V_{sy}. Other writers report comparisons of V_{sy} with both V_{st1} and V_{st2}. The majority of the sources do not present comparisons with V_{ran} in readily usable form, but it appears that in general V_{st2} gave gains in precision over V_{ran}.

TABLE 8.6
Natural Populations Used in Studies of Systematic Sampling

Reference	N	Type of Data
Yates (1948), table 13	288	Altitudes read at intervals of 0.1 mile from ordnance survey map.
Osborne (1942)	*	Per cent of area in (a) cultivated land, (b) shrub, (c) grass, (d) woodland on parallel lines drawn on a cover-type map.
Osborne (1942)	*	Per cent of area in Douglas fir on parallel lines drawn on a cover-type map.
Yates (1948)	192	Soil temperature (12 in. under grass) for 192 consecutive days.
Yates (1948)	192	Soil temperature (4 in. under bare soil) for 192 days.
Yates (1948)	192	Air temperature for 192 days.
Yates (1948)	96	Yields of 96 rows of potatoes.
Finney (1948)	160	Volume of saleable timber per strip, 3 chains wide and of varying length (Mt. Stuart forest).
Finney (1948)	288	Volume of virgin timber per strip, 2.5 chains wide, 80 chains long (Black's Mountain forest).
Finney (1950)	292	Volume of timber per strip, 2 chains wide and of varying length (Dehra Dun forest).
Johnson (1943)	400†	Number of seedlings per 1-ft-bed-width in 4 beds of hardwood seedbed stock.
Johnson (1943)	400†	Number of seedlings per 1-ft-bed-width in 3 beds of coniferous seedbed stock.
Johnson (1943)	400†	Number of seedlings per 1-ft-bed-width in 6 beds of coniferous transplant stock.

* Theoretically, N is infinite, if lines that are infinitely thin can be envisaged.
† Approximately. The number varied from bed to bed.

In the papers by Yates and Finney comparisons are given for a range of values of n and k within each finite population. In these cases the data in Table 8.7 are the geometric means of the variance ratios for the individual values of k. The other writers make computations for only one value of k per population but may give data for different items or for several populations of the same natural type. Here, again, geometric means of the variance ratios were taken.

Although the data are limited in extent, the results are impressive. In the studies that permit comparison with V_{st1} systematic sampling shows a consistent gain in precision which, although modest, is worth having. The median of the ratios V_{st1}/V_{sy} is 1.4. The gains in comparison with V_{st2} are substantial, the median ratio being 1.9.

TABLE 8.7

RELATIVE PRECISION OF SYSTEMATIC AND STRATIFIED RANDOM SAMPLING

Data	Range of k	Relative Precision of Systematic to Stratified	
		V_{st1}/V_{sy}	V_{st2}/V_{sy}
Altitudes	2–20	2.99	5.68
Per cent area (4 cover types)		—	4.42
Per cent area (Douglas fir)		—	1.83
Soil temperature (12 in.)	2–24	2.42	4.23
Soil temperature (4 in.)	4–24	1.45	2.07
Air temperature	4–24	1.26	1.65
Potatoes	3–16	1.37	1.90
Timber volume (Mt. Stuart)	2–32	1.07	1.35
Timber volume (Black's Mt.)	2–24	1.19	1.44
Timber volume (Dehra Dun)	2–32	1.39	1.89
Hardwood seedlings	14	—	1.89
Coniferous seedlings	14–24	—	2.22
Coniferous transplant	12–22	—	0.93

The internal trend of the results agrees with expectations, although not too much should be made of this in view of the small number of studies. The gains are largest for the types of data in which we would guess that variation would be nearest to continuous. The decline in V_{st1}/V_{sy} from soil to air temperatures would also be anticipated from this viewpoint. In the last three items (forest nursery data), the only one showing no gain is coniferous transplant stock, which is older and more uniform than seedling stock.

8.11 ESTIMATION OF THE VARIANCE FROM A SINGLE SAMPLE

From the results of a simple random sample with $n > 1$, we can calculate an unbiased estimate of the variance of the sample mean, the estimate being unbiased *whatever the form of the population*. Since a systematic sample can be regarded as a simple random sample with $n = 1$, this useful property does not hold for the systematic sample. As an illustration, consider the "sine curve" example. Let

$$y_i = m + a \sin \frac{\pi i}{2}$$

where $k = 4$ and $i = 1, 2, \ldots, 4n$. The successive observations in the population are

$$(m + a), m, (m - a), m, (m + a), m, (m - a), m, \ldots$$

If $i = 1$ is chosen as the first member, *all* members of the systematic sample have the value $(m + a)$. For the other three possible choices of i, all members have the values $m, (m - a)$, or m, respectively. Thus from a *single* sample we have no means of estimating the value of a. But the true sampling variance of the mean of the systematic sample is $a^2/2$. The illustration shows that it is impossible to construct an estimated variance that is unbiased if periodic variation is present.

These results do not mean that nothing can be done. Excluding the case of periodic variation, we might know enough about the structure of the population to be able to develop a mathematical model that adequately represents the type of variation present. We might then be able to manufacture a formula for the estimated variance that is approximately unbiased for this model, although it may be badly biased for other models. The decision to use one of these models must rest on the judgment of the sampler.

Some simple models with their corresponding estimated variances are presented below. No proofs are given.

The simplest models apply to populations in which y_i is composed of a trend plus a "random" component. Thus

$$y_i = \mu_i + e_i$$

where μ_i is some function of i. For the random component, we assume that there is a superpopulation in which

$$\mathscr{E}(e_i) = 0, \qquad \mathscr{E}(e_i^2) = \sigma_i^2, \qquad \mathscr{E}(e_i e_j) = 0 \qquad (i \neq j)$$

A proposed formula s_{sy}^2 for the estimated variance is called unbiased if

$$\mathscr{E}E(s_{sy}^2) = \mathscr{E}V_{sy}$$

that is, if it is unbiased over all finite populations that can be drawn from the superpopulation.

Population in "Random" Order

$$\mu_i = \text{constant} \qquad (i = 1, 2, \ldots, N)$$

$$s_{sy1}^2 = \frac{N - n}{Nn} \frac{\sum (y_i - \bar{y}_{sy})^2}{n - 1} \tag{8.43}$$

This case applies when we are confident that the order is essentially random with respect to the items being measured. The variance formula is the same as that for a simple random sample and is unbiased if the model is correct.

Stratification Effects Only

$$\mu_i \text{ constant} \qquad (rk + 1 \leq i \leq rk + k)$$

$$s_{sy2}^2 = \frac{N-n}{Nn} \frac{\sum (y_i - y_{i+k})^2}{2(n-1)} \tag{8.44}$$

In this case the mean is constant within each stratum of k units. The estimate s_{sy2}^2, which is based on the mean square successive difference, is not unbiased. It contains an unwanted contribution from the difference between μ's in neighboring strata, and the first and last strata carry too little weight in estimating the random component of the variance. With a reasonably large sample, this estimate would in general be too high, assuming that the model is correct.

Linear Trend

$$\mu_i = \mu + \beta i$$

$$s_{sy3}^2 = \frac{N-n}{N} \frac{n'}{n^2} \frac{\sum (y_i - 2y_{i+k} + y_{i+2k})^2}{6(n-2)} \qquad (1 \leq i \leq n-2) \tag{8.45}$$

The estimate is based on successive quadratic terms in the sequence y_i. The sum of squares contains $(n-2)$ terms. With a linear trend we have seen (section 8.7) that the trend can be eliminated by the use of end corrections. The term n'/n^2 is the sum of squares of the weights in $\bar{y}_{wsy}$. Unless n is small, n'/n^2 can be replaced by the usual factor $1/n$. Because the strata at the ends receive too little weight, the estimate is biased unless σ_i^2 is constant, but it should be satisfactory if n is large and the model is correct.

If continuous variation of a more complex type is present, the preceding formulas may give poor results. In Table 8.8 the second and third formulas are applied to six forest nursery beds (Johnson, 1943). The quadratic formula is slightly better than that based on successive differences, but both give consistently serious overestimates.

TABLE 8.8

VARIANCES OF SAMPLE MEAN NUMBERS OF SEEDLINGS (JOHNSON'S DATA)

	Bed	Actual V_{sy}	s_{sy2}^2	s_{sy3}^2
Silver maple	1	0.91	2.8	2.5
	2	0.74	3.6	2.9
American elm	1	4.8	28.4	12.6
	2	15.5	22.6	18.6
White spruce	1	5.5	17.2	11.2
	2	2.0	11.6	6.4
White pine	1	8.2	21.0	21.9

Formulas developed from simple assumptions about the nature of the correlogram have been given by Osborne (1942), Cochran (1946), and Matérn (1947). If the successive observations in the systematic sample are denoted by y_1', y_2', and so forth, Yates (1949) suggested an estimator based on differences d_u of which the first is

$$d_1 = (\tfrac{1}{2}y_1' + y_3' + y_5' + y_7' + \tfrac{1}{2}y_9') - (y_2' + y_4' + y_6' + y_8') \qquad (8.46)$$

The next difference, d_2, may start with y_9', and so on. For the estimated variance of $\bar{y}_{sy}$ we take

$$\hat{V}(\bar{y}_{sy}) = \frac{N-n}{Nn} \sum_{u=1}^{g} \frac{d_u^2}{7.5g} \qquad (8.47)$$

The factor 7.5 is the sum of squares of the coefficients in any d_u, and g is the number of differences that the sample provides ($g \doteq n/9$). In natural populations that Yates examined, a formula of this type was superior to s_{sy2}^2 based on successive differences but still overestimated $V(\bar{y}_{sy})$.

In summary, there is no dearth of formulas for the estimated variance, but all appear to have limited applicability.

With $N = nk$, suppose that n is divisible by an integer m (say 10). The following method uses systematic sampling in part and provides an unbiased sample estimate of $V(\bar{y}_{sy})$ based on $(m-1)$ degrees of freedom. Draw a simple random sample of size m from the units numbered 1 to mk. For every unit in this sample, take also every (mk)th thereafter. In effect, this method divides the population into mk clusters each of size $N/mk = n/m$, and chooses m clusters at random, so that we have a simple random sample of m clusters. For instance, suppose that we want a 20% sample from a population with $N = 2400$, so that $n = 480$, $k = 5$. Take a simple random sample of size $m = 10$ from units numbered 1 to $mk = 50$, and take every 50th unit thereafter. We then have 10 cluster samples each of size $2400/50 = 48$.

Gautschi (1957) has examined the accuracy of this method under the population structures considered in this chapter. As might be anticipated, the accuracy lies between those of simple random sampling and of systematic sampling with $m = 1$.

8.12 STRATIFIED SYSTEMATIC SAMPLING

We have seen that if the units are ordered appropriately, systematic sampling provides a kind of stratification with equal sampling fractions. If we stratify by some other criterion, we may draw a separate systematic sample within each stratum with starting points independently determined. This is suitable if separate estimates are wanted for each stratum or if unequal sampling fractions are to be used. This method will, of course, be more precise than stratified random sampling if systematic sampling within strata is more precise than simple random sampling within strata.

If $\bar{y}_{syh}$ is the mean of the systematic sample in stratum h, the estimate of the population mean $\bar{Y}$ and its variance are

$$\bar{y}_{stsy} = \sum W_h \bar{y}_{syh}, \qquad V(\bar{y}_{stsy}) = \sum W_h^2 V(\bar{y}_{syh})$$

With only a few strata, the problem of finding a sample estimate of this quantity amounts to that already discussed of finding a satisfactory sample estimate of $V(\bar{y}_{syh})$ in each stratum.

When the strata are more numerous, an estimate based on the method of "collapsed strata" (section 5A.12) may be preferable. From the results in that section, it follows that the estimate

$$v(\bar{y}_{stsy}) = \sum{}' W_h'^2 (\bar{y}_{syh} - \bar{y}_{syj})^2 \qquad (8.48)$$

where the sum extends over the pairs of strata, is on the average an overestimate, even if periodic variation is present within the strata.

An unbiased estimate of the error variance can be obtained if two systematic samples, with a different random start and an interval $2k$, are drawn within each stratum, one df being provided by each stratum. There will be some loss in precision if systematic sampling is effective. If there are many strata, one systematic sample can be used in most of them, drawing two in a random subsample of strata for the purposes of estimating the error.

8.13 SYSTEMATIC SAMPLING IN TWO DIMENSIONS

In sampling an area, the simplest extension of the one-dimensional systematic sample is the "square grid" pattern shown in Fig. 8.4a. The sample is completely determined by the choice of a pair of random numbers to fix the coordinates of the upper left unit. The performance of the square grid has been studied both on theoretical and natural populations. Matérn (1960) has investigated the best type of sample when the correlation between any two points in the area is a monotone decreasing concave upward function of their distance apart d. For correlograms like $e^{-\lambda d}$ the square grid does well, being superior to simple or stratified random sampling with one unit per stratum, although Matérn gives reasons for expecting that the best pattern for this situation is a triangular network in which the points lie at the vertices of equilateral triangles.

In 14 agricultural uniformity trials, Haynes (1948) found that the square grid had about the same precision as simple random sampling in two dimensions. Milne (1959) examined the *central* square grid, in which the point lies at the center of the square, in 50 uniformity trials. It performed better than simple random sampling and perhaps slightly better than stratified random sampling, although this difference was not statistically significant. These results suggest that, at least for data of this type, autocorrelation effects are weak. For estimating the area covered by forest or by water on a map, Matérn found the square grid superior to the random method in two examples.

Figure 8.4*b* shows an alternative systematic sample, called an *unaligned* sample. The coordinates of the upper left unit are selected first by a pair of random numbers. Two additional random numbers determine the *horizontal* coordinates

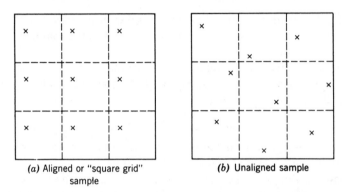

(a) Aligned or "square grid" sample

(b) Unaligned sample

Fig. 8.4   Two types of two-dimensional systematic sample.

of the remaining two units in the first *column* of strata. Another two are needed to fix the *vertical* coordinates of the remaining units in the first *row* of strata. The constant interval *k* (equal to the sides of the squares) then fixes the locations of all points. Investigations by Quenouille (1949) and Das (1950) for simple two-dimensional correlograms indicate that the unaligned pattern will often be superior both to the square grid and to stratified random sampling.

Further evidence of the superiority of an unaligned sample is obtained from experience in experimental design, in which the latin square has been found a precise method for arranging treatments in a rectangular field. The 5×5 latin square in Fig. 8.5*a* may be regarded as a division of the field into five systematic samples, one for each letter. There is some evidence that this particular square, which is called the "knight's move" latin square, is slightly more precise than a randomly chosen 5×5 square, probably because alignment is absent in the diagonals as well as in rows and columns.

The principle of the latin square has been used by Homeyer and Black (1946) in sampling rectangular fields of oats. Each field contained 21 plots. The three possible systematic samples are denoted by the letters *A*, *B*, and *C*, respectively, in Fig. 8.5*b*. This arrangement, with one of the letters chosen at random in each field, gave an increase in precision of around 25% over stratified random sampling with rows as strata. The arrangement does not quite satisfy the latin square property because each letter appears three times in one column and twice in the other columns, but it approaches this property as nearly as possible.

Yates (1960), who terms arrangements of this type *lattice sampling*, discusses their use in two- and three-dimensional sampling. In three dimensions each row,

```
 A   B   C   D   E           A   B   C
 D   E   A   B   C           B   C   A
 B   C   D   E   A           C   A   B
 E   A   B   C   D           A   B   C
 C   D   E   A   B           B   C   A
                             C   A   B
                             A   B   C
```

(a) "Knight's move" latin square   (b) Systematic design for a 3 × 7 rectangular field

Fig. 8.5 Two systematic designs based on the latin square.

column, and vertical level can be represented in the sample by choosing p units out of the p^3 in the population. With p^2 units in the sample, each of the p^2 combinations of levels of rows and columns, of rows and vertical heights, and of columns and vertical heights can be represented. Patterson (1954) has investigated the arrangements that provide an unbiased estimate of error.

8.14 SUMMARY

Systematic samples are convenient to draw and to execute. In most of the studies reported in this chapter, both on artificial and on natural populations, they compared favorably in precision with stratified random samples. Their disadvantages are that they may give poor precision when unsuspected periodicity is present and that no trustworthy method for estimating $V(\bar{y}_{sy})$ from the sample data is known.

In the light of these results systematic sampling can safely be recommended in the following situations.

1. Where the ordering of the population is essentially random or contains at most a mild stratification. Here systematic sampling is used for convenience, with little expectation of a gain in precision. Sample estimates of error that are reasonably unbiased are available (section 8.11).

2. Where a stratification with numerous strata is employed and an independent systematic sample is drawn from each stratum. The effects of hidden periodicities tend to cancel out in this situation, and an estimate of error that is known to be an overestimate can be obtained (section 8.12). Alternatively, we can use half the number of strata and draw two systematic samples, with independent random starts, from each stratum. This method gives an unbiased estimate of error.

3. For subsampling cluster units (Chapter 10). In this case an unbiased or almost unbiased estimate of the sampling error can be obtained in most practical situations. This is a common use of systematic sampling.

4. For sampling populations with variation of a continuous type, provided that an estimate of the sampling error is not regularly required. If a series of surveys of

Numbers of Seedlings

					Feet					Systematic Sample Totals
1–20	21–40	41–60	61–80	81–100	101–120	121–140	141–160	161–180	181–200	
1	2	3	4	5	6	7	8	9	10	
8	20	26	34	31	24	18	16	36	10	223
6	19	26	21	23	19	13	12	8	35	182
6	25	10	27	41	28	7	8	29	7	188
23	11	41	25	18	18	9	10	33	9	197
25	31	30	32	15	29	11	12	14	12	211
16	26	55	43	21	24	20	20	13	7	245
28	29	34	33	8	33	16	17	18	6	222
21	19	56	45	22	37	9	12	20	14	255
22	17	39	23	11	32	14	7	13	12	190
18	28	41	27	3	26	15	17	24	15	214
26	16	27	37	4	36	20	21	29	18	234
28	9	20	14	5	20	21	26	18	4	165
11	22	25	14	11	43	15	16	16	4	177
16	26	39	24	9	27	14	18	20	9	202
7	17	24	18	25	20	13	11	6	8	149
22	39	25	17	16	21	9	19	15	8	191
44	21	18	14	13	18	25	27	4	9	193
26	14	44	38	22	19	17	29	8	10	227
31	40	55	36	18	24	7	31	8	5	255
26	30	39	29	9	30	30	29	10	3	235
Strata Totals 410	459	674	551	325	528	303	358	342	205	4155

this type is being made, an occasional check on the sampling errors may be sufficient. Yates (1948) has shown how this may be done by taking supplementary observations.

EXERCISES

8.1 The data in the table (p.230) are the numbers of seedlings for each foot of bed in a bed 200 ft long.

Find the variance of the mean of a systematic sample consisting of every twentieth foot. Compare this with the variances for (a) a simple random sample, (b) a stratified random sample with two units per stratum, (c) a stratified random sample with one unit per stratum. All samples have $n = 10$. $\left[\sum (y_i - \bar{Y})^2 = 23{,}601.\right]$

8.2 A population of 360 households (numbered 1 to 360) in Baltimore is arranged alphabetically in a file by the surname of the head of the household. Households in which the head is nonwhite occur at the following numbers: 28, 31–33, 36–41, 44, 45, 47, 55, 56, 58, 68, 69, 82, 83, 85, 86, 89–94, 98, 99, 101, 107–110, 114, 154, 156, 178, 223, 224, 296, 298–300, 302–304, 306–323, 325–331, 333, 335–339, 341, 342. (The nonwhite households show some "clumping" because of an association between surname and color.)

Compare the precision of a 1-in-8 systematic sample with a simple random sample of the same size for estimating the proportion of households in which the head is nonwhite.

8.3 A neighborhood contains three compact communities, consisting, respectively, of people of Anglo-Saxon, Polish, and Italian descent. There is an up-to-date directory. In it the persons in a house are listed in the following order: husband, wife, children (by age), others. Houses are listed in order along streets. The average number of persons per house is five.

The choice is between a systematic sample of every fifth person in the directory and a 20% simple random sample. For which of the following variables do you expect the systematic sample to be more precise? (a) Proportion of people of Polish descent, (b) proportion of males, (c) proportion of children. Give reasons.

8.4 In a directory of 13 houses on a street the persons are listed as follows: $M =$ male adult, $F =$ female adult, $m =$ male child, $f =$ female child.

					Household							
1	2	3	4	5	6	7	8	9	10	11	12	13
M	M	M	M	M	M	M	M	M	M	M	M	M
F	F	F	F	F	F	F	F	F	F	F	F	F
f	f	m		m	f	f	m	m	m	f	f	
m	m	f		m	m	f	f			f	m	
f	f			f		m						

Compare the variances given by a systematic sample of one in five persons and a 20% simple random sample for estimating (a) the proportion of males, (b) the proportion of children, (c) the proportion of persons living in professional households (households 1, 2, 3, 12, and 13 are described as professional). Do the results support your answers to exercise 8.3? For the systematic sample, number down each column, then go to the top of the next column.

8.5 In exercise 8.1 we might estimate $V(\bar{y}_{sy})$ by (a) regarding each systematic sample as a simple random sample, (b) pretending that each 1-in-20 systematic sample is composed of two 1-in-40 systematic samples with a separate random start. For each method, compare the average of the estimated variances with the actual variance of $\bar{y}_{sy}$.

8.6   In a population consisting of a linear trend (section 8.6) show that a systematic sample is less precise than a stratified random sample with strata of size $2k$ and two units per stratum if $n > (4k+2)/(k+1)$.

8.7   A two-dimensional population with a linear trend may be represented by the relation

$$y_{ij} = i + j \qquad (i, j = 1, 2, ,\ldots, nk)$$

where y_{ij} is the value in the ith row and jth column. The population contains $N^2 = n^2 k^2$ units.

A systematic square grid sample is selected by drawing at random two independent starting coordinates i_0, j_0, each between 1 and k. The sample, of size n^2, contains all units whose coordinates are of the form

$$i_0 + \gamma k, j_0 + \delta k$$

where γ, δ are any two integers between 0 and $(n-1)$, inclusive.

Show that the mean of this sample has the same precision as the mean of a simple random sample of size n^2.

8.8   If the comparison in exercise 8.7 were made for a three-dimensional population with linear trend, what result would you expect?

8.9   In a population with $y_i = i^2$ $(i = 1, 2, \ldots, 16)$, compare the values of $E(\hat{\bar{Y}} - \bar{Y})^2$ given by every kth systematic sampling and by the Yates, Sethi, and Singh et al. methods for $n = 4$, $k = 4$, $N = 16$.

CHAPTER 9

Single-Stage Cluster Sampling: Clusters of Equal Sizes

9.1 REASONS FOR CLUSTER SAMPLING

Several references have been made in preceding chapters to surveys in which the sampling unit consists of a group or *cluster* of smaller units that we have called *elements* or *subunits*. There are two main reasons for the widespread application of cluster sampling. Although the first intention may be to use the elements as sampling units, it is found in many surveys that no reliable list of the elements in the population is available and that it would be prohibitively expensive to construct such a list. In many countries there are no complete and up-to-date lists of the people, the houses, or the farms in any large geographic region. From maps of the region, however, it can be divided into areal units such as blocks in the cities and segments of land with readily identifiable boundaries in the rural parts. In the United States these clusters are often chosen because they solve the problem of constructing a list of sampling units.

Even when a list of individual houses is available, economic considerations may point to the choice of a larger cluster unit. For a given size of sample, a small unit usually gives more precise results than a large unit. For example, a simple random sample of 600 houses covers a town more evenly than 20 city blocks containing an average of 30 houses apiece. But greater field costs are incurred in locating 600 houses and in travel between them than in locating 20 blocks and visiting all the houses in these blocks. When cost is balanced against precision, the larger unit may prove superior.

A rational choice between two types or sizes of unit may be made by the familiar principle of selecting the unit that gives the smaller variance for a given cost or the smaller cost for a prescribed variance. As in many practical decisions, there may be imponderable factors: one type of unit may have some special convenience or disadvantage that is difficult to include in a calculation of costs. In sampling a growing crop, some experiences suggest that a small unit may give biased estimates because of uncertainty about the exact boundaries of the unit. Homeyer and Black (1946) found that units 2×2 ft gave yields of oats about 8% higher than units 3×3 ft, possibly because samplers tend to place boundary plants inside the unit when there is doubt. Sukhatme (1947) cities similar results for wheat and rice.

This Chapter deals with the case in which every cluster unit contains the same number of elements or subunits.

9.2 A SIMPLE RULE

When the problem is to compare a few specific sizes or types of unit, the following result is helpful.

Theorem 9.1. This applies to simple random sampling in which the fpc is negligible. The quantity to be estimated is the population total. For the uth type of unit, let

$$M_u = \text{relative size of unit}$$

$$S_u^2 = \text{variance among the unit totals}$$

$$C_u = \text{relative cost of measuring one unit}$$

Then relative cost for specified variance or relative variance for specified cost $\propto C_u S_u^2 / M_u^2$.

Proof. Suppose V is the specified variance of the estimated population total. For the uth type of unit the estimate is $N_u \bar{y}_u$ with variance

$$V = \frac{N_u^2 S_u^2}{n_u}; \qquad n_u = \frac{N_u^2 S_u^2}{V} \tag{9.1}$$

The cost of taking n_u units is $C_u n_u = C_u N_u^2 S_u^2 / V$. Since $N_u M_u = \text{constant}$ for different types of unit, the cost is proportional to $C_u S_u^2 / M_u^2$. On the other hand, if the cost C is specified, $n_u = C/C_u$ and, in (9.1), $V \propto C_u S_u^2 / M_u^2$. This completes the proof.

Corollary 1. If we define the *relative net precision* of a unit as inversely proportional to the variance obtained for fixed cost, theorem 9.1 may be stated as

$$\text{relative net precision} \propto \frac{M_u^2}{C_u S_u^2} \tag{9.2}$$

Corollary 2. In the analysis of variance, the variances for units of different sizes are often computed on what is called a common basis—usually that applicable to the smallest unit. To put the variances on a common basis, the variance S_u^2 among the totals of units of size M_u is divided by M_u. Let

$$S_u'^2 = \frac{S_u^2}{M_u} = \text{variance among unit totals (on a common basis)}$$

$$C_u' \propto \frac{C_u}{M_u} = \text{relative cost of taking a given bulk of sample}$$

Then theorem 9.1 and corollary 1 may be stated as follows.

$$\text{relative cost for equal precision} \propto \frac{C_u S_u^{\,2}}{M_u^{\,2}} \propto C_u' S_u'^{\,2} \tag{9.3}$$

$$\text{relative net precision} \propto \frac{1}{C_u' S_u'^{\,2}} \tag{9.4}$$

If differences in the costs of taking the sample are ignored (i.e., if C_u' is constant), the relative net precision with the uth unit $\propto 1/S_u'^{\,2}$. Kish's *deff* factors for the different units (section 4.11) are therefore proportional to the $S_u'^{\,2} = S_u^{\,2}/M_u$.

Example. Johnson's data (1941) for a bed of white pine seedlings provide a simple example. The bed contained six rows, each 434 ft long. There are many ways in which the bed can be divided into sampling units. Data for four types of unit are shown in Table 9.1. Since the bed was completely counted, the data are correct population values.

TABLE 9.1

DATA FOR FOUR TYPES OF SAMPLING UNIT

Preliminary Data	Type of Unit			
	1-ft row	2-ft row	1-ft bed	2-ft bed
M_u = relative size of unit	1	2	6	12
N_u = number of units in pop.	2604	1302	434	217
$S_u^{\,2}$ = pop. variance per unit	2.537	6.746	23.094	68.558
Number of feet of row that can be counted in 15 min.	44	62	78	108

The units were

1 ft of a single row
2 ft of a single row
1 ft of the width of the bed
2 ft of the width of the bed

With the first two units it was assumed that sampling would be stratified by rows, so that the $S_u^{\,2}$ represent variances within rows. Simple random sampling was assumed for the last two units.

Since the principal cost is that of locating and counting the units, costs were estimated by a time study (last row of Table 9.1). With the larger units, a greater bulk of sample can be counted in 15 min, less time being spent in moving from one unit to another.

The quantity to be estimated is the total number of seedlings in the bed. In the notation of theorem 9.1, Table 9.1 gives the value of M_u and $S_u^{\,2}$. The relative values of C_u, expressed as the time required to count one unit, are as follows.

	1-ft row	2-ft row	1-ft bed	2-ft bed
C_u (in 15-min times)	$\frac{1}{44}$	$\frac{2}{62}$	$\frac{6}{78}$	$\frac{12}{108}$

By theorem 9.1, corollary 1, the relative net precisions are worked out in Table 9.2. The last line of Table 9.2 gives the relative precisions when that of the smallest unit is taken as 100. The 1-ft bed appears to be the best unit.

TABLE 9.2

RELATIVE NET PRECISIONS OF THE FOUR UNITS

	1-ft row	2-ft row	1-ft bed	2-ft bed
$\dfrac{M_u^2}{C_u S_u^2}$	$\dfrac{44}{2.537} = 17.34$	$\dfrac{(4)(62)}{(2)(6.746)} = 18.38$	$\dfrac{(36)(78)}{(6)(23.094)} = 20.27$	$\dfrac{(144)(108)}{(12)(68.558)} = 18.90$
	100	106	117	109

The variances among units, expressed on a common basis, are also worth looking at. The values of $S_u'^2 = S_u^2/M_u$, applicable to a single foot of row, are, respectively, 2.537, 3.373, 3.849, 5.713. Note that these variances increase steadily with increasing size of unit. This result is commonly found (although exceptions may occur). Since the relative net precision $\propto 1/C_u' S_u'^2$, the cost of taking a given bulk of sample must decrease with the larger units if they are to prove economical.

Theorem 9.1 and its corollaries remain valid for stratified sampling with proportional allocation if all strata are of the same size and if S_u^2, $S_u'^2$ represent average variances within strata. This is so, under the conditions stated, because the variance of the estimated population total, ignoring the fpc, is $N^2 S_u^2/n$, and therefore assumes the same form as in simple random sampling. Theorem 9.1 does not hold for more complex types of sampling.

The preceding results are intended merely as an illustration of the general procedure. *Comparisons among units should always be made for the kind of sampling that is to be used in practice or, if this has not been decided, for the kinds that are under consideration.* Changes in the method of sampling or of estimation will alter the relative net precisions of the different units. Even with a fixed method of sampling and estimation, relative net precisions vary with size of sample if the cost is not a linear function of size or if the size is large enough so that the fpc must be taken into account.

There is usually more than one item to consider. One approach is to fix the total cost and work out the relative net precisions for each type of unit and each item. Unless one type is uniformly superior, some compromise decision is made, giving principal weight to the most important items.

TABLE 9.3

ESTIMATED STANDARD ERRORS (%) FOR FOUR SIZES OF UNIT, WITH SIMPLE RANDOM SAMPLING

Items	$S/4$	$S/2$	S	$2S$	Best Unit
Number of swine	5.0	4.9	5.3	6.2	$S/2$
Number of horses	3.4	3.3	3.6	4.2	$S/2$
Number of sheep	17.4	15.7	14.9	14.3	$2S$
Number of chickens	3.0	3.0	3.3	3.8	$S/4, S/2$
Number of eggs yesterday	5.7	5.2	4.9	4.7	$2S$
Number of cattle	4.7	4.6	4.8	⁻5.5	$S/2$
Number of cows milked	3.7	3.6	3.8	4.4	$S/2$
Number of gallons of milk	4.4	4.2	4.4	4.9	$S/2$
Dairy products receipts	5.5	5.2	5.4	6.0	$S/2$
Number of farm acres	2.9	2.8	3.0	3.5	$S/2$
Number of corn acres	3.7	3.5	3.8	4.4	$S/2$
Number of oat acres	4.6	4.8	5.6	7.0	$S/4$
Corn yield	1.6	1.7	2.0	2.5	$S/4$
Oat yield	1.6	1.5	1.6	1.8	$S/2$
Commercial feed expenditures	12.6	13.6	16.7	21.8	$S/4$
Total expenditures, operator	7.8	8.1	9.6	12.0	$S/4$
Total receipts, operator	6.2	6.5	7.7	9.8	$S/4$
Net cash income, operator	6.8	6.9	7.8	9.5	$S/4$

In view of the numerous factors that influence the results, a study of optimum size of unit in an extensive survey is a large task. A good example for farm sampling is described by Jessen (1942). An excerpt from his results is given in Table 9.3. This compares four sizes of unit—a quarter-section, a half-section, a section, and a block consisting of two contiguous sections. The section is an area 1 mile square, containing on the average slightly under four farms. In this comparison the total field cost ($1000), the length of questionnaire (60 min to complete), and the travel cost (5 cents per mile) are all specified, because relative net precisions change if any of these variables is altered. Costs are at a 1939 level.

The data in the table are the relative standard errors (in per cent) of the estimated means per farm for 18 items. No unit is best for all items. The half-section and the quarter-section are, however, superior to the larger units for all except two items, with little to choose between the half- and quarter-sections. The half-section would probably be preferred, because the problem of identifying the boundaries accurately is easier.

9.3 COMPARISONS OF PRECISION MADE FROM SURVEY DATA

In the nursery seedling example the variances for the different types of unit were obtained from a complete count of the population. Except with small populations, however, it is seldom feasible to conduct a survey solely for the purpose of comparison. Information about the optimum unit is more usually procured as an ingenious by-product of a survey whose main purpose is to make estimates.

Suppose that in a survey each unit can be divided into M smaller units. Instead of recording only the totals for each "large" unit in the sample, we record data separately for each of the M small units. A comparison can then be made of the precision of the large and small units. A simple random sample of size n will be assumed at first.

The analysis of variance in Table 9.4 can be computed from the sample.

TABLE 9.4

ANALYSIS OF VARIANCE OF THE SAMPLE DATA (ON A SMALL-UNIT BASIS)

	df	ms
Between large units	$(n-1)$	s_b^2
Between small units within large	$n(M-1)$	s_w^2
Between small units in sample	$nM-1$	$s^2 = \dfrac{(n-1)s_b^2 + n(M-1)s_w^2}{nM-1}$

The estimated variance of a large unit (on a small-unit basis) is s_b^2. It might be thought that an appropriate estimate of the variance of a small unit would be the mean square between all small units in the sample; that is,

$$s^2 = \frac{(n-1)s_b^2 + n(M-1)s_w^2}{nM-1} \tag{9.5}$$

This estimate, although often satisfactory, is slightly biased because the sample is not a simple random sample of small units, since these are sampled in contiguous groups of M units.

An unbiased estimate is obtained from the sample by constructing an analysis of variance, as in Table 9.5, for the whole population, which contains N large units and NM small units.

By its definition, the population variance among small units is given by the last line of the table, that is,

$$S^2 = \frac{(N-1)S_b^2 + N(M-1)S_w^2}{NM-1} \tag{9.6}$$

TABLE 9.5

ANALYSIS OF VARIANCE FOR THE WHOLE POPULATION (ON A SMALL-UNIT BASIS)

	df	ms
Between large units	$N - 1$	S_b^2
Between small units within large units	$N(M - 1)$	S_w^2
Between small units in the population	$NM - 1$	$S^2 = \dfrac{(N - 1)S_b^2 + N(M - 1)S_w^2}{NM - 1}$

With simple random sampling, s_b^2 in Table 9.4 is an unbiased estimate of S_b^2 (this follows from section 2.3). It may be shown easily that s_w^2 is an unbiased estimate of S_w^2. Hence an unbiased estimate of the variance S^2 among all small units in the population is

$$\hat{S}^2 = \frac{(N-1)s_b^2 + N(M-1)s_w^2}{NM-1} \tag{9.7}$$

Clearly, this expression is almost the same as the simpler expression

$$\hat{S}^2 \doteq \frac{s_b^2 + (M-1)s_w^2}{M} \tag{9.8}$$

If $n > 50$, (9.5) for s^2 also reduces to (9.8), so that s^2 is a satisfactory approximation to S^2 for $n > 50$.

The two estimates s_b^2 (for the large unit) and $\hat{S}^2$ (for the small unit) are on a common basis and may be substituted in theorem 9.1, corollary 2.

If the sample is large, the small units may be measured for a random subsample of the large units (say 100 out of 600). Alternatively, two small units, chosen at random from each large unit, might be measured. More than one size of small unit may be investigated simultaneously, provided that we take data that give an unbiased estimate of S_w^2 for each small unit.

With stratified sampling, the variances for the large and small units can be estimated by these methods separately in each stratum and then substituted in the appropriate formula for the variance of the estimate from a stratified sample.

Example.   The data come from a farm sample taken in North Carolina in 1942 in order to estimate farm employment (Finkner, Morgan, and Monroe, 1943). The method of drawing the sample was to locate points at random on the map and to choose as sampling units the three farms that were nearest to each point. This method is not recommended because a large farm has a greater chance of inclusion in the sample than a small farm, and an isolated farm has a greater chance than another in a densely farmed area. Any effects of this bias will be ignored.

TABLE 9.6

SAMPLE ANALYSIS OF VARIANCE (NUMBER OF PAID WORKERS)
(SINGLE-FARM BASIS)

	df	ms
Between units within strata	825	6.218
Between farms within units	2768	2.918
Between farms within strata	3593	3.676

From the sample data for individual farms, the group of three farms can be compared with the individual farm as a sampling unit. The item chosen is the number of paid workers. The sample was stratified, the stratum being a group of townships similar in density of farm population and in ratio of cropland to farmland. Since the sampling fraction was 1.9%, the fpc can be ignored.

$$V(\hat{Y}_{st}) = \sum_h \frac{N_h^2 S_h^2}{n_h}$$

The correct procedure is to compute $N_h^2 S_h^2 / n_h$ separately within each stratum for the two types of unit, using an analysis of variance and expression (9.8). We will use a simpler procedure as an approximation.

The strata contained in general between 300 and 450 farms, and either two or three 3-farm units were taken in each stratum to make the sampling approximately proportional. Assuming proportionality, that is, $n_h/N_h = n/N$, we may write

$$V(\hat{Y}_{st}) = \frac{N}{n} \sum N_h S_h^2 \doteq \frac{N^2}{n} \bar{S}_h^2$$

if we assume further that the S_h^2 do not vary greatly among strata, so that they may be replaced by their average, $\bar{S}_h^2$.

Estimates of $\bar{S}_h^2$ are obtained from the analysis of variance in Table 9.6, which is on a single-farm basis.

For the group of three farms, the mean square $\bar{s}_{h3}^2 = 6.218$ serves as the estimate of $\bar{S}_h^2$ on a single-farm basis. For the individual farm, using (9.8), we have

$$\hat{\bar{S}}^2 = \frac{6.218 + 2(2.918)}{3} = 4.018$$

By theorem 9.1, corollary 2, the two figures, 6.218 for the group of three farms and 4.018 for the individual farm, indicate the relative variances obtained for a fixed total size of sample. The group of farms gives about two thirds the precision of the single farm. Consideration of costs would presumably make the result more favorable to the three-farm unit.

9.4 VARIANCE IN TERMS OF INTRACLUSTER CORRELATION

Variance formulas are sometimes expressed in terms of the correlation coefficient ρ between elements in the same cluster. This approach has already been used for systematic sampling (section 8.3).

Let y_{ij} be the observed value for the jth element within the ith unit, and let y_i be the unit total. In cluster sampling we need to distinguish between two kinds of average: the mean per unit $\bar{Y} = \sum y_i/N$ and the mean per element $\bar{\bar{Y}} = \sum y_i/NM = \bar{Y}/M$. The variance among elements is

$$S^2 = \frac{\sum_{i,j}(y_{ij} - \bar{\bar{Y}})^2}{NM - 1}$$

The intracluster correlation coefficient ρ was defined (section 8.3) as

$$\rho = \frac{E(y_{ij} - \bar{\bar{Y}})(y_{ik} - \bar{\bar{Y}})}{E(y_{ij} - \bar{\bar{Y}})^2} = \frac{2\sum_i \sum_{j<k}(y_{ij} - \bar{\bar{Y}})(y_{ik} - \bar{\bar{Y}})}{(M-1)(NM-1)S^2} \tag{9.9}$$

The number of terms (cross products) in the numerator E is $NM(M-1)/2$, and in the denominator E is $(NM-1)S^2/NM$.

Theorem 9.2. A simple random sample of n clusters, each containing M elements, is drawn from the N clusters in the population. Then the sample mean per element $\bar{\bar{y}}$ is an unbiased estimate of $\bar{\bar{Y}}$ with variance

$$V(\bar{\bar{y}}) = \frac{1-f}{n} \cdot \frac{NM-1}{M^2(N-1)} S^2[1 + (M-1)\rho]$$

$$\doteq \frac{1-f}{nM} S^2[1 + (M-1)\rho] \tag{9.10}$$

where ρ is the intracluster correlation coefficient.

Proof. Let y_i denote the total for the ith cluster and $\bar{y} = \sum_{}^{n} y_i/n$. By theorems 2.1 and 2.2, $\bar{y}$ is an unbiased estimate of $\bar{Y}$ with variance

$$V(\bar{y}) = \frac{(1-f)}{n} \frac{\sum(y_i - \bar{Y})^2}{N-1}$$

But $\bar{y} = M\bar{\bar{y}}$ and $\bar{Y} = M\bar{\bar{Y}}$. Hence $\bar{\bar{y}}$ is an unbiased estimate of $\bar{\bar{Y}}$ with variance

$$V(\bar{\bar{y}}) = \frac{1-f}{nM^2} \frac{\sum(y_i - \bar{Y})^2}{N-1} \tag{9.11}$$

But

$$(y_i - \bar{Y}) = (y_{i1} - \bar{\bar{Y}}) + (y_{i2} - \bar{\bar{Y}}) + \cdots + (y_{iM} - \bar{\bar{Y}})$$

Square and sum over all N clusters,

$$\sum_i^N (y_i - \bar{Y})^2 = \sum_i^N \sum_j^M (y_{ij} - \bar{\bar{Y}})^2 + 2\sum_i^N \sum_{j<k}^M (y_{ij} - \bar{\bar{Y}})(y_{ik} - \bar{\bar{Y}})$$

$$= (NM-1)S^2 + (M-1)(NM-1)\rho S^2$$

$$= (NM-1)S^2[1 + (M-1)\rho] \tag{9.12}$$

using the definition of ρ in (9.9). Substitute in (9.11) for $V(\bar{y})$. This gives

$$V(\bar{y}) = \frac{1-f}{n} \cdot \frac{NM-1}{M^2(N-1)} S^2 [1+(M-1)\rho]$$

This completes the proof.

If a simple random sample of nM elements is taken, the formula for $V(\bar{y})$ is the same as (9.10) except for the term in brackets. The factor

$$1+(M-1)\rho$$

shows by how much the variance is changed by the use of a cluster instead of an element as sampling unit. This factor is therefore Kish's (*deff*) for clusters of size M (section 4.11). If $\rho > 0$, the cluster is less precise for a given bulk of sample. If $\rho < 0$, as sometimes happens, the cluster is more precise. Theorem 9.2 is a simple extension of theorem 8.2, p. 209.

An alternative expression can be given for ρ. Let S_b^2 denote the variance among cluster totals, on a single unit basis. Then

$$\sum (y_i - \bar{Y})^2 = (N-1)MS_b^2$$

Equation 9.12 can be written as

$$(N-1)MS_b^2 = (NM-1)S^2[1+(M-1)\rho]$$

so that

$$\rho = \frac{(N-1)MS_b^2 - (NM-1)S^2}{(NM-1)(M-1)S^2} \doteq \frac{S_b^2 - S^2}{(M-1)S^2} \tag{9.13}$$

when terms in $1/N$ are negligible.

The value of the within-cluster mean square

$$S_w^2 = \sum_i^N \sum_j^M (y_{ij} - \bar{y}_i)^2 / N(M-1) \tag{9.14}$$

is worth noting. By the one-way analysis of variance,

$$(NM-1)S^2 = \sum_i^N (y_i - \bar{Y})^2 / M + \sum_i^N \sum_j^M (y_{ij} - \bar{y}_i)^2$$

$$= \frac{(NM-1)}{M} S^2 [1+(M-1)\rho] + N(M-1)S_w^2$$

by (9.12). Hence,

$$S_w^2 = \frac{NM-1}{NM} S^2(1-\rho) \doteq S^2(1-\rho) \tag{9.15}$$

A good discussion of the numerical values of ρ for different items and different sizes of cluster is given by Hansen, Hurwitz, and Madow (1953), who regard ρ as a "measure of homogeneity" of the cluster.

9.5   VARIANCE FUNCTIONS

In some types of surveys, for example, soil sampling, crop cutting, and surveys of farming that utilize an areal sampling unit, the size of the cluster unit may be capable of almost continuous variation. In the search for the best unit the problem is not that of choosing between two or three specific sizes that have been tried but of finding the optimum value of M regarded as a continuous variable. This problem requires a method of predicting the variance S_b^2 between units in the population as a function of M. By the analysis of variance, S_b^2 can be found if we know (a) the variance S^2 between all elements in the population and (b) the variance S_w^2 between elements that lie in the same unit. Our approach is to predict S_w^2 and S^2 and to find S_b^2 by the analysis of variance.

The sample data produce estimates of S^2 and S_w^2 for the size of unit actually used. Since S^2 is the variance among elements, it is not affected by the size of the unit. However, S_w^2 will be affected. It might be expected to increase as the size of the large unit increases. If the large units to be examined differ little in size from the unit actually used, a first approximation is to regard S_w^2 as constant, using the estimate given by the sample data. An investigation by McVay (1947) suggests that this approximation may often be satisfactory.

As a better approximation, attempts have been made (Jessen, 1942; Mahalanobis, 1944; Hendricks, 1944) to develop a general law to predict how S_w^2 changes with the size of unit. In several agricultural surveys, S_w^2 appeared to be related to M by the empirical formula

$$S_w^2 = AM^g \qquad (g > 0) \qquad (9.16)$$

where A and g are constants that do not depend on M. In this formula S_w^2 increases steadily as M increases. Usually g is small. A curve of this type might be expected when there are forces that exert a similar influence on elements close together. Climate, soil type, topography, and access to markets tend to give neighboring farms similar features.

Theoretically, the formula is open to objection, since it makes S_w^2 increase without bound as M increases. If we assume, as seem reasonable, that there is no correlation between elements that are far apart, a formula in which S_w^2 approaches an upper bound with large M would be more appropriate. However, a formula will suffice if it gives a good fit over the range of M that is under investigation.

If this formula fits, $\log S_w^2$ should plot as a straight line against $\log M$. Values of S_w^2 for at least two values of M are needed in order to estimate the constants $\log A$ and g. At least three values of M are necessary for any appraisal of the linearity of the fit.

From the analysis of variance in Table 9.5 (p. 239) we find

$$S_b^2 = \frac{(NM-1)S^2 - N(M-1)S_w^2}{N-1}$$

$$= \frac{(NM-1)S^2 - N(M-1)AM^g}{N-1}$$

$$\doteq MS^2 - (M-1)AM^g \qquad (9.17)$$

Hendricks (1944) has pointed out that the complete population might be regarded as a single large sampling unit containing NM elements. If (9.16) holds, then $S^2 = A(NM)^g$. The advantage of this device is that the values of A and g can now be estimated from the data for a survey in which only one value of M was used. The two equations that lead to the estimates are

$$\log S_w^2 = \log A + g \log M \qquad (9.18)$$

$$\log S^2 = \log A + g \log (NM) \qquad (9.19)$$

The formula for S_b^2 becomes [from (9.17)]

$$S_b^2 \doteq AM^g[MN^g - (M-1)] \qquad (9.20)$$

This method furnishes no check on the correctness of (9.16), which might hold well enough for small values of M but fail for a value as large as NM.

Formula 9.16 is presented as an example of the methodology rather than as a general law. The reader who faces a similar problem should construct and test whatever type of formula seems most appropriate to his material. In some cases $\log S_b^2$ might be a linear function of M, as Fairfield Smith (1938) suggested for yield data.

9.6 A COST FUNCTION

In an extensive survey the nature of the field costs plays a large part in determining the optimum unit. As an illustration of the role of cost factors, we will describe a cost function developed by Jessen (1942) for farm surveys in which the large units are clusters of neighboring farms.

Two components of field cost are distinguished. The component c_1Mn comprises costs that vary directly with the total number of elements (farms). Thus c_1 contains the cost of the interview and the cost of travel from farm to farm within the cluster.

The second component, $c_2\sqrt{n}$, measures the cost of travel between the clusters. Tests on a map showed that this cost, for a fixed population, varies approximately as the square root of the number of clusters. Total field cost is therefore

$$C = c_1Mn + c_2\sqrt{n} \qquad (9.21)$$

Assuming simple random sampling and ignoring the fpc, the variance of the mean per element $\bar{\bar{y}}$ is S_b^2/nM. From (9.17), this equals

$$V(\bar{\bar{y}}) = \frac{S^2 - (M-1)AM^{g-1}}{n} \tag{9.22}$$

To determine the optimum size of unit, we find M, and incidentally n, to minimize V for fixed C. The general solution is complicated, although its application in a numerical problem presents no great difficulty.

By some manipulation we can obtain the equation that gives the optimum M. First solve the cost equation (9.21) as a quadratic in $\sqrt{n}$. This gives

$$\frac{2c_1 M\sqrt{n}}{c_2} = \left(1 + \frac{4Cc_1 M}{c_2^2}\right)^{1/2} - 1 \tag{9.23}$$

The equation to be minimized is

$$C + \lambda V = c_1 Mn + c_2\sqrt{n} + \lambda V$$

Differentiating, and noting that $\partial V/\partial n = -V/n$, we obtain the equations

n:
$$c_1 M + \tfrac{1}{2}c_2 n^{-1/2} = -\frac{\lambda}{\partial n}\frac{\partial V}{} = \frac{\lambda V}{n} \tag{9.24}$$

M:
$$c_1 n = -\frac{\lambda}{\partial M}\frac{\partial V}{} \tag{9.25}$$

Divide (9.25) by (9.24) to eliminate λ. This leads to

$$\frac{n}{V}\frac{\partial V}{\partial M} = -\frac{c_1 n}{c_1 M + \tfrac{1}{2}c_2 n^{-1/2}}$$

or

$$\frac{M}{V}\frac{\partial V}{\partial M} = -\frac{1}{1 + c_2/2c_1 M\sqrt{n}} \tag{9.26}$$

If we substitute for $\sqrt{n}$ from (9.23), we obtain, after some simplification,

$$\frac{M}{V}\frac{\partial V}{\partial M} = \left(1 + \frac{4Cc_1 M}{c_2^2}\right)^{-1/2} - 1 \tag{9.27}$$

By writing out the left side of this equation in full and changing signs on both sides, we find

$$\frac{AM^{g-1}[gM - (g-1)]}{S^2 - (M-1)AM^{g-1}} = 1 - \left(1 + \frac{4Cc_1 M}{c_2^2}\right)^{-1/2} \tag{9.28}$$

This equation gives the optimum M. The left side does not involve any of the cost factors, being dependent only on the shape of the variance function. Both sides can be seen to be increasing functions of M, for $g > 0$, $M \geq 1$, within the region of interest. Suppose that the solution has been found for specified values of C, c_1, and c_2, and we wish to examine the effect of an increase in c_1 on this solution. The left side does not depend on c_1, but the right side increases as c_1 increases. However, the optimum M is found to decrease because of the term $c_1 M$ on the right. A decrease in c_2 produces a similar effect.

Now c_1 increases if the length of interview increases, whereas c_2 decreases if travel becomes cheaper or if the farms in a given area become denser. These facts lead to the conclusion that the optimum size of unit becomes *smaller* when

length of interview increases
travel becomes cheaper
the elements (farms) become more dense
total amount of money used (C) increases

This conclusion is a consequence of the type of cost function and would require reexamination with a different function. It illustrates the fact that the optimum unit is not a fixed characteristic of the population, but depends also on the type of survey and on the levels of prices and wages.

Hansen, Hurwitz, and Madow (1953) give an excellent discussion of the construction of cost functions for surveys involving cluster sampling.

9.7 CLUSTER SAMPLING FOR PROPORTIONS

The same techniques apply to cluster sampling for proportions. Suppose that the M elements in any cluster can be classified into two classes and that $p_i = a_i/M$ is the proportion in class C in the ith cluster. A simple random sample of n clusters is taken, and the average p of the observed p_i in the sample is used as the estimate of the population proportion P.

It will be recalled (section 3.12) that we cannot use binomial theory to find $V(p)$ but must apply the formula for continuous variates to the p_i. This gives

$$V(p) = \frac{N-n}{Nn} \frac{\sum_{i=1}^{N} (p_i - P)^2}{N-1} \doteq \frac{N-n}{N^2 n} \sum (p_i - P)^2 \tag{9.29}$$

Alternatively, if we take a simple random sample of nM elements, the variance of p is obtained by binomial theory (theorem 3.2) as

$$V_{bin}(p) = \frac{(NM - nM)}{NM - 1} \frac{PQ}{nM} \doteq \frac{N-n}{N} \frac{PQ}{nM} \tag{9.30}$$

if N is large. Consequently the factor, the *deff*,

$$\frac{V(p)}{V_{bin}(p)} \doteq \frac{M \sum (p_i - P)^2}{NPQ} \quad (N \text{ large}) \tag{9.31}$$

shows the relative change in the variance caused by the use of clusters. Numerical values of this factor are helpful in making preliminary estimates of sample size with cluster sampling. The required sample size is first estimated by the binomial formula and then multiplied by the factor to indicate the size that will be necessary with cluster sampling. For an illustration, see Cornfield (1951).

If the cluster sizes M_i are variable, the estimate $p = \sum a_i / \sum M_i$ is a ratio estimate. Its variance is given approximately by the formula (section 3.12)

$$V(p) \doteq \frac{N-n}{Nn\bar{M}^2} \frac{\sum\limits_{i=1}^{N} M_i^2 (p_i - P)^2}{N-1} \qquad (9.32)$$

where $\bar{M} = \sum M_i / N$ is the average size of cluster.

If this sample is compared with a simple random sample of $n\bar{M}$ elements, we find, as a generalization of (9.31),

$$\frac{V(p)}{V_{bin}(p)} \doteq \frac{\sum M_i^2 (p_i - P)^2}{N\bar{M}PQ} \qquad (9.33)$$

As with continuous variates, the relationship of size of cluster to between-cluster variance can be investigated, either by expressing the factor in (9.31) and (9.33) as a function of $\bar{M}$, or by seeking a relation between the within-cluster variance and $\bar{M}$. If we assign the value 1 to any unit that falls in class C and 0 to any other unit, the fundamental analysis of variance equation for fixed M is

$$NMP(1-P) = M \sum (p_i - P)^2 + M \sum p_i (1 - p_i) \qquad (9.34)$$
$$\text{total ss} = \text{ss between clusters} + \text{ss within clusters}$$

From this relation the mean square within clusters can be computed and plotted as a function of M. McVay (1947) describes how this analysis can be used to investigate optimum cluster size.

EXERCISES

9.1 For the data in Table 9.1 compare the relative net precisions of the four types of unit when the object is to estimate the total number of seedlings in the bed with a standard error of 200 seedlings. (Note that the fpc is involved.)

9.2 For the data in Table 3.5 (p. 67) estimate the relative precision of the household to the individual for estimating the sex ratio and the proportion of people who had seen a doctor in the past 12 months, assuming simple random sampling.

9.3 A population consisting of 2500 elements is divided into 10 strata, each containing 50 large units composed of five elements. The analysis of variance of the population for an item is as follows, on an element basis.

	df	ms
Between strata	9	30.6
Between large units within strata	490	3.0
Between elements within large units	2000	1.6

Ignoring the fpc, is the relative precision of the large to the small unit greater with simple random sampling than with stratified random sampling (proportional allocation)?

9.4   A population containing $L\bar{N}M$ elements is divided into L strata, each having $\bar{N}$ large units, each of which contains M small units. The following quantities come from the analysis of variance of the population, on an element basis.

$$S_1^2 = \text{mean square between strata}$$

$$S_2^2 = \text{mean square between large units within strata}$$

$$S_3^2 = \text{mean square between elements within strata}$$

If $\bar{N}$ is large and the fpc is ignored, show that the relative precision of the large to the small unit (element) is improved by stratification if

$$\frac{(M-1)}{S_1^2} < \frac{M}{S_2^2} - \frac{1}{S_3^2}$$

9.5   In a rural survey in which the sampling unit is a cluster of M farms, the cost of taking a sample of n units is

$$C = 4tMn + 60\sqrt{n}$$

where t is the time in hours spent getting the answers from a single farmer. If $2000 is spent on the survey, the values of n for $M = 1, 5, 10$; $t = \frac{1}{2}, 2$, work out as follows.

		M	
	1	5	10
$t = \frac{1}{2}$ hr	400	131	74
$t = 2$ hr	156	40	21

Verify two of these values to ensure that you understand the use of the formula.

The variance of the sample mean (ignoring the fpc) is

$$\frac{S^2}{Mn}[1 + (M-1)\rho]$$

If $\rho = 0.1$ for all M between 1 and 10, which size of unit is most precise for (a) $t = \frac{1}{2}$ hr, (b) $t = 2$ hr? How do you explain the difference in results?

9.6   If $5000 were available for the survey, would you expect the optimum size of unit to decrease or increase (relative to that for $2000)? Give reasons. You may, if you wish, find the optimum size in order to check your argument.

Single-Stage Cluster Sampling: Clusters of Unequal Sizes

9A.1 CLUSTER UNITS OF UNEQUAL SIZES

In most applications the cluster units (e.g., counties, cities, city blocks) contain different numbers of elements or subunits (areal units, households, persons). This chapter deals with some of the numerous methods of sample selection and estimation that have been produced for cluster units of unequal sizes. Let M_i be the number of elements in the ith unit. For the estimation of the population *total Y* of the y_{ij}, two methods are already familiar to us.

Simple random sample of clusters: unbiased estimate

As before, let

$$y_i = \sum_{i=1}^{M_i} y_{ij} = M_i \bar{y}_i$$

denote the item *total* for the ith cluster unit. Given a simple random sample of n of the N population units, an unbiased estimate of Y is (by theorem 2.1, corollary)

$$\hat{Y} = \frac{N}{n} \sum_{i=1}^{n} y_i \qquad (9A.1)$$

By theorem 2.2, its variance is

$$V(\hat{Y}) = \frac{N^2(1-f)}{n} \frac{\sum_{i=1}^{N} (y_i - \bar{Y})^2}{N-1} \qquad (9A.2)$$

where $\bar{Y} = Y/N$ is the population mean per cluster unit.

The estimate $\hat{Y}$ is often found to be of poor precision. This occurs when the $\bar{y}_i$ (means per *element*) vary little from unit to unit and the M_i vary greatly. In this event the $y_i = M_i \bar{y}_i$ also vary greatly from unit to unit and the variance (9A.2) is large.

Simple Random Sample of Clusters: Ratio-to-Size Estimate

Let

$$M_0 = \sum_{i=1}^{N} M_i = \text{total number of elements in the population.}$$

If the M_i and hence M_0 are all known, an alternative is a ratio estimate in which M_i is taken as the auxiliary variate x_i.

$$\hat{Y}_R = M_0 \frac{\sum_{i=1}^{n} y_i}{\sum_{i=1}^{n} M_i} = M_0 \text{ (sample mean per element)}$$

In the notation of the ratio estimate the population ratio $R = Y/X = Y/M_0 = \bar{\bar{Y}}$, the population mean per element. By theorem 6.1, assuming that the number of clusters in the sample is large,

$$V(\hat{Y}_R) \doteq \frac{N^2(1-f)}{n} \frac{\sum_{i=1}^{N} (y_i - M_i \bar{\bar{Y}})^2}{N-1} \tag{9A.3}$$

$$\doteq \frac{N^2(1-f)}{n} \frac{\sum_{i=1}^{N} M_i^2 (\bar{y}_i - \bar{\bar{Y}})^2}{N-1} \tag{9A.4}$$

As (9A.4) shows, the variance of $\hat{Y}_R$ depends on the variability among the means per element and is often found to be much smaller than $V(\hat{Y})$.

Note that $\hat{Y}_R$ requires a knowledge of the total M_0 of all the M_i, while $\hat{Y}$ does not. The reverse is true when we are estimating the population mean per element. In this case the corresponding estimates are

$$\hat{\bar{Y}} = \frac{\hat{Y}}{M_0} = \frac{N}{nM_0} \sum_{}^{n} y_i, \ \hat{\bar{Y}}_R = \frac{\hat{Y}_R}{M_0} = \frac{\sum_{}^{n} y_i}{\sum_{}^{n} M_i} = \text{sample mean per element}$$

Thus, $\hat{\bar{Y}}_R$ requires knowledge of only the M_i that fall in the selected sample.

9A.2 SAMPLING WITH PROBABILITY PROPORTIONAL TO SIZE

If all the M_i are known, another technique, developed by Hansen and Hurwitz (1943), is to select the units with probabilities proportional to their sizes M_i. One method of selecting a single unit is illustrated in the following small population of $N = 7$ units.

Unit	Size M_i	$\sum M_i$	Assigned Range
1	3	3	1–3
2	1	4	4
3	11	15	5–15
4	6	21	16–21
5	4	25	22–25
6	2	27	26–27
7	3	30	28–30

The cumulative sums of the M_i are formed. To select a unit, draw a random number between 1 and $M_0 = 30$. Suppose that this is 19. In the sum, number 19 falls in unit 4, which covers the range from numbers 16 to 21, inclusive. With this method of drawing, the probability that any unit another is selected is proportional to its size.

This method of selecting a unit is convenient when N is only moderate, or in stratified sampling when the N_h are moderate or small, but the cumulation of the M_i can be time-consuming with N large (e.g., $N = 20,000$). For this case, Lahiri (1951) has given an alternative method that avoids the cumulation. Let M_{max} be the largest of the M_i. Draw a random number between 1 and N; suppose this is i. Now draw another random number m between 1 and M_{max}. If m is less than or equal to M_i, the ith unit is selected. If not, try another pair of random numbers. Naturally, this method involves the fewest rejections when the M_i do not differ too much in size.

Now consider $n > 1$. Assume at present that sampling is *with replacement*. To select a second unit by the cumulative method, draw a new random number between 1 and 30. However, unlike sampling without replacement, we do not forbid the selection of unit 4 a second time. With this rule, the probabilities of selection remain proportional to the sizes at each draw. An advantage of selection with replacement is that the formulas for the true and estimated variances of the estimates are simple.

In sampling without replacement, on the other hand (section 9A.6), keeping the selection probabilities proportional to the chosen sizes is more difficult and sooner or later becomes impossible as n increases. This may be seen in the extreme (although impractical) case $n = 7$ in the preceding example. If selection were made without replacement, every unit would be certain to be chosen, irrespective of the original sizes M_i. However, for stratified sampling in which the N_h are small, much research has been done (section 9A.6 *ff*) to develop practical methods of sampling with unequal probabilities without replacement.

9A.3 SELECTION WITH UNEQUAL PROBABILITIES
WITH REPLACEMENT

Let the ith unit be selected with probability M_i/M_0 and with replacement, where $M_0 = \sum M_i$. We will show that an unbiased estimate of the population total Y is

$$\hat{Y} = \frac{M_0}{n}(\bar{y}_1 + \bar{y}_2 + \cdots + \bar{y}_n)$$

$$= M_0 \text{ (mean of the unit means per element)} \tag{9A.5}$$

For comparison with subsequent methods, this estimate will be denoted by $\hat{Y}_{pps}$. Furthermore,

$$V(\hat{Y}_{pps}) = \frac{M_0}{n} \sum_{i=1}^{N} M_i(\bar{y}_i - \bar{\bar{Y}})^2 \tag{9A.6}$$

so that the variance of $\hat{Y}_{pps}$, like that of $\hat{Y}_R$, depends on the variability of the unit *means per element.*

In some applications the sizes M_i are known only approximately. In others the "size" is not the number of elements in the unit but a measure of its bigness that is thought to be highly correlated with the unit total y_i. For instance, the "size" of a hospital might be measured by the total number of beds or by the average number of occupied beds over some time period. Similarly, various measures of the "size" of a restaurant, a bank, or an agricultural district can be devised. Consequently, we will consider a measure of size M_i' and a corresponding probability of selection $z_i = M_i'/M_0'$, where $M_0' = \sum_{i=1}^{N} M_i'$. As far as the theoretical results are concerned, the z_i can be any set of positive numbers that add to 1 over the population. It will be shown that

$$\hat{Y}_{ppz} = \frac{1}{n} \sum_{i=1}^{n} \frac{y_i}{z_i} \tag{9A.7}$$

is an unbiased estimate of Y with variance

$$V(\hat{Y}_{ppz}) = \frac{1}{n} \sum_{i=1}^{N} z_i\left(\frac{y_i}{z_i} - Y\right)^2 \tag{9A.8}$$

The proofs utilize a method introduced in section 2.10. Let t_i be the number of times that the ith unit appears in a specific sample of size, n, where t_i may have any of the values $0, 1, 2, \ldots, n$. Consider the joint frequency distribution of the t_i for all N units in the population.

The method of drawing the sample is equivalent to the standard probability problem in which n balls are thrown into N boxes, the probability that a ball goes into the ith box being z_i at every throw. Consequently the joint distribution of the t_i is the multinomial expression

$$\frac{n!}{t_1!t_2!\ldots t_N!}z_1{}^{t_1}z_2{}^{t_2}\ldots z_N{}^{t_N}$$

For the multinomial, the following properties of the distribution of the t_i are well known.

$$E(t_i)=nz_i, \qquad V(t_i)=nz_i(1-z_i), \qquad \text{Cov}\,(t_it_j)=-nz_iz_j \qquad (9A.9)$$

Theorem 9A.1. If a sample of n units is drawn with probabilities z_i and with replacement, then

$$\hat{Y}_{ppz}=\frac{1}{n}\sum_{i=1}^{n}\frac{y_i}{z_i} \qquad (9A.10)$$

is an unbiased estimate of Y with variance

$$V(\hat{Y}_{ppz})=\frac{1}{n}\sum_{i=1}^{N}z_i\left(\frac{y_i}{z_i}-Y\right)^2 \qquad (9A.11)$$

Proof.   We may write

$$\hat{Y}_{ppz}=\frac{1}{n}\left(t_1\frac{y_1}{z_1}+t_2\frac{y_2}{z_2}+\cdots+t_N\frac{y_N}{z_N}\right)=\frac{1}{n}\sum_{i=1}^{N}t_i\frac{y_i}{z_i}$$

where the sum extends over *all* units in the population. In repeated sampling the t's are the random variables, whereas the y_i and the z_i are a set of fixed numbers. Hence, since $E(t_i)=nz_i$ by (9A.9),

$$E(\hat{Y}_{ppz})=\frac{1}{n}\sum_{i=1}^{N}(nz_i)\frac{y_i}{z_i}=\sum_{i=1}^{n}y_i=Y$$

so that $\hat{Y}_{ppz}$ is unbiased. Also,

$$V(\hat{Y}_{ppz})=\frac{1}{n^2}\left[\sum_{i=1}^{N}\left(\frac{y_i}{z_i}\right)^2V(t_i)+2\sum_{i=1}^{N}\sum_{j>i}^{N}\frac{y_i}{z_i}\frac{y_j}{z_j}\text{Cov}\,(t_it_j)\right] \qquad (9A.12)$$

$$=\frac{1}{n}\left[\sum_{i=1}^{N}\left(\frac{y_i}{z_i}\right)^2z_i(1-z_i)-2\sum_{i=1}^{N}\sum_{j>i}^{N}\frac{y_i}{z_i}\frac{y_j}{z_j}z_iz_j\right] \qquad (9A.13)$$

$$=\frac{1}{n}\left(\sum_{i=1}^{N}\frac{y_i^2}{z_i}-Y^2\right)=\frac{1}{n}\sum_{i=1}^{N}z_i\left(\frac{y_i}{z_i}-Y\right)^2 \qquad (9A.11)$$

since $\sum z_i=1$. This completes the proof.

Taking $z_i=M_i/M_0$ in theorem 9A.1 gives the corresponding results for sampling with probability proportional to size.

An alternative expression for $V(\hat{Y}_{ppz})$ can be given. From (9A.13),

$$nV(\hat{Y}_{ppz}) = \sum_{i=1}^{N} \frac{y_i^2(1-z_i)}{z_i} - 2 \sum_{i=1}^{N} \sum_{j>i}^{N} y_iy_j \qquad (9A.14)$$

Since $(1-z_i)$ equals the sum of all other z's in the population, the coefficient of y_i^2/z_i in (9A.14) contains a term z_j for any $j \neq i$. Similarly, the coefficient of y_j^2/z_j contains a term z_i. Hence,

$$V(\hat{Y}_{ppz}) = \frac{1}{n} \sum_{i=1}^{N} \sum_{j>i}^{N} \left(\frac{y_i^2 z_j}{z_i} + \frac{y_j^2 z_i}{z_j} - 2y_iy_j \right)$$

$$= \frac{1}{n} \sum_{i=1}^{N} \sum_{j>i}^{N} z_iz_j \left(\frac{y_i}{z_i} - \frac{y_j}{z_j} \right)^2 \qquad (9A.15)$$

Theorem 9A.2. If a sample of n units is drawn with probability proportional to z_i with replacement, an unbiased sample estimate of $V(\hat{Y}_{ppz})$ is, for any $n > 1$,

$$v(\hat{Y}_{ppz}) = \sum_{i=1}^{n} \left(\frac{y_i}{z_i} - \hat{Y}_{ppz} \right)^2 \Big/ n(n-1) \qquad (9A.16)$$

Proof. By the usual algebraic identity,

$$\sum_{i=1}^{n} \left(\frac{y_i}{z_i} - \hat{Y}_{ppz} \right)^2 = \sum_{i=1}^{n} \left(\frac{y_i}{z_i} - Y \right)^2 - n(\hat{Y}_{ppz} - Y)^2 \qquad (9A.17)$$

Hence, from (9A.16)

$$E[n(n-1)v(\hat{Y}_{ppz})] = E \sum_{i=1}^{n} \left(\frac{y_i}{z_i} - Y \right)^2 - nV(\hat{Y}_{ppz}) \qquad (9A.18)$$

by definition of $V(\hat{Y}_{ppz})$. Introducing the variables t_i, we get

$$n(n-1)E[v(\hat{Y}_{ppz})] = E \sum_{i=1}^{N} t_i \left(\frac{y_i}{z_i} - Y \right)^2 - nV(\hat{Y}_{ppz})$$

$$= n \sum_{i=1}^{n} z_i \left(\frac{y_i}{z_i} - Y \right)^2 - nV(\hat{Y}_{ppz})$$

$$= n(n-1)V(\hat{Y}_{ppz}) \qquad (9A.19)$$

using (9A.11) in theorem 9A.1. This completes the proof.

Since $\hat{Y}_{ppz}$ is the mean of the n values y_i/z_i, formula (9A.16) has a very simple form.

When selection is strictly proportional to size (i.e., $z_i = M_i/M_0$), theorems 9A.1, 9A.2 are expressed as follows.

Theorem 9A.3. If a sample of n units is drawn with probabilities proportional to size, $z_i = M_i/M_0$ and, with replacement,

$$\hat{Y}_{pps} = \frac{M_0}{n} \sum_{i=1}^{n} \left(\frac{y_i}{M_i} \right) = \frac{M_0}{n} \sum_{i=1}^{n} (\bar{y}_i) = M_0\bar{\bar{y}} \qquad (9A.20)$$

where $\bar{y}$ is the unweighted mean of the unit means, is an unbiased estimate of Y with variance

$$V(\hat{Y}_{pps}) = \frac{M_0}{n} \sum_{i=1}^{N} M_i(\bar{y}_i - \bar{\bar{Y}})^2 \qquad (9A.21)$$

These results follow from theorem 9A.1, since $\bar{y}_i = y_i/M_i$ and $\bar{\bar{Y}} = Y/M_0$.

Theorem 9A.4.   Under the conditions of theorem 9A.3, an unbiased sample estimate of $V(\hat{Y}_{pps})$ is

$$v(\hat{Y}_{pps}) = M_0^2 \sum_{i=1}^{n} (\bar{y}_i - \bar{y})^2/n(n-1) \qquad (9A.22)$$

The result follows by substituting $z_i = M_i/M_0$ in (9A.16), since $\bar{y}_i = y_i/M_i$ and $\hat{Y}_{ppz} = M_0\bar{y}$.

9A.4   THE OPTIMUM MEASURE OF SIZE

In cases in which the measure of size M_i' is some estimate of the bigness of the unit, a question of interest is: what measure of size minimizes the variance of $\hat{Y}_{ppz}$? Now,

$$V(\hat{Y}_{ppz}) = \frac{1}{n} \sum_{i=1}^{N} z_i\left(\frac{y_i}{z_i} - Y\right)^2 = \frac{1}{n}\left(\sum_{i=1}^{N} \frac{y_i^2}{z_i} - Y^2\right)$$

This expression becomes zero if $z_i \propto y_i$: that is, $z_i = y_i/Y$. If the y_i are all positive, this set of z_i is an acceptable set of probabilities. Consequently, the best measures of size are numbers proportional to the item totals y_i for the units.

This result is not of direct practical application; if the y_i were known in advance for the whole population the sample would be unnecessary. The result suggests that if the y_i are relatively stable through time, the most recently available previous values of the y_i may be the best measures of size for this item. In practice, of course, a *single* measure of size must be used for all items in selecting the sample. If there is a choice between different measures of size, the measure most nearly proportional to the unit totals of the principal items is likely to be best.

9A.5   RELATIVE ACCURACIES OF THREE TECHNIQUES

In this section we compare the accuracies of three preceding techniques for estimating the population total with cluster units of unequal sizes (assuming that the M_i are known if the technique requires them).

1. Selection: equal probabilities. Estimate; $\hat{Y}_u$.
2. Selection: equal probabilities. Estimate; $\hat{Y}_R$.
3. Selection: probability $\propto$ size. Estimate; $\hat{Y}_{pps}$.

There is no simple rule for deciding which is most accurate. The issue depends on the relation between $\bar{y}_i$ and M_i and on the variance of $\bar{y}_i$ as a function of M_i. The situation favorable to the ratio and *pps* estimates is that in which $\bar{y}_i$ is unrelated to M_i. The situation favorable to $\hat{Y}_u$ is that in which the unit *total* y_i is unrelated to M_i.

Some guidance can be obtained by expressing the variances of the three estimates in a comparable form. We assume that $(N-1) \doteq N$ and write $E(y_i - \bar{Y})^2 = \sum_{}^{N} (y_i - \bar{Y})^2/N$. We also assume that the bias of $\hat{Y}_R$ is negligible.

For $\hat{Y}_u$ we have, from (9A.2),

$$nV(\hat{Y}_u) = N^2(1-f)E(y_i - \bar{Y})^2 = (1-f)E(Ny_i - Y)^2 \qquad (9A.23)$$

For $\hat{Y}_R$, from (9A.4),

$$nV(\hat{Y}_R) = N^2(1-f)EM_i^2(\bar{y}_i - \bar{\bar{Y}})^2 = (1-f)E\left(\frac{M_i}{\bar{M}}\right)^2(M_0\bar{y}_i - Y)^2$$

$$(9A.24)$$

where $\bar{M} = \sum M_i/N = M_0/N$.

From (9A.21), for $\hat{Y}_{pps}$,

$$nV(\hat{Y}_{pps}) = NM_0EM_i(\bar{y}_i - \bar{\bar{Y}})^2 = M_0^2E\left(\frac{M_i}{\bar{M}}\right)(\bar{y}_i - \bar{\bar{Y}})^2$$

$$= E\left(\frac{M_i}{\bar{M}}\right)(M_0\bar{y}_i - Y)^2 \qquad (9A.25)$$

From (9A.23), (9A.24), (9A.25), we see that $V(\hat{Y}_u)$ depends on the accuracy of the quantities $Ny_i = NM_i\bar{y}_i$ as estimates of Y, while $V(\hat{Y}_R)$ and $V(\hat{Y}_{pps})$ depend on the accuracy of the quantities $M_0\bar{y}_i = M_0y_i/M_i$ as estimates of Y. If $\bar{y}_i$ is unrelated to M_i, we expect the $M_0\bar{y}_i$ to be more accurate than the $NM_i\bar{y}_i$, and the reverse if y_i is unrelated to M_i.

As regards $\hat{Y}_R$ and $\hat{Y}_{pps}$, note from (9A.24) and (9A.25) that $V(\hat{Y}_R)$ gives relatively greater weight to large units than $V(\hat{Y}_{pps})$. Note also that $\hat{Y}_u$ and $\hat{Y}_R$ benefit from the fpc term, which can become substantial in small strata (e.g., with $n_h = 2$, $N_h = 10$). This point stimulated the development of unequal probability selection without replacement. Formula (9A.24) for $\hat{Y}_R$, of course, holds only in large samples.

Further comparisons among the methods have been made from an infinite population model by Cochran, 2nd ed., Des Raj (1954, 1958), Yates (1960), Zarcovic (1960), and Foreman and Brewer (1971). Most writers assume that the finite population is a random sample from an infinite superpopulation in which

$$y_i = \alpha + \beta M_i + e_i; \qquad E(e_i|M_i) = 0 \qquad (9A.26)$$

which is hoped to approximate the relation that holds in many surveys. Some assumption must also be made about the variance of e_i in clusters of given size.

From (9.12) in Section 9.4, we get on dividing by $(N-1) \doteq N$,

$$V(e_i) = V(y_i|M_i) \doteq M_i S^2 [M_i \rho + (1-\rho)]$$

As an approximation this suggests $V(e_i) = c M_i^g$ where $1 < g < 2$ in most applications. From the model (9A.26) we get

$$\bar{Y} = \alpha + \beta \bar{M} + \bar{e}_N: \qquad \bar{\bar{Y}} = \frac{\alpha}{\bar{M}} + \beta + \frac{\bar{e}_N}{\bar{M}} \qquad (9A.27)$$

We assume here that $\bar{e}_N$ is negligible; this amounts to ignoring the fpc. It follows that,

$$\frac{n V(\hat{Y}_u)}{N^2} = E(y_i - \bar{Y})^2 = E[\beta(M_i - \bar{M}) + e_i]^2$$

$$= \beta^2 V(M_i) + c E(M_i^g) \qquad (9A.28)$$

$$\frac{n V(\hat{Y}_R)}{N^2} \doteq E M_i^2 (\bar{y}_i - \bar{\bar{Y}})^2 = E(y_i - M_i \bar{\bar{Y}})^2$$

$$= E[\alpha(1 - M_i/\bar{M}) + e_i]^2 = \frac{\alpha^2 V(M_i)}{\bar{M}^2} + c E(M_i^g) \qquad (9A.29)$$

$$\frac{n V(\hat{Y}_{pps})}{N^2} = \bar{M} E M_i (\bar{y}_i - \bar{\bar{Y}})^2 = \bar{M} E M_i \left[\alpha \left(\frac{1}{M_i} - \frac{1}{\bar{M}} \right) + \frac{e_i}{M_i} \right]^2$$

$$= \alpha^2 E \left[\frac{(M_i - \bar{M})^2}{M_i \bar{M}} \right] + v \bar{M} E(M_i^{g-1}) \doteq \frac{\alpha^2 V(M_i)}{\bar{M}^2} + c \bar{M} E(M_i^{g-1}) \qquad (9A.30)$$

Hence, approximately comparable variances under this model are

$$\frac{n V_u}{N^2} = \beta^2 V(M_i) + c E(M_i^g) \qquad (9A.31)$$

$$\frac{n V_R}{N^2} \doteq \alpha^2 \frac{V(M_i)}{\bar{M}^2} + c E(M_i^g) \qquad (9A.32)$$

$$\frac{n V_{pps}}{N^2} \doteq \frac{\alpha^2 V(M_i)}{\bar{M}^2} + c \bar{M} E(M_i^{g-1}) \qquad (9A.33)$$

Consider first $\alpha = 0$, the case in which $\bar{y}_i$ is unrelated to M_i. Clearly,

$$V_R < V_u$$

If β (which in this case becomes $\bar{\bar{Y}}$) is large, the superiority of the ratio estimate may be great. With $\alpha = 0$,

$$V_{pps} < V_u \qquad \text{for} \qquad g \geq 1$$

This holds because the covariance of M_i and M_i^{g-1} is positive if $g > 1$ and zero if

$g = 1$, so that with $g \geq 1$,

$$E(M_i^g) = E[(M_i)(M_i^{g-1})] \geq \bar{M}E(M_i^{g-1})$$

If $\beta = 0$ and $\alpha \neq 0$, so that the unit *total* y_i is unrelated to M_i, $\hat{Y}_u$ always beats $\hat{Y}_R$ and beats $\hat{Y}_{pps}$ except possibly in the unlikely case $g = 2$ with $\beta = 0$.

When neither α nor β vanishes, the relative performances of $\hat{Y}_u$ and $\hat{Y}_R$, $\hat{Y}_{pps}$ depend on the relative sizes of $|\alpha|$ and $|\beta|$. For instance, $\hat{Y}_R$ beats $\hat{Y}_u$ if $\beta^2 > \alpha^2/\bar{M}^2$, as noted by Foreman and Brewer (1971).

As regards V_R versus V_{pps}, the coefficients of α^2 are approximately the same in V_R and V_{pps}. From the terms in c, $V_{pps} < V_R$ in the case $g > 1$ expected to apply to most applications that have $\beta \neq 0$, while $V_{pps} > V_R$ if $g < 1$.

The results from this model agree with the conclusions suggested earlier in this section. If $\bar{y}_i$ shows no trend or only a moderate trend as M_i increases, the ratio method with equal probabilities and the *pps* method are more precise than unbiased estimation with equal probabilities and may be much more precise. $\hat{Y}_u$ is superior if the unit total y_i is unrelated to M_i. There is less to choose between the ratio and the *pps* estimates. Since g is expected to lie between 1 and 2, the *pps* estimate is usually more precise and its results are not restricted to large samples.

The estimates $\hat{Y}_u$ and $\hat{Y}_R$ are helped by the fpc term when this is appreciable. To anticipate later work (section 9A.12), the evidence suggests that the best *pps* methods without replacement benefit from about the same size of fpc term as in equal-probability sampling.

9A.6 SAMPLING WITH UNEQUAL PROBABILITIES WITHOUT REPLACEMENT

Much of this work was produced for extensive surveys in which the cluster units had first been stratified by some other principle (e.g., geographic location) into a substantial number of relatively small strata, only a small number of cluster units being drawn from each stratum. The case $n_h = 2$, which provides one degree of freedom from each stratum for estimating sampling errors, is of particular interest.

Suppose that two units are to be drawn from a stratum. The first unit is drawn with probabilities z_j, proportional to some measure of size. Let the ith unit be selected. If we follow the most natural method, at the second draw one of the remaining units is selected with assigned probabilities $z_j/(1 - z_i)$. Hence the total probability π_i that the ith unit will be selected at either the first or the second draw is

$$\pi_i = z_i + \sum_{j \neq i}^{N} \frac{z_j z_i}{(1 - z_j)} = z_i\left(1 + \sum_{j \neq i}^{N} \frac{z_j}{1 - z_j}\right) \tag{9A.34}$$

$$= z_i\left(1 + A - \frac{z_i}{1 - z_i}\right) \tag{9A.35}$$

where $A = \sum z_j/(1 - z_j)$ taken over all N units.

Suppose that $\pi_i = 2z_i$, the relative probabilities of selection of the units remaining proportional to the measure of size z_i. The simple estimator introduced by Horvitz and Thompson (1952),

$$\hat{Y}_{HT} = \sum_i^2 \frac{y_i}{\pi_i} = \frac{1}{2} \sum_i^2 \frac{y_i}{z_i}$$

will then have zero variance if the z_i are proportional to the y_i, since $z_i = y_i/Y$ and every selected unit will give the correct estimate $y_i/z_i = Y$. However, in (9A.35), the quantities $z_i' = \pi_i/2$ are always closer to equality than the original z_i because of the second factor in (9A.35). In the example given by Yates and Grundy (1953), with $N = 4$, $n = 2$, $z_i = 0.1$, 0.2, 0.3, and 0.4, the z_i' are found to be 0.1173, 0.2206, 0.3042, and 0.3579.

Three approaches have been used to cope with this distortion of the intended probabilities of selection. One is to retain the preceding natural method of sample selection, but make appropriate changes in the method of estimating the population total. One is to use a different method of sample selection that keeps $\pi_i = nz_i$ subject to certain restrictions on the values of the nz_i. One is to accept some distortion of the probabilities if this has compensating advantages (e.g. in simplicity or generality). Examples of the three methods will be given in section 9A.8 and following sections.

First, we present the best-known general estimate of the population total for unequal probability sampling without replacement.

9A.7 THE HORVITZ–THOMPSON ESTIMATOR

A sample of n units is selected, without replacement, by some method. Let

π_i = probability that the ith unit is in the sample

π_{ij} = probability that the ith and jth units are both in the sample

The following relations hold:

$$\sum_i^N \pi_i = n, \qquad \sum_{j \neq i}^N \pi_{ij} = (n-1)\pi_i, \qquad \sum_i \sum_{j>i}^N \pi_{ij} = \tfrac{1}{2} n(n-1) \qquad (9A.36)$$

To establish the second relation, let $P(s)$ denote the probability of a sample consisting of n specified units. Then $\pi_{ij} = \Sigma P(s)$ over all samples containing the ith and jth units, and $\pi_i = \Sigma P(s)$ over all samples containing the ith unit. When we take $\Sigma \pi_{ij}$ for $j \neq i$, every $P(s)$ for a sample containing the ith unit is counted $(n-1)$ times in the sum, since there are $(n-1)$ other values of j in the sample. This proves the second relation. The third relation follows from the second.

The Horvitz–Thompson (1952) estimator of the population total is

$$\hat{Y}_{HT} = \sum_i^n \frac{y_i}{\pi_i} \qquad (9A.37)$$

where y_i is the measurement for the ith unit.

Theorem 9A.5. If $\pi_i > 0$, $(i = 1, 2, \cdots, N)$,

$$\hat{Y}_{HT} = \sum_i^n \frac{y_i}{\pi_i}$$

is an unbiased estimator of Y, with variance

$$V(\hat{Y}_{HT}) = \sum_{i=1}^N \frac{(1-\pi_i)}{\pi_i} y_i^2 + 2 \sum_{i=1}^N \sum_{j>i}^N \frac{(\pi_{ij} - \pi_i\pi_j)}{\pi_i\pi_j} y_i y_j \qquad (9\text{A}.38)$$

where π_{ij} is the probability that units i and j both are in the sample.

Proof. Let t_i $(i = 1, 2, \ldots, N)$ be a random variable that takes the value 1 if the ith unit is drawn and zero otherwise. Then t_i follows the binomial distribution for a sample of size 1, with probability π_i. Thus

$$E(t_i) = \pi_i, \qquad V(t_i) = \pi_i(1-\pi_i) \qquad (9\text{A}.39)$$

The value of Cov $(t_i t_j)$ is also required. Since $t_i t_j$ is 1 only if both units appear in the sample,

$$\text{Cov}(t_i t_j) = E(t_i t_j) - E(t_i)E(t_j) = \pi_{ij} - \pi_i\pi_j \qquad (9\text{A}.40)$$

Hence, regarding the y_i as fixed and the t_i as random variables,

$$E(\hat{Y}_{HT}) = E\left(\sum_{i=1}^N \frac{t_i y_i}{\pi_i}\right) = \sum_{i=1}^N y_i = Y$$

$$V(\hat{Y}_{HT}) = \sum_i^N \left(\frac{y_i}{\pi_i}\right)^2 V(t_i) + 2\sum_i^N \sum_{j>i}^N \frac{y_i}{\pi_j} \frac{y_j}{\pi_j} \text{Cov}(t_i t_j)$$

$$= \sum_i^N \frac{(1-\pi_i)}{\pi_i} y_i^2 + 2\sum_i^N \sum_{j>i}^N \frac{(\pi_{ij} - \pi_i\pi_j)}{\pi_i\pi_j} y_i y_j \qquad (9\text{A}.41)$$

This proves the theorem.

This variance may be expressed in another form by using the first two of the relations in (9A.36). These give

$$\sum_{j\neq i} (\pi_{ij} - \pi_i\pi_j) = (n-1)\pi_i - \pi_i(n - \pi_i) = -\pi_i(1 - \pi_i)$$

Substituting for $(1 - \pi_i)$ in the first term in (9A.41),

$$\sum_i^N \frac{(1-\pi_i)}{\pi_i} y_i^2 = \sum_i^N \sum_{j\neq i}^N (\pi_i\pi_j - \pi_{ij})\left(\frac{y_i}{\pi_i}\right)^2 = \sum_i^N \sum_{j>i}^N (\pi_i\pi_j - \pi_{ij})\left[\left(\frac{y_i}{\pi_i}\right)^2 + \left(\frac{y_j}{\pi_j}\right)^2\right]$$

Hence,

$$V(\hat{Y}_{HT}) = \sum_i^N \sum_{j>i}^N (\pi_i\pi_j - \pi_{ij})\left[\left(\frac{y_i}{\pi_i}\right)^2 + \left(\frac{y_j}{\pi_j}\right)^2 - 2\frac{y_i}{\pi_i}\frac{y_j}{\pi_j}\right]$$

$$= \sum_i^N \sum_{j>i}^N (\pi_i\pi_j - \pi_{ij})\left(\frac{y_i}{\pi_i} - \frac{y_j}{\pi_j}\right)^2 \qquad (9\text{A}.42)$$

Corollary. From (9A.41), using the t_i method, an unbiased sample estimator of $V(\hat{Y}_{HT})$ is seen to be

$$v_1(\hat{Y}_{HT}) = \sum_{i}^{n} \frac{(1-\pi_i)}{\pi_i^2} y_i^2 + 2\sum_{i}^{n}\sum_{j>i}^{n} \frac{(\pi_{ij}-\pi_i\pi_j)}{\pi_i\pi_j\pi_{ij}} y_i y_j \qquad (9A.43)$$

provided that none of the π_{ij} in the population vanishes.

A different sample estimator has been given by Yates and Grundy (1953) and by Sen (1953). From (9A.42), this estimator is

$$v_2(\hat{Y}_{HT}) = \sum_{i}^{n}\sum_{j>i}^{n} \frac{(\pi_i\pi_j-\pi_{ij})}{\pi_{ij}} \left(\frac{y_i}{\pi_i} - \frac{y_j}{\pi_j}\right)^2 \qquad (9A.44)$$

with the same restriction on the π_{ij}.

Since the terms $(\pi_i\pi_j - \pi_{ij})$ often vary widely, being sometimes negative, v_1 and v_2 tend to be unstable quantities. Both estimators can assume negative values for some sample selection methods. Rao and Singh (1973) compared the coefficients of variation of v_1 and v_2 in samples of $n = 2$ from 34 small natural populations found in books and papers on sample surveys, using Brewer's sample selection method (section 9A.8), for which $\pi_i = 2z_i$ as desired. The estimator v_2 was considerably more stable as well as being always ≥ 0 for this method, while v_1 frequently took negative values.

We turn now to methods of sample selection. The bibliography by Brewer and Hanif (1969) mentions over 30 methods, of which 4 methods will be described. Most methods become steadily more complex as n increases beond 2, a few extend fairly easily. The presentation will give most attention first to $n = 2$, both for simplicity and because this is a common situation in nationwide samples using many small strata with $n_h = 2$.

9A.8 BREWER'S METHOD

For $n = 2$ this method of sample selection keeps $\pi_i = 2z_i$ and uses the Horvitz-Thompson estimator

$$\hat{Y}_{HT} = \frac{y_i}{\pi_i} + \frac{y_j}{\pi_j} = \frac{1}{2}\left(\frac{y_i}{z_i} + \frac{y_j}{z_j}\right)$$

Using different approaches, methods produced by Brewer (1963), Rao (1965), and Durbin (1967) all gave the same π_i and π_{ij} values. We assume every $z_i < 1/2$.

Brewer draws the first unit with what he calls revised probabilities *proportional* to $z_i(1-z_i)/(1-2z_i)$, and the second unit with probabilities $z_i/(1-z_j)$, where j is the unit drawn first. The divisor needed to convert the $z_i(1-z_i)/(1-2z_i)$ into actual probabilities is their sum

$$D = \sum_{i=1}^{N} \frac{z_i(1-z_i)}{1-2z_i} = \frac{1}{2}\sum_{i=1}^{N} \frac{z_i(2-2z_i)}{1-2z_i} = \frac{1}{2}\left(1+\sum_{i}^{N} \frac{z_i}{1-2z_i}\right) \qquad (9A.45)$$

With $n = 2$ the probability that the ith unit is drawn is the sum of the probabilities that it was drawn first and drawn second. Thus

$$\pi_i = \frac{z_i(1-z_i)}{D(1-2z_i)} + \frac{1}{D} \sum_{j \neq i}^{N} \frac{z_j(1-z_j)}{(1-2z_j)} \frac{z_i}{(1-z_j)}$$

$$= \frac{z_i}{D} \left[\frac{(1-2z_i)+z_i}{1-2z_i} + \sum_{j \neq i}^{N} \frac{z_j}{1-2z_j} \right] = \frac{z_i}{D} \left(1 + \sum_{j=1}^{N} \frac{z_j}{1-2z_j} \right) = 2z_i \quad (9A.46)$$

noting (9A.45) for D. Similarly,

$$\pi_{ij} = \frac{z_i z_j}{D} \left(\frac{1}{1-2z_i} + \frac{1}{1-2z_j} \right) = \frac{2z_i z_j}{D} \frac{(1-z_i-z_j)}{(1-2z_i)(1-2z_j)} \quad (9A.47)$$

Since this method uses the HT estimate, theorem 9A.5 and its corollary provide formulas for the variance and estimated variance of $\hat{Y}_B$.

The method has two desirable properties. Brewer (1963) has shown that its variance is always less than that of the estimate $\hat{Y}_{ppz}$ in sampling with replacement. Second, some algebra shows (Rao, 1965) that $(\pi_i \pi_j - \pi_{ij}) > 0$ for all $i \neq j$, so that the Yates–Grundy estimate v_2 of the variance is always positive.

Durbin's (1967) approach draws the first unit (i) with probability z_i. If unit i was drawn first, the probability that unit j is drawn second is made *proportional* to

$$z_j \left[\frac{1}{(1-2z_i)} + \frac{1}{(1-2z_j)} \right] \quad (9A.48)$$

In this case the divisor of the proportions is

$$\sum_{j \neq i}^{N} z_j \left[\frac{1}{(1-2z_i)} + \frac{1}{(1-2z_j)} \right] = \frac{(1-z_i)}{1-2z_i} + \sum_{j \neq i}^{N} \frac{z_j}{(1-2z_j)} = 1 + \sum_{j=1}^{N} \frac{z_j}{1-2z_j} \quad (9A.49)$$

Thus the divisor is equal to $2D$ in Brewer's (9A.45). The probability that the ith unit was drawn *first* and the jth unit *second* is, therefore,

$$P(i)P(j|i) = \frac{z_i z_j}{2D} \left[\frac{1}{(1-2z_i)} + \frac{1}{(1-2z_j)} \right] \quad (9A.50)$$

By symmetry, this equals $P(j)P(i|j)$, so that Durbin's π_{ij} is the same as Brewer's in (9A.47).

Sampford (1967) has extended this method to samples of size n, provided $nz_i < 1$ for all units in the population. With his method of sample selection, the probability that the sample consists, for example, of units $1, 2, \ldots, n$ is a natural extension of (9A.47), being proportional to

$$\left(1 - \sum_{i=1}^{n} z_i \right) \prod_{i=1}^{n} z_i \Big/ \prod_{i=1}^{n} (1-nz_i) \quad (9A.51)$$

For this method it can be shown that $\pi_i = nz_i$. A formula for π_{ij} is given, with advice on its calculation by computer. The HT estimator of Y is used, so that formulas are available for its variance and estimated variance. The Yates–Grundy estimator v_2 in (9A.44) is always positive. Several methods of actually drawing the sample so as to satisfy (9A.51) are suggested by Sampford. One is to draw the first unit with probabilities z_i and all subsequent units with probabilities proportional to $z_j/(1 - nz_j)$ *with replacement*. If a sample with n *distinct* units is obtained, this is accepted. An attempt at a sample is rejected as soon as a unit appears twice. This method can be seen to lead to (9A.51). As a guide to its speed, a formula is given for the expected number of attempts required to obtain a sample.

For $n = 2$ Durbin's (1967) method of drawing the sample, unlike the Brewer method, has the property that the unconditional probability of drawing unit i is z_i at both the first and the second draws. In multistage sampling in surveys repeated at regular intervals, Felligi (1963) pointed out earlier that it is necessary or advisable to drop units and replace them from time to time on some regular pattern called a rotation scheme, because of the undesirability of long-continued questioning of the same persons. He produced a method of selection of successive units that also has the Durbin property. His method, based on iterative calculations, is similar to the Brewer–Durbin method, but has slightly different π_{ij}.

9A.9 MURTHY'S METHOD

This method uses the first selection technique suggested (section 9A.6), the successive units being drawn with probabilities z_i, $z_j/(1 - z_i)$, $z_k/(1 - z_i - z_j)$, and so on. Murthy's estimator (1957) follows earlier work by Des Raj (1956a), who produced ingenious unbiased estimates based on the specific order in which the n units in the sample were drawn. Murthy showed that corresponding to any ordered estimate of this class we can construct an unordered estimate that is also unbiased and has smaller variance.

His proposed estimator is

$$\hat{Y}_M = \frac{\sum\limits_{i}^{n} P(s|i)y_i}{P(s)} \tag{9A.52}$$

where

$P(s|i) =$ conditional probability of getting the set of units that was drawn, given that the ith unit was drawn *first*

$P(s) =$ unconditional probability of getting the set of units that was drawn

We now prove that the estimate $\hat{Y}_M$ is unbiased. For any unit i in the population, $\sum P(s|i) = 1$, taken over *all* samples having unit i drawn first. To show this for $n = 2$, where j is the other unit in the sample,

$$P(s|i) = \frac{z_j}{1 - z_i}; \qquad \sum P(s|i) = \frac{\sum\limits_{j \neq i}^{N} z_j}{1 - z_i} = 1$$

For $n = 3$, with j and k the second and third units,

$$\sum P(s|i) = \sum_{j \neq i}^{N} \sum_{k \neq i,j}^{N} z_j z_k / (1 - z_i)(1 - z_i - z_j) = \sum_{j \neq i}^{N} z_j / (1 - z_i) = 1$$

and so on for $n > 3$. Hence, when we sum $\sum P(s) \hat{Y}_M$ over all samples of size n, the coefficient of y_i in the sum is 1, so that

$$E(\hat{Y}_M) = \sum P(s) \hat{Y}_M = \sum_{i}^{N} y_i = Y \qquad (9A.53)$$

General expressions for $V(\hat{Y}_M)$ and $v(\hat{Y}_M)$ for any n have been given by Murthy (1957).

When $n = 2$, the sample consisting of units i and j,

$$P(s|i) = \frac{z_j}{1 - z_i}; \qquad P(s|j) = \frac{z_i}{1 - z_j} \qquad (9A.54)$$

$$P(s) = \pi_{ij} = z_i P(s|i) + z_j P(s|j) = z_i z_j (2 - z_i - z_j) / (1 - z_i)(1 - z_j) \qquad (9A.55)$$

The estimate is therefore, using (9A.52), (9A.54), (9A.55),

$$\hat{Y}_M = \frac{1}{2 - z_i - z_j} \left[(1 - z_j) \frac{y_i}{z_i} + (1 - z_i) \frac{y_j}{z_j} \right] \qquad (9A.56)$$

By definition, $V(\hat{Y}_M)$ is (for $n = 2$)

$$\sum_{s} P(s) \hat{Y}_M^2 - Y^2 = \sum_{i}^{N} \sum_{j > i}^{N} \frac{z_i z_j (2 - z_i - z_j)}{(1 - z_i)(1 - z_j)} \hat{Y}_M^2 - Y^2 \qquad (9A.57)$$

After substituting for $\hat{Y}_M^2$ from (9A.56) and for Y^2 and rearranging, we find

$$V(\hat{Y}_M) = \sum_{i}^{N} \sum_{j > i}^{N} \frac{z_i z_j (1 - z_i - z_j)}{2 - z_i - z_j} \left(\frac{y_i}{z_i} - \frac{y_j}{z_j} \right)^2 \qquad (9A.58)$$

By comparison with $V(\hat{Y}_{ppz})$ for sampling with replacement, formula (9A.15), clearly $V(\hat{Y}_M) < V(\hat{Y}_{ppz})$.

Since for any proposed $v(\hat{Y}_M)$ its average is $\sum P(s) v(\hat{Y}_M)$, an unbiased sample estimate of variance for $n = 2$ is

$$v(\hat{Y}_M) = \frac{(1 - z_i)(1 - z_j)(1 - z_i - z_j)}{(2 - z_i - z_j)^2} \left(\frac{y_i}{z_i} - \frac{y_j}{z_j} \right)^2 \qquad (9A.59)$$

which is always positive.

The corresponding formula for samples of size n is

$$v(\hat{Y}_M) = \frac{1}{[P(s)]^2}\left\{\sum_i^n \sum_{j>i}^n [P(s)P(s|i,j) - P(s|i)P(s|j)]z_i z_j \left(\frac{y_i}{z_i} - \frac{y_j}{z_j}\right)^2\right\} \quad (9A.60)$$

where $P(s|ij)$ is the conditional probability of getting the sample given that units i and j were selected (in either order) in the first two draws. A computer program for calculating the $P(s|i)$, and $P(s|ij)$ has been given by Bayless (1968).

9A.10 METHODS RELATED TO SYSTEMATIC SAMPLING

A method suggested by Madow (1949), which Murthy (1967) notes has been used in Indian surveys, is an extension of systematic sampling. Its advantages are that the sample is easy to draw for any n, the method keeps $\pi_i = nz_i$, and it provides an unbiased estimate of Y. Furthermore, the sampler may sometimes have enough advance knowledge to arrange the units at least roughly in an order under which this systematic sampling performs well. A drawback of the systematic method is, as usual, the absence of an unbiased $v(\hat{Y}_{\text{sys}})$.

A sample of size n may be drawn using either the z_i or the measures of size M_i' from which the $z_i = M_i'/\sum M_i' = M_i'/M_0'$ were derived. If the M_i' are used, form a column of the cumulative totals T_i of the quantities nM_i'. From this column, within the interval 1 to nM_0' assign a range of size nM_i' to unit i. For sample selection, draw a random number r between 1 and M_0' and select the n units for which the assigned range includes the numbers $r, r+M_0', \ldots, r+(n-1)M_0'$.

This method is illustrated by the data used in Section 9A.2 with $N = 7$. We suppose $n = 3$. Having formed the T_i and the assigned ranges, draw a random number between 1 and $M_0' = 30$, say $r = 17$. The units selected are those whose ranges assigned include the numbers 17, 47, 77, that is, units 3, 4, 6.

Unit	Size M_i'	$T_i = 3\sum M_i'$	Assigned Range	Units Selected
1	3	9	1–9	
2	1	12	10–12	
3	11	45	13–45	17 (unit 3)
4	6	63	46–63	47 (unit 4)
5	4	75	64–75	
6	2	81	76–81	77 (unit 6)
7	3	90	82–90	
$M_0' =$	30			

If $nz_i < 1$ (i.e., $nM_i' \le M_0'$) for all i, any unit has a probability nz_i of being selected, and no unit is selected more than once. For instance, unit 5 is selected if $4 \le r \le 15$, giving probability $12/30 = 3z_5$. If $nz_i > 1$ for one or more units, such units may be selected more than once in the sample, but the *average* frequency of selection is $\pi_i = nz_i$. This happens in the example for unit 3, since $3M_3' = 33 > 30 = M_0'$. Unit 3 is chosen *once* if $1 \le r \le 12$ or $16 \le r \le 30$ (in all 27 choices), and *twice* if $13 \le r \le 15$ (3 choices). The average frequency of selection is $(1 \times 27 + 2 \times 3)/30 = 33/30 = 3z_3$.

It follows that

$$\hat{Y}_{\text{sys}} = \sum_i^n \frac{y_i}{\pi_i} = \frac{1}{n} \sum_i^n \frac{y_i}{z_i} \tag{9A.61}$$

is an unbiased estimate of Y.

Hartley and Rao (1962) examined this method with the units first arranged in random order. With the restriction $nz_i < 1$, (all i), they obtained approximate expressions for $V(\hat{Y}_{\text{sys}})$ and $v(\hat{Y}_{\text{sys}})$.

9A.11 THE RAO, HARTLEY, COCHRAN METHOD

For a sample of size n, this method first forms n random groups of units, *one* unit to be drawn from each group. The numbers of units $N_1, N_2, \ldots, N_n$ in the respective groups may be chosen in advance: we will see that it is advantageous to make the N_g as equal as possible. If Z_g is the total measure of size for group g, the ith unit in the group is given probability of selection z_i/Z_g. The estimate of the population total, Rao, Hartley, and Cochran (1962), is

$$\hat{Y}_{RHC} = \sum_{g=1}^n Z_g \frac{y_g}{z_g} = \sum_{g=1}^n \hat{Y}_g \tag{9A.62}$$

where y_g, z_g refer to the unit drawn from group g.

Since the Z_g will not be equal, this method does not keep the probabilities of selection proportional to the sizes, and there is some evidence that its estimator suffers a slight loss in precision. Its advantages are its simplicity and generality.

In developing $V(\hat{Y}_{RHC})$ we average over two stages. Stage 1 is the randomization into groups, stage 2 the selection of a unit within each group. For any specific split into groups, $\hat{Y}_g$ in (9A.62) is an unbiased estimator of the group total Y_g, and hence $E_2(\hat{Y}_{RHC}) = Y$. A well-known formula for finding a variance over two stages of sampling, proved in Chapter 10, is

$$V(\hat{Y}_{RHC}) = E_1[V_2(\hat{Y}_{RHC})] + V_1[E_2(\hat{Y}_{RHC})] \tag{10.2}$$

Since $E_2(\hat{Y}_{RHC}) = Y$ and is a constant, it has zero variance and the second term in (10.2) disappears in this application.

For $V_2(\hat{Y}_{RHC})$ we can use, within a group, the variance formula for sampling with replacement, since only one unit is selected from each group. Within a group the probabilities of selection are z_i/Z_g. By (9A.15) in section 9A.3 we get, for any specific split (with $n_g = 1$),

$$V(\hat{Y}_g) = \sum_{i<j}^{N_g} z_i z_j \left(\frac{y_i}{z_i} - \frac{y_j}{z_j} \right)^2 \tag{9A.63}$$

taken over the pairs of units in group g. Over the set of random subdivisions into groups, the probability that any pair of units falls in group g is $N_g(N_g-1)/N(N-1)$. Hence the average value of $V_2(\hat{Y}_g)$ over the randomization is

$$E_1 V_2(\hat{Y}_g) = \frac{N_g(N_g-1)}{N(N-1)} \sum_{i}^{N} \sum_{j>i}^{N} z_i z_j \left(\frac{y_i}{z_i} - \frac{y_j}{z_j}\right)^2 = \frac{N_g(N_g-1)}{N(N-1)} \left(\sum_{i=1}^{N} \frac{y_i^2}{z_i} - Y^2\right) \quad (9A.64)$$

from (9A.63). Since $\hat{Y}_{RHC} = \sum \hat{Y}_g$,

$$E_1[V_2(\hat{Y}_{RHC})] = \frac{\left(\sum_{g=1}^{N} N_g^2 - N\right)}{N(N-1)} \left(\sum_{1}^{N} \frac{y_i^2}{z_i} - Y^2\right) = \frac{n\left(\sum_{g=1}^{N} N_g^2 - N\right)}{N(N-1)} V(\hat{Y}_{ppz}) \quad (9A.65)$$

Thus $V(\hat{Y}_{RHC})$ is simply a multiple of the variance of the estimate $\hat{Y}_{ppz}$ in sampling with replacement. If N/n is integral, the choice $N_g = N/n$ minimizes the multiplier. In this case

$$V(\hat{Y}_{RHC}) = \left(1 - \frac{n-1}{N-1}\right) V(\hat{Y}_{ppz}) \quad (9A.66)$$

If $N = nR + k$, with R integral and $k < n$, the best choice is to make k groups of size $(R+1)$, the remaining $(n-k)$ of size R. This gives

$$V(\hat{Y}_{RHC}) = \left[1 - \frac{n-1}{N-1} + \frac{k(n-k)}{N(N-1)}\right] V(\hat{Y}_{ppz}) \quad (9A.67)$$

If N/n is integral, (9A.66) gives $V(\hat{Y}_{RHC})/V(\hat{Y}_{ppz}) = (N-n)/(N-1)$, the same ratio as obtained in simple random sampling.

An unbiased variance estimator can be shown to be

$$v(\hat{Y}_{RHC}) = \frac{\left(\sum_{g=1}^{n} N_g^2 - N\right)}{\left(N^2 - \sum_{g=1}^{n} N_g^2\right)} \sum_{g=1}^{n} Z_g \left(\frac{y_g}{z_g} - \hat{Y}_{RHC}\right)^2 \quad (9A.68)$$

With $N = nR + k$, and k groups of size $(R+1)$, formula (9A.68) becomes

$$v(\hat{Y}_{RHC}) = \frac{N^2 + k(n-k) - Nn}{N^2(n-1) - k(n-k)} \sum_{g}^{n} Z_g \left(\frac{y_g}{z_g} - \hat{Y}_{RHC}\right)^2 \quad (9A.69)$$

9A.12 NUMERICAL COMPARISONS

In the literature, comparisons of the performances of some of the methods have been made, particularly by Rao and Bayless (1969, 1970), in three situations: (a) on small artificial populations [e.g., the populations with $N = 4$, $n = 2$ constructed by Yates and Grundy (1953)], (b) under the linear regression model used in section 9A.5, and (c) on 20 natural populations.

Seven methods are compared here on three artificial populations with $N = 5$, $n = 2$. The relative sizes z_i of the units are the same in all three populations (A, B, C) (Table 9A.1). In A the *mean per element*, which is proportional to y_i/z_i, is uncorrelated with z_i. In B the mean per element rises as the sizes increase. In C the unit *totals* have little relation to the sizes.

TABLE 9A.1

THREE SMALL ARTIFICIAL POPULATIONS

Relative sizes	(z_i)	0.1	0.1	0.2	0.3	0.3
Population A	y_i	0.3	0.5	0.8	0.9	1.5
	y_i/z_i	3	5	4	3	5
Population B	y_i	0.3	0.3	0.8	1.5	1.5
	y_i/z_i	3	3	4	5	5
Population C	y_i	0.7	0.6	0.4	0.9	0.6
	y_i/z_i	7	6	2	3	2

The plans compared are as follows.

1. Simple random sampling of the units (equal probability). Estimate: $\hat{Y}_{SRS} = N\bar{y}$.
2. Simple random sampling of the units. Estimate: Ratio to size, $\hat{Y}_R = \sum y_i / \sum z_i$.
3. Sampling with probability proportional to size with replacement. Estimate: $\hat{Y}_{ppz} = (1/n) \sum (y_i/z_i)$.
4. Brewer's method, sampling without replacement. Estimate: $\hat{Y}_B = (1/n) \sum (y_i/z_i)$.
5. Murthy's method. Estimate: $\hat{Y}_M = \left[(1 - z_j)\dfrac{y_i}{z_i} + (1 - z_i)\dfrac{y_j}{z_j} \right] \Big/ (2 - z_i - z_j)$.
6. The RHC method with one group of three units, one of two units. Estimate: $\hat{Y}_{RHC} = \sum Z_g(y_g/z_g)$.
7. Madow's systematic method, with units arranged in increasing order of y_i/z_i. Estimate: $\hat{Y}_{sys} = (1/n) \sum (y_i/z_i)$.

To obtain a rough idea of what to expect in the relative performances of methods like B, M, and RHC versus simple random sampling with equal probabilities, note that the relative variances of the newer methods to $V(\hat{Y}_{SRS})$ are approximately the relative coefficients of variation of the quantities y_i/z_i and y_i. The ratios $cv(y_i/z_i)/cv(y_i)$ are 0.19 in population A, 0.11 in B, and 6.7 in C. Thus the unequal probability methods should have a relative efficiency of around 5 in A, 9 in B, but only 0.15 in C. This comparison is somewhat unfair to the unequal-probability methods, which weight the squared errors in y_i/z_i by the measures of size. The *MSE*'s of $\hat{Y}_R$ should be fairly close to the variances of the unequal-probability methods.

In a choice of a method in practice, major considerations are the ease with which the sample can be drawn, the simplicity of the estimate, the accuracy of the estimate, and the availability of an estimator of the variance of the estimate. With $n = 2$, all methods are simple. As regards accuracy, the systematic method is shown here only in its most favorable case, which the sampler rarely has the knowledge to use. All methods except $\hat{Y}_{sys}$ and $\hat{Y}_R$ provide unbiased sample estimates of error variance. Table 9A.2 presents the variances (with the MSE for $\hat{Y}_R$).

TABLE 9A.2

VARIANCES OF THE ESTIMATED POPULATION TOTALS

Estimate	Population		
	A	B	C
$\hat{Y}_{SRS}$	1.575	2.715	0.248
$\hat{Y}_R(MSE)$	0.344	0.351	1.421
$\hat{Y}_{ppz}$	0.400	0.320	1.480
$\hat{Y}_B$	0.246	0.248	1.251
$\hat{Y}_M$	0.267	0.237	1.130
$\hat{Y}_{RHC}$	0.320	0.256	1.184
$\hat{Y}_{sys}$	0.150	0.140	0.760

In equal-probability selection, the ratio of $V(\hat{Y})$ in sampling without replacement to $V(\hat{Y})$ in sampling with replacement is $(N-n)/(N-1) = 0.75$ for these populations. For $\hat{Y}_{RHC}$, the ratio $V(\hat{Y}_{RHC})/V(\hat{Y}_{ppz})$, where $\hat{Y}_{ppz}$ denotes sampling with replacement is 0.8 in Table 9A.2. For the other unequal-probability methods, the ratio varies from population to population. The average of the three ratios in A, B, and C is 0.74 for Brewer's method and 0.73 for Murthy's method, about the same ratio as for equal-probability methods. Results for $n = 3$, 4 on small natural populations by Bayless and Rao (1970) agree in general with $(N-n)/(N-1)$.

In populations A and B, all other methods are much more accurate than $\hat{Y}_{SRS}$ with simple random sampling. The ratio-to-size estimate with *SRS* performs roughly similarly to the unequal probability methods, although not quite as well. In the latter, there is little to choose among the Brewer, Murthy, or *RHC* methods. The systematic method at its best performs very well.

In population *C*, in which the unit totals bear little relation to the sizes, simple random sampling with equal probabilities is much the best. As noted at the end of section 9A.13, this superiority is probably due to the estimator $N\bar{y}$, not the *SRS* feature.

Rao and Bayless (1969) compared 10 unequal-probability methods in 20 natural populations found in books and papers on sampling, with N ranging from 9 to 35. They confined themselves to methods (a) known to have smaller variances than $\hat{Y}_{ppz}$ and (b) providing a positive unbiased variance estimator. Among the methods presented here, they compared the efficiencies of $\hat{Y}_M$, $\hat{Y}_{RHC}$, and $\hat{Y}_{ppz}$ with that of $\hat{Y}_B$. For $n = 2$, there was little to choose among the three "without replacement" methods, with $\hat{Y}_M$ slightly ahead, beating $\hat{Y}_{RHC}$ whenever the two methods differed in precision. Also, we have noted (section 9A.7) that the variance estimator may be unstable for methods using the Horvitz–Thompson estimate. The Rao–Bayless results compare the coefficients of variation of $v(\hat{Y}_M)$, $v(\hat{Y}_{RHC})$, and $v(\hat{Y}_B)$ in the Yates–Grundy form with that of $v(\hat{Y}_{ppz})$, as measures of the stabilities of the variance estimators. Relative to $v(\hat{Y}_B)$ as 100%, the median efficiencies of the other variance estimators were: $v(\hat{Y}_{RHC}) = 109\%$, $v(\hat{Y}_M) = 104\%$, $v(\hat{Y}_{ppz}) = 97\%$, the three methods all showing a few large individual gains.

Bayless and Rao (1970) give similar comparisons for $n = 3$ (14 populations) and $n = 4$ (10 populations). Sampford's extension of the Brewer method was used. For n much beyond 2, both Sampford's and Murthy's methods require computer aid in calculating the needed probabilities. The variances of $\hat{Y}_M$ and $\hat{Y}_S$ agreed closely in nearly all populations, with $\hat{Y}_{RHC}$ slightly behind, its median efficiency relative to $\hat{Y}_S$ dropping to 92% for $n = 4$. In stability of the variance estimators, on the other hand, the superiority of $\hat{Y}_{RHC}$ and $\hat{Y}_M$ increased, the median relative efficiencies being 118% ($n = 3$) and 129% ($n = 4$) for $v(\hat{Y}_{RHC})$ and 110% ($n = 3$) and 120% ($n = 4$) for $v(\hat{Y}_M)$ relative to 100% for $v(\hat{Y}_S)$.

Rao and Bayless also compared the efficiencies of $\hat{Y}$ and $v(\hat{Y})$ for some of the estimators under the linear regression model of section 9A.5 with $\alpha = 0$. While comparative results depended on the power g, the general trend was similar to that in the natural populations.

9A.13 STRATIFIED AND RATIO ESTIMATES

The preceding formulas have been presented as they would apply to a single stratum, although the concentration on small n implied previous stratification by another principle (e.g., geographic location, urban-rural). The extension of the formulas to this stratification is as usual. For any method, if n_h, z_{hi}, π_{hi}, y_{hi}, and so forth, refer to stratum h,

$$\hat{Y} = \sum_{h}^{L} \hat{Y}_h: \quad V(\hat{Y}) = \sum_{h}^{L} V(\hat{Y}_h): \quad v(\hat{Y}) = \sum v(\hat{Y}_h) \qquad (9A.70)$$

Ratio estimates enter either when the variable of interest is a ratio (e.g., unemployed females/females eligible for work), or when they are used to increase precision. In unequal-probability sampling a *single* choice of the z_i or z_{hi} must be made for all variables for which estimates are required. For instance, the z_i or z_{hi} may be proportional to the *total* recent sales of a type of business, where the survey has to estimate current sales of individual classes of items as well as current total sales. For some classes, sales may not be closely proportional to the z_i. In such cases, use of the familiar "ratio to the same variable last time" estimate may bring substantial increases in precision.

With unequal-probability sampling the change in formulas to those for ratio estimators is easily made. In an unstratified population, replace $\hat{Y}$ for any method by $X\hat{Y}/\hat{X}$. For instance, with the *HT* estimator, we use $X(\sum^{n} y_i/\pi_i)/\sum^{n} (x_i/\pi_i)$ for the ratio form instead of $\sum (y_i/\pi_i)$. For the standard approximations to the *MSE* and estimated *MSE* of a ratio estimate, replace y_i by $d_i = (y_i - Rx_i)$ in $V(\hat{Y})$ and by $d_i' = (y_i = \hat{R}x_i)$ in $v(\hat{Y})$.

For instance, with the ratio form of the Horvitz–Thompson estimator, the approximation to v_2, the estimated variance in the Yates–Grundy form, is from (9A.44), p. 261,

$$v_2[\hat{Y}_{HT(R)}] \doteq \sum_{i}^{N} \sum_{j>i}^{N} \frac{(\pi_i\pi_j - \pi_{ij})}{\pi_{ij}} \left(\frac{d_i'}{\pi_i} - \frac{d_j'}{\pi_j}\right)^2 \tag{9A.71}$$

In a stratified population it is likely with small n_h that the combined ratio estimate (section 6.11) will be used [i.e., $\hat{Y} = X(\sum \hat{Y}_h)/(\sum \hat{X}_h)$]. For approximate variance formulas, replace y_{hi} by $d_{hi} = y_{hi} - Rx_{hi}$ in V and by $d_{hi}' = y_{hi} - \hat{R}x_{hi}$ in v.

When the y_i for an item are not related to the z_i and no suitable ratio estimate is feasible for such an item, Rao (1966) has investigated alternative estimators. These are produced from any of the unequal-probability estimators (*HT*, *M*, *RHC*, etc.) by replacing y_i/z_i by Ny_i, wherever y_i appears in the estimator. Thus, for the Horvitz–Thompson estimator, the alternative form is

$$\hat{Y}^*_{HT} = \frac{N}{n} \sum_{i}^{n} y_i \tag{9A.72}$$

while for the *RHC* method

$$\hat{Y}^*_{RHC} = N \sum_{g}^{n} Z_g y_g \tag{9A.73}$$

The estimators are biased but intuition suggests that if the y_i have no relation to the z_i, the biases should be relatively small. By the same method as used in finding $V(\hat{Y}_{HT})$ in (9A.42), we get

$$V[\hat{Y}^*_{HT}] = \frac{N^2}{n^2} \sum_{i}^{N} \sum_{j>i}^{N} (\pi_i\pi_j - \pi_{ij})(y_i - y_j)^2 \tag{9A.74}$$

Thus $V[\hat{Y}^*_{HT}]$ depends on the amount of variation in the unit *totals*, as with $V(\hat{Y}_{SRS})$. In population C, Table 9A.2, this method gave 0.266 and 0.280 for the

MSE's of the alternative forms of $\hat{Y}_{HT}$ and $\hat{Y}_M$, these forms doing almost as well as $\hat{Y}_{SRS}$. For both methods the (Bias)^2 term was about 4% of the MSE.

EXERCISES

9A.1   Horvitz and Thompson (1952) give the following data for eye estimates M_i of the numbers of households and for the actual numbers y_i in 20 city blocks in Ames, Iowa. To assist in the calculations, values of $\bar{y}_i$ and $\bar{y}_i^2/M_i$ are also given. A sample of $n = 1$ block is chosen. Compute the variances of the total number of households Y, as obtained by (a) the unbiased estimate in sampling with equal probabilities, (b) the ratio estimate in sampling with equal probabilities, (c) sampling with probability proportional to M_i. (For the ratio estimate, compute the true mean square error, not the approximate formula.)

M_i	y_i	$\bar{y}_i$	y_i^2/M_i	M_i	y_i	$\bar{y}_i$	y_i^2/M_i
9	9	1.0000	9.000	19	19	1.0000	19.000
9	13	1.4444	18.778	21	25	1.1905	29.762
12	12	1.0000	12.000	23	27	1.1739	31.696
12	12	1.0000	12.000	24	21	0.8750	18.375
12	14	1.1667	16.333	24	35	1.4583	51.042
14	17	1.2143	20.643	25	22	0.8800	19.360
14	15	1.0714	16.071	26	25	0.9615	24.038
17	20	1.1765	23.529	27	27	1.0000	27.000
18	19	1.0556	20.056	30	47	1.5667	73.633
18	18	1.0000	18.000	40	37	0.9250	34.225

Do the results agree with the discussion in section 9A.5?

9A.2   A questionnaire is to be sent to a sample of high schools to find out which schools provide certain facilities, for example, a course in Russian or a swimming pool. If M_i is the number of students in the ith school, the quantity to be estimated for any given facility is the proportion P of high-school students who are in schools having the facility, that is,

$$P = \frac{\sum_w M_i}{\sum_{i=1}^{N} M_i}$$

where $\sum_w$ is a sum over those schools *with* the facility.

A sample of n schools is drawn with probability proportional to M_i with replacement. For one facility, a schools out of n are found to possess it. (a) Show that $\hat{P} = a/n$ is an unbiased estimate of P and that its true variance is $P(1-P)/n$. (*Hint.* In the corollary to theorem 9.4 let $y_i = M_i$ if the school has the facility and 0 otherwise.) (b) Show that an unbiased estimate of $V(\hat{P})$ is $v(\hat{P}) = \hat{P}(1-\hat{P})/(n-1)$.

9A.3   The large units in a population arrange themselves into a finite number of size classes: all units in class h contain M_h small units. (a) Under what conditions does sampling with *pps* give, on the average, the same distribution of the size classes in the sample as stratification by size of unit, with optimum allocation for fixed sample size? (b) If the variance among large units in class h is kM_h, where k is a constant for all classes, what system of probabilities of selection of the units gives a sample in which the sizes have approximately the same distribution as a stratified random sample with optimum allocation for fixed sample size?

9A.4 For a population with $N = 3$, $z_i = \frac{1}{2}, \frac{1}{3}, \frac{1}{6}$ and $y_i = 7, 5, 2$, two units are drawn without replacement, the first with probability proportional to z_i, the second with probability proportional to the remaining sizes. (a) Verify that $\pi_1 = \frac{51}{60}$, $\pi_2 = \frac{44}{60}$, $\pi_3 = \frac{25}{60}$ and that $\pi_{12} = \frac{35}{60}$, $\pi_{13} = \frac{16}{60}$, $\pi_{23} = \frac{9}{60}$. (b) For this method of sample selection, compare the variances of $\hat{Y}_{HT}$ and $\hat{Y}_M$ and also compare them with the variance of $\hat{Y}_{RHC}$ using its method of sample selection. You may either construct all three possible estimates or use the variance formulas. (c) Show that the ratio of $V(\hat{Y}_M)$ to $V(\hat{Y}_{ppz})$ in sampling with replacement is close to the value $\frac{1}{2}$ that applies for equal probability sampling.

9A.5 For the population in exercise 9A.4, a second variable had values $y_{2i} = 8, 5, 9$, not at all closely related to the z_i, so that with the sampling method used in exercise 9A.4, $\hat{Y}_{HT}$ and $\hat{Y}$ would be expected to perform poorly for this variable. For Rao's estimator $\hat{Y}_{HT}^* = 1.5(y_{2i} + y_{2j})$, compare its MSE with the variance of $\hat{Y}_{SRS} = 1.5(y_{2i} + y_{2j})$ in equal-probability sampling. How much does bias contribute to the MSE?

9A.6 For Brewer's method with $n = 2$, section 9A.8 showed that

$$\pi_{ij} = \frac{4z_i z_j (1 - z_i - z_j)}{(1 - 2z_i)(1 - 2z_j)} \bigg/ \left(1 + \sum_i^N \frac{z_i}{1 - 2z_i}\right)$$

(a) Show that if every $z_i < \frac{1}{2}$,

$$0 < \pi_{ij} < 4z_i z_j \qquad (\text{every } i \neq j)$$

(b) Show that this result makes the Yates–Grundy estimator of variance always positive for this method.

Hint. For (a) it is sufficient to show that

$$(1 - z_i - z_j) = (1 - 2z_i)(1 - 2z_j) \left[1 + \frac{z_i}{1 - 2z_i} + \frac{z_j}{1 - 2z_j}\right]$$

9A.7 (a) For Durbin's method with $n = 2$, verify directly that the probability that the jth unit is drawn second is z_j as stated on p. 263.

(b) With $N = 4$, $z_i = 0.1$, 0.2, 0.3, 0.4, calculate the probability that with Brewer's method unit 1 is drawn first and the probability that it is drawn second. Verify that the two probabilities add to $0.2 = 2z_i$.

9A.8 In Madow's systematic method, a unit may be chosen more than once in the sample if $nz_i > 1$ (i.e., $nM_i' > M_0'$). Show (as stated on p. 266) that for such units the average frequency of selection is nz_i, so that the Horvitz–Thompson estimator of Y remains unbiased for $nz_i > 1$.

Subsampling with Units
of Equal Size

10.1 TWO-STAGE SAMPLING

Suppose that each unit in the population can be divided into a number of smaller units, or subunits. A sample of n units has been selected. If subunits within a selected unit give similar results, it seems uneconomical to measure them all. A common practice is to select and measure a sample of the subunits in any chosen unit. This technique is called *subsampling*, since the unit is not measured completely but is itself sampled. Another name, due to Mahalanobis, is *two-stage sampling*, because the sample is taken in two steps. The first is to select a sample of units, often called the *primary units*, and the second is to select a sample of second-stage units or subunits from each chosen primary unit.

Subsampling has a great variety of applications, which go far beyond the immediate scope of sample surveys. Whenever any process involves chemical, physical, or biological tests that can be performed on a small amount of material, it is likely to be drawn as a subsample from a larger amount that is itself a sample.

In this chapter we consider the simplest case in which every unit contains the same number M of subunits, of which m are chosen when any unit is subsampled. A schematic representation of a two-stage sample, in which $M = 9$ and $m = 2$, is shown in Fig. 10.1.

The principal advantage of two-stage sampling is that it is more flexible than one-stage sampling. It reduces to one-stage sampling when $m = M$ but, unless this is the best choice for m, we have the opportunity of taking some smaller value that appears more efficient. As usual, the issue reduces to a balance between statistical precision and cost. When subunits in the same unit agree very closely, considerations of precision suggest a small value of m. On the other hand, it is sometimes almost as cheap to measure the whole of a unit as to subsample it, for example, when the unit is a household and a single respondent can give accurate data about all members of the household.

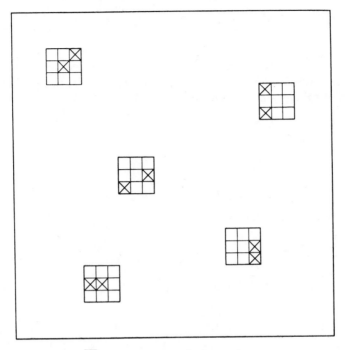

$\boxtimes$ denotes an element in the sample

Fig. 10.1 Schematic representation of two-stage sampling $(N = 81, n = 5, M = 9, m = 2)$.

10.2 FINDING MEANS AND VARIANCES IN TWO-STAGE SAMPLING

In two-stage sampling the sampling plan gives first a method for selecting n units. Then, for each selected unit, a method is given for selecting the specified number of subunits from it. In finding the mean and variance of an estimate, averages must be taken over all samples that can be generated by this two-stage process. One way of calculating this average is first to average the estimate over all second-stage selections that can be drawn from a *fixed* set of n units that the plan selects. Then we average over all possible selections of n units by the plan. For an estimate $\hat{\theta}$, this method can be expressed as

$$E(\hat{\theta}) = E_1[E_2(\hat{\theta})] \tag{10.1}$$

where E denotes expected or average value over all samples, E_2 denotes averaging over all possible second-stage selections from a fixed set of units, and E_1 denotes averaging over all first-stage selections.

For $V(\hat{\theta})$ this method gives the following easily remembered result.

$$V(\hat{\theta}) = V_1[E_2(\hat{\theta})] + E_1[V_2(\hat{\theta})] \tag{10.2}$$

where $V_2(\hat{\theta})$ is the variance over all possible subsample selections for a given set of units. To show this, let $\theta = E(\hat{\theta})$ (where θ is not necessarily the quantity that $\hat{\theta}$ is designed to estimate, since $\hat{\theta}$ may be biased). By definition,

$$V(\hat{\theta}) = E(\hat{\theta} - \theta)^2 = E_1 E_2(\hat{\theta} - \theta)^2 \tag{10.3}$$

But

$$E_2(\hat{\theta} - \theta)^2 = E_2(\hat{\theta}^2) - 2\theta E_2(\hat{\theta}) + \theta^2 \tag{10.4}$$

$$= [E_2(\hat{\theta})]^2 + V_2(\hat{\theta}) - 2\theta E_2(\hat{\theta}) + \theta^2 \tag{10.5}$$

Average now over first-stage selections. Since $E_1 E_2(\hat{\theta}) = \theta$,

$$V(\hat{\theta}) = E_1[E_2(\hat{\theta})]^2 - \theta^2 + E_1[V_2(\hat{\theta})] \tag{10.6}$$

$$= V_1[E_2(\hat{\theta})] + E_1[V_2(\hat{\theta})] \tag{10.7}$$

Formula (10.7) extends naturally to three or more stages. For three-stage sampling,

$$V(\hat{\theta}) = V_1\{E_2[E_3(\hat{\theta})]\} + E_1\{V_2[E_3(\hat{\theta})]\} + E_1\{E_2[V_3(\hat{\theta})]\} \tag{10.7'}$$

10.3 VARIANCE OF THE ESTIMATED MEAN IN TWO-STAGE SAMPLING

The following notation is used.

y_{ij} = value obtained for the jth subunit in the ith primary unit

$$\bar{y}_i = \sum_{j=1}^{m} \frac{y_{ij}}{m} = \text{sample mean per subunit in the } i\text{th primary unit}$$

$$\bar{\bar{y}} = \sum_{i=1}^{n} \frac{\bar{y}_i}{n} = \text{over-all sample mean per subunit}$$

$$S_1^2 = \frac{\sum_{i=1}^{N} (\bar{Y}_i - \bar{\bar{Y}})^2}{N-1} = \text{variance among primary unit means}$$

$$S_2^2 = \frac{\sum_{i=1}^{N} \sum_{j=1}^{M} (y_{ij} - \bar{Y}_i)^2}{N(M-1)} = \text{variance among subunits within primary units}$$

Note that Y_i denotes the total over-all subunits in the ith unit (denoted by y_i in Chapters 9 and 9A).

Theorem 10.1. If the n units and the m subunits from each chosen unit are selected by simple random sampling, $\bar{y}$ is an unbiased estimate of $\bar{Y}$ with variance

$$V(\bar{y}) = \left(\frac{N-n}{N}\right)\frac{S_1^{\,2}}{n} + \left(\frac{M-m}{M}\right)\frac{S_2^{\,2}}{mn} \qquad (10.8)$$

Proof. With simple random sampling at both stages,

$$E(\bar{y}) = E_1[E_2(\bar{y})] = E_1\left(\frac{1}{n}\sum_{i}^{n}\bar{Y}_i\right) = \left(\frac{1}{N}\sum_{i}^{N}\bar{Y}_i\right) = \bar{Y} \qquad (10.9)$$

For $V(\bar{y})$, we use formula (10.2).

$$V(\bar{y}) = V_1[E_2(\bar{y})] + E_1[V_2(\bar{y})] \qquad (10.10)$$

Since $E_2(\bar{y}) = \sum^{n} \bar{Y}_i/n$, the first term on the right is the variance of the mean per subunit for a one-stage simple random sample of n units. Hence, by Theorem 2.2,

$$V_1[E_2(\bar{y})] = \left(\frac{N-n}{N}\right)\frac{S_1^{\,2}}{n} \qquad (10.11)$$

Furthermore, with $\bar{y} = \sum^{n} \bar{y}_i/n$ and simple random sampling used at the second stage,

$$V_2(\bar{y}) = \frac{(M-m)}{Mn^2}\sum_{i}^{n}\frac{S_{2i}^{\,2}}{m} \qquad (10.12)$$

where $S_{2i}^{\,2} = \sum_{j}^{M}(y_{ij} - \bar{Y}_i)^2/(M-1)$ is the variance among subunits for the ith primary unit. When we average over the first-stage samples, $\sum_{i}^{n} S_{2i}^{\,2}/n$ averages to $\sum_{i}^{N} S_{2i}^{\,2}/N = S_2^{\,2}$.

Hence

$$E_1[V_2(\bar{y})] = \left(\frac{M-m}{M}\right)\frac{S_2^{\,2}}{mn} \qquad (10.13)$$

The theorem follows from formula (10.10), on adding (10.11) and (10.13).

If $f_1 = n/N$ and $f_2 = m/M$ are the sampling fractions in the first and second stages, an alternative form of the result is

$$V(\bar{y}) = \frac{1-f_1}{n}S_1^{\,2} + \frac{1-f_2}{mn}S_2^{\,2} \qquad (10.14)$$

10.3 SAMPLE ESTIMATION OF THE VARIANCE

Theorem 10.2. Under the conditions of theorem 10.1, an unbiased estimate of $V(\bar{\bar{y}})$ is

$$v(\bar{\bar{y}}) = \frac{1-f_1}{n}s_1^2 + \frac{f_1(1-f_2)}{mn}s_2^2 \qquad (10.15)$$

where $f_1 = n/N$, $f_2 = m/M$, and

$$s_1^2 = \frac{\sum\limits_{i}^{n}(\bar{y}_i - \bar{\bar{y}})^2}{n-1} \qquad s_2^2 = \frac{\sum\limits_{i}\sum\limits_{j}(y_{ij} - \bar{y}_i)^2}{n(m-1)} \qquad (10.16)$$

Proof.

$$(n-1)s_1^2 = \sum\limits^{n}(\bar{y}_i - \bar{\bar{y}})^2 = \sum\limits^{n}\bar{y}_i^2 - n\bar{\bar{y}}^2 \qquad (10.17)$$

Hence

$$(n-1)E_2(s_1^2) = \sum\limits^{n}\bar{Y}_i^2 + \sum\limits^{n}\frac{(1-f_2)}{m}S_{2i}^2 - n\bar{Y}_n^2 - \frac{1}{n}\sum\limits^{n}\frac{(1-f_2)}{m}S_{2i}^2 \qquad (10.18)$$

where $\bar{Y}_n = \sum\limits^{n} Y_i/n$. The last term on the right holds because subsampling is independent in different units and $\bar{\bar{y}} = \sum \bar{y}_i/n$. Thus,

$$(n-1)E_2(s_1^2) = \sum\limits^{n}(\bar{Y}_i - \bar{Y}_n)^2 + \frac{(n-1)(1-f_2)}{nm}\sum\limits^{n}\frac{S_{2i}^2}{n} \qquad (10.19)$$

Multiplying by $(1-f_1)/n(n-1)$ and averaging over the first stage of simple random sampling,

$$E\frac{(1-f_1)}{n}s_1^2 = \frac{(1-f_1)}{n}S_1^2 + \frac{(1-f_1)(1-f_2)}{mn}S_2^2 \qquad (10.20)$$

By comparison with (10.14) for $V(\bar{\bar{y}})$, note that the term in S_2^2 is too small by the amount $f_1(1-f_2)S_2^2/mn$. Since $E_1E_2(s_2^2) = S_2^2$, an unbiased estimate of $V(\bar{\bar{y}})$ is therefore

$$v(\bar{\bar{y}}) = \frac{1-f_1}{n}s_1^2 + \frac{f_1(1-f_2)}{mn}s_2^2 \qquad (10.21)$$

Corollary. A result used later is that from (10.20).

$$E(s_1^2) = S_1^2 + \frac{(1-f_2)}{m}S_2^2 = S_1^2 + \frac{S_2^2}{m} - \frac{S_2^2}{M} \qquad (10.22)$$

It follows that an unbiased estimate of S_1^2 is

$$s_1^2 - \frac{s_2^2(1-f_2)}{m}$$

Notes on Theorem 10.2. If $m = M$, that is, $f_2 = 1$, formula (10.15) becomes that appropriate to simple random sampling of the units. If $n = N$, the formula is that for proportional stratified random sampling, since primary units may then be regarded as strata, all of which are sampled. In this connection, two-stage sampling is a kind of incomplete stratification, with the units as strata.

When $f_1 = n/N$ is negligible, we obtain the simple result,

$$v(\bar{\bar{y}}) = \frac{s_1^2}{n} = \frac{\sum\limits_{i=1}^{n} (\bar{y}_i - \bar{\bar{y}})^2}{n(n-1)} \tag{10.23}$$

Thus the estimated variance can be computed from a knowledge of the unit means only. This result is helpful when subsampling is systematic, because in this event we cannot compute an unbiased estimate of S_2^2. But (10.23) still applies, provided that n/N is small. If n/N is not small, (10.23) overestimates by the amount $f_1 S_1^2/n$, as seen from (10.20) and (10.14).

10.5 THE ESTIMATION OF PROPORTIONS

If the subunits are classified into two classes and we estimate the proportion that falls in the first class, the preceding formulas can be applied by the usual device of defining y_{ij} as 1 if the corresponding subunit falls into this class and as zero otherwise. Let $p_i = a_i/m$ be the proportion falling in the first class in the subsample from the ith unit. The two estimated variances s_1^2 and s_2^2 required for theorem 10.2 work out as follows:

$$s_1^2 = \frac{\sum\limits_{i=1}^{n} (p_i - \bar{p})^2}{n-1}$$

$$s_2^2 = \frac{m}{n(m-1)} \sum\limits_{i=1}^{n} p_i q_i$$

where $\bar{p} = \Sigma p_i/n$. Consequently, by theorem 10.2,

$$v(\bar{p}) = \frac{1-f_1}{n(n-1)} \sum\limits_{i}^{n} (p_i - \bar{p})^2 + \frac{f_1(1-f_2)}{n^2(m-1)} \sum\limits_{i}^{n} p_i q_i \tag{10.24}$$

Example. In a study of plant disease the plants were grown in 160 small plots containing nine plants each. A random sample of 40 plots was chosen and three random plants in each sampled plot were examined for the presence of disease. It was found that 22 plots had no diseased plants (out of three), 11 had one, 4 had two, and 3 had three. Estimate the proportion of diseased plants and its s.e. The symbol ϕ denotes the frequencies 22, 11, 4, 3.

We have $N = 160$, $M = 9$, $n = 40$, $m = 3$. In finding s_1^2 and s_2^2, it is convenient to work at first with the numbers of diseased plants ($3p_i$) and the numbers of healthy plants ($3q_i$). The calculations are set out as follows.

$3p_i$	Frequency ϕ	$9p_iq_i$	$9\phi p_iq_i$	$3\phi p_i$	$9\phi p_i^2$
0	22	0	0	0	0
1	11	2	22	11	11
2	4	2	8	8	16
3	3	0	0	9	27
	40		30	28	54

$$\bar{p} = \frac{3 \sum \phi p_i}{3 \sum \phi} = \frac{28}{120} = 0.233$$

$$\sum \phi (p_i - \bar{p})^2 = \frac{1}{(9)}\left(54 - \frac{(28)^2}{40}\right) = 3.822$$

$$\sum \phi p_i q_i = \frac{30}{9} = 3.333$$

Hence, from the formula immediately before this example,

$$v(\bar{p}) = \frac{(3)(3.822)}{(4)(40)(39)} + \frac{(2)(3.333)}{(4)(3)(1600)(2)} = 0.00201$$

The proportion diseased is 0.233 with s.e. 0.045. The approximate formula $s_1/\sqrt{n}$, from (10.23), gives 0.049, a reasonably good estimate considering that $f_1 = \frac{1}{4}$.

10.6 OPTIMUM SAMPLING AND SUBSAMPLING FRACTIONS

These depend on the type of cost function. If travel costs between units are unimportant, one form that has proved useful is

$$C = c_1 n + c_2 nm$$

The first component of cost, $c_1 n$, is proportional to the number of primary units in the sample; the second, $c_2 nm$, is proportional to the total number of second-stage units or elements. From theorem 10.1, $V(\bar{y})$ may be written

$$V(\bar{y}) = \frac{1}{n}\left(S_1^2 - \frac{S_2^2}{M}\right) + \frac{1}{mn}S_2^2 - \frac{1}{N}S_1^2 \tag{10.25}$$

The last term on the right does not depend on the choice of n and m. Minimizing V

for fixed C, or C for fixed V, is equivalent to minimizing the product

$$\left(V + \frac{1}{N}S_1{}^2\right)C = \left[\left(S_1{}^2 - \frac{S_2{}^2}{M}\right) + \frac{S_2{}^2}{m}\right](c_1 + c_2 m)$$

By the Cauchy–Schwarz inequality,

$$m_{opt} = \frac{S_2}{\sqrt{S_1{}^2 - S_2{}^2/M}}\sqrt{c_1/c_2} \tag{10.26}$$

provided $S_1{}^2 > S_2{}^2/M$. Round m_{opt} to the nearest integer. If m is an integer such that $m < m_{opt} < m + 1$, a slightly better rule for m_{opt} small is (Cameron, 1951): round up if $m_{opt}^2 > m(m + 1)$, otherwise round down. If $m_{opt} > M$ or if $S_1{}^2 < S_2{}^2/M$, take $m = M$, using one-stage sampling. (The product $(V + S_1{}^2/N)C$ is a strictly decreasing function of m when $S_1{}^2 < S_2{}^2/M$.)

The value of n is found by solving either the cost equation or the variance equation, depending on which has been preassigned.

In most practical situations the optimum is relatively flat. An error of a few units in the choice of m produces only a small loss of precision, as the following example illustrates. Write

$$S_u{}^2 = S_1{}^2 - \frac{S_2{}^2}{M} \tag{10.27}$$

Example. Let

$$c_1 = 10c_2, \qquad S_2 = 1.3S_u$$

then

$$m_{opt} = 1.3\sqrt{10} = 4.1$$

We will regard total cost as fixed and see how the variance of $\bar{y}$ changes with m. N is assumed large. From (10.14).

$$V(\bar{y}) = \frac{S_u{}^2}{n} + \frac{S_2{}^2}{nm}$$

$$= \left(S_u{}^2 + \frac{S_2{}^2}{m}\right)\frac{c_1 + mc_2}{C}$$

eliminating n by means of the cost equation. This gives

$$V(\bar{y}) = \frac{S_u{}^2 c_2}{C}\left(1 + \frac{S_2{}^2}{mS_u{}^2}\right)\left(\frac{c_1}{c_2} + m\right) = \frac{S_u{}^2 c_2}{C}\left(1 + \frac{1.69}{m}\right)(10 + m)$$

Omitting the constant factor, the relative variance can be calculated for different values of m. Table 10.1 shows these variances and the relative precisions (with the maximum precision for $m = 4$ taken as the standard).

For any value of m between 2 and 9, the loss of precision relative to the optimum is less than 12%.

TABLE 10.1

RELATIVE VARIANCES AND PRECISIONS FOR DIFFERENT VALUES OF m

$m =$	1	2	3	4	5	6	7	8	9	10
Rel. variance	29.59	22.14	20.32	19.92	20.07	20.51	21.10	21.80	22.56	23.38
Rel. precision	0.67	0.90	0.98	1.00	0.99	0.97	0.94	0.91	0.88	0.85

In practice, the choice of m requires estimates of c_1/c_2 and S_2/S_1 or equivalently S_2/S_u. Because of the flatness of the optimum, these ratios need not be obtained with high accuracy. If c_1/c_2 is known reasonably well and a value of m, say m_0, has been selected, a useful table (Brooks, 1955) shows the range of values of S_2^2/S_u^2 within which this m_0 gives a precision at least 90% of the optimum.

The table was obtained as follows. For given cost, assuming N large, the relative precision of m_0 to m_{opt} is found to be

$$\frac{V(\bar{y}|m_{opt})}{V(\bar{y}|m_0)} = \frac{(S_u\sqrt{c_1} + S_2\sqrt{c_2})^2}{S_u^2 c_1 + S_2^2 c_2 + m_0 c_2 S_u^2 + c_1 S_2^2/m_0} \tag{10.28}$$

The set of values of S_2/S_u for which this expression exceeds some assigned level L are those lying between the two roots

$$\frac{S_2}{S_u} = \frac{\gamma \pm \sqrt{L(1-L)}(\sqrt{m_0} + \gamma^2/\sqrt{m_0})}{(L\gamma^2/m_0) - (1-L)} \tag{10.29}$$

where $\gamma^2 = c_1/c_2$.

Table 10.2, adapted from Brooks (1955), shows the lower and upper limits of S_2^2/S_u^2 for $L = 0.9$. The wide interval between the lower and upper limits is striking in nearly all cases. Note that the range of m_0 changes in different parts of the table.

If we have a rough idea about the values of S_2^2/S_u^2 for the principal items in a survey, Table 10.2 may be used to select a value m_0. Note that if ρ is the correlation between elements in the same primary unit, as defined in section 9.4, the ratio S_2^2/S_u^2 is nearly equal to $(1-\rho)/\rho$. A value of S_2^2/S_u^2 as low as 1 corresponds to $\rho = 0.5$. This would be an unusually high degree of intraunit correlation. Similarly, $\rho = 0.1$ gives $S_2^2/S_u^2 = 9$, whereas $\rho = 0.01$ gives $S_2^2/S_u^2 = 99$.

Example. Suppose that c_1/c_2 is about 1 and that S_2^2/S_u^2 is expected to lie between 5 and 100 for the principal items. The columns $c_1/c_2 = 1$ give $m_0 = 4$ as a satisfactory choice, since this covers ratios from 4 to more than 100 (actually to 196). With $c_1/c_2 = 16$ and the same desired range, the table suggests a value of m_0 somewhere between 15 and 20. Further calculation from (10.29) shows that $m_0 = 18$ is best. This covers the range from 5.2 to 84—not quite so wide as desired.

TABLE 10.2

LIMITS FOR S_2^2/S_u^2 WITHIN WHICH m_0 GIVES AT LEAST 90% OF THE MAXIMUM PRECISION

$c_1/c_2 =$ m_0	$\frac{1}{2}$		1		$c_1/c_2 =$ m_0	2		4	
	L	U	L	U		L	U	L	U
1	0.0	11	0.0	4	2	0.5	8	0.2	4
2	2.0	98	1.1	22	3	1.2	21	0.5	8
3	4.1	>*	2.4	72	4	2.2	44	1.0	16
4	6.6	>	4.0	>	5	3.3	82	1.6	27
5	9.5	>	5.9	>	6	4.7	>	2.4	42
6	13	>	8.1	>	7	6.3	>	3.3	61
7	16	>	11	>	8	8.0	>	4.3	87
8	20	>	13	>	9	10	>	5.4	>

$c_1/c_2 =$ m_0	8		16		$c_1/c_2 =$ m_0	32		64	
	L	U	L	U		L	U	L	U
6	1.0	17	0.3	8	5	0.1	3	0.0	2
7	1.5	24	0.5	11	10	0.4	12	0.1	7
8	2.0	32	0.7	15	15	1.2	26	0.3	14
9	2.6	42	1.0	19	20	2.7	46	0.7	24
10	3.3	53	1.3	23	25	4.5	74	1.5	37
15	7.6	>	3.5	55	30	6.9	>	2.5	52
20	13.3	>	6.6	>	35	9.7	>	3.7	71
25	20.4	>	10.5	>	40	13	>	5.2	93

* > denotes "> 100."

When the cost of travel between primary units is substantial, a more accurate cost function may be

$$C = c_1 n + c_t \sqrt{n} + c_2 nm \qquad (10.30)$$

since travel costs tend to be proportional to $\sqrt{n}$. If a desired value of $V(\bar{y})$ has been specified, pairs of values of (n, m) that give this variance are easily computed from (10.25) for $V(\bar{y})$. The costs for different combinations are then computed from (10.30) and the combination giving the smallest cost is found. When cost is fixed in advance, Hansen, Hurwitz, and Madow (1953) give a method for determining the (n, m) combination that minimizes the variance and a table that facilitates a rapid choice. Note that their n is our m and vice versa.

10.7 ESTIMATION OF m_{opt} FROM A PILOT SURVEY

Sometimes estimates of S_2^2 and S_1^2 are obtained from a pilot survey in which n' primary units are chosen, with m' elements taken from each unit. This section deals with the choice of n' and m'. If s_1^2 is the variance between unit means and s_2^2 is the variance between subunits within units, as defined in section 10.4, (10.22) gives

$$E(s_1^2) = \left(S_1^2 - \frac{S_2^2}{M}\right) + \frac{S_2^2}{m'} = S_u^2 + \frac{S_2^2}{m'} \qquad (10.31)$$

For the simple cost function $c_1 n + c_2 nm$, we had

$$m_{opt} = \frac{S_2}{\sqrt{S_1^2 - S_2^2/M}}\sqrt{c_1/c_2}$$

As an estimate of m_{opt} from the pilot survey, (10.31) suggests that we take

$$\hat{m}_{opt} = \frac{S_2}{\sqrt{s_1^2 - s_2^2/m'}}\sqrt{c_1/c_2} = \frac{\sqrt{m'}}{\sqrt{(m's_1^2/s_2^2) - 1}}\sqrt{c_1/c_2} \qquad (10.32)$$

The estimate $\hat{m}_{opt}$ is subject to a sampling error that depends on the sampling error of the ratio s_1^2/s_2^2. From the analysis of variance it is known that $m's_1^2/s_2^2$ is distributed as

$$F\left(1 + m'\frac{S_u^2}{S_2^2}\right)$$

where F has $(n'-1)$ and $n'(m'-1)$ degrees of freedom, provided that the y_{ij} are normally distributed. This result leads to the sampling distribution of $\hat{m}_{opt}$ for given values of n' and m', that is,

$$\hat{m}_{opt} = \frac{\sqrt{m'c_1/c_2}}{\sqrt{F\left(1 + \frac{m'S_u^2}{S_2^2}\right) - 1}} \qquad (10.33)$$

Example. For the example in section 10.6, in which

$$c_1 = 10c_2, \qquad S_2 = 1.3S_u, \qquad m_{opt} = 1.3\sqrt{10} = 4.1$$

consider how well m_{opt} is estimated from a pilot sample with $n' = 10$ and $m' = 4$. From (10.33),

$$\hat{m}_{opt} = \frac{6.324}{\sqrt{F[1 + (4/1.69)] - 1}} = \frac{6.324}{\sqrt{3.367F - 1}}$$

where F has 9 and 30 df. To find the limits within which $\hat{m}_{opt}$ will lie 80% of the time, we have, from the 10% one-tailed significance levels of F,

$$F_{.10}(9, 30) = 1.8490, \qquad F_{.90}(9, 30) = 1/F_{.10}(30, 9) = 1/2.2547 = 0.4435$$

Substitution of these values of F gives

$$\text{lower limit,} \qquad \hat{m}_{opt} = 2.8; \qquad \text{upper limit,} \qquad \hat{m}_{opt} = 9.0$$

As shown in Table 10.1, any m in this range gives a degree of precision close to the optimum. Thus, with $n' = 10$, $m' = 4$, the chances are 8 in 10 that the loss of precision is small with normal data.

The 80 and 95% limits for $n' = 5, 10, 20$ and $m' = 4$ appear in Table 10.3. With $n' = 20$, we are almost certain to estimate m_{opt} with precision close to the optimum. This is not so with $n' = 5$.

TABLE 10.3

LOWER AND UPPER LIMITS FOR $\hat{m}_{opt}$

n'	80%	90%
5	2.5, ∞	1.8, ∞
10	2.8, 9.0	2.3, ∞
20	3.1, 6.4	2.7, 9.1

If the ratio c_1/c_2 is the same in the pilot survey as in the main survey, the cost of the pilot survey will be proportional to $c_1 n' + c_2 n' m'$. Brooks (1955) gives a table of the values of (n', m') in the most economical pilot survey that provides an expected relative precision of 90% in the estimation of m_{opt}. Table 10.4 shows part of this table.

TABLE 10.4

PILOT SAMPLE DESIGNS HAVING AN EXPECTED RELATIVE PRECISION OF 90%

c_1/c_2	≤ 1		2		4		8		16		32		64	
S_2^2/S_u^2	n'	m'	n'	m'	n'	m'	n'	m'	n'	m'	n'	m'	n'	m'
1	7	3	6	4	6	5	5	6	5	7	4	10	4	12
2	8	5	7	7	6	9	6	9	5	13	5	14	4	20
4	9	9	8	11	8	12	7	14	7	15	5	25	5	27
8	10	14	10	15	9	17	9	18	8	22	6	32	5	44
16	10	25	10	27	10	27	10	28	8	37	7	46	6	60
32	10	46	10	47	10	48	10	49	9	58	8	69	6	102
64	10	92	10	93	10	96	10	100	10	104	8	137	7	169

The computations assume that N and M are large: the designs are conservative if fpc terms are taken into account. Note that no more than 10 primary units are required and that the designs are relatively insensitive to the ratio c_1/c_2.

10.8   THREE-STAGE SAMPLING

The process of subsampling can be carried to a third stage by sampling the subunits instead of enumerating them completely. For instance, in surveys to estimate crop production in India (Sukhatme, 1947), the village is a convenient sampling unit. Within a village, only some of the fields growing the crop in question are selected, so that the field is a subunit. When a field is selected, only certain parts of it are cut for the determination of yield per acre; thus, the subunit itself is sampled. If physical or chemical analyses of the crop are involved, an additional subsampling may be used, since these determinations are often made on a part of the sample cut from a field.

The results are a straightforward extension of those for two-stage sampling and are given briefly. The population contains N first-stage units, each with M second-stage units, each of which has K third-stage units. The corresponding numbers for the sample are n, m, and k, respectively. Let y_{iju} be the value obtained for the uth third-stage unit in the jth second-stage unit drawn from the ith primary unit. The relevant population means per third-stage unit are as follows.

$$\bar{Y}_{ij} = \frac{\sum\limits_{u}^{K} y_{iju}}{K}, \qquad \bar{\bar{Y}}_{i} = \frac{\sum\limits_{j}^{M}\sum\limits_{u}^{K} y_{iju}}{MK}, \qquad \bar{\bar{\bar{Y}}} = \frac{\sum\limits_{i}^{N}\sum\limits_{j}^{M}\sum\limits_{u}^{K} y_{iju}}{NMK}$$

The following population variances are required.

$$S_1^2 = \frac{\sum\limits_{i}^{N}(\bar{\bar{Y}}_i - \bar{\bar{\bar{Y}}})^2}{N-1}$$

$$S_2^2 = \frac{\sum\limits_{i}^{N}\sum\limits_{j}^{M}(\bar{Y}_{ij} - \bar{\bar{Y}}_i)^2}{N(M-1)}$$

$$S_3^2 = \frac{\sum\limits_{i}^{N}\sum\limits_{j}^{M}\sum\limits_{u}^{K}(y_{ijk} - \bar{Y}_{ij})^2}{NM(K-1)}$$

Theorem 10.3. If simple random sampling is used at all three stages, the sample mean $\bar{\bar{y}}$ per third-stage unit is an unbiased estimate of $\bar{\bar{\bar{Y}}}$ with variance

$$V(\bar{\bar{y}}) = \frac{1-f_1}{n}S_1^2 + \frac{1-f_2}{nm}S_2^2 + \frac{1-f_3}{nmk}S_3^2 \qquad (10.34)$$

where $f_1 = n/N, f_2 = m/M, f_3 = k/K$ are the sampling fractions at the three stages.

Proof. Only the principal steps are indicated. Write

$$\bar{\bar{y}} - \bar{\bar{\bar{Y}}} = (\bar{\bar{y}} - \bar{\bar{\bar{Y}}}_{nm}) + (\bar{\bar{\bar{Y}}}_{nm} - \bar{\bar{\bar{Y}}}_{n}) + (\bar{\bar{\bar{Y}}}_{n} - \bar{\bar{\bar{Y}}}) \qquad (10.35)$$

where $\bar{\bar{\bar{Y}}}_{nm}$ is the population mean of the nm second-stage units that were selected and $\bar{\bar{\bar{Y}}}_n$ is the population mean of the n primary units that were selected. When we square and take the average, the cross-product terms vanish. The contributions of the squared terms turn out to be as follows.

$$E(\bar{\bar{y}} - \bar{\bar{Y}}_{nm})^2 = \frac{1-f_3}{nmk}S_3^{\,2}$$

$$E(\bar{\bar{Y}}_{nm} - \bar{\bar{Y}}_n)^2 = \frac{1-f_2}{nm}S_2^{\,2}$$

$$E(\bar{\bar{Y}}_n - \bar{\bar{Y}})^2 = \frac{1-f_1}{n}S_1^{\,2}$$

When these three terms are added, the theorem is obtained.

Theorem 10.4. An unbiased estimate of $V(\bar{\bar{y}})$ from the sample is

$$v(\bar{\bar{y}}) = \frac{1-f_1}{n}s_1^{\,2} + \frac{f_1(1-f_2)}{nm}s_2^{\,2} + \frac{f_1 f_2(1-f_3)}{nmk}s_3^{\,2} \tag{10.36}$$

where $s_1^{\,2}$, $s_2^{\,2}$, $s_3^{\,2}$ are the sample analogues of $S_1^{\,2}$, $S_2^{\,2}$, $S_3^{\,2}$, respectively.

Proof. This may be proved by the methods in section 10.4 or alternatively by showing that

$$E(s_1^{\,2}) = S_1^{\,2} + \frac{1-f_2}{m}S_2^{\,2} + \frac{1-f_3}{mk}S_3^{\,2} \tag{10.37}$$

$$E(s_2^{\,2}) = S_2^{\,2} + \frac{1-f_3}{k}S_3^{\,2}$$

and $E(s_3^{\,2}) = S_3^{\,2}$. To obtain the first result, let $\bar{\bar{y}}_{iK}$ denote the mean over the m second-stage units in the ith primary unit, given that all K elements were enumerated at the third stage. Let $\bar{\bar{y}}_K$ be the mean of the n values $\bar{\bar{y}}_{iK}$. Then, from (10.22) for two-stage sampling, it follows that

$$E\left[\frac{\sum\limits_{i}^{n}(\bar{\bar{y}}_{iK} - \bar{\bar{y}}_K)^2}{n-1}\right] = S_1^{\,2} = \frac{1-f_2}{m}S_2^{\,2} \tag{10.38}$$

Now, if $\bar{\bar{y}}_i$ is the sample mean for the ith primary unit, write

$$(\bar{\bar{y}}_i - \bar{\bar{y}}) = (\bar{\bar{y}}_{iK} - \bar{\bar{y}}_K) + [(\bar{\bar{y}}_i - \bar{\bar{y}}_{iK}) - (\bar{\bar{y}} - \bar{\bar{y}}_K)] \tag{10.39}$$

By first averaging over samples in which the first-stage and second-stage units are fixed, it may be shown that

$$\frac{1}{(n-1)}E\sum_{}^{n}[(\bar{\bar{y}}_i - \bar{\bar{y}}_{iK}) - (\bar{\bar{y}} - \bar{\bar{y}}_K)]^2 = \frac{(1-f_3)S_3^{\,2}}{mk} \tag{10.40}$$

and that the cross-product terms from (10.39) contribute nothing. This establishes the result for $E(s_1^2)$. That for $E(s_2^2)$ is found similarly. Hence

$$E[v(\bar{\bar{y}})] = \frac{1-f_1}{n}\left(S_1^2 + \frac{1-f_2}{m}S_2^2 + \frac{1-f_3}{mk}S_3^2\right)$$

$$+ \frac{f_1(1-f_2)}{nm}\left(S_2^2 + \frac{1-f_3}{k}S_3^2\right) + \frac{f_1 f_2(1-f_3)}{nmk}S_3^2$$

$$= \frac{1-f_1}{n}S_1^2 + \frac{1-f_2}{nm}S_2^2 + \frac{1-f_3}{nmk}S_3^2 = V(y) \qquad (10.41)$$

As with two-stage sampling, it is clear from (10.36) that if f_1 is negligible $v(\bar{\bar{y}})$ reduces to

$$v(\bar{\bar{y}}) = \frac{s_1^2}{n} = \frac{\sum\limits_{}^{n}(\bar{y}_i - \bar{\bar{y}})^2}{n(n-1)} \qquad (10.42)$$

This estimate is conservative if f_1 is not negligible.

With a cost function of the form

$$C = c_1 n + c_2 nm + c_3 nmk \qquad (10.43)$$

the optimum values of k and m are

$$k_{opt} = \frac{S_3}{\sqrt{S_2^2 - S_3^2/K}}\sqrt{c_2/c_3}, \qquad m_{opt} = \frac{\sqrt{S_2^2 - S_3^2/K}}{\sqrt{S_1^2 - S_2^2/M}}\sqrt{c_1/c_2} \qquad (10.44)$$

The extension of the results in this section to additional stages of sampling should be clear from the structure of the formulas.

10.9 STRATIFIED SAMPLING OF THE UNITS

Subsampling may be combined with any type of sampling of the primary units. The subsampling itself may employ stratification or systematic sampling. Variance formulas for these modifications can be built up from the formulas for the simpler methods.

Results are given for stratified sampling of the primary units in a two-stage sample. The primary unit sizes are assumed constant for a given stratum but may vary from stratum to stratum. This situation occurs when primary units are stratified by size so that sizes within a stratum become constant or nearly so.

The hth stratum contains N_h primary units, each with M_h second-stage units; the corresponding sample numbers are n_h and m_h. The estimated population mean per second-stage unit is

$$\bar{y}_{st} = \frac{\sum\limits_{h} N_h M_h \bar{\bar{y}}_h}{\sum\limits_{h} N_h M_h} = \sum\limits_{h} W_h \bar{\bar{y}}_h \qquad (10.45)$$

where $W_h = N_h M_h / \Sigma N_h M_h$ is the relative size of the stratum in terms of *second-stage units* and $\bar{\bar{y}}_h$ is the sample mean in the stratum. By applying theorem 10.1 within each stratum, we have

$$V(\bar{\bar{y}}_{st}) = \sum_h W_h^2 \left(\frac{1-f_{1h}}{n_h} S_{1h}^2 + \frac{1-f_{2h}}{n_h m_n} S_{2h}^2 \right) \tag{10.46}$$

where $f_{1h} = n_h/N_h$, $f_{2h} = m_h/M_h$.

From theorem 10.2, an unbiased sample estimate is

$$v(\bar{\bar{y}}_{st}) = \sum_h W_h^2 \left[\frac{1-f_{1h}}{n_h} s_{1h}^2 + \frac{f_{1h}(1-f_{2h})}{n_h m_h} s_{2h}^2 \right] \tag{10.47}$$

Corresponding variances for the estimated population total are obtained by multiplying formulas (10.46) and (10.47) by $(\Sigma N_h M_h)^2$.

10.10 OPTIMUM ALLOCATION WITH STRATIFIED SAMPLING

This deals with the best choice of the n_h and the m_h. If travel costs between units are not a major factor, the cost may be represented adequately by the formula

$$C = \sum_h c_{1h} n_h + \sum_h c_{2h} n_h m_h \tag{10.48}$$

From (10.46), the variance may be rewritten as

$$V(\bar{\bar{y}}_{st}) = \sum_h W_h^2 \left[\frac{1}{n_h} \left(S_{1h}^2 - \frac{S_{2h}^2}{M_h} \right) + \frac{1}{n_h m_h} S_{2h}^2 - \frac{1}{N_h} S_{1h}^2 \right]$$

The quantity

$$V(\bar{\bar{y}}_{st}) + \lambda \left(\sum_h c_{1h} n_h + \sum_h c_{2h} n_h m_h - C \right)$$

where λ is a Lagrange multiplier, is a function of the variables n_h and $(n_h m_h)$. Hence, to minimize V for fixed C, or vice versa, we have

$$n_h \sqrt{\lambda} = \frac{W_h}{\sqrt{c_{1h}}} \sqrt{S_{1h}^2 - S_{2h}^2/M_h} \tag{10.49}$$

$$n_h m_h \sqrt{\lambda} = \frac{W_h S_{2h}}{\sqrt{c_{2h}}} \tag{10.50}$$

These give

$$m_h = \frac{S_{2h}}{\sqrt{S_{1h}^2 - S_{2h}^2/M_h}} \sqrt{c_{1h}/c_{2h}} \tag{10.51}$$

The formula for optimum m_h is exactly the same as in unstratified sampling [(10.26) in section 10.6].

From (10.49), since $W_h \propto N_h M_h$

$$n_h \propto \frac{N_h M_h S_{uh}}{\sqrt{c_{1h}}} \qquad \text{where } S_{uh}^2 = S_{1h}^2 - \frac{S_{2h}^2}{M_h} \qquad (10.52)$$

Since self-weighting estimates are convenient, we consider under what circumstances the optimum allocation leads to a self-weighting estimate. From (10.45), it follows that $\bar{y}_{st}$ is self-weighting if $n_h m_h / N_h M_h = f_0 = \text{constant}$ since, in this event,

$$\bar{y}_{st} = \frac{\sum_h (N_h M_h / n_h m_h) \sum_i^{n_h} \sum_j^{m_h} y_{hij}}{\sum N_h M_h} = \frac{\sum_h \sum_i \sum_j y_{hij}}{f_0 \sum N_h M_h}$$

$$= \frac{\sum_h \sum_i \sum_j y_{hij}}{\sum n_h m_h} = \bar{\bar{y}}$$

The condition is, as might be expected, that the probability f_0 of selecting a subunit be the same in all strata.

From (10.50), the optimum allocation gives

$$f_{0h} = \frac{n_h m_h}{N_h M_h} \propto \frac{S_{2h}}{\sqrt{c_{2h}}} \qquad (10.53)$$

Frequently c_{2h}, the cost per second-stage unit, will be approximately the same in large and small primary units; but S_{2h} may be greater in large units than in small. However, since the optimum is flat, a self-weighting sample will often be almost as precise as the optimum. Note that this result holds even if the optimum sampling of primary units is far from proportional.

EXERCISES

10.1　A set of 20,000 records are stored in 400 file drawers, each containing 50 records. In a two-stage sample, five records are drawn at random from each of 80 randomly selected drawers. For one item, the estimates of variance were $s_1^2 = 362$, $s_2^2 = 805$, as defined in section 10.4. (a) Compute the standard error of the mean per record from this sample. (b) Compare this with the standard error given by the approximate formula (10.23) in section 10.4.

10.2　From the results of a pilot two-stage sample, in which m' subunits were chosen from each of n' units, it is useful to be able to estimate the value of $V(\bar{y})$ that would be given by a subsequent sample having m subunits from each of n units. Show that an unbiased estimate of $V(\bar{y})$ is

$$\hat{V}(\bar{y}) = \left(\frac{N-n}{N}\right)\frac{s_1^2}{n} + \frac{s_2^2}{mn}\left(1 - \frac{m}{m'} + \frac{mn}{m'N} - \frac{mn}{MN}\right)$$

where s_1^2 and s_2^2 are computed from the preliminary sample. *Hint.* Use theorem 10.1 and the result (10.22):

$$E(s_1^2) = S_1^2 - \frac{S_2^2}{M} + \frac{S_2^2}{m}$$

10.3   In sampling wheat fields in Kansas, with the field as a primary unit, King and McCarty (1941) report the following mean squares for yield in bushels per acre: $s_1^2 = 165$, $s_2^2 = 66$. Two subsamples were taken per field. For a sample of n fields, compare the variances of the sample mean as given by (*a*) the sample as actually taken, (*b*) four subsamples per field from n fields, (*c*) completely harvesting n fields.

N and M may be assumed large and constant. In (*c*) assume that complete harvesting is equivalent to single-stage sampling (i.e., to having $m = M$).

10.4   In the same survey, with two subsamples per field, the mean squares for the percentage of protein were $s_1^2 = 7.73$, $s_2^2 = 1.43$. How many fields are required to estimate the mean yield to within ± 1 bushel and the mean protein percentage to within $\pm \frac{1}{4}$, apart from a 1-in-20 chance in each case? Perform the calculations (*a*) assuming that two subsamples per field are taken in the main survey, (*b*) assuming complete harvesting of a field in the main survey.

10.5   For the wheat-yield data in exercise 10.3, what is the value of c_1/c_2 in a linear cost function if the estimated optimum m is 2?

10.6   If m/M and n/N are both small and the cost function is linear, show that $m = 2$ gives a smaller value of $V(\bar{y})$ than $m = 1$ if

$$\frac{c_1}{c_2} > 2\frac{S_1^2}{S_2^2}$$

10.7   A large department store handles about 20,000 accounts receivable per month. A 2% sample ($m = 400$) was verified each month over a 2-year period ($n = 24$). The numbers of accounts found to be in error per month (out of 400) were (in order of magnitude) 0, 0, 1, 1, 2, 4, 4, 5, 5, 5, 5, 6, 6, 6, 7, 7, 8, 9, 9, 10, 10, 13, 14, 17, the time pattern being erratic. From the results in section 10.5, compute s_1^2 and s_2^2. Hence compute the standard error of $\bar{p}$, as an estimate of the percentage of accounts that are in error over a period of a year, that would be obtained from verifying (*a*) 1200 accounts from a single month, chosen at random, (*b*) 300 accounts from each of four random months, (*c*) 100 accounts each month. *Hint.* Either use the formula in exercise 10.2 with $m' = 400$ or obtain unbiased estimates of S_1^2 and S_2^2 and use theorem 10.1.

10.8   In planning a two-stage survey it was expected that c_1/c_2 would be about 4 and that S_2^2/S_u^2 would lie between 5 and 50. (*a*) What value of m would you choose from Table 10.2? (*b*) Suppose that after the survey was completed it was found that c_1/c_2 was close to 8 and S_2^2/S_u^2 was about 25. Compute the relative precision given by your m to that given by the optimum m. (*c*) Make the same computation for $c_1/c_2 = 4$, $S_2^2/S_u^2 = 100$.

10.9   If ρ is the correlation coefficient between second-stage units in the same primary unit, prove that

$$\frac{1-\rho}{\rho} = \frac{S_2^2}{[(N-1)/N]S_1^2 - S_2^2/M} \doteq \frac{S_2^2}{S_u^2}$$

(This establishes a result used in section 10.6.)

10.10   Show that if $S_u^2 > 0$, in the notation of section 10.6, a simple random sample of n primary units, with 1 element chosen per unit, is more precise than a simple random sample of n elements ($n > 1$, $M > 1$). Show that the precision of the two methods is equal if n/N is negligible. Would you expect this intuitively?

CHAPTER 11

Subsampling with Units of Unequal Sizes

11.1 INTRODUCTION

In sampling extensive populations, primary units that vary in size are encountered frequently. Moreover, considerations of cost often dictate the use of multistage sampling, so that the problems discussed in this chapter are of common occurrence. If the sizes do not vary greatly, one method is to stratify by size of primary unit, so that the units within a stratum become equal, or nearly so. The formulas in section 10.9 may then be an adequate approximation. Often, however, substantial differences in size remain within some strata, and sometimes it is advisable to base the stratification on other variables. In a review of the British Social Surveys, which are nationwide samples with districts as primary units, Gray and Corlett (1950) point out that size was at first included as one of the variables for stratification but that another factor was found more desirable when the characteristics of the population became better known.

Some concentrated effort is required in order to obtain a good working knowledge of multistage sampling when the units vary in size, because the technique is flexible. The units may be chosen either with equal probabilities or with probabilities proportional to size or to some estimate of size. Various rules can be devised to determine the sampling and subsampling fractions, and various methods of estimation are available. The advantages of the different methods depend on the nature of the population, on the field costs, and on the supplementary data that are at our disposal.

The first part of this chapter is devoted to a description of the principal methods that are in use. We will begin with a population that consists of a single stratum. The extension to stratified sampling can be made, as in preceding chapters, by summing the appropriate variance formulas over the strata. For simplicity, we assume at first that only a single primary unit is chosen, that is, that $n = 1$. This case is not so impractical as it might appear at first sight, because when there is a large number of strata we may achieve satisfactory precision in estimation even though $n_h = 1$. In the monthly surveys taken by the U.S. Census Bureau to estimate numbers of employed people, the primary unit is a county or a group of neighboring counties. This is a large unit, but it has administrative advantages that decrease costs. Since counties are far from uniform in their characteristics, stratification is extended to the point at which only one is selected from each stratum. Consequently, the theory to be discussed is applicable to a single stratum in this sampling plan.

As in preceding chapters, the quantities to be estimated may be the population total Y, the population mean (usually the mean per subunit $\bar{\bar{Y}}$), or a ratio of two variates.

Notation.   The observation for the jth subunit within the ith unit is denoted by y_{ij}. The following symbols refer to the ith unit.

	Population	Sample
Number of subunits	M_i	m_i
Mean per subunit	$\bar{Y}_i$	$\bar{y}_i$
Total	$Y_i = M_i \bar{Y}_i$	$y_i = m_i \bar{y}_i$

The following symbols refer to the whole population or sample.

	Population	Sample
Number of subunits	$M_0 = \sum\limits^{N} M_i$	$\sum\limits^{n} m_i$
Total	$Y = \sum\limits^{N} Y_i$	$\sum\limits^{n} y_i$
Mean per subunit	$\bar{\bar{Y}} = Y/M_0$	$\bar{\bar{y}} = \sum y_i / \sum m_i$
Mean per primary unit	$\bar{Y} = Y/N$	$\bar{y} = \sum y_i / n$

11.2   SAMPLING METHODS WHEN $n = 1$

Suppose that the ith unit is selected and that it contains M_i subunits, of which m_i are sampled at random. We consider three methods of estimating $\bar{\bar{Y}}$, the mean per subunit, or *second-stage unit.* as it is often called.

I. Units Chosen with Equal Probability

$$\text{Estimate} = \bar{y}_\mathrm{I} = \bar{y}_i.$$

The estimate is the sample mean per subunit. It is biased, for in repeated sampling from the same unit the average of $\bar{y}_i$ is $\bar{Y}_i$, and, since every unit has an equal chance of being selected, the average of $\bar{Y}_i$ is

$$\frac{1}{N} \sum_{i=1}^{N} \bar{Y}_i = \bar{\bar{Y}}_a \quad \text{(say)}$$

But the population mean is

$$\bar{\bar{Y}} = \frac{\sum\limits_{i=1}^{N} M_i \bar{Y}_i}{M}, \qquad \text{where } M = \sum\limits_{i=1}^{N} M_i$$

Hence the bias equals $(\bar{\bar{Y}}_a - \bar{\bar{Y}})$. Since the method is biased, we will compute the mean square error (MSE) about $\bar{\bar{Y}}$. Write

$$\bar{y}_i - \bar{\bar{Y}} = (\bar{y}_i - \bar{Y}_i) + (\bar{Y}_i - \bar{\bar{Y}}_a) + (\bar{\bar{Y}}_a - \bar{\bar{Y}})$$

Square and take the expectation over all possible samples. All contributions from cross-product terms vanish. The expectations of the squared terms follow easily by the methods given in Chapter 10. We find

$$\text{MSE}(\bar{y}_I) = \frac{1}{N} \sum\limits_{i=1}^{N} \frac{(M_i - m_i)}{M_i} \frac{S_{2i}^2}{m_i} + \frac{1}{N} \sum\limits_{i=1}^{N} (\bar{Y}_i - \bar{\bar{Y}}_a)^2 + (\bar{\bar{Y}}_a + \bar{\bar{Y}})^2 \qquad (11.1)$$

where

$$S_{2i}^2 = \frac{1}{M_i - 1} \sum\limits_{j=1}^{M_i} (y_{ij} - \bar{Y}_i)^2$$

is the variance among subunits in the ith unit.

The MSE of $\bar{y}_I$ contains three components: one arising from variation within units, one from variation between the true means of the units, and one from the bias.

The values of the m_i have not been specified. The most common choice is either to take all m_i equal or to take m_i proportional to M_i, that is, to subsample a fixed proportion of whatever unit is selected. The choice of the m_i affects only the *first* of the three components of the variance—the component that arises from variation within units.

II. Units Chosen with Equal Probability

$$\text{Estimate} = \bar{y}_{II} = \frac{NM_i \bar{y}_i}{M_0}$$

This estimate is *unbiased*. Since $\bar{y}_i$ is an unbiased estimate of $\bar{Y}_i$, the product $M_i \bar{y}_i$ is an unbiased estimate of the unit total Y_i. Hence $NM_i \bar{y}_i$ is an unbiased estimate of the population total Y. Dividing by M_0, the total number of subunits in the population, we obtain an unbiased estimate of $\bar{\bar{Y}}$.

To find $V(\bar{y}_{II})$ which, of course, equals its MSE, we have

$$\bar{y}_{II} - \bar{\bar{Y}} = \frac{NM_i \bar{y}_i}{M_0} - \bar{\bar{Y}}$$

$$= \frac{NM_i}{M_0} (\bar{y}_i - \bar{Y}_i) + \left(\frac{NM_i}{M_0} \bar{Y}_i - \bar{\bar{Y}} \right)$$

Now $M_i \bar{Y}_i = Y_i$, the total for the unit, and $\bar{\bar{Y}} = N\bar{Y}/M_0$, where $\bar{Y}$ is the population mean per unit. This gives

$$\bar{y}_{II} - \bar{\bar{Y}} = \frac{NM_i}{M_0}(\bar{y}_i - \bar{Y}_i) + \frac{N}{M_0}(Y_i - \bar{Y})$$

Hence

$$V(\bar{y}_{II}) = \frac{N}{M_0^2} \sum_{i=1}^{N} M_i(M_i - m_i)\frac{S_{2i}^2}{m_i} + \frac{N}{M_0^2} \sum_{i=1}^{N} (Y_i - \bar{Y})^2 \tag{11.2}$$

The between-units component of this variance (second term on the right) represents the variation among the unit *totals* Y_i. This component is affected both by variations in the M_i from unit to unit by variations in the means $\bar{Y}_i$ per element. If the units vary considerably in size, this component is large, even though the means per element $\bar{Y}_i$ are almost constant from unit to unit. Frequently this component is so large that $\bar{y}_{II}$ has a much higher MSE than the biased estimate $\bar{y}_I$. Thus neither method I nor method II is fully satisfactory.

III. Units Chosen with Probability Proportional to Size

$$\text{Estimate} = \bar{y}_{III} = \bar{y}_i = \text{sample mean}$$

This technique is due to Hansen and Hurwitz (1943). It gives a sample mean that is unbiased and is not subject to the inflation of the variance in method II.

In repeated sampling, the ith unit appears with relative frequency M_i/M_0. Hence

$$E(\bar{y}_{III}) = \sum_{i=1}^{N} \frac{M_i}{M_0} \bar{Y}_i = \bar{\bar{Y}}$$

Furthermore,

$$\bar{y}_{III} - \bar{\bar{Y}} = (\bar{y}_{III} - \bar{Y}_i) + (\bar{Y}_i - \bar{\bar{Y}})$$

Average first over samples in which the ith unit is selected.

$$\underset{i}{E}(\bar{y}_{III} - \bar{\bar{Y}})^2 = \left(\frac{M_i - m_i}{M_i}\right)\frac{S_{2i}^2}{m_i} + (\bar{Y}_i - \bar{\bar{Y}})^2$$

Now average over all possible selections of the unit. Since the ith unit is selected with relative frequency M_i/M_0,

$$V(\bar{y}_{III}) = \frac{1}{M_0}\left[\sum_{i=1}^{N}(M_i - m_i)\frac{S_{2i}^2}{m_i} + \sum_{i=1}^{N} M_i(\bar{Y}_i - \bar{\bar{Y}})^2\right] \tag{11.3}$$

Note that, as in method I, the between-units component arises from differences among the means per subunit $\bar{Y}_i$ in the successive units. If these means per subunit are nearly equal, this component is small.

TABLE 11.1

ARTIFICIAL POPULATION WITH UNITS OF UNEQUAL SIZES

Unit	y_{ij}	M_i	Y_i	S_{2i}^2	$\bar{Y}_i$	$\bar{Y}_i - \bar{\bar{Y}}$
1	0, 1	2	1	0.500	0.5	−2.25
2	1, 2, 2, 3	4	8	0.667	2.0	−0.75
3	3, 3, 4, 4, 5, 5	6	24	0.800	4.0	+1.25
	Totals	12	33			

Example. Let us apply these results to a small population, artificially constructed. The data are presented in Table 11.1. There are three units, with 2, 4, and 6 elements, respectively. The reader may verify the figures given for Y_i, S_{2i}^2, and $\bar{Y}_i$. The population mean $\bar{\bar{Y}}$ is $\frac{33}{12}$, or 2.75. The unweighted mean of the $\bar{Y}_i$ is $2.167 = \bar{\bar{Y}}_a$, so that the bias in method I is −0.583. Its square, the contribution to the MSE, is 0.340.

One unit is to be selected and two subunits sampled from it. We consider four methods, two of which are variants of method I.

Method Ia.
  Selection: unit with equal probability, $m_i = 2$.
  Estimate: $\bar{y}_i$ (biased).
Method Ib.
  Selection: unit with equal probability, $m_i = \frac{1}{2}M_i$.
  Estimate: $\bar{y}_i$ (biased).
Method II.
  Selection: unit with equal probability, $m_i = 2$.
  Estimate: $NM_i\bar{y}_i/M_0$ (unbiased).
Method III.
  Selection: unit with probability M_i/M_0, $m_i = 2$.
  Estimate: $\bar{y}_i$ (unbiased).

Method *Ib* (proportional subsampling) does not guarantee a sample size of 2 (it may be 1, 2, or 3), but the average sample size is 2.

By application of the sampling error formulas (11.1), (11.2), and (11.3), we obtain the results in Table 11.2.

TABLE 11.2

MSE'S OF SAMPLE ESTIMATES OF $\bar{Y}$

Method	Contribution to MSE from			Total MSE
	Within Units	Between Units	Bias	
Ia	0.144	2.056	0.340	2.540
Ib	0.183	2.056	0.340	2.579
II	0.256	5.792	0.000	6.048
III	0.189	1.813	0.000	2.002

Although the example is artificial, the results are typical of those found in comparisons made on many populations. Method III gives the smallest MSE because it has the smallest contribution from variation between units. Method II, although unbiased, is very inferior. Method Ia (equal size of subsample) is slightly better than method Ib (proportional subsampling).

Some comparisons of these methods have also been made on actual populations. For six items (total workers, total agricultural workers, total nonagricultural workers, estimated separately for males and females), Hansen and Hurwitz (1943) found that method III produced large reductions in the contribution from variation between units as compared with the unbiased method II, and reductions that averaged 30% as compared with method I. (They assumed the contribution from variation within units to be negligible.) In estimating typical farm items for the state of North Carolina, Jebe (1952) reported reductions in the total variance of the order of 15% as compared with methods of type I. In both studies the primary unit was a county.

11.3 SAMPLING WITH PROBABILITY PROPORTIONAL TO ESTIMATED SIZE

As mentioned in Chapter 9, the sizes M_i of the units are sometimes known only approximately from previous data, and in other surveys several possible measures of the size of a unit may be available. Let z_i be the probability or relative size assigned to the ith unit, where the z_i are any set of positive numbers that add to unity. We still assume $n = 1$.

Method IV.   An unbiased estimate of $\bar{\bar{Y}}$ is

$$\bar{y}_{\text{IV}} = \frac{M_i \bar{y}_i}{z_i M_0} \tag{11.4}$$

This follows because, in repeated sampling, the ith unit appears with relative frequency z_i, so that

$$E(\bar{y}_{\text{IV}}) = \sum_{i=1}^{N} z_i \left(\frac{M_i \bar{y}_i}{z_i M_0} \right) = \sum_{i=1}^{N} \frac{M_i \bar{y}_i}{M_0} = \bar{\bar{Y}}$$

The variance of $\bar{y}_{\text{IV}}$ is obtained in the usual way. Write

$$\bar{y}_{\text{IV}} - \bar{\bar{Y}} = \frac{M_i \bar{y}_i}{z_i M_0} - \bar{\bar{Y}} \qquad \text{[by (11.4)]}$$

$$= \frac{1}{M_0} \left[\frac{M_i}{z_i} (\bar{y}_i - \bar{Y}_i) + \left(\frac{M_i}{z_i} \bar{Y}_i - M_0 \bar{\bar{Y}} \right) \right]$$

In the variance, each square receives a weight z_i. Hence

$$V(\bar{y}_{\text{IV}}) = \frac{1}{M_0^2} \left[\sum_{i=1}^{N} \frac{M_i (M_i - m_i)}{z_i} \frac{S_{2i}^2}{m_i} + \sum_{i=1}^{N} z_i \left(\frac{M_i \bar{Y}_i}{z_i} - M_0 \bar{\bar{Y}} \right)^2 \right] \tag{11.5}$$

If $z_i = M_i/M_0$, (11.5) reduces to (11.3) for $V(\bar{y}_{III})$. If $z_i = 1/N$ (initial probabilities equal), (11.5) reduces to (11.2) for the variance of the unbiased estimate when probabilities are equal.

Unless $z_i = M_i/M_0$, the between-units component in (11.5) is affected to some extent by variations in the sizes M_i as well as by variations in the means per element $\bar{Y}_i$.

TABLE 11.3

COMPUTATION OF $V(\bar{y}_{IV})$

Unit	M_i	M_i/M_0	z_i	m_i	$\dfrac{M_i(M_i - m_i)}{z_i m_i}$	$S_{2i}^{\ 2}$	Y_i	$\dfrac{Y_i}{z_i}$	$\dfrac{Y_i}{z_i} - Y$
1	2	0.17	0.2	2	0	0.500	1	5	-28
2	4	0.33	0.4	2	10	0.667	8	20	-13
3	6	0.50	0.4	2	30	0.800	24	60	$+27$

Example. Table 11.3 shows the computations for finding $V(\bar{y}_{IV})$ in the artificial population in Table 11.1. The z_i have been taken as 0.2, 0.4, and 0.4, and the $m_i = 2$. From (11.5), the variance comes out as follows:

$$\text{within-units contribution} = \sum \frac{M_i(M_i - m_i)S_{2i}^{\ 2}}{z_i m_i}/M_0^{\ 2} = 0.213$$

$$\text{between-units contribution} = \sum z_i\left(\frac{Y_i}{z_i} - Y\right)^2/M_0^{\ 2} = 3.583$$

Comparison with Table 11.2 shows that method IV has a lower variance than the unbiased method II in which the primary unit is chosen with equal probabilities, but method IV is decidedly inferior to method I or method III. In this example method IV pays too high a price in order to obtain an unbiased estimate.

Consequently, it is natural to consider whether the sample mean (as in method I) would be better than the estimate adopted in method IV.

V. Units Chosen with Probability Proportional To Estimated Size

$$\text{Estimate} = \bar{y}_V = \bar{y}_i = \text{sample mean}$$

The estimate is biased since, for example,

$$E(\bar{y}_i) = \sum z_i \bar{Y}_i = \bar{\bar{Y}}_z$$

If the z_i are good estimates, $\bar{\bar{Y}}_z$ is close to the correct mean $\bar{Y} = \sum M_i \bar{Y}_i/M_0$ and the bias is small.

If we write

$$\bar{y}_V - \bar{Y} = (\bar{y}_i - \bar{Y}_i) + (\bar{Y}_i - \bar{\bar{Y}}_z) + (\bar{\bar{Y}}_z - \bar{Y})$$

the three components of the MSE work out as follows.

$$\text{MSE}(\bar{y}_V) = \sum_{i=1}^{N} \frac{z_i(M_i - m_i)}{M_i} \frac{S_{2i}^2}{m_i} + \sum_{i=1}^{N} z_i(\bar{Y}_i - \bar{Y}_z)^2 + (\bar{Y}_z - \bar{Y})^2 \qquad (11.6)$$

Example. If the values of z_i and m_i are chosen as in Table 11.3, the reader may verify that the components of the variance of $\bar{y}_V$ are as shown in Table 11.4.

TABLE 11.4

CONTRIBUTIONS TO THE MSE IN METHOD V

Within Units	Between Units	Bias	Total MSE
0.173	1.800	0.062	2.035

This is superior to all methods except method III (pps) and is almost as good as method III.

11.4 SUMMARY OF METHODS FOR $n = 1$

The five methods of estimating the mean per element $\bar{\bar{Y}}$ and their MSE's in the numerical example are summarized in Table 11.5.

For estimating the population total $Y = M_0 \bar{\bar{Y}}$, the preceding estimates are multiplied by M_0 and their MSE's in Table 11.5 by $M_0^2 = 144$. Their relative performances remain the same.

TABLE 11.5

TWO-STAGE SAMPLING METHODS $(n = 1)$

Method	Probabilities in Selecting Units	Estimate of $\bar{\bar{Y}}$	Bias Status	MSE in Example
I	Equal	$\bar{y}_i$	Biased	Ia: 2.541 Ib: 2.579
II	Equal	$\dfrac{NM_i\bar{y}_i}{M}$	Unbiased	6.048
III	$\dfrac{M_i}{M_0} \propto$ size	$\bar{y}_i$	Unbiased	2.002
IV	$z_i \propto$ estimated size	$\dfrac{M_i\bar{y}_i}{z_iM_0}$	Unbiased	3.796
V	$z_i \propto$ estimated size	$\bar{y}_i$	Biased	2.035

11.5 SAMPLING METHODS WHEN $n > 1$

For $n > 1$ the principal sampling methods are natural extensions both of the preceding methods I to IV and of the methods discussed in Chapter 9A for one-stage sampling with cluster units of unequal sizes. The quantities most commonly estimated are population totals, population means per subunit, and means or proportions that have the structure of ratio estimates.

In many applications the quantity M_0, the total number of subunits in the population, is not known, but only the M_i for the sample primary units that are drawn, since these M_i can be counted by listing. It is worth noting, therefore, that methods II and IV and their extensions to $n > 1$ do not require knowledge of M_0 for estimation of the population *total*. Method I and its extensions do not require knowledge of M_0 for estimating the mean per subunit. Method III, probability proportional to size, requires knowledge of M_0.

Estimates are called *self-weighting* when they are simply a multiple of the total over all subunits in the sample. In view of the practical convenience of self-weighting estimates in large-scale surveys, the condition under which each estimate becomes self-weighting will be noted. For unbiased estimates the condition is, as will be seen, that every second-stage unit in the population or, more generally, every unit of the final stage of sampling, has an equal chance of being drawn.

11.6　TWO USEFUL RESULTS

In finding variances and sample estimates, two general results are useful. They show how to extend one-stage sampling results to two-stage or more generally multistage sampling. They were first proved by Durbin (1953) for estimators that are sums of estimators for the sample primary units, extended by Des Raj (1966) to linear functions of the primary unit estimators, and extended to still more complex cases by Rao (1975b). We will follow the approach used by Des Raj.

Primary units are chosen without replacement, with equal or unequal probabilities. In the ith primary unit, let $\hat{Y}_i$ be an unbiased estimator of the unit total Y_i, with second-stage variance σ_{2i}^2. Subsampling is independent in different primary units. Consider an unbiased estimator of the population total Y of the form

$$\hat{Y} = \sum_{i=1}^{n} w_{is} \hat{Y}_i \qquad (11.7)$$

where the weight w_{is} is known for every sample s, and may depend on other primary units that are in the sample as well as on unit i.

We will adopt the device used in section 2.9 of letting $w_{is}{}'$ be a random variable that equals w_{is} if unit i appears in the sample and equals zero otherwise. This gives

$$\hat{Y} = \sum_{i=1}^{N} w_{is}{}' \hat{Y}_i \qquad (11.8)$$

Now

$$E(\hat{Y}) = E_1 E_2(\hat{Y}) = E_1\left(\sum_{i=1}^{N} w_{is}' Y_i\right) = Y \qquad (11.9)$$

if and only if $E_1(w_{is}') = 1$ for every i.

Theorem 11.1.

$$V(\hat{Y}) = V\left(\sum_{i=1}^{n} w_{is}\hat{Y}_i\right) = V\left(\sum_{i=1}^{n} w_{is} Y_i\right) + \sum_{i=1}^{N} E_1(w_{is}'^2)\sigma_{2i}^2 \qquad (11.10)$$

Proof.   By formula (10.2),

$$V(\hat{Y}) = V_1[E_2(\hat{Y})] + E_1[V_2(\hat{Y})]$$

$$= V\left(\sum_{i=1}^{n} w_{is} Y_i\right) + E_1\left[\sum_{i=1}^{N} w_{is}'^2 V_2(\hat{Y}_i)\right] \qquad (11.11)$$

since the second-stage covariance between $\hat{Y}_i$ and $\hat{Y}_j$ $(i \neq j)$ is zero, because subsampling is independent. Hence

$$V(\hat{Y}) = V\left(\sum_{i=1}^{n} w_{is} Y_i\right) + \sum_{i=1}^{N} [E_1(w_{is}'^2)\sigma_{2i}^2] \qquad (11.12)$$

This completes the proof.

The first term in (11.12) is the variance of the corresponding estimator when every primary sample unit is completely enumerated, and it is obtained from results in single-stage sampling.

Example.   For the two-stage analogue of the Horvitz–Thompson estimator, $\hat{Y}_{HT} = \sum^{n} \hat{Y}_i/\pi_i$, the weight $w_{is}' = 1/\pi_i$ if unit i is in the sample and zero otherwise. Hence, $E_1(w_{is}'^2) = \pi_i/\pi_i^2 = 1/\pi_i$, where π_i is the probability that primary unit i is drawn. Furthermore, if m_i subunits are drawn from the M_i by simple random sampling whenever unit i is drawn,

$$\sigma_{2i}^2 = V_2(\hat{Y}_i) = \frac{M_i(M_i - m_i)}{m_i} S_{2i}^2 \qquad (11.13)$$

Hence, by theorem 11.1, using formula (9A.42) for the single-stage analogue of $V(\hat{Y}_{HT})$, we have

$$V(\hat{Y}_{HT}) = \sum_{i=1j>i}^{N}\sum^{N} (\pi_i\pi_j - \pi_{ij})\left(\frac{Y_i}{\pi_i} - \frac{Y_j}{\pi_j}\right)^2 + \sum_{i=1}^{N} \frac{M_i(M_i - m_i)S_{2i}^2}{m_i\pi_i} \qquad (11.14)$$

Theorem 11.2.   Suppose that we have an unbiased estimate $\hat{\sigma}_{2i}^2$ of the second-stage variance σ_{2i}^2 of $\hat{Y}_i$, and an unbiased sample estimate of $V(\sum^{n} w_{is} Y_i) = V(\sum^{N} w_{is}' Y_i)$ from one-stage sampling. The latter will be a quadratic of the form

$$v\left(\sum_{i}^{n} w_{is} Y_i\right) = \sum_{i}^{n} a_{is} Y_i^2 + 2\sum_{i}^{n}\sum_{j>i}^{n} b_{ijs} Y_i Y_j \qquad (11.15)$$

Then an unbiased sample estimator of $V(\overset{n}{\underset{}{\sum}} w_{is}\hat{Y}_i)$ is

$$v\left(\overset{n}{\underset{}{\sum}} w_{is}\hat{Y}_i\right) = \left(\overset{n}{\underset{i}{\sum}} a_{is}\hat{Y}_i^2 + 2\overset{n}{\underset{i}{\sum}}\overset{n}{\underset{j>i}{\sum}} b_{ijs}\hat{Y}_i\hat{Y}_j\right) + \overset{n}{\underset{i}{\sum}} w_{is}\hat{\sigma}_{2i}^2 \qquad (11.16)$$

Thus the rule for constructing the sample estimator of $V(\overset{n}{\underset{}{\sum}} w_{is}\hat{Y}_i)$ is: In an unbiased estimator of $V(\overset{n}{\underset{}{\sum}} w_{is}Y_i)$ from one-stage sampling, insert $\hat{Y}_i$ wherever Y_i appears. To this add the term $\overset{n}{\underset{}{\sum}}(w_{is}\hat{\sigma}_{2i}^2)$, where $\overset{n}{\underset{}{\sum}} w_{is}\hat{Y}_i = \hat{Y}$, and $\hat{\sigma}_{2i}^2$ is an unbiased estimator of $V_2(\hat{Y}_i)$.

Proof.

$$V\left(\overset{N}{\underset{i}{\sum}} w_{is}{}'Y_i\right) = \overset{N}{\underset{i}{\sum}} Y_i^2 V(w_{is}{}') + 2\overset{N}{\underset{i}{\sum}}\overset{N}{\underset{j>i}{\sum}} Y_i Y_j \text{ cov }(w_{is}{}'w_{js}{}') \qquad (11.17)$$

Introduce the random variable, $a_{is}{}'$, where $a_{is}{}' = a_{is}$ if unit i is in the sample and $a_{is}{}' = 0$ otherwise. Similarly, let $b_{ijs}{}' = b_{ijs}$ if units i and j are in the sample and $b_{ijs}{}' = 0$ otherwise. From (11.15) for one-stage sampling,

$$v\left(\overset{N}{\underset{}{\sum}} w_{is}{}'Y_i\right) = \overset{N}{\underset{i}{\sum}} a_{is}{}'Y_i^2 + 2\overset{N}{\underset{i}{\sum}}\overset{N}{\underset{j>i}{\sum}} b_{ijs}{}'Y_i Y_j \qquad (11.18)$$

If this is to be unbiased, comparison of (11.18) and (11.17) shows that we must have $E_1(a_{is}{}') = V(w_{is}{}')$. Now for our variance estimator (11.16)

$$E_1 E_2\left(\overset{N}{\underset{i}{\sum}} a_{is}{}'\hat{Y}_i^2 + 2\overset{N}{\underset{i}{\sum}}\overset{N}{\underset{j>i}{\sum}} b_{ijs}{}'\hat{Y}_i\hat{Y}_j\right) + E_1 E_2\left(\overset{N}{\underset{}{\sum}} w_{is}{}'\hat{\sigma}_{2i}{}'^2\right)$$

$$= E_1\left(\overset{N}{\underset{i}{\sum}} a_{is}{}'Y_i^2 + 2\overset{N}{\underset{i}{\sum}}\overset{N}{\underset{j>i}{\sum}} b_{ijs}{}'Y_i Y_j\right) + \overset{N}{\underset{i}{\sum}}[V(w_{is}{}') + E^2(w_{is}{}')]\sigma_{2i}^2 \qquad (11.19)$$

In (11.19) we have used the results that $E(a_{is}{}') = V(w_{is}{}')$ and that for any i, $E_1(w_{is}{}') = 1 = E_1^2(w_{is}{}')$. Continuing,

$$E\left[v\left(\overset{n}{\underset{}{\sum}} w_{is}\hat{Y}\right)\right] = V\left(\overset{N}{\underset{}{\sum}} w_{is}Y_i\right) + \overset{N}{\underset{i}{\sum}} E_1(w_{is}{}'^2)\sigma_{2i}^2 = V\left(\overset{n}{\underset{}{\sum}} w_{is}\hat{Y}_i\right) \qquad (11.20)$$

by (11.12). This completes the proof.

We consider now some specific estimators of the population total Y. In extensive surveys, selection of primary units with unequal probabilities has become the most commonly used technique. Sections 11.9 (sampling with replacement) and 11.10 (sampling without replacement) deal with these methods; section 11.13 presents the ratio estimator in *ppz* sampling. Other sections (11.7 and 11.8) give the corresponding methods when primary units are chosen with equal probabilities.

11.7   UNITS SELECTED WITH EQUAL PROBABILITIES: UNBIASED ESTIMATOR

Except where noted otherwise, the m_i sample subunits in the ith unit are chosen by simple random sampling. The unbiased estimator of the population total is

$$\hat{Y}_u = \frac{N}{n}\sum^n M_i\bar{y}_i = \frac{N}{n}\sum^n \hat{Y}_i \qquad (11.21)$$

To apply theorem 11.1, note that

$$w_{is} = \frac{N}{n}; \qquad E(w_{is}') = \frac{n}{N}\frac{N}{n} = 1; \qquad E(w_{is}'^2) = \frac{n}{N}\frac{N^2}{n^2} = \frac{N}{n}$$

Hence, by theorem 11.1 and earlier results,

$$V(\hat{Y}_u) = \frac{N^2}{n}(1-f_1)\frac{\sum(Y_i-\bar{Y})^2}{(N-1)} + \frac{N}{n}\sum^N \frac{M_i^2(1-f_{2i})S_{2i}^2}{m_i} \qquad (11.22)$$

where $f_{2i} = m_i/M_i$. The estimator becomes self-weighting if f_{2i} is constant, $(=f_2,$ say). We then have

$$\hat{Y}_u = \frac{N}{nf_2}\sum_i^m\sum_j^{m_i} y_{ij} \qquad (11.23)$$

The quantity nf_2/N is, of course, the probability that any second-stage unit is drawn.

For an unbiased sample estimate of variance, theorem 11.2 gives, from (11.16),

$$v(\hat{Y}_u) = \frac{N^2(1-f_1)}{n}\frac{\sum_i^n(\hat{Y}_i-\hat{\bar{Y}}_u)^2}{n-1} + \frac{N}{n}\sum_i^n \frac{M_i^2(1-f_{2i})s_{2i}^2}{m_i} \qquad (11.24)$$

11.8   UNITS SELECTED WITH EQUAL PROBABILITIES: RATIO TO SIZE ESTIMATE

This estimator of the population total Y is

$$\hat{Y}_R = M_0\frac{\sum^n M_i\bar{y}_i}{\sum^n M_i} = M_0\frac{\sum^n \hat{Y}_i}{\sum^n M_i} \qquad (11.25)$$

This is a typical ratio estimator, since both numerator and denominator vary from sample to sample. It is used mainly for estimating means per subunit, for which knowledge of M_0 is not required, and for which it is an extension of method I with $n = 1$.

To find its approximate MSE, write

$$\hat{Y}_R - Y = M_0 \frac{\sum\limits_{}^{n} M_i \bar{y}_i}{\sum\limits_{}^{n} M_i} - Y \doteq \frac{N}{n} \sum\limits_{}^{n} M_i(\bar{y}_i - \bar{\bar{Y}}) \qquad (11.26)$$

where $\bar{\bar{Y}} = Y/M_0$.

Since $\hat{Y}_u = (N/n) \sum\limits_{}^{n} M_i \bar{y}_i$, we can obtain the approximate $\text{MSE}(\hat{Y}_R)$ from formula (11.22) for $V(\hat{Y}_u)$ by substituting $M_i(\bar{y}_i - \bar{\bar{Y}})$ for $M_i\bar{y}_i$ or, more generally, $(y_{ij} - \bar{\bar{Y}})$ for y_{ij}. Now (11.22) is

$$V(\hat{Y}_u) = \frac{N^2}{n}(1-f_1)\frac{\sum\limits_{}^{N}(Y_i - \bar{Y})^2}{N-1} + \frac{N}{n}\sum\limits_{}^{n}\frac{M_i^2(1-f_{2i})S_{2i}^2}{m_i}$$

In the substitution, $Y_i = M_i\bar{Y}_i$ becomes $M_i(\bar{Y}_i - \bar{\bar{Y}})$ while $\bar{Y} = \sum\limits_{}^{N} y_i/N$ is replaced by 0. Also $S_{2i}^2 = \sum\limits_{j}(y_{ij} - \bar{Y}_i)^2/(M_i - 1)$ remains unchanged when $(y_{ij} - \bar{Y})$ replaces y_{ij}. This gives the result

$$\text{MSE}(\hat{Y}_R) \doteq \frac{N^2}{n}(1-f_1)\frac{\sum\limits_{}^{N}M_i^2(\bar{Y}_i - \bar{\bar{Y}})^2}{N-1} + \frac{N}{n}\sum\limits_{}^{N}\frac{M_i^2(1-f_{2i})S_{2i}^2}{m_i} \qquad (11.27)$$

As with $\hat{Y}_u$, this estimator becomes self-weighting if

$$f_{2i} = \frac{m_i}{M_i} = \text{constant} = f_2 = \frac{\bar{m}}{\bar{M}} = \frac{N\bar{m}}{M_0}$$

In this event the within-units contribution may be expressed more simply by putting $m_i = N\bar{m}M_i/M_0$, giving

$$\text{MSE}(\hat{Y}_R) \doteq \frac{N^2}{n}(1-f_1)\frac{\sum\limits_{}^{N}M_i^2(\bar{Y}_i - \bar{\bar{Y}})^2}{(N-1)} + \frac{M_0^2(1-f_2)}{n\bar{m}}\sum\limits_{}^{N}\left(\frac{M_i}{M_0}\right)S_{2i}^2$$

$$(11.28)$$

The resemblance to the corresponding formula when primary units are of equal sizes may be noted. From (10.14) section 10.3, multiplying by M_0^2 since $\hat{Y} = M_0\bar{y}$, we have

$$V(\hat{Y}) = \frac{M_0^2(1-f_1)}{n}\sum\limits_{}^{N}\frac{(\bar{Y}_i - \bar{\bar{Y}})^2}{(N-1)} + \frac{M_0^2(1-f_2)}{nm}\sum\limits_{}^{N}\left(\frac{1}{N}\right)S_{2i}^2 \qquad (11.29)$$

The difference is that in (11.28) the contributions to the MSE from the primary units are weighted, larger units receiving greater weight.

By theorem 11.2 an approximate sample estimate of $\text{MSE}(\hat{Y}_R)$ in (11.27) is given by

$$v(\hat{Y}_R) = \frac{N^2}{n}(1-f_1)\frac{\sum\limits^{n}M_i^2(\bar{y}_i-\hat{\bar{Y}}_R)^2}{n-1} + N\frac{n}{n}\sum\frac{M_i^2(1-f_{2i})s_{2i}^2}{m_i} \qquad (11.30)$$

where $\hat{\bar{Y}}_R = \sum\limits^{n} M_i\bar{y}_i / \sum\limits^{n} M_i = \hat{Y}_R/M_0$.

When this estimator is used to estimate the population mean per subunit, we have $\hat{\bar{Y}}_R = \sum\limits^{n} \hat{Y}_i / \sum\limits^{n} M_i$ and $v(\hat{\bar{Y}}_R) = v(\hat{Y}_R)'/M_0^2$. When M_0 is not known, we substitute $\hat{M}_0 = N(\sum\limits^{n} M_i/n)$ for M_0 in calculating $v(\hat{\bar{Y}}_R)$.

When primary units are selected with equal probabilities, an alternative estimate of the population mean per subunit (another extension of method I) is

$$\frac{1}{n}(\bar{y}_1 + \bar{y}_2 + \cdots + \bar{y}_n)$$

This estimate is self-weighting if m_i = constant, as in the following example. When m_i and $\bar{Y}_i$ are uncorrelated, this estimate may be satisfactory, but it is liable to a bias that does not vanish even with n large.

Example. From the volume *American Men of Science*, 20 pages were selected at random. On each page the ages of two scientists, from two biographies also selected at random, were recorded. The total number of biographies per page varies in general from about 14 to 21. Estimate the average age and its standard error from the data in Table 11.6, using the ratio estimate.

From the extreme right column,

$$\hat{\bar{Y}}_R = \frac{\sum M_i\bar{y}_i}{\sum M_i} = \frac{17,121.5}{359} = 47.7 \text{ years}$$

Since n/N is negligible, we have from (11.30), dividing by M_0^2,

$$v(\hat{\bar{Y}}_R) \doteq \frac{\sum\limits^{n} M_i^2(\bar{y}_i - \hat{\bar{Y}}_R)^2}{n\bar{M}^2(n-1)}$$

The numerator is computed as

$$\sum(M_i\bar{y}_i)^2 - 2\hat{\bar{Y}}_R \sum(M_i\bar{y}_i)M_i + \hat{\bar{Y}}_R^2 \sum M_i^2$$
$$= 15,375,020 - (95.3844)(309,747.5) + (2274.55)(6481) = 571,300$$

Since $\bar{M}_n = 359/20$, as estimated from the sample, this gives

$$v(\hat{\bar{Y}}_R) = \frac{(20)(571,300)}{(19)(359)^2} = 4.67$$

$$s(\hat{\bar{Y}}_R) = 2.16 \text{ years}$$

TABLE 11.6

AGES OF 40 SCIENTISTS IN *American Men of Science* ($n = 20$, $m = 2$)

Unit No.	M_i	Ages y_{i1}	Ages y_{i2}	Total y_i	$M_i \bar{y}_i$
1	15	47	30	77	577.5
2	19	38	51	89	845.5
3	19	43	45	88	836.0
4	16	55	41	96	768.0
5	16	59	45	104	832.0
6	19	39	38	77	731.5
7	18	43	43	86	774.0
8	18	49	51	100	900.0
9	18	45	35	80	720.0
10	18	46	59	105	945.0
11	20	71	64	135	1,350.0
12	18	35	46	81	729.0
13	19	61	54	115	1,092.5
14	19	45	87	132	1,254.0
15	18	31	38	69	621.0
16	16	64	39	103	824.0
17	16	63	47	110	880.0
18	19	36	33	69	655.5
19	19	61	39	100	950.0
20	19	54	34	88	836.0
Totals	359			1,904	17,121.5

11.9 UNITS SELECTED WITH UNEQUAL PROBABILITIES WITH REPLACEMENT: UNBIASED ESTIMATOR

Primary units are selected with probabilities proportional to z_i *with replacement*. Results for $z_i = M_i/M_0$ (probability proportional to size) follow as a special case.

The subsample of m_i subunits from the ith unit is assumed to be randomly drawn *without replacement*. If the ith unit is selected more than once, we suppose that on each selection the whole subsample is replaced, a new independent drawing of m_i units being made without replacement from the complete unit.

An unbiased estimate of the population total is

$$\hat{Y}_{ppz} = \frac{1}{n} \sum_{i=1}^{n} \frac{M_i \bar{y}_i}{z_i} = \frac{1}{n} \sum_{i=1}^{n} \frac{\hat{Y}_i}{z_i} \qquad (11.31)$$

For $n = 1$, it was shown in section 11.3 that this estimator, $M_0 \bar{y}_{IV} = \hat{Y}_{IV}$, is unbiased. Its variance is obtained from formula (11.5) on multiplying by M_0^2 as

$$V(\hat{Y}_{IV}) = \sum_{i=1}^{N} z_i \left(\frac{Y_i}{z_i} - Y \right)^2 + \sum_{i=1}^{N} \frac{M_i(M_i - m_i)S_{2i}^2}{z_i m_i} \tag{11.32}$$

With this method of sampling, the estimator $\hat{Y}_{ppz}$ is the mean of n independent estimates of the form $\hat{Y}_{IV}$. Consequently, from classical sampling theory $\hat{Y}_{ppz}$ is unbiased and

$$V(\hat{Y}_{ppz}) = \frac{1}{n} V(\hat{Y}_{IV}) = \frac{1}{n} \sum_{i=1}^{N} z_i \left(\frac{Y_i}{z_i} - Y \right)^2 + \frac{1}{n} \sum_{i=1}^{N} \frac{M_i^2(1 - f_{2i})S_{2i}^2}{m_i z_i} \tag{11.33}$$

Furthermore, given n independent estimates $\hat{Y}_{IV} = Y_i / z_i$, an unbiased sample estimator of $V(\hat{Y}_{IV})$ is, of course,

$$v(\hat{Y}_{IV}) = \frac{\sum_{i=1}^{n} \left(\frac{\hat{Y}_i}{z_i} - \hat{Y}_{ppz} \right)^2}{(n-1)} \tag{11.34}$$

Consequently, an unbiased sample estimator of $V(\hat{Y}_{ppz})$ is the very simple expression

$$v(\hat{Y}_{ppz}) = \frac{\sum_{i=1}^{n} \left(\frac{\hat{Y}_i}{z_i} - \hat{Y}_{ppz} \right)^2}{n(n-1)} \tag{11.35}$$

These results hold also for multistage sampling, provided that $\hat{Y}_i$ is an unbiased estimator of Y_i and that subsampling is independent whenever a primary unit is drawn.

To discover when $\hat{Y}_{ppz}$ becomes self-weighting in two-stage sampling, write

$$\hat{Y}_{ppz} = \frac{1}{n} \sum_{i}^{n} \frac{M_i}{z_i m_i} \sum_{j}^{m_i} y_{ij} \tag{11.36}$$

The condition is therefore

$$\frac{n z_i m_i}{M_i} = \text{constant} = f_0 \tag{11.37}$$

This expression is the probability that any specific second-stage unit is drawn. From (11.37), $m_i / M_i = f_0 / n z_i$. If the probability f_0 is chosen in advance, the field worker can be told in advance what second-stage fraction m_i / M_i to take from any primary unit that has been chosen. For instance, suppose $f_0 = 1/50 = 0.02$ and $n = 60$ primary units have been chosen. If $z_i = 0.0026$ for one selected unit, we must have $m_i / M_i = 0.02 / (60)(0.0026)$, or 1 in 7.8.

With a self-weighting sample, the variance estimator (11.35) takes the simpler form

$$v(\hat{Y}_{ppz}) = \frac{n}{(n-1)f_0^2} \sum_i^n (y_i - \bar{y})^2 \qquad (11.38)$$

where $y_i = \sum_j y_{ij}$ is the sample total in the ith unit. The simplicity of the estimated variances (11.35) and (11.38) is an attractive feature of with-replacement sampling.

With this method there are other ways in which the subsamples may be drawn. If the ith unit is selected t_i times, one variant is to draw a *single* subsample of size $t_i m_i$ without replacement, provided that $M_i > m_i t_i$. Sukhatme (1954) has shown that with this method, $V(\hat{Y}_{ppz})$ is reduced by $(n-1) \sum^N M_i S_{2i}^2 / n$. Another method is to draw a single subsample of size m_i no matter how many times the ith unit is selected. The estimate $M_i \bar{y}_i / z_i$ from this unit receives a weight t_i (the number of times that the unit has been drawn) with either method. The effect is to increase $V(\hat{Y}_{ppz})$ by

$$\left(\frac{n-1}{n}\right) \sum^N \frac{M_i^2(1 - f_{2i})S_{2i}^2}{m_i}$$

For the same cost, the differences in precision among these methods are seldom likely to be substantial.

If $z_i = M_i/M_0$ the unbiased estimate (11.31) reduces to

$$\hat{Y}_{pps} = \frac{M_0}{n} \sum_i^n \bar{y}_i \qquad (11.39)$$

Clearly this estimate becomes self-weighting when $m_i = m$, so that $\hat{Y}_{pps} = M_0 \bar{y}$.

An unbiased sample estimator is

$$v(\hat{Y}_{pps}) = \frac{M_0^2}{n(n-1)} \sum_i^n \left(\bar{y}_i - \frac{\hat{Y}_{pps}}{M_0}\right)^2 \qquad (11.40)$$

11.10 UNITS SELECTED WITHOUT REPLACEMENT

For any of the "without-replacement" methods studied in Chapter 9A, the formulas for the variance and estimated variance in two-stage sampling are obtained from theorems 11.1, 11.2. With the Brewer or Durbin methods,

$$\hat{Y}_B = \sum^n \frac{M_i \bar{y}_i}{\pi_i} = \frac{1}{n} \sum^n \frac{M_i \bar{y}_i}{z_i} = \frac{1}{n} \sum^n \frac{\hat{Y}_i}{z_i} \qquad (11.41)$$

with variance

$$V(\hat{Y}_B) = \sum_i^N \sum_{j>i}^N (\pi_i \pi_j - \pi_{ij}) \left(\frac{Y_i}{\pi_i} - \frac{Y_j}{\pi_j}\right)^2 + \sum_i^N \frac{M_i^2(1 - f_{2i})S_{2i}^2}{m_i \pi_i} \qquad (11.42)$$

For the corresponding estimator $\hat{Y}_{ppz}$ in sampling with replacement, we have, from (11.33), with $\pi_i = nz_i$,

$$V(\hat{Y}_{ppz}) = \frac{1}{n}\sum_{1}^{N} z_i\left(\frac{Y_i}{z_i} - Y\right)^2 + \sum^{N} \frac{M_i^2(1-f_{2i})S_{2i}^2}{m_i\pi_i} \tag{11.43}$$

Since the "within-units" contributions to the variance are the same in (11.33) and (11.43), any relative gain in precision from selecting units without replacement is watered down in two-stage samples by the within-units contribution to the variance. For instance, in a stratified multistage sample of $n = 44$ provinces out of $N = 147$ provinces, Des Raj (1964) found that the ratio $V(\hat{Y}_{WOR})/V(\hat{Y}_{WR})$ averaged about 0.79 over seven items for the between-units component, but the average ratio over both components was 0.92.

With $n = 2$ an unbiased sample estimate of $V(\hat{Y}_B)$ is

$$v(\hat{Y}_B) = (\pi_1\pi_2\pi_{12}^{-1} - 1)\left(\frac{M_1\bar{y}_1}{\pi_1} - \frac{M_2\bar{y}_2}{\pi_2}\right)^2 + \sum_{i=1}^{2} \frac{M_i^2(1-f_{2i})s_{2i}^2}{m_i\pi_i} \tag{11.44}$$

where suffixes 1, 2 denote the chosen units.

The extension to the Sampford estimator for $n > 2$ is straightforward. The alternative *RHC* estimator

$$\hat{Y}_{RHC} = \sum_{g}^{n} \frac{Z_g M_g \bar{y}_g}{z_g} \tag{11.45}$$

where $Z_g = \sum z_i$ over the gth group and M_g, $\bar{y}_g$, and z_g refer to the *unit* drawn from the group. For a sample of n units an unbiased estimator of $V(\hat{Y}_{RHC})$ is

$$v(\hat{Y}_{RHC}) = \frac{\left(\sum_{g}^{n} N_g^2 - N\right)}{(N^2 - \sum N_g^2)}\sum_{g}^{n} Z_g\left(\frac{M_g\bar{y}_g}{z_g} - \hat{Y}_{RHC}\right)^2 + \sum_{g}^{n} Z_g\frac{M_g^2}{z_g}\frac{(1-f_{2g})s_{2g}^2}{m_g} \tag{11.46}$$

Formulas (11.44) and (11.46) require separate calculation of the between-units and the within-units contributions to the estimated variance. In extensive surveys with many strata and numerous items, the complexity of such formulas makes the estimation of variances a task requiring more computer time than can be devoted to it in practice. recall the much simpler form of the variance formula (11.35) when units are drawn with replacement; that is,

$$v(\hat{Y}_{ppz}) = \frac{1}{n(n-1)}\sum^{n}\left(\frac{Y_i}{z_i} - \hat{Y}_{ppz}\right)^2$$

where $y_i'' = M_i\bar{y}_i/z_i$. This result makes with-replacement sampling appealing if it does not involve too much loss of precision.

Among without-replacement methods, Platek and Singh (1972), in planning the redesign of the Canadian Labor Force Survey for strata in which $n_h = 6$ or a multiple of 6, point out the advantages of $\hat{Y}_{RHC}$ and the Hartley–Rao version of systematic sampling (section 9A.10) in simplicity and variance estimation.

For many small strata with $n_h = 2$, Durbin (1967) has produced methods that give simple variance estimation when units are selected without replacement. He uses $\hat{Y}_{HT}$ and seeks methods for which $\pi_i = 2z_i$. Within a stratum, the strategy is to use selection methods that make $(\pi_i\pi_j\pi_{ij}^{-1} - 1)$ either 1 or 0. These choices enable $v(\hat{Y}_{HT})$ in (11.44) to reduce either to the term

$$\left(\frac{\hat{Y}_1}{\pi_1} - \frac{\hat{Y}_2}{\pi_2}\right)^2$$

or to the second term in (11.44). Brewer and Hanif (1969) have improved and extended the methods to estimators other than $\hat{Y}_{HT}$.

11.11 COMPARISON OF THE METHODS

In one-stage sampling (section 9A.12), equal-probability sampling with the unbiased estimate $\hat{Y}_u$ was compared with the corresponding ratio estimates and with various unequal-probability sampling methods. Roughly, the ratio of the variance of $\hat{Y}_u$ to that of another method that sampled without replacement was equal to the ratio of the coefficient of variation of the unit total Y_i to that of Y_i/z_i. As regards unequal-probability sampling with and without replacement, the ratio $V(\hat{Y}_{WOR})/V(\hat{Y}_{WR})$ roughly equaled $(N-n)/(N-1)$, the same figure as in equal-probability selection.

In two-stage and multistage sampling the net effect of the within-units contribution to the variance is to dilute the differences created by the between-units contributions. With the subsampling methods used in practice, the within-units contribution is often approximately the same for different methods. The net effect is that the overall relative precisions of different methods move toward equality. In practice the within-units component is often at least as large as the between-units component.

Even with large computers the self-weighting forms of the estimates are a convenience and are widely used. Use of a self-weighting plan should incur only minor loss of precision, since the choice of the m_i affects only the within-units contribution to the variance. The self-weighting form requires $m_i \propto M_i/z_i$, while the m_i that minimizes V_2 for a given expected total number of subunits is $m_i \propto M_i S_{2i}/z_i$. These two allocations differ little unless the S_{2i} vary over a wide range.

In surveys with many items we also noted (section 9A.13) that the z_i used for sample selection may not be a good choice for some items, Y_i having little relation to z_i. If an auxiliary variate x_i can be found such that y_i/x_i is reasonably constant, one possibility is to switch to a ratio estimate (section 11.13). If no such x_i is available, the alternative estimate suggested by Rao (1966) is of the form

$$\hat{Y}_{ppz}^* = \frac{N}{n} \sum_i^N M_i \bar{y}_i \qquad (11.47)$$

Preliminary evidence by Rao suggests that if Y_i and z_i are unrelated, $\hat{Y}_{ppz}^*$ may perform about as well as $\hat{Y}_u$ would if sampling were with equal probabilities (as illustrated in section 9A.13). If primary units are chosen with replacement, an unbiased sample estimate of $V(\hat{Y}_{ppz}^*)$ is

$$v(\hat{Y}_{ppz}^*) = \frac{N^2}{n(n-1)} \sum^n (M_i \bar{y}_i - \hat{\bar{Y}}_{ppz}^*)^2 \qquad (11.48)$$

where $\hat{\bar{Y}}_{ppz}^* = \hat{Y}_{ppz}^*/N$. This expression does not include the contribution of the bias in this estimate to its MSE, which it is hoped will be small if Y_i and z_i are unrelated.

11.12 RATIOS TO ANOTHER VARIABLE

In two-stage sampling the quantity to be estimated is often a ratio Y/X. This happens for two different reasons. As mentioned previously, if x is the value of y at a recent census, the ratio y/x may be relatively stable. An estimate of the population total or mean of y that is based on this ratio may be more precise than the estimates considered thus far in this chapter.

Ratio estimates of this type are encountered also in the estimation of proportions or means over parts of the population. In an urban survey with the city block as primary unit, an example of a proportion of this type is

$$\frac{\text{number of employed males over 16 years}}{\text{total number of males over 16 years}}$$

If $y_{ij} = 1$ for any employed male over 16 and $y_{ij} = 0$ otherwise, and $x_{ij} = 1$ for any male over 14 and $x_{ij} = 0$ otherwise, the population proportion is Y/X. Other examples for this type of survey are the average income of families that subscribe to a certain magazine or the average amount of pocket money per teen-aged child.

With any of the preceding methods of selection of the units (equal or unequal probabilities, with or without replacement), the standard approximate formula for the MSE or variance of $\hat{Y}_R$ and $\hat{R}$ is easily obtained from the formula for $V(\hat{Y})$ for the same method of sample selection by the technique which we have used repeatedly.

To obtain $V(\hat{Y}_R)$, replace y_{ij} by $d_{ij} = y_{ij} - Rx_{ij}$ in $V(\hat{Y})$ for the sampling method used. For $V(\hat{R})$, also divide by X^2.

For the estimated MSE or variance $v(\hat{Y}_R)$, replace y_{ij} by $d_{ij}' = y_{ij} - \hat{R}x_{ij}$ in $v(\hat{Y})$. For $v_2(\hat{R})$, divide also by $\hat{X}^2$.

This follows by the method used in Theorem 2.5. For

$$\hat{Y}_R - Y = X\frac{\hat{Y}}{\hat{X}} - Y = \frac{X}{\hat{X}}(\hat{Y} - R\hat{X}) \doteq (\hat{Y} - R\hat{X}) = \hat{D} \qquad (11.49)$$

where $d_{ij} = y_{ij} - Rx_{ij}$. Since $E(\hat{D}) = 0$ with any sampling plan for which $\hat{Y}$ and $\hat{X}$ are unbiased, we find $E(\hat{Y}_R - Y)^2$ by taking the formula for $V(\hat{Y})$ for this plan and substituting $d_{ij} = y_{ij} - Rx_{ij}$ in place of y_{ij}.

For example, consider $\hat{Y}_u = (N/n) \sum^n M_i\bar{y}_i = (N/n) \sum^n \hat{Y}_i$ in sampling with equal probabilities. From formula (11.22) in section 11.7,

$$V(\hat{Y}_u) = \frac{N^2}{n}(1 - f_1)\frac{\sum(Y_i - \bar{Y})^2}{N - 1} + \frac{N}{n}\sum^N \frac{M_i^2(1 - f_{2i})S_{2i}^2}{m_i} \qquad (11.50)$$

Hence, with equal-probability sampling and $\hat{Y}_R = X\hat{Y}_u/\hat{X}_u$,

$$V(\hat{Y}_R) \doteq \frac{N^2}{n}(1 - f_1)\frac{\sum^N(Y_i - RX_i)^2}{N - 1} + \frac{N}{n}\sum^N \frac{M_i^2(1 - f_{2i})S_{d2i}^2}{m_i} \qquad (11.51)$$

where

$$S_{d2i}^2 = \frac{1}{M_i - 1}\sum_{i=1}^{M_i}[(y_{ij} - Rx_{ij}) - (\bar{Y}_i - R\bar{X}_i)]^2$$

The conditions under which $\hat{Y}_R$ reduces to a multiple of the ratio of the sample totals $\Sigma\Sigma y_{ij}/\Sigma\Sigma x_{ij}$ are always those under which the corresponding $\hat{Y}$ becomes self-weighting; in this case $f_{2i} = m_i/M_i = $ constant.

For the estimated variance, substitution of $d_{ij}' = y_{ij} - \hat{R}x_{ij}$ for y_{ij} in (11.24) for $v(\hat{Y}_u)$ gives

$$v(\hat{Y}_R) \doteq \frac{N^2(1 - f_1)}{n}\frac{\sum(\hat{Y}_i - \hat{R}\hat{X}_i)^2}{n - 1} + \frac{N}{n}\sum^n \frac{M_i^2(1 - f_{2i})s_{d'2i}^2}{m_i} \qquad (11.52)$$

Similarly, with *ppz* selection with replacement we had for

$$\hat{Y}_{ppz} = \frac{1}{n}\sum^n M_i\bar{y}_i/z_i$$

(from (11.33) in section 11.9)

$$V(\hat{Y}_{ppz}) = \frac{1}{n}\sum^N z_i\left(\frac{y_i}{z_i} - Y\right)^2 + \frac{1}{n}\sum^N \frac{M_i^2(1 - f_{2i})S_{2i}^2}{z_im_i} \qquad (11.53)$$

Hence, for $\hat{Y}_R = X\hat{Y}/\hat{X}$ in sampling with replacement,

$$V(\hat{Y}_R) \doteq \frac{1}{n}\sum^N \frac{1}{z_i}(Y_i - RX_i)^2 + \frac{1}{n}\sum^N \frac{M_i^2(1 - f_{2i})S_{d2i}^2}{z_im_i} \qquad (11.54)$$

From (11.35) a sample estimate $v(\hat{Y}_R)$ that is slightly biased is

$$v(\hat{Y}_R) = \frac{1}{n(n-1)} \sum_{i}^{n} \left(\frac{\hat{Y}_i - \hat{R}\hat{X}_i}{z_i} \right)^2 \tag{11.55}$$

With Brewer's method of sampling without replacement, $\hat{Y}_{RB} = X\hat{Y}_B/\hat{X}_B$, where $\hat{Y}_B = \sum M_i \bar{y}_i / \pi_i$. For $n = 2$, formulas (11.42) and (11.44) give, for the ratio estimate,

$$V(\hat{Y}_{RB}) = \sum_{i}^{N} \sum_{j>i}^{N} (\pi_i \pi_j - \pi_{ij}) \left(\frac{D_i}{\pi_i} - \frac{D_j}{\pi_j} \right)^2 + \sum_{i=1}^{N} \frac{M_i^2 (1 - f_{2i}) S_{d2i}^2}{m_i \pi_i} \tag{11.56}$$

where $D_i = Y_i - RX_i$ and units i and j are the two selected.

$$v(\hat{Y}_{RB}) = (\pi_i \pi_j \pi_{ij}^{-1} - 1) \left(\frac{\hat{D}_i'}{\pi_i} - \frac{\hat{D}_j'}{\pi_j} \right)^2 + \sum_{i=1}^{2} \frac{M_i^2 (1 - f_{2i}) s_{d'2i}^2}{m_i \pi_i} \tag{11.57}$$

11.13 CHOICE OF SAMPLING AND SUBSAMPLING FRACTIONS. EQUAL PROBABILITIES

This problem is discussed first for the ratio-to-size estimate when units are chosen with equal probabilities. The subsampling fraction $f_2 = m_i/M_i$ is assumed constant, so that the estimate is the sample mean per subunit.

The simplest cost function contains three terms.

c_u = fixed cost per primary unit
c_2 = cost per subunit
c_l = cost of listing per subunit in a selected unit

The third term is included because the sampler must usually list the elements in any selected unit and verify their number in order to draw a subsample. Hence

$$\text{cost} = c_u n + c_2 \sum^{n} m_i + c_l \sum^{n} M_i$$

This formula is not usable as it stands, since the cost depends on the particular set of units that is chosen. Instead, consider the average cost over n units, which equals

$$E(C) = c_u n + c_2 n \bar{m} + c_l n \bar{M} = (c_u + c_l \bar{M}) n + c_2 n \bar{m} = c_1 n + c_2 n \bar{m} \tag{11.58}$$

where c_1 now includes the average cost of listing a unit.

We determine n and $\bar{m} = f_2 \bar{M}$ so as to minimize $V(\bar{y})$ for given cost or *vice versa*. From (11.28) in section 11.8, dividing by $M_0^2 = (\bar{M}N)^2$,

$$\text{MSE}(\bar{y}) \doteq \frac{1-f_1}{n} \frac{\sum_{i}^{N} M_i^2 (\bar{Y}_i - \bar{Y})^2}{\bar{M}^2 (N-1)} + \frac{1-f_2}{n\bar{m}} \sum^{N} \frac{M_i}{M_0} S_{2i}^2$$

Write

$$S_b^{\ 2} = \frac{\sum\limits_{}^{N} M_i^2 (\bar{Y}_i - \bar{Y})^2}{\bar{M}^2(N-1)}$$

This is a weighted variance among unit means per element. It is analogous to the variance $S_1^{\ 2}$ in section 10.3 and reduces to $S_1^{\ 2}$ if all M_i are equal. We may also write

$$S_2^{\ 2} = \sum^{N} \frac{M_i}{M_0} S_{2i}^{\ 2}$$

This is a weighted mean of the within-unit variances. It reduces to the $S_2^{\ 2}$ of section 10.3 if all M_i are equal.

In this notation, since $f_2 = \bar{m}/\bar{M}$,

$$\text{MSE}(\bar{y}) = \frac{1}{n}\left(S_b^{\ 2} - \frac{S_2^{\ 2}}{\bar{M}}\right) + \frac{1}{n\bar{m}}S_2^{\ 2} - \frac{1}{N}S_b^{\ 2} \qquad (11.59)$$

Applying the Cauchy–Schwarz inequality as usual to (11.58) and (11.59) we get

$$\bar{m}_{opt} \doteq \frac{S_2}{\sqrt{S_b^{\ 2} - S_2^{\ 2}/\bar{M}}}\sqrt{\frac{c_1}{c_2}} \qquad (11.60)$$

The methods given in section 10.8 for utilizing knowledge about the ratios S_2/S_b and c_1/c_2 to guide the selection of $\bar{m}_{opt}$ are applicable here. The unbiased estimate when units are drawn with equal probabilities can be handled similarly.

The next section presents a more general analysis of this problem.

11.14 OPTIMUM SELECTION PROBABILITIES AND SAMPLING AND SUBSAMPLING RATES

An important early analysis by Hansen and Hurwitz (1949) determines at the same time the optimum z_i as functions of the M_i and the optimum sampling and subsampling fractions. Selection of units is with replacement. The analysis is presented for $\hat{Y}_R$ in the self-weighting form, so that $m_i = f_0 M_i/n z_i = f_0 M_i/\pi_i$.

As in Section 11.13, the cost function is

$$C = c_u n + c_2 \sum^{n} m_i + c_l \sum^{n} M_i$$

Although the M_i are known, a listing cost is included for those units that are in the sample, since listing may be needed to provide a frame for subsampling.

The average cost of sampling n units is

$$E(C) = c_u n + C_2 f_0 M_0 + c_l \sum^{N} \pi_i M_i \qquad (11.61)$$

By (11.54) in section 11.12,

$$V(\hat{Y}_R) = \frac{1}{n}\left[\sum_{i}^{N}\frac{1}{z_i}(Y_i - RX_i)^2 + \frac{M_i(M_i - m_i)}{z_i m_i}S_{d2i}^2\right] \qquad (11.62)$$

Since $d_{ij} = y_{ij} - Rx_{ij}$, we may write $(Y_i - RX_i) = M_i\bar{D}_i$. Noting that $\pi_i = nz_i$, $M_i/nz_i m_i = 1/f_0$, and combining the first and third terms in (11.62), we get

$$V = V(\hat{Y}_R) = \sum_{i}^{N}\left[\frac{M_i^2}{\pi_i}\left(\bar{D}_i^2 - \frac{S_{d2i}^2}{M_i}\right) + \frac{M_i}{f_0}S_{d2i}^2\right] \qquad (11.63)$$

The problem is to choose n, f_0, and the $\pi_i = nz_i$ to minimize V subject to fixed average cost and to the restriction

$$\sum_{i}^{N} z_i = 1, \qquad \sum_{i}^{N} \pi_i = n$$

Take λ and μ as Lagrangian multipliers and minimize

$$V + \lambda\left[c_u n + c_2 f_0 M_0 + c_l\sum_{i}^{N}\pi_i M_i - E(C)\right] + \mu\left(n - \sum_{i}^{N}\pi_i\right) \qquad (11.64)$$

Differentiation gives

$$n: \quad \lambda c_u + \mu = 0; \quad \mu = -\lambda c_u \qquad (11.65)$$

$$\pi_i: \quad \frac{-M_i^2}{\pi_i^2}\left(\bar{D}_i^2 - \frac{S_{d2i}^2}{M_i}\right) + \lambda c_l M_i - \mu = 0 \qquad (11.66)$$

Relations (11.65) and (11.66) lead to

$$z_i = \frac{\pi_i}{n} \propto M_i\left(\bar{D}_i^2 - \frac{S_{d2i}^2}{M_i}\right)^{1/2} \Big/ (c_u + c_l M_i)^{1/2} \qquad (11.67)$$

Since the individual values of $(\bar{D}_i^2 - S_{d2i}^2/M_i)$ will not be known, we consider how the average value may depend on the size of unit M_i, using the following rough argument. Suppose a population were divided into units of size M. Since $E(\bar{D}_i) = 0$, formula (9.10) on p. 241 gives

$$E(\bar{D}_i^2) = V(\bar{D}_i) \doteq \frac{S_d^2}{M}[1 + (M-1)\rho_M]$$

where S_d^2 is the population variance of the d_{ij} and ρ_M is the intraunit correlation for units of size M. Also, from (9.15), p. 242,

$$E(S_{d2i}^2) \doteq S_d^2(1 - \rho_M)$$

hence

$$E\left(\bar{D}_i^2 - \frac{S_{d2i}^2}{M_i}\right) \doteq \frac{S_d^2}{M}[1 + (M-1)\rho_M - (1 - \rho_M)] = \rho_M S_d^2$$

From (11.67) this gives as an approximation

$$z_i \propto \frac{M_i \sqrt{\rho_{M_i}}}{\sqrt{c_u + c_l M_i}} \tag{11.68}$$

With ρ positive, ρ_{M_i} may be expected to decrease as M_i increases, since subunits far apart are less subject to common influences, but this decrease may be only slight for $\sqrt{\rho_{M_i}}$. Deductions from (11.68) are as follows.

1. If the cost of listing $c_l M_i$ is unimportant, $z_i \propto M_i$ (i.e., *pps* selection) is best if $\sqrt{\rho_{M_i}}$ changes little over the range of sizes in the population. If $\sqrt{\rho_{M_i}}$ decreases noticeably, optimum probabilities lies between $z_i \propto M_i$ and $z_i \propto \sqrt{M_i}$.

2. If listing cost predominates, z_i should lie between $\sqrt{M_i}$ and constant (equal probabilities).

3. If listing costs and fixed costs are of the same order of magnitude, $z_i \propto \sqrt{M_i}$ may be a good compromise.

Differentiation of (11.64) with respect to the overall sampling fraction f_0 gives

$$f_0^2 = \frac{\sum\limits_{}^{N} M_i S_{d2i}^2}{\lambda c_2 M_0} \tag{11.69}$$

The value of λ is found in terms of the known π's by adding (11.66) over all units. This step leads to the result

$$f_0 = \left[\frac{(c_u + c_l \bar{M}) \sum\limits_{}^{N} (M_i/M_0) S_{d2i}^2}{c_2 \sum \frac{M_i^2}{\pi_i^2} \left(\bar{D}_i^2 - \frac{S_{d2i}^2}{M_i} \right)} \right]^{1/2} \tag{11.70}$$

Comparison with (11.60) will show that f_0 has the same structure as in sampling with equal probabilities, remembering that in (11.60), $f_0 = n\bar{m}_{opt}/M_0$ and $c_1 = c_u + c_l \bar{M}$.

The optimum n is found from the average cost equation (11.61).

11.15 STRATIFIED SAMPLING. UNBIASED ESTIMATORS

For the unbiased methods ($\hat{Y}_u$, $\hat{Y}_{ppz}$, $\hat{Y}_B$, etc.) the extension to stratified sampling is straightforward. The subscript h denotes the stratum.

The estimated population total is $\hat{Y}_{st} = \sum \hat{Y}_h$, with

$$V(\hat{Y}_{st}) = \sum\limits_{h}^{L} V(\hat{Y}_h), \qquad v(\hat{Y}_{st}) = \sum\limits_{h}^{L} v(\hat{Y}_h) \tag{11.71}$$

These variances are obtainable from the formulas already given.

We may note the conditions under which the estimates become self-weighting. For $\hat{Y}_{ppz}$ (section 11.9) or $\hat{Y}_B$ (section 11.10),

$$\hat{Y}_{st} = \sum_h^L \frac{1}{n_h} \sum_i^{n_h} \frac{M_{hi} y_{hi}}{m_{hi} z_{hi}} \tag{11.72}$$

where y_{hi} is the *total* over the m_{hi} subunits from the ith unit in stratum h. These estimates are seen to be self-weighting within strata if the probability $f_{0h} = n_h z_{hi} m_{hi}/M_{hi}$ of selecting any subunit in the stratum is constant within the stratum. In this event, the estimate becomes

$$\hat{Y}_{st} = \sum_h^L \frac{1}{f_{0h}} \sum_i^{n_h} y_{hi} \tag{11.73}$$

which is completely self-weighting if f_{0h} is the same in all strata, as might be anticipated intuitively.

If units are of the same size within a given stratum (i.e., $M_{hi} = M_h$), it was shown (section 10.10) that sample allocation leading to a completely self-weighting estimate is close to optimum, provided that $S_{2h}/\sqrt{c_{2h}}$ is reasonably constant. When the M_{hi} vary within strata, a corresponding result for estimators like $\hat{Y}_{ppz}$ and $\hat{Y}_B$ is as follows. Suppose that the cost function is linear as in sections 11.13 and 11.14, and that the estimator is self-weighting within strata (i.e., $f_{0h} = n_h z_{hi} m_{hi}/M_{hi}$). Then the optimum f_{0h} for given expected cost may be shown to be of the form

$$f_{0h} \propto \frac{1}{\sqrt{c_{2h}}} \sqrt{\sum_i (M_{hi}/M_{0h}) S_{2hi}^2} \tag{11.74}$$

where $M_{0h} = \Sigma M_{hi}$. Thus, choice of $f_{0h} = f_0$, which makes these estimators completely self-weighting, will be near-optimal as regards precision unless either the S_{2hi}^2 or the c_{2h} vary widely from stratum to stratum. especially since this choice affects only the second-stage contribution to the variance.

11.16 STRATIFIED SAMPLING. RATIO ESTIMATES

The formulas for the separate estimate $\hat{Y}_{Rs}$ follow from those in section 11.12 for a single stratum, assuming a large sample in each stratum and independent sampling in different strata.

For the combined estimate, $\hat{Y}_{Rc} = X\hat{Y}_{st}/\hat{X}_{st}$,

$$\hat{Y}_{Rc} - Y = X \sum_h^L \hat{Y}_h / \sum_h^L \hat{X}_h - Y$$

$$= \frac{X}{\hat{X}_{st}} \sum_h^L (\hat{Y}_h - R\hat{X}_h) \doteq \sum_h^L (\hat{Y}_h - R\hat{X}_h)$$

Hence, we substitute $d_{hij} = y_{hij} - Rx_{hij}$ for y_{hij} to obtain the approximate formulas for $V(\hat{Y}_{Rc})$ from those for $V(\hat{Y}_{st})$ by the sampling plan used. For $v(\hat{Y}_{Rc})$ we substitute $d'_{hij} = y_{hij} - \hat{R}_c x_{hij}$ in $v(\hat{Y}_{st})$.

For example, with unequal probabilities with replacement (section 11.9), formula (11.33) leads to

$$V(\hat{Y}_{Rc}) \doteq \sum_{h}^{L} \frac{1}{n_h} \sum_{i}^{N_h} \left[z_{hi}\left(\frac{D_{hi}}{z_{hi}} - D_h\right)^2 + \frac{M_{hi}^2(1-f_{2hi})S_{d2hi}^2}{z_{hi}m_{hi}} \right] \tag{11.75}$$

where

$$D_{hi} = Y_{hi} - RX_{hi}$$

$$S_{d2hi}^2 = \frac{1}{(M_{hi}-1)} \sum_{j}^{M_{hi}} [(y_{hij} - Rx_{hij}) - (\bar{Y}_{hi} - R\bar{X}_{hi})]^2$$

For the estimated variance, formula (11.35) gives, after the substitution,

$$v(\hat{Y}_{Rc}) \doteq \frac{\sum_{h}^{L} (D_{hi}' - \bar{D}_h')^2}{n_h(n_h-1)} \tag{11.76}$$

where

$$D_{hi}' = \frac{M_{hi}\bar{d}_{hi}'}{z_{hi}}, \qquad \bar{D}_h' = \frac{\sum_{i}^{n_h} D_{hi}'}{n_h}, \qquad \bar{d}_{hi}' = \bar{y}_{hi} - \hat{R}_c\bar{x}_{hi}$$

11.17 NONLINEAR ESTIMATORS IN COMPLEX SURVEYS

In addition to the estimation of totals, means, ratios and differences among them, analyses of survey data may involve estimates of more complex mathematical structure (e.g. simple and partial correlation coefficients, medians, and other percentiles). The objectives of the analyses may include attaching a confidence interval to the quantity estimated or performing a test of significance.

Given a random sample from an infinite population, statistical theory has produced a variety of methods for meeting these objectives—exact small-sample methods based on assumptions about the nature of the distributions followed by the observations, and approximate large-sample methods requiring fewer assumptions. With the more complex types of sample studied in this book, on the other hand, we have been able to give methods for computing unbiased estimators of the variances of unbiased *linear* estimators like $\hat{Y}_{st}$, $\hat{Y}_{ppz}$, or $\hat{Y}_{HT}$. Assuming the sample large enough so that these estimators have distributions close to normality, the normal tables supply confidence limits and tests of significance. There remain many problems with nonlinear estimators in complex surveys.

Three approximate methods for estimating the standard errors of nonlinear estimators have been produced. They will be described for stratified random samples with $n_h = 2$ in all strata, the case for which they have received the most study. All methods give the usual unbiased variance estimates when the estimator is linear. The sample can be a multistage sample, primary units being drawn either with equal probabilities or with unequal probabilities with replacement.

11.18   TAYLOR SERIES EXPANSION

This is the method that produced the approximate formulas for the estimated variances of $\hat{R}$, and $\hat{R}_c$, $\hat{R}_s$ in stratified sampling. The function to be estimated, $f(Y_1, Y_2, \ldots, Y_k) = f(\mathbf{Y})$ is expressed as a function of the population totals of certain variables, the estimate $f(\hat{\mathbf{Y}})$ being the corresponding function of unbiased sample estimates of $\hat{\mathbf{Y}}$. For variable j with $n_h = 2$, when units are chosen with ppz with replacement, the population total and its linear estimator are

$$Y_j = \sum_h^L \sum_i^{N_h} Y_{jhi}; \qquad \hat{Y}_j = \sum_h^L \sum_i^2 \frac{\hat{Y}_{jhi}}{2z_{hi}} = \sum_h^L (y'_{jh1} + y'_{jh2}) \tag{11.77}$$

where $y'_{jhi} = \hat{Y}_{jhi}/2z_{hi}$ and $\hat{Y}_{jhi}$ is an unbiased sample estimate of Y_{jhi} from the subsample in this unit.

From (11.35) an unbiased estimate of $V(\hat{Y}_j)$ works out as

$$v(\hat{Y}_j) = \sum_h \sum_i^2 2(y'_{jhi} - \bar{y}'_{jhi})^2 \equiv \sum_h (y'_{jh1} - y'_{jh2})^2 \tag{11.78}$$

Hence, by Woodruff's (1971) extension in section (6.13) of the Keyfitz short-cut method for $n_h = 2$, a Taylor series approximation to the variance of $f(\hat{\mathbf{Y}})$ is

$$v[f(\hat{\mathbf{Y}})] \doteq \sum_h \left[\sum_j^k \left(\frac{\partial f}{\partial \hat{Y}_j} \right) (y'_{jh1} - y'_{jh2}) \right]^2 \tag{11.79}$$

Expression of a nonlinear function of the measured variables in the form $f(\mathbf{Y})$ may require some work and care and is not possible with some functions of interest. Consider a simple example—simple random single-stage sampling with primary units of equal sizes, where the correlation between the unit totals of two variables U_1 and U_2 is to be estimated. The population value $f(\mathbf{Y})$ is

$$\rho = \frac{\displaystyle\sum_h^L \sum_i^N U_{1hi} U_{2hi} - \frac{\left(\sum_h \sum_i U_{1hi} \right)\left(\sum_h \sum_i U_{2hi} \right)}{N}}{\left[\sum_h \sum_i U_{1hi}^2 - \frac{\left(\sum_h \sum_i U_{1hi} \right)^2}{N} \right]^{1/2} \left[\sum_h \sum_i U_{2hi}^2 - \frac{\left(\sum_h \sum_i U_{2hi} \right)^2}{N} \right]^{1/2}}$$

In terms of our variables Y_{jhi}, this is a function of five such variables.

$$Y_{1hi} = U_{1h_\bullet} {}_{1hi}; \qquad Y_{2hi} = U_{1hi}^2; \qquad Y_{3hi} = U_{2hi}^2$$

$$Y_{4hi} = U_{1hi}; \qquad Y_{5hi} = U_{2hi}$$

with $y'_{1hi} = \hat{U}_{1hi}\hat{U}_{2hi}/2z_{hi}$, $i = 1, 2$, $z_{hi} = 1/N_h$ in this case.

11.19 BALANCED REPEATED REPLICATIONS

We consider this method first for a linear function of a single variable. Let $f(Y) = Y$, with its estimate $f(\hat{Y}) = \hat{Y} = \sum_h (y_{h1}' + y_{h2}')$, where $y_{hi}' = \hat{Y}_{hi}/2z_{hi}$. An unbiased variance estimate $v(\hat{Y})$ is, from (11.78),

$$v(\hat{Y}) = \sum_h (y_{h1}' - y_{h2}')^2 \tag{11.80}$$

Select a half-sample H by choosing one unit from each stratum. The estimate of $f(Y) = Y$ from this half-sample is $f(H) = 2 \sum_h y_{hi}'$, where hi is the unit chosen in stratum h. Hence, if $f(S)$ denotes the estimate of $f(Y)$ made from the whole sample,

$$f(H) - f(S) = 2 \sum_h y_{hi}' - \sum_h (y_{h1}' + y_{h2}') = \sum_h \pm (y_{h1}' - y_{h2}') \tag{11.81}$$

the signs depending on whether unit 1 or unit 2 was selected in stratum h.

Comparison of (11.81) and (11.80) shows that for any of the 2^L possible selections, $[f(H) - f(S)]^2$ contains the correct squared terms $(y_{h1}' - y_{h2}')^2$ in $v(\hat{Y})$ in every stratum. However, for any specific half-sample every pair of strata contributes a cross-product term with a $\pm$ sign to $[f(H) - f(S)]^2$. McCarthy (1966) noted that a set of balanced half-samples can be found in which, for every pair of strata, half the cross-product signs in the set are $+$ and half $-$. If this set contains g different half-samples, the unwanted cross-product terms cancel when we sum over the set giving

$$\frac{1}{g} \sum_i^g [f(H_i) - f(S)]^2 \equiv \sum_h^L (y_{h1}' - y_{h2}')^2 = v(\hat{Y}) \tag{11.82}$$

If C_i is the complementary half-sample to H_i, second and third forms are

$$v(\hat{Y}) \equiv \frac{1}{g} \sum_i^g [f(C_i) - f(S)]^2 \equiv \frac{1}{4g} \sum_i^g [f(H_i) - f(C_i)]^2 \tag{11.83}$$

A fourth estimator is the average of the first two above.

Balanced sets of this type were given earlier by Plackett and Burman (1946) for use in experimental design for g any multiple of 4. For L strata the smallest balanced set has g equal to the smallest multiple of 4 that is $\geq L$. With $L = 5$ strata, $g = 8$. Table 11.7 shows a balanced set for $L = 5$, $+$ denoting unit 1, $-$ denoting unit 2 in a stratum.

TABLE 11.7

BALANCED HALF-SAMPLES FOR $L = 5$ STRATA

	HALF-SAMPLE							
Stratum	1	2	3	4	5	6	7	8
1	+	+	+	−	+	−	−	−
2	−	+	+	+	−	+	−	−
3	−	−	+	+	+	−	+	−
4	+	−	−	+	+	+	−	−
5	−	+	−	−	+	+	+	−

When this balanced repeated replication (BRR) method is applied to a non-linear function $f(\mathbf{Y})$ of several variables, the four estimators

$$\frac{1}{g}\sum_{i}^{g}[f(H_i) - f(S)]^2; \qquad \frac{1}{g}\sum_{i}^{g}[f(C_i) - f(S)]^2; \qquad \frac{1}{4g}\sum_{i}^{g}[f(H_i) - f(C_i)]^2$$

and the average of the first two, all differ to some extent. Also, $f(S) = f(\hat{\mathbf{Y}})$ is biased, and the variance estimates do not include the correct bias contribution to $\mathrm{MSE}[f(\hat{\mathbf{Y}})]$. However, note that if we drew m independent samples S_i, each with a complementary H_i, C_i, the quantity

$$\frac{1}{2m}\sum_{i}^{m}[f(H_i) - f(C_i)]^2 \tag{11.84}$$

would be an unbiased estimate of $V[f(H)]$. Thus, roughly speaking, the BRR method amounts to assuming, for $f(S)$, that (a) its bias is negligible, (b) $V[f(S)] \doteq (1/2)V[f(H)]$, and (c) calculation of $v[f(S)]$ from BRR for a *single* stratified S agrees well enough with its calculation from independent samples S_i to be useful.

The repeated replications method has been extensively used both for sample selection and variance estimation by the U.S. Census Bureau in the 1950s and by Deming (1956, 1960). The idea goes back to the Mahalanobis technique of interpenetrating subsamples (Mahalanobis, 1944). McCarthy (1969) reviews the properties of the BRR method, with further discussion by Kish and Frankel (1974).

11.20   THE JACKNIFE METHOD

For $\hat{R}$ in simple random samples this method was described in section 6.17. It was proposed as a means of estimating $V(\hat{R}_O)$, but could also be tried for estimating $V(\hat{R})$. Omit unit j from the sample and calculate $\hat{R}_j$, the ratio estimate from the rest of the sample. Ignoring the fpc, an approximate estimator of $V(\hat{R})$ can be obtained from (6.86) with $g = n$. If we assume $\hat{R} \doteq \hat{R}_-$, the mean of the $\hat{R}_j$, this estimator becomes

$$v(\hat{R}) \doteq \frac{(n-1)}{n} \sum_i^n (\hat{R}_j - \hat{R})^2 = \frac{(n-1)}{n}[f(S_j) - f(S)]^2 \qquad (11.85)$$

the expression on the extreme right indicating how the method might be applied to other nonlinear estimation.

For a linear estimator like $\bar{y}$ it is easily verified that formula (11.85) reduces to the usual $v(\bar{y}) = \sum (y_i - \bar{y})^2 / n(n-1)$.

In extending this method to stratified samples with $n_h = 2$, Frankel (1971) suggests omitting one unit at random from stratum h for $h = 1, 2, \ldots L$ in turn, calculating $f(S_h)$ from the remaining sample of size $(2L-1)$. One form of the Jacknife estimate of $V[f(S)]$ is then

$$v[f(S)] = \sum_h^L [f(S_h) - f(S)]^2 \qquad (11.86)$$

For a linear estimator $f(\hat{\mathbf{Y}})$, this variance estimator reduces to the usual unbiased estimator in *ppz* sampling with replacement.

As with *BRR*, there are four analogous versions of the Jacknife estimator of $v[f(S)]$.

11.21 COMPARISON OF THE THREE APPROACHES

The Taylor, *BRR*, and *J* methods have had very limited rigorous theoretical justification. Appraisal of their performance with different estimators and types of surveys has relied thus far primarily on Monte Carlo studies. One study is described by Frankel (1971) and Kish and Frankel (1974).

The sample was a single-stage sample of cluster units of slightly differing sizes, a unit having on the average 14.1 households. The population of $N = 3240$ units or 45,737 households was divided in turn into 6, 12, and 30 equal-sized strata, with $n_h = 2$ in each stratum. Simple random sampling within strata was used. The data were taken from the Current Population Survey (U.S. Census Bureau). The eight original variables for a household were the number of persons, number under 18, number in labor force, income, age, sex, and years of schooling of household head, and total income. Various means (ratios), differences of means, and simple, partial and multiple regression or correlation coefficients (three independent variables) were computed. In numbers of primary units the samples were quite small, 12, 24, and 60, being about 170, 340, and 850 in numbers of households.

In computing variance estimators, all four forms of the *BRR* and the *J* methods were compared. The Taylor method was not used for partial and multiple regressions, because of difficulty in expressing the partial derivatives $\partial f / \partial Y_j$ in manageable form. Since the sampling method was proportional stratification without replacement, the fpc $(1-f)$ was included in all variance estimates, with $f = n/N$, although scarcely needed with these small samples.

Except for the multiple correlation coefficients, the estimators $f(\mathbf{Y})$ had relatively unimportant biases for all three sample sizes, the ratios |Bias|/s.e. being

<0.1 for ratios, differences of ratios, and simple regression coefficients, and around 0.2 for simple and partial correlation coefficients. When the approximate $v(\hat{\mathbf{Y}})$ are regarded as estimators of $\mathrm{MSE}(\hat{\mathbf{Y}})$, all three methods—Taylor, *BRR*, and *J*—performed well for ratios and differences of ratios, the average |Bias in $v(\hat{\mathbf{Y}})|/\mathrm{MSE}(\hat{\mathbf{Y}})$ being under 5%. For simple correlation coefficients, the *BRR* variance estimators had substantially smaller biases than the *T* or *J* estimators. The reverse was true for simple regression coefficients. With both *BRR* and *J* the best of the four $v(\hat{Y})$ methods was the average of the *H* and *C* estimates (i.e. with *BRR*)

$$\frac{1}{2g}\left\{\sum_i^g [f(H_i)-f(S)]^2 + \sum_i^g [f(C_i)-f(S)]^2\right\} \tag{11.87}$$

In confidence interval statements and tests of significance, the important issue is how well the tail probabilities of the variate $[f(\hat{\mathbf{Y}})-f(\mathbf{Y})]/\mathrm{s.e.}[f(\hat{\mathbf{Y}})]$ agree with those of Student's t with L d.f. For selected t-values Frankel (1971) gives the Monte Carlo estimates of the tail frequencies for the related variate $[f(\hat{\mathbf{Y}})-Ef(\hat{\mathbf{Y}})]/\mathrm{s.e.}[f(\hat{\mathbf{Y}})]$. Some results for 6 and 30 strata are shown in Table 11.8 for the t-values that Frankel presents nearest to the 5 and 10% two-tailed levels. Version (11.87) of the *BRR* and *J* methods is the one shown.

TABLE 11.8

AVERAGE TAIL PROBABILITIES OF $[f(\hat{\mathbf{Y}})-Ef(\hat{\mathbf{Y}})]/\mathrm{s.e.}[f(\hat{\mathbf{Y}})]$
COMPARED WITH THOSE OF STUDENTS' t

	6 strata					
	$P(t)=.042, t=2.576$			$P(t)=.098, t=1.960$		
	BRR	J	Taylor	BRR	J	Taylor
Ratios	.044	.049	.052	.096	.106	.112
b's[a]	.034	.048	.058	.085	.117	.127
r's[b]	.052	.069	.084	.114	.137	.163
Partial r's	.043	.063	—	.092	.132	—
Multiple R's	.065	.088	—	.105	.160	—

	30 strata					
	$P(t)=.059, t=1.960$			$P(t)=.110, t=1.645$		
	BRR	J	Taylor	BRR	J	Taylor
Ratios	.056	.057	.057	.109	.111	.112
b's	.062	.067	.068	.110	.116	.116
r's	.089	.098	.102	.138	.153	.164
Partial r's	.103	.121	—	.156	.181	—
Multiple R's	.175	.207	—	.265	.297	—

[a] b = simple regression coefficient.
[b] r = simple correlation coefficient.

To summarize, *BRR* consistently performs best in Table 11.8. Except for multiple R's, it can be regarded as adequate for practical use if one has the view that in data analysis a tabular 5% tail value represents an actual tail value somewhere between 3 and 8%. J does slightly better than Taylor. Except for *BRR* with ratios and simple regressions, all methods give actual tail frequencies higher than the t-tables, so that confidence probabilities are overstated. A puzzling feature is that for correlation coefficients the increase in sample size from 12 to 60 has not brought a corresponding improvement in the closeness of the actual to the t tail frequencies.

This study opens up a wide area for investigation of the methods with different survey plans and different types of estimator $f(\mathbf{Y})$.

In a Monte Carlo study of a larger, more complex sample (two-stage *pps* sampling with replacement, including both stratification and poststratification) Bean (1975) compared the Taylor and *BRR* methods for estimators of the ratio type. Both methods gave satisfactory variance estimates and adequate two-sided confidence probabilities calculated from the normal distribution. Sufficient skewness remained, however, so that one-sided confidence intervals could not be trusted.

EXERCISES

11.1   By working out the estimates for all possible samples that can be drawn from the artificial population in Table 11.1, by methods Ia, Ib, II, and III, verify the total MSE's given in Table 11.2.

11.2   For methods II (equal probabilities, unbiased estimate) and III(pps selection), recompute the variances of $\hat{\bar{Y}}$ for the example in Table 11.1 when $m_i = 1$. Show that the precision of method III in relation to method II is lower for $m_i = 1$ than for $m_i = 2$. What general result does this illustrate?

11.3   For the population in Table 11.1, if the estimated sizes z_i are 0.1, 0.3 and 0.6, with $m_i = 2$, show that the unbiased estimate (method IV) gives a smaller variance than *pps* sampling. What is the explanation of this result?

11.4   The elements in a population with three primary units are classified into two classes. The unit sizes M_i and the proportions P_i of elements that belong to the first class are as follows.

$$M_1 = 100, \quad M_2 = 200, \quad M_3 = 300, \quad P_1 = 0.40, \quad P_2 = 0.45, \quad P_3 = 0.35$$

For a sample consisting of 50 elements from one primary unit, compare the MSE's of methods Ia, II, and III for estimating the proportion of elements in the first class in the population. (In the variance formulas in section 11.2, S_i^2 is approximately P_iQ_i.)

11.5   A sample of n primary units is selected with equal probabilities. From each chosen unit, a constant fraction f_2 of the subunits is taken. If a_i out of the m_i subunits in the ith unit fall in class C, show that the ratio-to-size estimate (section 11.8) of the population proportion in class C is $\bar{p} = \Sigma a_i / \Sigma m_i$. From formula (11.36), show that an estimate of MSE($\bar{p}$) is

$$v(\bar{p}) = \frac{1-f_1}{n\bar{M}^2} \frac{\sum\limits^{n} M_i^2(p_i - \bar{p})^2}{n-1} + \frac{f_1(1-f_2)}{n^2\bar{m}\bar{M}} \sum\limits^{n} \frac{M_im_i}{m_1 - 1} p_iq_i$$

where $p_i = a_i / m_i$.

11.6   A firm with 36 factories decides to check the condition of some equipment of which $M_0 = 25,012$ pieces are in use. A random sample of 12 factories is taken, a 10% subsample being checked in each selected factory. The numbers of pieces checked (m_i) and the numbers found with signs of deterioration (a_i) are as follows.

Factory	m_i	a_i	$p_i = \dfrac{a_i}{m_i}$	Factory	m_i	a_i	$p_i = \dfrac{a_i}{m_i}$
1	65	8	0.123	7	85	18	0.212
2	82	21	0.256	8	73	11	0.151
3	52	4	0.077	9	50	7	0.140
4	91	12	0.132	10	76	9	0.118
5	62	1	0.016	11	64	20	0.312
6	69	3	0.043	12	50	2	0.040

Estimate the percentage and the total number of defective pieces in use and give estimates of their standard errors.

Note.   Since $M_i/\bar{M} \doteq m_i/\bar{m}$, the between-units component of $v(\bar{p})$ may be computed as

$$\frac{1-f_1}{n\bar{m}^2(n-1)}\left(\sum a_i^2 - 2\bar{p}\sum a_i m_i + \bar{p}^2\sum m_i^2\right)$$

and, since the m_i are fairly large, the within-units component as

$$\frac{f_1(1-f_2)}{(n\bar{m})^2}\sum a_i q_i$$

11.7   If primary units are selected with equal probabilities and f_2 is constant, show that in the notation of exercise 11.5 the unbiased estimate of a population proportion is $p = N\Sigma a_i/nM_0 f_2$ and that, if terms in $1/m_i$ are negligible, its variance may be computed as

$$v(p) = \frac{1-f_1}{n(n-1)\bar{m}^2}\sum_{}^{n}(a_i - \bar{a})^2 + \frac{f_1(1-f_2)}{(n\bar{m})^2}\sum_{}^{n} a_i q_i$$

Calculate p and its standard error for the data in exercise 11.6.

11.8   A sample of n primary units is chosen with probabilities proportional to estimated sizes z_i (with replacement) and with a constant expected over-all sampling fraction f_0. Show that the unbiased and the ratio-to-size estimates of the population total are, respectively,

T/f_0 and $TM_0/\sum_{}^{n} m_i$, where T is the sample total. (It follows that if M_0 is not known the unbiased estimate can be used, but not the ratio to size. For estimating the population mean per subunit, the situation is reversed.)

11.9   In a study of overcrowding in a large city one stratum contained 100 blocks of which 10 were chosen with probabilities proportional to estimated size (with replacement). An expected over-all sampling fraction $f_o = 2\%$ was used. Estimate the total number of persons and the average persons per room and their s.e.'s from the data below.

Block	1	2	3	4	5	6	7	8	9	10
Rooms	60	52	58	56	62	51	72	48	71	58
Persons	115	80	82	93	105	109	130	93	109	95

11.10   For Durbin's method (section 11.10) of simplifying variance estimation in *ppz* sampling without replacement, a simple method of sample selection, due essentially to Kish

(1965), is as follows. The subscript h to denote the stratum will be omitted and the number of primary units is assumed to be even.

Arrange the units in order of increasing z_i and mark them off in pairs. The method is exact only if $z_i = z_j$ for members of the same pair; this will be assumed here. Select two units *ppz with replacement*. If two different units are drawn, accept both. If the same unit is drawn twice, let the sample consist of the two members of the pair to which this unit belongs. Show that for this method: (a) $\pi_i = 2z_i$, (b) for units not in the same pair, $\pi_{ij} = 2z_i z_j = \pi_i \pi_j / 2$, so that $\pi_i \pi_j \pi_{ij}^{-1} - 1 = 1$, and (c) for units in the same pair, $\pi_{ij} = 4z_i z_j = \pi_i \pi_j$, so that $\pi_i \pi_j \pi_{ij}^{-1} - 1 = 0$.

11.11   In section 11.9, formula (11.33) for $V(\hat{Y}_{ppz})$ in sampling with replacement was proved under the plan that whenever the ith unit was selected, an independent simple random subsample of size m_i was drawn from the whole of the unit. Prove the following results for two alternative plans.

(a)   When the ith unit is selected t_i times, a simple random subsample of size $m_i t_i$ is drawn from it (assume $m_i t_i \le M_i$). Under this plan, $V(\hat{Y}_{ppz})$ in (11.41) is reduced by $(n-1)\sum^{N} M_i S_{2i}^2 / n$ (Sukhatme, 1954).

(b)   When the ith unit is selected t_i times, a simple random subsample of size m_i is drawn. Then $V(\hat{Y}_{ppz})$ in (11.41) is increased by

$$\frac{(n-1)}{n} \sum^{N} M_i^2 (1 - f_{2i}) S_{2i}^2 / m_i$$

In both (a) and (b), $\hat{Y}_{ppz} = \sum^{N} t_i M_i \bar{y}_i / n z_i$, the ith unit receiving weight t_i.

CHAPTER 12

Double Sampling

12.1 DESCRIPTION OF THE TECHNIQUE

As we have seen, a number of sampling techniques depend on the possession of advance information about an auxiliary variate x_i. Ratio and regression estimates require a knowledge of the population mean $\bar{X}$. If it is desired to stratify the population according to the values of the x_i, their frequency distribution must be known.

When such information is lacking, it is sometimes relatively cheap to take a large preliminary sample in which x_i alone is measured. The purpose of this sample is to furnish a good estimate of $\bar{X}$ or of the frequency distribution of x_i. In a survey whose function is to make estimates for some other variate y_i, it may pay to devote part of the resources to this preliminary sample, although this means that the size of the sample in the main survey on y_i must be decreased. This technique is known as *double sampling* or *two-phase sampling*. As the discussion implies, the technique is profitable only if the gain in precision from ratio or regression estimates or stratification more than offsets the loss in precision due to the reduction in the size of the main sample.

Double sampling may be appropriate when the information about x_i is on file cards that have not been tabulated. For instance, in surveys of the German civilian population in 1945, the sample from any town was usually drawn from rationing registration lists. In addition to geographic stratification within the town, for which data were usually already available, stratification by sex and age was proposed. Since the sample had to be drawn in a hurry and since the lists were in constant use, tabulation of the complete age and sex distribution was not feasible. A moderately large systematic sample could, however, be drawn quickly. Each person drawn was classified into the appropriate age-sex class. From these data the much smaller list of persons to be interviewed was selected.

12.2 DOUBLE SAMPLING FOR STRATIFICATION

The theory was first given by Neyman (1938). The population is to be stratified into L classes (strata). The first sample is a simple random sample of size n'.

Let

$$W_h = N_h/N = \text{proportion of population falling in stratum } h$$

$$w_h = n_h'/n' = \text{proportion of first sample falling in stratum } h$$

Then w_h is an unbiased estimate of W_h.

The second sample is a stratified random sample of size n in which the y_{hi} are measured: n_h units are drawn from stratum h. Usually the second sample in stratum h is a random subsample from the n_h' in the stratum. The objective of the first sample is to estimate the strata weights; that of the second sample is to estimate the strata means $\bar{Y}_h$.

The population mean $\bar{Y} = \sum W_h \bar{Y}_h$. As an estimate we use

$$\bar{y}_{st} = \sum_{h=1}^{L} w_h \bar{y}_h \tag{12.1}$$

The problem is to choose n' and the n_h to minimize $V(\bar{y}_{st})$ for given cost.

We must then verify whether the minimum variance is smaller than can be attained by a single simple random sample in which y_i alone is measured. In presenting the theory, we assume that the n_h are a random subsample of the n_h'. Thus, $n_h = \nu_h n_h'$, where $0 < \nu_h \le 1$ and the ν_h are chosen in advance. Repeated sampling implies a fresh drawing of both the first and the second samples, so that the w_h, n_h, and $\bar{y}_h$ are all random variables. The problem is therefore one of stratification in which the strata sizes are not known exactly (section 5A.2).

Two approximations will be made for simplicity. The first sample size n' is assumed large enough so that every $w_h > 0$. Second, when we come to discuss optimum strategy, every optimum ν_h as found by the formula is assumed ≤ 1.

Theorem 12.1. The estimate $\bar{y}_{st}$ is unbiased.

Proof. Average first over samples in which the w_h are fixed. Since $\bar{y}_h$ is the mean of a simple random sample from the stratum, $E(\bar{y}_h) = \bar{Y}_h$. Furthermore, when we average over different selections of the first sample, $E(w_h) = W_h$, since the first sample is itself a simple random sample. Hence

$$E(\bar{y}_{st}) = E[E(\sum w_h \bar{y}_h | w_h)] = E(\sum w_h \bar{Y}_h) = \sum W_h \bar{Y}_h = \bar{Y} \tag{12.2}$$

Theorem 12.2. If the first sample is random and of size n', the second sample is a random subsample of the first, of size $n_h = \nu_h n_h'$, where $0 < \nu_h \le 1$, and the ν_h are fixed,

$$V(\bar{y}_{st}) = S^2\left(\frac{1}{n'} - \frac{1}{N}\right) + \sum_{h}^{L} \frac{W_h S_h^2}{n'}\left(\frac{1}{\nu_h} - 1\right) \tag{12.3}$$

where S^2 is the population variance.

Proof. The proof is easily obtained by the following device. Suppose that the y_{hi} were measured on all n_h' first-sample units in stratum h, not just on the random subsample of n_h. Then, since $w_h = n_h'/n'$,

$$\sum_h^L w_h \bar{y}_h' = \bar{y}'$$

is the mean of a simple random sample of size n' from the population. Hence, averaging over repeated selections of the sample of size n',

$$V\left(\sum_h^L w_h \bar{y}_h'\right) = S^2\left(\frac{1}{n'} - \frac{1}{N}\right) \tag{12.4}$$

But

$$\bar{y}_{st} = \sum_h^L w_h \bar{y}_h = \sum_h^L w_h \bar{y}_h' + \sum_h^L w_h(\bar{y}_h - \bar{y}_h') \tag{12.5}$$

Let the subscript 2 refer to an average over all random subsamples of n_h units that can be drawn from a given n_h' units. Clearly, $E_2(\bar{y}_h) = \bar{y}_h'$. Results that follow immediately are (see exercise 2.16):

$$\text{cov}\,[\bar{y}_h', (\bar{y}_h - \bar{y}_h')] = 0:$$
$$\text{cov}\,(\bar{y}_h', \bar{y}_h) = V(\bar{y}_h'): \quad V(\bar{y}_h - \bar{y}_h') = V(\bar{y}_h) - V(\bar{y}_h') \tag{12.6}$$

Hence, for fixed w_h,

$$V_2\left[\sum w_h(\bar{y}_h - \bar{y}_h')\right] = \left] = \sum w_h^2 S_h^2\left(\frac{1}{n_h} - \frac{1}{n_h'}\right) = \sum \frac{w_h S_h^2}{n'}\left(\frac{1}{v_h} - 1\right) \tag{12.7}$$

since $n_h = v_h n_h' = v_h w_h n'$.

Averaging over the distribution of the w_h obtained by repeated selections of the first sample, we have, from (12.4), (12.5) and (12.7),

$$V(\bar{y}_{st}) = S^2\left(\frac{1}{n'} - \frac{1}{N}\right) + \sum_h^L \frac{W_h S_h^2}{n'}\left(\frac{1}{v_h} - 1\right) \tag{12.8}$$

Corollary 1. The result can be expressed in a number of different forms. By the analysis of variance,

$$(N-1)S^2 = \sum(N_h - 1)S_h^2 + \sum N_h(\bar{Y}_h - \bar{Y})^2 \tag{12.9}$$

Hence, if $g' = (N-n')/(N-1)$, multiplying by $g'/n'N$ gives

$$\frac{(N-n')S^2}{n'N} = S^2\left(\frac{1}{n'} - \frac{1}{N}\right) = \frac{g'}{n'}\sum(W_h - N^{-1})S_h^2 + \frac{g'}{n'}\sum W_h(\bar{Y}_h - \bar{Y})^2 \tag{12.10}$$

From (12.3) this gives

$$V(\bar{y}_{st}) = \sum_h^L \frac{W_h S_h^2}{n'}\left(\frac{1}{v_h} - 1\right) + \frac{g'}{n'}\sum_h^L (W_h - N^{-1})S_h^2 + \frac{g'}{n'}\sum_h^L W_h(\bar{Y}_h - \bar{Y})^2 \tag{12.11}$$

Furthermore, by the definition of $g' = (N - n')/(N - 1)$, it follows that

$$-\frac{1}{n'} + \frac{g'}{n'} = -\frac{1}{N} + \frac{g'}{n'N} \tag{12.12}$$

Hence, in (12.11), the second and third terms in $\sum W_h S_h^2$, which have coefficients $-1/n'$ and g'/n', may be written alternatively, giving

$$V(\bar{y}_{st}) = \sum_h W_h S_h^2 \left(\frac{1}{n'\nu_h} - \frac{1}{N}\right) + \frac{g'}{n'N} \sum_h (W_h - 1)S_h^2 + \frac{g'}{n'} \sum W_h (\bar{Y}_h - \bar{Y})^2 \tag{12.13}$$

If classification is cheap, it may not be reasonable to assume n'/N negligible, since a sizable proportion of the population may be classified. For most applications, however, the term in $g'/n'N$ in (12.13) can be neglected. In this event (12.13) simplifies to

$$V(\bar{y}_{st}) \doteq \sum_h^L W_h S_h^2 \left(\frac{1}{n'\nu_h} - \frac{1}{N}\right) + \frac{g'}{n'} \sum_h^L W_h (\bar{Y}_h - \bar{Y})^2 \tag{12.14}$$

The results in theorem (12.2) were given by Rao (1973).

Corollary 2. If a proportion is being estimated in the second sample, the expressions: $S^2 \doteq NPQ/(N - 1)$ and

$$S_h^2 = \frac{N_h P_h Q_h}{(N_h - 1)}; \qquad (\bar{Y}_h - \bar{Y})^2 = (P_h - P)^2$$

are substituted in (12.3), (12.11), and (12.13). In (12.14) we have, for $n'\nu_h/N$ also negligible,

$$V(p_{st}) \doteq \sum_h^L \frac{W_h P_h Q_h}{n'\nu_h} + \frac{g'}{n'} \sum_h^L W_h (P_h - P)^2 \tag{12.15}$$

Corollary 3. Results for the case where the second sample is drawn *independently* of the first, so that the n_h do not depend on the n_h' (except for the assumption $n_h \leq n_h'$), were given in the second edition, p. 329. For n_h/N_h negligible, the leading term in the variance has the same structure as (12.14), being,

$$V(\bar{y}_{st}) \doteq \sum_h^L \frac{W_h^2 S_h^2}{n_h} + \frac{g'}{n'} \sum_h^L W_h (\bar{Y}_h - \bar{Y})^2 \tag{12.16}$$

Papers by Robson (1952) and Robson and King (1953) extend the stratification theory to two-stage sampling, applying it to the estimation of magazine readership.

12.3 OPTIMUM ALLOCATION

The objective is to choose the n' and ν_h so as to minimize $V(\bar{y}_{st})$ for specified cost. Let c' be the cost of classification per unit and c_h the cost of measuring a unit in stratum h. For a specific sample,

$$C = c'n' + \sum_h c_h n_h \qquad (12.17)$$

Since the n_h are random variables, we minimize the expected cost for chosen n' and ν_h.

$$E(C) = C^* = c'n' + n' \sum c_h \nu_h W_h \qquad (12.18)$$

For $V = V(\bar{y}_{st})$, formula (12.3) leads to

$$n'(V + S^2/N) = (S^2 - \sum W_h S_h^2) + \sum_h \frac{W_h S_h^2}{\nu_h} \qquad (12.19)$$

The product $C^*(V + S^2/N)$ does not involve n'. Application of the Cauchy–Schwarz inequality to this product shows that the product is minimized if, for every h,

$$\frac{\nu_h^2 c_h}{S_h^2} = \frac{c'}{(S^2 - \sum W_h S_h^2)} \qquad (12.20)$$

This gives

$$\nu_h = S_h [c'/c_h (S^2 - \sum W_h S_h^2)]^{1/2} \qquad (12.21)$$

The value of n' is obtained from the expected cost equation (12.18).

By substitution of the optimum ν_h in the formula for $C^*(V + S^2/N)$, the minimum variance is found to be

$$V_{min}(\bar{y}_{st}) = \frac{1}{C^*} [\sum W_h S_h \sqrt{c_h} + (S^2 - \sum W_h S_h^2)^{1/2} \sqrt{c'}]^2 - \frac{S^2}{N} \qquad (12.22)$$

Use of formula (12.21) for sample allocation in practice demands more knowledge of the population than the sampler is likely to have. Fortunately, errors in guessing have compensating features. Thus, if the ν_h in (12.21) are too high, n' from (12.18) will be too low and the stratum weights will not be determined as well as they should be. However, in partial compensation, the stratum means $\bar{y}_h$ will be determined more precisely than under the optimum solution. When there is little advance knowledge of the stratum weights W_h, Srinath (1971) and Rao (1973) suggest a slightly different method of choosing the n_h, thought to be more robust against poor guesses at the W_h.

The case that presents the easiest allocation problem is that in which c_h and S_h are constant. Then (12.21) becomes

$$\nu_h = \nu = \left[\left(\frac{c'}{c} \right) \frac{S_w^2}{(S^2 - S_w^2)} \right]^{1/2} = \left[\left(\frac{c'}{c} \right) \frac{1}{(\phi - 1)} \right]^{1/2} \qquad (12.23)$$

where $\phi = S^2/S_w^2$ is the relative efficiency of proportional stratification to simple random sampling. Thus, if we guessed that stratification would reduce $V(\bar{y})$ in half so that $\phi = 2$, then $\nu_h = (c'/c)^{1/2}$.

Example. This example did not arise from a double sampling problem, but illustrates some features of the solution. We use the Jefferson data from p. 168. In estimating corn acres, assume that we could either take a simple random sample of farms or devote some resources to classifying farms into two strata by farm size. Relevant population data are for corn acres.

Strata	W_h	S_h^2	S_h	$\bar{Y}_h$
1	0.786	312	17.7	19.404
2	0.214	922	30.4	51.626
Population		620		26.300

Suppose that $C^* = 100$, $c_h = 1 = c$, and that S^2/N is negligible. This implies that if double sampling is not used, we can afford to take a sample of $n = 100$ farms, giving $V(\bar{y}) = 6.20$. Let c' be the cost per farm of classifying farms into stratum 1 (≤ 160 acres) and stratum 2 (> 160 acres). Consider the questions:

1. For what values of c'/c does double sampling bring an increase in precision?
2. What is the optimum double sampling plan if $c' = c/100$, and what is the resulting $V(\bar{y}_{st})$?
3. In problem 2, how do the plan and the value of $V(\bar{y}_{st})$ change if the ν_h are guessed as twice the optimum fractions?

1. From the population data, $\Sigma W_h S_h = 20.4$, $(S^2 - \Sigma W_h S_h^2) = 177$. Hence from (12.22)

$$V_{min}(\bar{y}_{st}) = 0.01 (20.4 + 13.3\sqrt{c'})^2$$

If this is to be less than 6.20 for simple random sampling, $c' < 0.11$; that is, $c'/c < 1/9$.

2. If $c'/c = 1/100$, then with the optimum plan, (12.22) gives

$$V_{min}(\bar{y}_{st}) = 0.01(20.4 + 1.3)^2 = 4.71$$

Note that if classification by farm size cost nothing, we would have

$$V_{min}(\bar{y}_{st}) = 0.01(20.4)^2 = 4.16$$

As regards the details of the plan with $c'/c = 1/100$, from (12.21), we get

$$\nu_h = S_h/133; \qquad \nu_1 = 0.133; \qquad \nu_2 = 0.229$$

Since $\Sigma W_h \nu_h = 0.1535$, we find, from (12.18), that $n' = 612$. In return this gives expected values of 64, 30 for n_1, n_2. Thus nearly all the money is spent on measurement: only 6% on classification.

3. If we guess $\nu_1 = 0.266$, $\nu_2 = 0.458$, then $\Sigma W_h \nu_h = 0.307$ and from (12.18), $n' = 315$, leading to an expected $n_1 = 66$, $n_2 = 31$. From (12.3), $V(\bar{y}_{st})$ for this plan will be found to be 4.85, only a 3% increase over the optimum 4.71 in problem 2.

12.4 ESTIMATED VARIANCE IN DOUBLE SAMPLING FOR STRATIFICATION

If $1/n'$ and $1/N$ are both negligible with respect to 1 (e.g., <0.02), an almost unbiased sample estimate of $V(\bar{y}_{st})$ in (12.14) is simply the sample copy of this formula.

$$v(\bar{y}_{st}) = \sum_h^L w_h s_h^2 \left(\frac{1}{n' \nu_h} - \frac{1}{N} \right) + \frac{g'}{n'} \sum_h^L w_h (\bar{y}_h - \bar{y}_{st})^2 \tag{12.24}$$

$$= \sum_h^L \frac{w_h^2 s_h^2}{n_h} - \sum_h^L \frac{w_h s_h^2}{N} + \frac{g'}{n'} \sum_h^L w_h (\bar{y}_h - \bar{y}_{st})^2 \tag{12.24'}$$

where $g' = (N - n')/(N - 1)$. This formula will suffice for almost all applications. With $1/N$ and $1/n'$ not negligible a more complex algebraic expression is needed.

Theorem 12.3. An unbiased sample estimate of $V(\bar{y}_{st})$ in double sampling is

$$v(\bar{y}_{st}) = \frac{n'(N-1)}{(n'-1)N} \left[\sum_h w_h s_h^2 \left(\frac{1}{n' \nu_h} - \frac{1}{N} \right) + \frac{g'}{n'} \sum_h s_h^2 \left(\frac{w_h}{N} - \frac{1}{n' \nu_h} \right) + \frac{g'}{n'} \sum_h w_h (\bar{y}_h - \bar{y}_{st})^2 \right] \tag{12.25}$$

Proof. From (12.13) the general form of the variance to be estimated is

$$V(\bar{y}_{st}) = \sum_h W_h S_h^2 \left(\frac{1}{n' \nu_h} - \frac{1}{N} \right) + \frac{g'}{n' N} \sum_h (W_h - 1) S_h^2 + \frac{g'}{n'} \sum_h W_h (\bar{Y}_h - \bar{Y})^2 \tag{12.26}$$

By averaging first for fixed n' and w_h and then over variations in the w_h, the average of $w_h s_h^2$ in (12.25) is $W_h S_h^2$, while that of s_h^2 is S_h^2. These results will be used after equation (12.31).

In the last term in (12.25)

$$\sum w_h (\bar{y}_h - \bar{y}_{st})^2 = \sum w_h \bar{y}_h^2 - \bar{y}_{st}^2 \tag{12.27}$$

Averaging first for fixed w_h,

$$E(\sum w_h \bar{y}_h^2) = \sum w_h \bar{Y}_h^2 + \sum w_h S_h^2 \left(\frac{1}{\nu_h w_h n'} - \frac{1}{w_h N} \right) \tag{12.28}$$

Furthermore,

$$E_w E(\sum w_h \bar{y}_h^2) = \sum W_h \bar{Y}_h^2 + \sum S_h^2 \left(\frac{1}{\nu_h n'} - \frac{1}{N} \right) \tag{12.29}$$

Also,

$$E(\bar{y}_{st}^2) = \bar{Y}^2 + V(\bar{y}_{st}) \tag{12.30}$$

Subtracting (12.30) from (12.29) and multiplying by g'/n' gives

$$\frac{g'}{n'}E \sum w_h(\bar{y}_h - \bar{y}_{st})^2 = \frac{g'}{n'}\left[\sum W_h(\bar{Y}_h - \bar{Y})^2 + \sum S_h^2\left(\frac{1}{\nu_h n'} - \frac{1}{N}\right) - V(\bar{y}_{st})\right]$$

(12.31)

Substitute (12.31) in finding $(n'-1)NEv(\bar{y}_{st})/n'(N-1)$ from (12.25). We get

$$\frac{(n'-1)N}{n'(N-1)}Ev(\bar{y}_{st}) = \left(1 - \frac{g'}{n'}\right)V(\bar{y}_{st}) = \frac{(n'-1)N}{n'(N-1)}V(\bar{y}_{st})$$

This proves the result. Note that the two middle terms in (12.25) are of order $1/n'N$ and $1/n'^2\nu_h$ and are negligible relative to terms retained if $1/N$ and $1/n'$ are negligible. This supports the simpler form (12.24).

Rao (1973) has given the result (12.25) in terms of the n_h and n_h' as follows.

$$v(\bar{y}_{st}) = \frac{N-1}{N}\sum_h \left(\frac{n_h'-1}{n'-1} - \frac{n_h-1}{N-1}\right)\frac{w_h s_h^2}{n_h} + \frac{(N-n')}{N(n'-1)}\sum_h w_h(\bar{y}_h - \bar{y}_{st})^2$$

(12.32)

Corollary. To use (12.24) in the estimation of a proportion, put p_h for $\bar{y}_h$ and $n_h p_h q_h/(n_h-1)$ for s_h^2.

Example. In a simple random sample of 374 households from a large district, 292 were occupied by white families and 82 by nonwhite families. A sample of about one in four households gave the following data on ownership.

	Owned	Rented	Total
White	31	43	74
Nonwhite	4	14	18

Estimate the proportion of rented households in the area from which the sample was drawn and its standard error.

If the first stratum consists of the white-occupied households.

$$w_1 = \frac{292}{374} = 0.78, \qquad w_2 = \frac{82}{374} = 0.22$$

$$p_1 = \frac{43}{74} = 0.58, \qquad P_2 = \frac{14}{18} = 0.78$$

$$p_{st} = w_1 p_1 + w_2 p_2 = 0.624$$

$$n' = 374, \qquad n_1 = 74, \qquad n_2 = 18$$

It is readily found that only the first term in (12.24) is of importance. Hence

$$v(p_{st}) = \sum_h \frac{n_h}{(n_h-1)} \frac{w_h p_h q_h}{v_h n'} = \sum_h \frac{w_h^2 p_h q_h}{(n_h-1)} = \frac{(0.78)^2(0.2436)}{73} + \frac{(0.22)^2(0.1716)}{17}$$

$$= 0.00252$$

$$s(p_{st}) = 0.050$$

The estimated proportion of rented households is 0.62 ± 0.050.

12.5 DOUBLE SAMPLING FOR ANALYTICAL COMPARISONS

Section 5A.13 dealt with the determination of sample sizes in subgroups of the population for certain analytical comparisons among L subgroup means. If the objective is to have the variances of the differences between the means of every pair of subgroups all equal to V, it was noted that a simple compromise allocation, often adequate, is to specify or minimize only the *average* variance of the $L(L-1)/2$ differences.

$$\frac{2}{L} \sum_i^L \frac{S_i^2}{n_i} = V \tag{12.33}$$

In another application, Booth and Sedransk (1969), the subgroups formed a 2×2 factorial classification, the problem being to estimate row and column effects with equal precision. In this case the variance to be specified or minimized took the form

$$\frac{1}{2} \sum_i^2 \sum_j^2 (\theta_{i.}^2 + \theta_{.j}^2) \frac{S_{ij}^2}{n_{ij}} = V \tag{12.34}$$

where the $\theta_{i.}$ and $\theta_{.j}$ are weights chosen by the investigator, the symbols (ij) denoting the 4 subgroups.

In the discussion in section 5A.13, subgroups could be sampled directly. The problem becomes one of double sampling when members of the subgroups cannot be identified in advance but units can be classified relatively cheaply into the subgroups from an initial simple random sample of size n'. If n_i' of the initial sample fall in subgroup i, the n_i for the final sample that is measured to provide the desired comparisons are drawn from the n_i'. The simplest cost function is

$$C = c'n' + c \sum n_i = c'n' + cn \tag{12.35}$$

where we assume $c' \ll c$.

For fixed n_i, both relations (12.33) and (12.34) are of the form

$$\sum_i^L \frac{a_i^2}{n_i} = V \tag{12.36}$$

where we assume that the S_i^2, S_{ij}^2 and hence the a_i^2 are known in advance.

When the objective is to minimize V for given C, Sedransk (1965) and Booth and Sedransk (1969) try different values of n' in turn. For given n', the value of n is then known from (12.35). For given n, the n_i that minimize V are $n_i = na_i/(\sum a_i)$. Difficulty may arise in using these n_i, however, because the n_i' provided by the initial sample are random variables. In some subgroups we may find $n_i' < na_i/(\sum a_i)$ so that the minimizing n_i exceeds the available n_i'. Allocation rules to handle this situation will be illustrated for $L = 3$ groups, from Sedransk (1965). Number the subgroups in increasing order of $n_i'(\sum a_i)/na_i$, and let $(\sum_1 a_i)$ denote the sum over all classes except the first. Take

$$n_1 = \frac{na_1}{(\sum a_i)} \qquad \text{if} \qquad n_1' \geq \frac{na_1}{(\sum a_i)}$$

$$n_1 = n_1' \qquad \text{if} \qquad n_1' < \frac{na_1}{(\sum a_i)}$$

$$n_2 = \frac{(n-n_1)a_2}{\sum_1 a_i} \qquad \text{if} \qquad n_2' \geq \frac{(n-n_1)a_2}{\sum_1 a_i} \qquad (12.37)$$

$$n_2 = n_2' \qquad \text{if} \qquad n_2' < \frac{(n-n_1)a_2}{\sum_1 a_i}$$

$$n_3 = n - n_1 - n_2$$

where $\sum_1 a_i = \sum a_i - a_1$.

These rules are not complete, but will cover most n' likely to be near optimal. The principle is to keep close to the $n_i \propto a_i$ allocation. See Booth and Sedransk (1969) for more detail.

Example. An example in which double sampling should perform well is as follows: $C = 2,000$, $c' = 1$, $c = 10$. The three subgroups are of relative sizes $W_1 = 0.05$, $W_2 = 0.25$, $W_3 = 0.70$, $a_i^2 = S_i^2 = 10$ $(i = 1, 2, 3)$. Consider first single sampling. Since it costs 11 monetary units to select, classify, and measure a sampling unit, we can afford $n = 182$ with single sampling.

Optimum allocation would require equal n_i, but on the average the values of n_i from single sampling with $n = 182$ are 9.1, 45.5, 127.4. Assuming $E(1/n_i) \doteq 1/E(n_i)$, the average V from single sampling is approximately

$$E(V) \doteq 10\left(\frac{1}{9.1} + \frac{1}{45.5} + \frac{1}{127.4}\right) = 1.40$$

With double sampling, calculations for different n' show that the optimum n' is close to 620, giving $n = 138$. With $n = 138$ the optimal n_i is 46 in each class. However, $n' = 620$ provides only an expected 31 in class 1. Hence, on the average, the allocation rule (12.37) for double sampling gives $n_1 = E(n_1') = (0.05)(620) = 31$, $n_2 = n_3 = 53.5$, leading to

$$E(V) \doteq 10\left(\frac{1}{31} + \frac{2}{53.5}\right) = 0.70$$

a 50% reduction from $E(V)$ with single sampling.

Rao (1973) handles these problems by the method used with stratified sampling: this method specifies the fraction v_i of the n_i' in the ith subgroup that are to be measured. An advantage is that the optimum n' can be determined analytically. With $C = c'n' + cn$ as before, the expected cost is

$$C^* = c'n' + cn' \sum v_i W_i \tag{12.38}$$

where W_i is the relative size of subgroup i. Assuming as before that $E(1/n_i) \doteq 1/E(n_i)$, the average V is

$$E(V) = \sum_i^L \frac{a_i^2}{n' W_i v_i} \tag{12.39}$$

By the Cauchy–Schwarz inequality, the optimal v_i for fixed n' is given by

$$n' W_i v_i = \frac{a_i}{(\sum a_i)} \frac{(C^* - n'c')}{c} \tag{12.40}$$

provided all $v_i \leq 1$. Substitution of $n' W_i v_i$ from (12.40) into (12.39) gives, for the minimal $E(V)$,

$$E(V) = \frac{c(\sum a_i)^2}{C^* - n'c'} \tag{12.41}$$

Since $E(V)$ in (12.41) decreases as n' decreases, the optimum n' will decrease at least until it reaches the value m_1, say, at which one of the v_i, say v_1, becomes 1. At this point (12.40) and (12.41) no longer apply. From (12.40), v_1 becomes 1 when

$$n' W_1 = \frac{a_1(C^* - n'c')}{c(\sum a_i)} \tag{12.42}$$

Solving (12.42) for m_1, the value of n' at which $v_1 = 1$, we get

$$m_1 = \frac{C^*}{[c' + cW_1(\sum a_i)/a_1]} \tag{12.43}$$

To examine values of n' smaller than m_1, set $v_1 = 1$ and use the Cauchy–Schwarz inequality to obtain the remaining v_i. We find, for $i > 1$,

$$n' W_i v_i = \frac{a_i}{(\sum_1 a_i)} \frac{C^* - n'(c' + cW_1)}{c} \tag{12.44}$$

where $\sum_1 a_i = \sum a_i - a_1$. The resulting minimal $E(V)$ is now

$$E(V) = \frac{a_1^2}{n'W_1} + \frac{c(\sum_1 a_i)^2}{C^* - n'(c' + cW_1)} \tag{12.45}$$

From this $E(V)$ the derivative $dE(V)/dn'$ vanishes at

$$n' = C^* \Big/ \left\{ (c' + cW_1) + \frac{(\sum a_i)}{a_1} [cW_1(c' + cW_1)]^{1/2} \right\} \tag{12.46}$$

The value m_2 at which v_1 and v_2 are both 1, so that (12.45) ceases to hold, is

$$m_2 = C^* \Big/ \left[(c' + cW_1) + \frac{cW_2(\sum_1 a_i)}{a_2} \right] \tag{12.47}$$

Thus expressions (12.45) and (12.46) apply only over the range $m_1 \geq n' \geq m_2$. If $dE(V)/dn'$ does not vanish for $n' \geq m_2$, we need to set v_1, v_2, and so forth, $= 1$ in turn until the turning point of $E(V)$ is found. In many situations for which double sampling is economical, however, the turning value of $E(V)$ occurs for $m_1 \geq n' \geq m_2$.

Example.   For the worked example in this section, $C^* = 2000$, $c' = 1$, $c = 10$, $a_i^2 = S_i^2 = 10$, $W_i = 0.05, 0.25, 0.70$. We have, from (12.43) and (12.47),

$$m_1 = \frac{2000}{[1 + (10)(0.05)(3)]} = 800$$

$$m_2 = \frac{2000}{[1.5 + (10)(.25)2]} = 308$$

Furthermore, $dE(V)/dn'$ in formula (12.46) vanishes at

$$n' = \frac{2000}{\{1.5 + (2)[(0.5)(1.5)]^{1/2}\}} = 619$$

Since this value lies between 800 and 308, it gives the required minimum. Formula (12.44) gives $v_2 = .346$, $v_3 = .124$. Numerically this solution is essentially the same as that found by Sedransk's method, which had $n' = 620$ and similar values of n_i in the three subgroups.

Both methods extend easily to the case of differential costs of measurement in different subgroups. Suggestions are also given (Rao, 1973) for the case where $E(1/n_i)$ is substantially larger than $1/E(n_i)$.

12.6   REGRESSION ESTIMATORS

In some applications of double sampling the auxiliary variate x_i has been used to make a regression estimate of $\bar{Y}$. In the first (large) sample of size n', we measure only x_i; in the second, a random subsample of size $n = \nu n' = n'/k$ where the fraction ν is chosen in advance, we measure both x_i and y_i. The estimate of $\bar{Y}$ is

$$\bar{y}_{lr} = \bar{y} + b(\bar{x}' - \bar{x}) \tag{12.48}$$

where $\bar{x}'$, $\bar{x}$ are the means of the x_i in the first and second samples and b is the least squares regression coefficient of y_i on x_i, computed from the second sample.

If no assumption is made about the presence of a linear regression in the population, $\bar{y}_{lr}$ will be biased, just as in one-stage sampling (Chapter 7). An approximation to $V(\bar{y}_{lr})$ can be given, assuming random sampling and $1/n$ and $1/n'$ negligible with respect to 1.

$$V(\bar{y}_{lr}) \doteq \frac{S_y^2(1-\rho^2)}{n} + \frac{\rho^2 S_y^2}{n'} - \frac{S_y^2}{N} \tag{12.49}$$

Proof. In finding the sampling error of $\bar{y}_{lr}$ in simple random sampling, we showed that if b in $\bar{y}_{lr}$ is replaced by the finite population regression coefficient $B = S_{yx}/S_x^2$, the error in the approximation is of order $1/\sqrt{n}$ relative to that in $\bar{y}_{lr}$. The same device applies here. We therefore examine the variance of the approximation

$$\tilde{y}_{lr} = \bar{y} + B(\bar{x}' - \bar{x})$$

The subscripts 1, 2 will denote variations over the first and second phases of sampling. Let $u_i = y_i - Bx_i$. In the second phase, regard the large sample as a finite population. Then, since the small sample was drawn at random from the large sample,

$$E_2(\tilde{y}_{lr}) = \bar{y}': \qquad V_2(\tilde{y}_{lr}) = \left(\frac{1}{n} - \frac{1}{n'}\right)s_u'^2$$

where $s_u'^2$ is the variance of u within the large sample. It follows that

$$V(\bar{y}_{lr}) \doteq V(\tilde{y}_{lr}) = V_1(\bar{y}') + E_1\left(\frac{1}{n} - \frac{1}{n'}\right)s_u'^2 \tag{12.50}$$

$$= \left(\frac{1}{n'} - \frac{1}{N}\right)S_y^2 + \left(\frac{1}{n} - \frac{1}{n'}\right)S_y^2(1-\rho^2) \tag{12.51}$$

since $s_u'^2$ is an unbiased estimate of $S_u^2 = S_y^2(1-\rho^2)$. Hence

$$V(\bar{y}_{lr}) \doteq \frac{S_y^2(1-\rho^2)}{n} + \frac{\rho^2 S_y^2}{n'} - \frac{S_y^2}{N} \tag{12.49}$$

This completes the proof.

As in section 7.8, suppose we assume, following Royall (1970), that the finite population is itself a random sample from an infinite superpopulation in which a linear regression model holds. Then $\bar{y}_{lr}$ becomes model-unbiased and exact small-sample results for its variance can be obtained. Let the regression model in the superpopulation be

$$y = \alpha + \beta x + \varepsilon \qquad (12.52)$$

where, for given x's, the ε's are independent with means 0 and variance $\sigma_y^2(1-\rho^2)$, where σ_y and ρ are now parameters of the superpopulation.

On substituting for $\bar{y}$, $\bar{Y}$, and b from (12.52), straightforward algebra gives

$$\bar{y}_{lr} - \bar{Y} = \bar{\varepsilon}_n - \bar{\varepsilon}_N + \beta(\bar{x}' - \bar{X}) + \frac{(\bar{x}' - \bar{x})\sum_{i}^{n}\varepsilon_i(x_i - \bar{x})}{\sum(x_i - \bar{x})^2} \qquad (12.53)$$

By averaging over the distribution of the ε's, it follows from (12.53) that $\bar{y}_{lr}$ is model-unbiased for fixed x's in the finite population and the two samples. Furthermore, from (12.53),

$$E[(\bar{y}_{lr} - \bar{Y})^2|x] = \sigma_y^2(1-\rho^2)\left(\frac{1}{n} - \frac{1}{N}\right) + \beta^2(\bar{x}' - \bar{X})^2 + \sigma_y^2(1-\rho^2)\frac{(\bar{x}' - \bar{x})^2}{\sum(x_i - \bar{x})^2} \qquad (12.54)$$

The last term on the right in (12.54) arises from the sampling error of b and is of order $1/n$ relative to the first two terms on the right. Averaging the first two terms on the right over the distribution of the x's created by repeated random selections of the finite population and the two samples, we get

$$EV(\bar{y}_{lr}) \doteq \sigma_y^2(1-\rho^2)\left(\frac{1}{n} - \frac{1}{N}\right) + \rho^2\sigma_y^2\left(\frac{1}{n'} - \frac{1}{N}\right) \qquad (12.55)$$

$$= \frac{\sigma_y^2(1-\rho^2)}{n} + \frac{\rho^2\sigma_y^2}{n'} - \frac{\sigma_y^2}{N} \qquad (12.56)$$

This expression has the same form as (12.49) except that in (12.56) σ_y^2 and ρ refer to the superpopulation.

Double sampling with regression has been extended by Khan and Tripathi (1967) to the case where p auxiliary x variables are measured in the second sample, $\bar{Y}$ being estimated by the multiple linear regression of y on these variables. With the second sample a random subsample of the first and with multivariate normality assumed for y and the x's, the extension of (12.54) for $p > 1$ gives for the average variance

$$V(\bar{y}_{lr}) = \frac{S_y^2(1-R^2)}{n}\left[1 + \frac{(n'-n)}{n'}\frac{p}{(n-p-2)}\right] + \frac{R^2 S_y^2}{n'} - \frac{S_y^2}{N} \qquad (12.57)$$

where R is the multiple correlation coefficient between y and the x's.

The case where the second sample is drawn independently of the first was considered by Chameli Bose (1943).

12.7 OPTIMUM ALLOCATION AND COMPARISON WITH SINGLE SAMPLING

From the variance formula (12.49), which assumes $1/n$ negligible, double sampling with a regression estimate can be compared with a single simple random sample with no regression adjustment. We have

$$V + \frac{S_y^2}{N} = \frac{S_y^2(1-\rho^2)}{n} + \frac{\rho^2 S_y^2}{n'}: \qquad C = cn + c'n' \qquad (12.58)$$

By the Cauchy–Schwarz inequality, the product VC is minimized when

$$\frac{cn^2}{S_y^2(1-\rho^2)} = \frac{c'n'^2}{\rho^2 S_y^2} \qquad \text{so that} \qquad \frac{n}{n'} = \left[\frac{c'}{c} \frac{(1-\rho^2)}{\rho^2} \right]^{1/2} \qquad (12.59)$$

Substitution in VC gives

$$(VC)_{min} = S_y^2 (\sqrt{c(1-\rho^2)} + \sqrt{c'\rho^2})^2 - \frac{CS_y^2}{N} \qquad (12.60)$$

Thus, for a specified cost C,

$$V_{min} = S_y^2 \frac{(\sqrt{c(1-\rho^2)} + \sqrt{c'\rho^2})^2}{C} - \frac{S_y^2}{N} \qquad (12.61)$$

If all resources are devoted instead to a single sample with no regression adjustment, this sample has size C/c and the variance of its mean is

$$V(\bar{y}) = \frac{cS_y^2}{C} - \frac{S_y^2}{N} \qquad (12.62)$$

Hence, optimum use of double sampling gives a smaller variance if

$$c > (\sqrt{c(1-\rho^2)} + \sqrt{c'\rho^2})^2 \qquad (12.63)$$

This inequality may be expressed in two ways.

$$\frac{c}{c'} > \frac{(1+\sqrt{1-\rho^2})^2}{\rho^2} \qquad (12.64)$$

or

$$\rho^2 > \frac{4(c/c')}{(1+c/c')^2} \qquad (12.65)$$

Equations (12.64) and (12.65) give the critical ranges of c/c' for given ρ and of ρ for given c/c' that make double sampling profitable.

Figure 12.1 plots the values of the ratio c/c' (on a log scale) against ρ. Curve I is the relationship when double and single sampling are equally precise; curve II holds when $V_{opt} = 0.8 \, V(\bar{y})$, that is, when double sampling gives a 25% increase in precision; and curve III refers to a 50% increase in precision. For example, when $\rho = 0.8$, double sampling equals single sampling in precision if c/c' is 4, gives a 25% increase in precision if c/c' is about $7\frac{1}{2}$, and a 50% increase if c/c' is about 13.

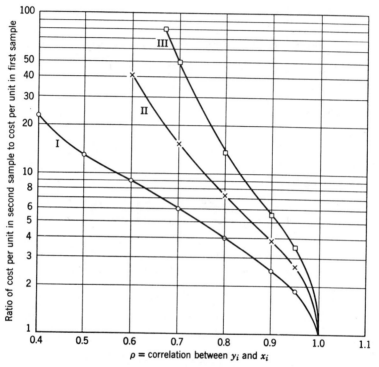

Fig. 12.1 Relation between c/c' and ρ for three fixed values of the relative precision of double and single sampling.

Curve I: double and single sampling equally precise.
Curve II: double sampling gives 25 per cent increase in precision.
Curve III: double sampling gives 50 per cent increase in precision.

For practical use, the curves overestimate the gains to be achieved from double sampling, because the best values of n and n' must either be estimated from previous data or be guessed. Some allowance for errors in these estimations should be made before deciding to adopt double sampling.

For any ρ, there is an upper limit to the gain in precision from double sampling. This occurs when information on $\bar{x}'$ is obtained free ($c' = 0$). The upper limit to the relative precision is $1/(1-\rho^2)$.

12.8 ESTIMATED VARIANCE IN DOUBLE SAMPLING FOR REGRESSION

If terms in $1/n$ are negligible, $V(\bar{y}_{lr})$ is given by (12.49):

$$V(\bar{y}_{lr}) \doteq \frac{S_y^2(1-\rho^2)}{n} + \frac{\rho^2 S_y^2}{n'} - \frac{S_y^2}{N}$$

With a linear regression model, the quantity

$$s_{y.x}^2 = \frac{1}{n-2}\left[\sum_{i=1}^n (y_i - \bar{y})^2 - b^2 \sum_{i=1}^n (x_i - \bar{x})^2\right] \tag{12.66}$$

is an unbiased estimate of $S_y^2(1-\rho^2)$. Since

$$s_y^2 = \frac{\sum (y_i - \bar{y})^2}{n-1}$$

is an unbiased estimate of S_y^2, it follows that

$$s_y^2 - s_{y.x}^2$$

is an unbiased estimate of $\rho^2 S_y^2$.

Thus a sample estimate of $V(\bar{y}_{lr})$ is

$$v(\bar{y}_{lr}) = \frac{s_{y.x}^2}{n} + \frac{s_y^2 - s_{y.x}^2}{n'} - \frac{s_y^2}{N} \tag{12.67}$$

If the second sample is small and terms in $1/n$ are not negligible relative to 1, an estimate of variance suggested for simple random samples from (12.54) is

$$v(\bar{y}_{lr}) = s_{y.x}^2\left[\frac{1}{n} + \frac{(\bar{x}'-\bar{x})^2}{\sum (x_i - \bar{x})^2}\right] + \frac{s_y^2 - s_{y.x}^2}{n'} - \frac{s_y^2}{N} \tag{12.68}$$

This is a hybrid of the conditional variance and the average variance.

12.9 RATIO ESTIMATORS

If the first sample is used to obtain $\bar{x}'$ as an estimate of $\bar{X}$ in a ratio estimate of $\bar{Y}$, the estimator of $\bar{Y}$ is

$$\bar{y}_R = \frac{\bar{y}}{\bar{x}}\bar{x}' \tag{12.69}$$

To find the approximate variance, write

$$\bar{y}_R - \bar{Y} = \frac{\bar{y}}{\bar{x}}\bar{x}' - \bar{Y}$$

$$= \left(\frac{\bar{y}}{\bar{x}}\bar{X} - \bar{Y}\right) + \frac{\bar{y}}{\bar{x}}(\bar{x}' - \bar{X})$$

$$= \frac{\bar{X}}{\bar{x}}(\bar{y} - R\bar{x}) + \frac{\bar{y}}{\bar{x}}(\bar{x}' - \bar{X})$$

The first component is the error of the ordinary ratio estimate (section 2.11). In obtaining the appropriate error variance in section 2.11, we replaced the factor $\bar{X}/\bar{x}$ by unity in this term. To the same order of approximation, we replace the factor $\bar{y}/\bar{x}$ in the second component by the population ratio $R = \bar{Y}/\bar{X}$. Thus

$$\bar{y}_R - \bar{Y} \doteq (\bar{y} - R\bar{x}) + R(\bar{x}' - \bar{X}) \tag{12.70}$$

If the second sample is a random subsample of the first,

$$E_2(\bar{y}_R - \bar{Y}) \doteq \bar{y}' - \bar{Y}: \qquad V_2(\bar{y}_R - \bar{Y}) \doteq \left(\frac{1}{n} - \frac{1}{n'}\right)s_d'^2 \tag{12.71}$$

where $s_d'^2$ is the variance within the second sample of the variate $d_i = y_i - Rx_i$. Averaging now over repeated random selections of the first sample,

$$V(\bar{y}_R) = V_1 E_2 + E_1 V_2$$

$$\doteq \left(\frac{1}{n'} - \frac{1}{N}\right)S_y^2 + \left(\frac{1}{n} - \frac{1}{n'}\right)(S_y^2 - 2RS_{yx} + R^2 S_x^2) \tag{12.72}$$

since $s_d'^2$ is an unbiased estimate of $S_d^2 = S_y^2 - 2RS_{yx} + R^2 S_x^2$.

Separating the terms in $1/n$ and $1/n'$ we get

$$V(\bar{y}_R) \doteq \frac{S_y^2 - 2RS_{yx} + R^2 S_x^2}{n} + \frac{2RS_{yx} - R^2 S_x^2}{n'} - \frac{S_y^2}{N} \tag{12.72'}$$

12.10   REPEATED SAMPLING OF THE SAME POPULATION

The practice of relying on samples for the collection of important series of data that are published at regular intervals has become common. In part, this is due to a realization that with a dynamic population a census at infrequent intervals is of limited use. Highly precise information about the characteristics of a population in July 1960 and July 1970 may not help much in planning that demands a knowledge of the population in 1976. A series of small samples at annual or even shorter intervals may be more serviceable.

When the same population (apart from the changes that the passage of time introduces) is sampled repeatedly, the sampler is in an ideal position to make realistic estimates both of costs and of variances and to apply the techniques that lead to optimum efficiency of sampling. One important question is how frequently and in what manner the sample should be changed as time progresses. Many considerations affect the decision. People may be unwilling to give the same type of information time after time. The respondents may be influenced by information which they receive at the interviews, and this may make them progressively less representative as time proceeds. Sometimes, however, cooperation is better in a second interview than in the first, and when the information is technical or confidential the second visit may produce more accurate data than the first.

The remainder of this chapter considers the question of replacement of the sample and the related question of making estimates from the series of repeated samples. The topic is appropriate to the present chapter because double sampling techniques can be utilized.

Given the data from a series of samples, there are three kinds of quantity for which we may wish estimates.

1. The change in $\bar{Y}$ from one occasion to the next.
2. The average value of $\bar{Y}$ over all occasions.
3. The average value of $\bar{Y}$ for the most recent occasion.

In most surveys, interest centers on the current average (3), particularly if the characteristics of the population are likely to change rapidly with time. With a population in which time changes are slow, on the other hand, an annual average (2) taken over 12 monthly samples or four quarterly samples may be adequate for the major uses. This would be the situation in a study of the prevalence of chronic diseases of long duration. With a disease whose prevalence shows marked seasonal variation, the current data are of major interest, but annual averages are also useful for comparisons between different regions and different years. Estimates of change (1) are wanted mainly in attempts to study the effects of forces that are known to have acted on the population. For instance, if a bill is passed that is supposed to stimulate the building of houses, it is interesting to know whether the building rate of new houses has increased in the succeeding year (with a realization that an increase may not be entirely due to the bill).

Suppose that we are free to alter or retain the composition of the sample and that the total size of sample is to be the same on all occasions. If we wish to maximize precision, the following statements can be made about replacement policy:

1. For estimating change, it is best to retain the same sample throughout all occasions.
2. For each estimating the average over all occasions, it is best to draw a new sample on each occasion.
3. For current estimates, equal precision is obtained either by keeping the same sample or by changing it on every occasion. Replacement of part of the sample on each occasion may be better than these alternatives.

Statements 1 and 2 hold because there is nearly always a *positive* correlation between the measurements on the same unit on two successive occasions. The estimated change on a unit has variance $S_1^2 + S_2^2 - 2\rho S_1 S_2$, where the subscripts refer to the occasions. If the change is estimated from two *different* units, the variance is $S_1^2 + S_2^2$. In estimating the over-all mean for the two occasions, the variance is $(S_1^2 + S_2^2 + 2\rho S_1 S_2)/4$ if the same unit is retained and $(S_1^2 + S_2^2)/4$ if a new unit is chosen.

Statement 3, which is less obvious, is investigated in succeeding sections.

12.11 SAMPLING ON TWO OCCASIONS

Suppose that the samples are of the same size n on both occasions and that the current estimates are of primary interest. Replacement policy has been examined by Jessen (1942). For simplicity, we assume that simple random sampling is used.

The mean of the first sample has varience S_1^2/n, there being no previous information to utilize. In selecting the second sample, m of the units in the first sample are retained (m for matched). The remaining u units (u for unmatched) are discarded and replaced by a new selection from the units not previously selected.

Notation.

$\bar{y}_{hu}$ = mean of unmatched portion on occasion h

$\bar{y}_{hm}$ = mean of matched portion on occasion h

$\bar{y}_h$ = mean of whole sample on occasion h

The unmatched and matched portions of the second sample provide independent estimates $\bar{y}_{2u}{}'$, $\bar{y}_{2m}{}'$ of $\bar{Y}_2$, as shown in Table 12.1. In the matched portion we use a double-sampling regression estimate, where the "large" sample is the first sample and the auxiliary variate x_i is the value of y_i on the first occasion. The variance of $\bar{y}_{2m}{}'$ comes from (12.49), p. 339: note that our m and n correspond to n and n', respectively, in (12.49). The fpc is ignored.

TABLE 12.1
ESTIMATES FROM THE UNMATCHED AND MATCHED PORTIONS

	Estimate	Variance
Unmatched:	$\bar{y}_{2u}{}' = \bar{y}_{2u}$	$\dfrac{S_2^2}{u} = \dfrac{1}{W_{2u}}$
Matched:	$\bar{y}_{2m}{}' = \bar{y}_{2m} + b(\bar{y}_1 - \bar{y}_{1m})$	$\dfrac{S_2^2(1-\rho^2)}{m} + \rho^2\dfrac{S_2^2}{n} = \dfrac{1}{W_{2m}}$

The best combined estimate of $\bar{Y}_2$ is found by weighting the two independent estimates inversely as their variances. If W_{2u}, W_{2m} are the inverse variances, this estimate is

$$\bar{y}_2' = \phi_2\bar{y}_{2u}' + (1-\phi_2)\bar{y}_{2m}' \tag{12.73}$$

where

$$\phi_2 = \frac{W_{2u}}{W_{2u} + W_{2m}}$$

By least squares theory, the variance of $\bar{y}_2'$ is

$$V(\bar{y}_2') = \frac{1}{W_{2u} + W_{2m}}$$

From Table 12.1, this works out after simplification as

$$V(\bar{y}_2') = \frac{S_2^2(n-u\rho^2)}{n^2-u^2\rho^2}$$ (12.74)

Note that if $u = 0$ (complete matching) or if $u = n$ (no matching) this variance has the same value, S_2^2/n.

The optimum value of u is found by minimizing (12.74) with respect to variation in u. This gives

$$\frac{u}{n} = \frac{1}{1+\sqrt{1-\rho^2}}, \qquad \frac{m}{n} = \frac{\sqrt{1-\rho^2}}{1+\sqrt{1-\rho^2}}$$ (12.75)

When the optimum u is substituted in (12.74), the minimum variance works out as

$$V_{opt}(\bar{y}_2') = \frac{S_2^2}{2n}(1+\sqrt{1-\rho^2})$$ (12.76)

Table 12.2 shows for a series of values of ρ the optimum percent that should be matched and the relative gain in precision compared with no matching. The best

TABLE 12.2

Optimum % Matched

ρ	Optimum % matched	% gain in precision	% gain with $\frac{m}{n}=\frac{1}{3}$	% gain with $\frac{m}{n}=\frac{1}{4}$
0.5	46	7	7	6
0.6	44	11	11	9
0.7	42	17	17	15
0.8	38	25	25	23
0.9	30	39	39	39
0.95	24	52	50	52
1.0	0	100	67	75

percentage to match never exceeds 50% and decreases steadily as ρ increases. When $\rho = 1$, the formula suggests $m = 0$, which lies outside the range of our assumptions, since m has been assumed reasonably large. The correct procedure in this case is to take $m = 2$. The two matched units are sufficient to determine the regression line exactly.

The greatest attainable gain in precision is 100% when $\rho = 1$. Unless ρ is high, the gains are modest.

Although the optimum percentage to match varies with ρ, only a single percentage can be used in practice for all items in a survey. The right-hand columns of Table 12.2 show the percent gains in precision when one third and one fourth of the units are matched. Both are good compromises, except for items in which ρ exceeds 0.95.

Kulldorff (1963) gives an extensive discussion of this model. He considers the case where on the second occasion the cost of measuring a matched unit (i.e., one previously measured) may differ from that of measuring a new unmatched unit and does not assume equal sample sizes on two occasions. Thus, apart from fixed costs, his cost on the second occasion is

$$C_2 = mc_m + uc_u: \qquad \frac{C_2}{c_u} = m\,\delta + u \qquad (12.77)$$

where $\delta = c_m/c_u$. If sample sizes are the same on the two occasions so that $m + u = n$, the optimum unmatched proportion on the second occasion is found by minimizing

$$\frac{VC_2}{c_u S_2^2} = [n\delta + u(1-\delta)]\frac{(n-u\rho^2)}{(n^2-u^2\rho^2)} = [\delta + \mu(1-\delta)]\frac{(1-\mu\rho^2)}{(1-\mu^2\rho^2)} \qquad (12.78)$$

where $\mu = u/n$ and V comes from (12.74). If $\delta < 1$, matching being cheaper, the optimum proportion matched is, of course, greater than the values in Table 12.2. He also deals with the case where the costs are to be the same on the two occasions.

In some applications the data for occasion 1 provide several auxiliary variables correlated with y_2, one of which will, of course, usually be y_1. For example, in estimating the kill y_2 of waterfowl per hunter in Ontario from 1968 to 1969, Sen (1973a) found that the kill per hunter and the number of days hunted in 1967 and 1968 were both correlated with y_2. In this paper he extended the preceding analysis to the case where $\bar{y}_{2m}$ is adjusted by its multiple linear regression on the auxiliary variables and where the samples on the two occasions are of unequal sizes. With large samples of equal size the only change in (12.76) for $V_{opt}(\bar{y}_2')$ is to replace ρ^2 by the square R^2 of the multiple correlation coefficient between y_2 and the auxiliary variables (assuming multivariate normality). The corresponding theory for the case where $\bar{y}_{2m}$ is adjusted by the multivariate ratio estimate is given in Sen (1972) for equal sample sizes and in Sen (1973b) for unequal sample sizes.

12.12 SAMPLING ON MORE THAN TWO OCCASIONS

The general problem of replacement has been studied by Yates (1960) and Patterson (1950), with respect to both current estimates and estimates of change. When there are more than two occasions, the opportunities for a flexible use of the data are increased. On occasion h we may have parts of the sample that are matched with occasion $h-1$, parts that are matched with both occasions $h-1$ and $h-2$, and so on. In attempting to improve the current estimate, we might try a multiple regression involving all matchings to previous occasions. It is also possible to *revise* the current estimate for occasion $h-1$ after the data for occasion h are known. In the revised estimate the regression of occasion $h-1$ on both occasion $h-2$ and occasion h could be utilized, assuming that suitably matched portions of the sample were available.

TABLE 12.3

ESTIMATES OF $\bar{Y}_h$ ON THE hth OCCASION

	Estimate	Variance
Unmatched:	$\bar{y}_{hu}' = \bar{y}_{hu}$	$\dfrac{S^2}{u} = \dfrac{1}{W_{hu}}$
Matched:	$\bar{y}_{hm}' = \bar{y}_{hm} + b(\bar{y}_{h-1}' - \bar{y}_{h-1,m})$	$\dfrac{S^2(1-\rho^2)}{m} + \rho^2 V(\bar{y}_{h-1}') = \dfrac{1}{W_{hm}}$

The present section contains an introduction to the subject. Attention will be restricted to current estimates in which only the regression on the sample immediately preceding is used. This results in some loss of precision but, since the correlation ρ usually decreases as the time interval between the occasions is increased, the loss of precision will seldom be great. The variance S^2 and the correlation coefficient ρ between the item values on the same unit on two successive occasions are assumed constant throughout.

On the hth occasion let m_h and u_h be the numbers of units that are matched and unmatched, respectively, with the $(h-1)$th occasion. The two estimates of $\bar{Y}_h$ that can be made are given in Table 12.3. The only change in procedure from the second occasion (Table 12.1) is that in the regression adjustment of the estimate from the matched portion we use the improved estimate $\bar{y}_{h-1}'$ instead of the sample mean $\bar{y}_{h-1}$.

The variance of the matched estimate $\bar{y}_{hm}'$ in Table 12.3 is derived from (12.49) on page 339. Note that (a) our m corresponds to the n in (12.49) and (b) the term $\rho^2 S_y^2/n'$ in (12.49), which equals $B^2 V(\bar{x}')$, is replaced by $\rho^2 V(\bar{y}_{h-1}')$, since $B = \rho$ when S is constant on successive occasions and $\bar{y}_{h-1}'$ corresponds to $\bar{x}'$ in the earlier analysis.

We now examine the precision obtained if the optimum m_h and u_h and the optimum weights are used on every occasion. It will be found that the optimum m_h/n increases steadily on successive occasions, rapidly approaching a limiting value of $\frac{1}{2}$.

Weighting inversely as the variance, the best estimate of $\bar{y}_h$ is

$$\bar{y}_h' = \phi_h \bar{y}_{hu}' + (1 - \phi_h) \bar{y}_{hm}' \qquad (12.79)$$

where $\phi_h = W_{hu}/(W_{hu} + W_{hm})$. This gives

$$V(\bar{y}_h') = \frac{1}{W_{hu} + W_{hm}} = \frac{g_h S^2}{n}$$

where g_h denotes the ratio of the variance on occasion h to that on the first occasion. Substituting for W_{hu}, W_{hm} from Table 12.3, we have

$$\frac{S^2}{V(\bar{y}_h')} = \frac{n}{g_h} = S^2(W_{hu} + W_{hm}) = u_h + \frac{1}{\dfrac{(1 - \rho^2)}{m_h} + \dfrac{\rho^2 g_{h-1}}{n}} \qquad (12.80)$$

We now choose m_h and u_h to maximize this quantity and therefore to minimize $V(\bar{y}_h')$. Writing $u_h = n - m_h$ and differentiating the right side of (12.80) with respect to m_h we obtain

$$\frac{1 - \rho^2}{m_h^2} = \left(\frac{1 - \rho^2}{m_h} + \frac{\rho^2 g_{h-1}}{n} \right)^2$$

This gives, on solving for the optimum $\hat{m}_h$, say,

$$\frac{\hat{m}_h}{n} = \frac{\sqrt{1 - \rho^2}}{g_{h-1}(1 + \sqrt{1 - \rho^2})} \qquad (12.81)$$

When this value is substituted in (12.80), the relation becomes, after some algebraic manipulation,

$$\frac{1}{g_h} = 1 + \frac{1 - \sqrt{1 - \rho^2}}{g_{h-1}(1 + \sqrt{1 - \rho^2})} \qquad (12.82)$$

This relation may be written

$$r_h = 1 + b r_{h-1}$$

where $r_h = 1/g_h$ and $r_1 = 1/g_1 = 1$. Repeated use of this recurrence relation gives

$$\frac{1}{g_h} = r_h = 1 + b + b^2 + \cdots + b^{h-1} = \frac{1 - b^h}{1 - b}$$

where, from (12.82), $b = (1 - \sqrt{1 - \rho^2})/(1 + \sqrt{1 - \rho^2})$. Since $0 < b < 1$, the limiting variance factor g_∞ is

$$g_\infty = 1 - b = \frac{2\sqrt{1 - \rho^2}}{1 + \sqrt{1 - \rho^2}} \qquad (12.83)$$

Hence the variance of $\bar{y}_h{}'$ tends to

$$V(\bar{y}_\infty') = \frac{S^2}{n}\left(\frac{2\sqrt{1-\rho^2}}{1+\sqrt{1-\rho^2}}\right) \tag{12.84}$$

Finally, the limiting value of $\hat{m}_h$ is obtained from (12.81) as

$$\frac{\hat{m}_\infty}{n} = \frac{\sqrt{1-\rho^2}}{g_\infty(1+\sqrt{1-\rho^2})} = \frac{1}{2}$$

irrespective of the value of ρ.

Table 12.4 shows the optimum percentage to match—$100\hat{m}_h/n$, as found from (12.81)—and the resulting variances for $\rho = 0.7, 0.8, 0.9$ and 0.95 and for a series of values of h.

TABLE 12.4

OPTIMUM % MATCHED AND VARIANCES

	\% matched $100\hat{m}_h/n$				$g_h = nV(\bar{y}_h')/S^2$			
	$\rho =$				$\rho =$			
h	0.7	0.8	0.9	0.95	0.7	0.8	0.9	0.95
2	42	38	30	24	0.857	0.800	0.718	0.656
3	49	47	42	36	0.837	0.762	0.646	0.556
4	50	49	47	43	0.834	0.753	0.622	0.515
5	50	50	49	46	0.833	0.751	0.613	0.495
∞	50	50	50	50	0.833	0.750	0.607	0.476

By the fourth occasion, the optimum percent matched is close to 50 for all the values of ρ shown, although a smaller amount of matching is indicated for the second and third occasions. The reductions in variance, $(1 - g_h)$, are modest if ρ is less than 0.8.

12.13 SIMPLIFICATIONS AND FURTHER DEVELOPMENTS

In practical application the preceding analysis may need modification. We assumed that all replacement policies cost the same and are equally feasible. With human populations, field costs are likely to be lower if the same units are retained for a number of occasions. If estimates of the change in the population total or mean are of interest, this factor also points toward matching more than half the units from one occasion to the next.

It is convenient also to keep the weights and the proportion matched constant, instead of changing them on every occasion. Consequently, we will investigate the variances of $\bar{y}_h{}'$ and of the estimated change $(\bar{y}_h{}' - \bar{y}_{h-1}')$ when m, u, and ϕ are held constant. We continue to write $V(\bar{y}_h{}) = g_h S^2/n$, although the actual value of g_h will be different from that in the preceding section.

The estimate is now

$$\bar{y}_h{}' = \phi \bar{y}_{hu}{}' + (1 - \phi) \bar{y}_{hm}{}'$$

Substituting the expressions for the two variances (from Table 12.3), we have

$$V(\bar{y}_h{}') = \frac{g_h S^2}{n} = \phi^2 V(\bar{y}_{hu}{}') + (1 - \phi)^2 V(\bar{y}_{hm}{}')$$

$$S^2 \left[\frac{\phi^2}{u} + \frac{(1-\phi)^2(1-\rho^2)}{m} \right] + \frac{S^2 \rho^2 (1-\phi)^2 g_{h-1}}{n}$$

Hence

$$g_h = \left[\frac{\phi^2}{\mu} + \frac{(1-\phi)^2(1-\rho^2)}{\lambda} \right] + \rho^2 (1-\phi)^2 g_{h-1} \qquad (12.85)$$

where $\mu = u/n$, $\lambda = m/n$. Write this relation as

$$g_h = a + b g_{h-1}$$

By repeated application, we have, since $g_1 = 1$,

$$g_h = \frac{a(1 - b^{h-1})}{1 - b} + b^{h-1}$$

Since $b = \rho^2(1-\phi)^2$ is less than 1, the limiting value is

$$g_\infty = \frac{a}{1 - b} = \frac{\lambda \phi^2 + \mu(1-\phi)^2(1-\rho^2)}{\lambda \mu [1 - \rho^2 (1-\phi)^2]} \qquad (12.86)$$

The value of the weight ϕ that minimizes the limiting variance may be found by differentiating (12.86). This leads to a quadratic equation whose appropriate root is

$$\phi_{opt} = \frac{\sqrt{1-\rho^2}[\sqrt{1-\rho^2 + 4\lambda\mu\rho^2} - \sqrt{1-\rho^2}]}{2\lambda\rho^2}$$

In practice, the value of ρ will not be known exactly and will differ from item to item. A compromise value can usually be chosen. Clearly, ϕ_{opt} will be less than $\mu = u/n$, since the matched part of the sample gives higher precision per unit than the unmatched part. For example, with $\mu = 0.25$, that is, $\frac{1}{4}$ of the sample unmatched, ϕ_{opt} turns out to be 0.216, 0.198, and 0.164 for $\rho = 0.7, 0.8, 0.9$. The choice of $\phi = 0.2$ would be adequate for this range of ρ.

For the estimate of change, we have

$$V(\bar{y}_h{}' - \bar{y}_{h-1}') = V(\bar{y}_h{}') + V(\bar{y}_{h-1}') - 2 \operatorname{cov}(\bar{y}_h{}' \bar{y}_{h-1}') \qquad (12.87)$$

To find the covariance term, note that if y_{hi}, $y_{h-1,i}$ are the values for the ith unit in the matched set on occasions h and $(h-1)$, our model is

$$y_{hi} = \bar{Y}_h + \rho(y_{h-1,i} - \bar{Y}_{h-1}) + e_{hi}$$

where the e_{hi} are independent of the y's. From this model it is found by substitution that

$$\bar{y}_{hm}' = \bar{y}_{hm} + \rho(\bar{y}_{h-1}' - \bar{y}_{h-1,m}) = \bar{Y}_h + \rho(\bar{y}_{h-1}' - \bar{Y}_{h-1}) + \bar{e}_{hm}$$

Hence the covariance of $\bar{y}_{hm}'$ and $\bar{y}_{h-1}'$ is $\rho V(\bar{y}_{h-1}')$. But

$$\mathrm{cov}\,(\bar{y}_h'\bar{y}_{h-1}') = \mathrm{cov}\,\{[\phi\bar{y}_{hu} + (1-\phi)\bar{y}_{hm}']\bar{y}_{h-1}'\} = \rho(1-\phi)V(\bar{y}_{h-1}')$$

since $\bar{y}_{hu}$ is independent of $\bar{y}_{h-1}'$. From (12.87), this gives

$$V(\bar{y}_h' - \bar{y}_{h-1}') = \frac{S^2}{n}\{g_h + g_{h-1}[1 - 2\rho(1-\phi)]\} \tag{12.88}$$

From (12.85) and (12.87), the variances of $\bar{y}_h'$ and $(\bar{y}_h' - \bar{y}_{h-1}')$ may be computed for any values of m, ϕ, and ρ. Table 12.5 shows the resulting percent gains in efficiency for these estimates relative to the estimates obtained from independent

TABLE 12.5

PERCENT GAINS IN EFFICIENCY FOR THE CURRENT ESTIMATE $\bar{y}_h'$ AND THE
ESTIMATE OF CHANGE $(\bar{y}_h' - \bar{y}_{h-1}')$. PROPORTIONS MATCHED: $\frac{1}{2}$ AND $\frac{3}{4}$

Current estimate: percent gain $= 100(1 - g_h)/g_h$

h	$\rho = 0.7$		$\rho = 0.8$		$\rho = 0.9$		$\rho = 0.95$	
	$\frac{1}{2}$	$\frac{3}{4}$	$\frac{1}{2}$	$\frac{3}{4}$	$\frac{1}{2}$	$\frac{3}{4}$	$\frac{1}{2}$	$\frac{3}{4}$
2	14	10	22	14	33	19	41	22
3	16	14	30	20	52	32	67	39
4	17	15	32	24	59	40	79	52
∞	17	15	33	26	62	50	89	74
RG^a	15	10	27	18	56	40	—	—

h	Estimate of change: percent gain $= 100(2 - g_h')/g_h'$							
2	106	153	156	233	245	399	326	565
3	113	160	170	245	277	415	365	588
4	115	160	174	251	285	424	388	603
∞	115	163	178	251	292	440	397	624
RG	101	163	166	269	381	605	—	—

^{a} Described on p. 355.

samples on each occasion. The proportions matched are $\lambda = m/n = \frac{1}{2}$ and $\frac{3}{4}$. The weight ϕ was taken as 0.35 for $\lambda = \frac{1}{2}$ and 0.2 for $\lambda = \frac{3}{4}$.

Not surprisingly, the major features of Table 12.5 are the large gains in efficiency for the estimates of change when ρ is at least 0.7. Moreover, increase in the proportion matched from $\frac{1}{2}$ to $\frac{3}{4}$ produces substantial gains in efficiency for the estimates of change at the expense of smaller losses in efficiency for the current estimates. The results suggest that retention of $\frac{2}{3}$, $\frac{3}{4}$ or $\frac{4}{5}$ from one occasion to the next may be a good practical policy if current estimates and estimates of change are both important.

Comparison of the gains in efficiency for the current estimate $\bar{y}_h{}'$ in Table 12.5 with the optimum gains from Table 12.4 suggests that after the second occasion little precision is lost by using a constant weight and a fixed proportion matched, unless $\rho \geq 0.95$.

If ρ exceeds 0.8, the regression coefficient $b = \rho$ may be replaced by 1 with only a small additional loss of precision. This gives an estimate $\bar{y}_h{}''$ of the form

$$\bar{y}_h{}'' = \phi \bar{y}_{hu} + (1 - \phi)(\bar{y}''_{h-1} + \bar{y}_{hm} - \bar{y}_{h-1,m}) \qquad (12.89)$$

In the important Current Population Survey taken monthly by the U.S. Bureau of the Census, one quarter of the second-stage units are replaced each month, so that an individual household remains in the sample during four consecutive months. The household is omitted for the eight succeeding months but is then brought back for another four months, thus increasing slightly the precision of year-to-year comparisons.

The composite estimate used in this survey is of a form related to (12.89) but slightly different.

$$\bar{y}_h{}'' = (1 - K)\bar{y}_h + K(\bar{y}''_{h-1} + \bar{y}_{hm} - \bar{y}_{h-1,m}) \qquad (12.90)$$

where K is a constant weighting factor. The difference is that $\bar{y}_h$, the current estimate for the whole sample, takes the place of the $\bar{y}_{hu}$ in (12.89). The quantities $\bar{y}_{hm}, \bar{y}_{h-1,m}, \bar{y}_h$ in (12.90) are ratio estimates of a fairly complex type.

The variance of $\bar{y}_h{}''$ (due to Bershad) is given in Hansen, Hurwitz, and Madow (1953); see also the Appendix in Hansen et al. (1955). The estimator of the month-to-month change is

$$\bar{d}_h = \bar{y}_h{}'' = \bar{y}''_{h-1}$$

Since the primary units remain unchanged, only the within-units components of $V(\bar{y}_h{}'')$ and $V(\bar{d}_h)$ are affected by the sample rotation policy.

Rao and Graham (1964) have examined the performance of composite estimators in rotation policies of this type, in which a respondent remains in the sample for r months and then drops out for m months. They used as models an exponential correlogram and a linear correlogram in time, descending to zero. A more complex correlogram, needed if there is a high correlation between months h and $(h - 12)$, has been studied by Graham (1973).

Their gains in efficiency for $r = 2$, 4 and $m = \infty$ correspond to the results for $\lambda = \frac{1}{2}, \frac{3}{4}$ in Table 12.5. For an exponential correlogram in which the correlation between results on the same unit on occasions h, $h - i$ is ρ^i, the lines labeled RG in Table 12.5 show for $\rho = 0.7$, 0.8, 0.9, their percent gains in efficiency. For each ρ they use in (12.90) the optimum K for the current estimates as $h \to \infty$. As Table 12.5 shows, they also find that the gains in efficiency from the composite estimators are much greater in estimating change than current level.

In a more general framework, Scott and Smith (1974) have discussed the role of time series methods in making estimates in repeated surveys of various types.

In another rotation policy a new sample is drawn on each occasion, with no matching. With weekly or monthly sampling, this plan is appropriate when annual estimates, and to a lesser extent semiannual or quarterly estimates, are of primary importance, for example, in an illness survey with emphasis on chronic diseases. If the questionnaire obtains for any unit the results for the preceding month as well as for the current month, we can consider composite estimates of the form

$$\bar{y}_h' = \bar{y}_h + \phi_h(\bar{y}_{h-1}' - \bar{y}_{h-1,h}) \tag{12.91}$$

where $\bar{y}_h$ = estimate made from current data in the current sample

$\bar{y}_{h-1,h}$ = estimate made from previous month's data in the current sample

$\bar{y}_{h-1}'$ = composite estimate for the previous month

The theory is discussed by Hansen, Hurwitz, and Madow (1953) and Woodruff (1959), who apply it to a survey of retail sales, and by Eckler (1955). In the Retail Trade Survey the composite estimate involves a ratio estimate, being of the form

$$\bar{y}_h'' = (1 - W)\bar{y}_h + W\left(\frac{\bar{y}_h}{\bar{y}_{h-1,h}}\right)\bar{y}_{h-1}''$$

where W is a weighting factor. Since month-to-month correlations are very high, averaging around 0.98, the gains in precision are substantial. One month later a revised composite estimate for month h is computed, using the results for month h from the new sample taken in month $(h + 1)$.

With this method, it is essential that the data obtained for the preceding month from the current sample be accurate. This may not be so when the data depend on the unrecorded memory of the respondent, although the method may work successfully if the data are of a type that the respondent records carefully as a routine matter.

EXERCISES

12.1　$3000 is allocated for a survey to estimate a proportion. The main survey will cost $10 per sampling unit. Information is available in files, at a cost of $0.25 per sampling unit, that enables the units to be classified into two strata of about equal sizes. If the true proportion is 0.2 in stratum 1 and 0.8 in stratum 2, estimate the optimum n, n', and the resulting value of $V(p_{st})$. Does double sampling produce a gain in precision over single sampling? (The ratios n'/N, n_h/N_h may be ignored.)

12.2   For the W_h, P_h in exercise 12.1, find the cost ratios c_n/c_n' for which double sampling is more economical than single sampling.

12.3   A population contains L strata of equal size. If V_{ran} denotes the variance of the mean of a simple random sample and V_{st}, V_{ds} are the corresponding variances for stratified random sampling with proportional allocation and for double sampling with stratification, show that, approximately,

$$n V_{ran} = \bar{S}_h{}^2 + \frac{\sum_h (\bar{Y}_h - \bar{Y})^2}{L}$$

$$n V_{st} = \bar{S}_h{}^2$$

$$n V_{ds} = \bar{S}_h{}^2 + \frac{n}{n'} \frac{\sum_h (\bar{Y}_h - \bar{Y})^2}{L}$$

where $\bar{S}_h{}^2$ is the average variance within strata. (N and n' may both be assumed large relative to L, and the n_h in double sampling may be assumed equal to n/L.)

Hence, if $(RP)_{st}$ denotes the relative precision of the stratified sample to the simple random sample, with a corresponding definition for $(RP)_{ds}$, show that

$$(RP)_{ds} = \frac{(RP)_{st}}{1 + (n/n')[(RP)_{st} - 1]}$$

For $(RP)_{st} = 2$, plot $(RP)_{ds}$ against n/n'. How small must this ratio be in order that $(RP)_{ds} = 1.9$?

12.4   If $\rho = 0.8$ in double sampling for regression, how large must n' be relative to n, if the loss in precision due to sampling errors in the mean of the large sample is to be less than 10%.

12.5   In an application of double sampling for regression, the small sample was of size 87 and the large sample of size 300. The following computations apply to the small sample.

$$\sum (y_i - \bar{y})^2 = 17,283, \qquad \sum (y_i - \bar{y})(x_i - \bar{x}) = 5114, \qquad \sum (x_i - \bar{x})^2 = 3248$$

Compute the standard error of the regression estimate of $\bar{Y}$.

12.6   For $\rho = 0.95$, verify the data given in Table 12.4 for the optimum percentage that should be matched and for the gain in precision relative to no matching. Compute the corresponding percent gains in precision if one third of the units are retained from the first to the second occasion and one half of the units are retained on each subsequent occasion.

12.7   In simple random sampling on two occasions, suppose that the estimate on the second occasion is, in the notation of section 12.11,

$$\bar{y}_2'' = (1 - \phi)(\bar{y}_1 + \bar{y}_{2m} - \bar{y}_{1m}) + \phi \bar{y}_{2u}$$

(a) Ignoring the fpc, show that

$$V(\bar{y}_2'') = \frac{S^2}{n} \left\{ (1 - \phi)^2 \frac{[1 + \mu(1 - 2\rho)]}{\lambda} + \frac{\phi^2}{\mu} \right\}$$

where $\lambda = m/n$, $\mu = u/n$. (b) For given ρ, λ, μ, find the value of ϕ that minimizes $V(\bar{y}_2'')$. Show that if ρ exceeds $\frac{1}{2}$ the best weight ϕ lies between μ and $\mu/(1 + \mu)$.

12.8   For $\mu = \frac{1}{4}$, $\mu = \frac{1}{2}$, $\rho = 0.8$, and $\rho = 0.9$, compare $V(\bar{y}_2'')$ in the preceding exercise with the variance of the optimum composite regression estimate $\bar{y}_2'$, as given by equation 12.74. (In $\bar{y}_2''$ take $\phi = 0.2$ when $\mu = \frac{1}{4}$ and $\phi = 0.4$ when $\mu = \frac{1}{2}$.) Verify that for these values of ρ the estimate $\bar{y}_2''$ is almost as precise as $\bar{y}_2'$ for both $\mu = \frac{1}{4}$ and $\mu = \frac{1}{2}$.

12.9   An independent sample of size n is drawn each month. From the sample taken in any month, data are obtained for the current and the preceding month. A composite estimate $\bar{y}_h'$ is made as in (12.91), section 12.13.

$$\bar{y}_h' = \bar{y}_h + \phi_h(\bar{y}_{h-1}' - \bar{y}_{h-1,h})$$

The model is

$$y_{hi} = \bar{Y}_h + \rho(y_{h-1,i} - \bar{Y}_{h-1}) + e_{hi}$$

where e_{hi} is independent of the y's and has variance $(1 - \rho^2)$. Show that (a)

$$\bar{y}_h' - \bar{Y}_h = \bar{e}_h + \phi_h(\bar{y}_{h-1}' - \bar{Y}_{h-1}) + (\rho - \phi_h)(\bar{y}_{h-1,h} - \bar{Y}_{h-1})$$

(b) If $V(\bar{y}_h') = g_h S^2/n$, where S^2 is constant on all occasions,

$$g_h = (1 - \rho^2) + \phi_h^2 g_{h-1} + (\rho - \phi_h)^2$$

(c) The optimum $\phi_h = \rho/(1 + g_{h-1})$ and the resulting optimum g_h is

$$g_h = 1 - \frac{\rho^2}{1 + g_{h-1}}$$

(d) The limiting g_h is $g_\infty = \sqrt{1 - \rho^2}$. These results were given by Eckler (1955).

12.10   If $E_{lr} = V(\bar{y})/V(\bar{y}_{lr})$ and $E_R = V(\bar{y})/V(\bar{y}_R)$ are the relative efficiencies of the linear regression and ratio estimates of the sample mean of a simple random sample, show that for both $\bar{y}_{lr}$ and $\bar{y}_R$ the corresponding relative efficiency to $\bar{y}$ in double sampling with optimum choice of n/n' is (ignore $1/N$),

$$E_{ds} = E \left/ \left(1 + \sqrt{\frac{c'}{c}}\sqrt{E - 1}\right)^2 \right.$$

Hence note that with either of these estimators, double sampling will not be highly effective unless c'/c is small (e.g., $< 1/10$). For example, with $c'/c = 1/10$, $E = 6$ gives $E_{ds} = 2.1$.

12.11   In sampling on two occasions, suppose that $S_1 = S_2 = S$ and that the samples are large, so that the regression coefficients of y_{2i} on y_{1i} and of y_{1i} on y_{2i} in the matched part of the samples on the two occasions are both effectively equal to ρ. The estimate $\bar{y}_2'$ in section 12.10 is constructed and an analogous estimate $\bar{y}_1'$ using the regression of y_{1i} on y_{2i}. Show that

(i)
$$V(\bar{y}_2' - \bar{y}_1') = \frac{2S^2(1 - \rho)}{(n - u\rho)}$$

(ii)
$$V(\bar{y}_2' + \bar{y}_1') = \frac{2S^2(1 + \rho)}{(n + u\rho)}$$

(One way of doing this is to express $(\bar{y}_2' \pm \bar{y}_1')$ as linear functions of $(\bar{y}_{2m} \pm \bar{y}_{1m})$ and $(\bar{y}_{2u} \pm \bar{y}_{1u})$, which are uncorrelated).

Note that, as intuition suggests, (i) is minimized when $u = 0$, while (ii) is minimized when $u = n$.

12.12   The most favorable case for the application of the method in exercise 12.1 occurs when the true proportion is 0 in stratum 1. In estimating the total number of units Y_1 in the population that possess an attribute costly to measure, this happens when there is a second attribute, cheap to measure, such that only units having the second attribute can possess the first attribute. In a simple random sample of size n', count the number m' who have attribute 2. Draw a subsample of size $vm' = m'/k$ from these, and count the number r who have attribute 1.

(a) From theorems 12.1 and 12.2, show that $\hat{Y}_1 = Nkr/n$ is an unbiased estimate of Y_1 with variance

$$V(\hat{Y}_1) = \frac{N^2 P_1}{n'} \left[Q_1 \left(1 - \frac{n'}{N} \right) + \frac{P_2(k-1)}{P_1 + P_2} \right]$$

where $P_1, (P_1 + P_2)$ are the population proportions having attributes 1, 2. Assume $1/N(P_1 + P_2)$ negligible.

(b) With $c = 10$, $c' = 1$, the investigator guesses that $P_1 \doteq 0.25$, $P_2 \doteq 0.15$. Is double sampling profitable in this case?

CHAPTER 13

Sources of Error in Surveys

13.1 INTRODUCTION

The theory presented in preceding chapters assumes throughout that some kind of probability sampling is used and that the observation y_i on the ith unit is the correct value for that unit. The error of estimate arises solely from the random sampling variation that is present when n of the units are measured instead of the complete population of N units.

These assumptions hold reasonably well in the simpler types of surveys in which the measuring devices are accurate and the quality of work is high. In complex surveys, particularly when difficult problems of measurement are involved, the assumptions may be far from true. Three additional sources of error that may be present are as follows.

1. Failure to measure some of the units in the chosen sample. This may occur by oversight or, with human populations, because of failure to locate some individuals or their refusal to answer the questions when located.
2. Errors of measurement on a unit. The measuring device may be biased or imprecise. With human populations the respondents may not possess accurate information or they may give biased answers.
3. Errors introduced in editing, coding and tabulating the results.

These sources of error necessitate a modification of the standard theory of sampling. The principal aims of such a modification are to provide guidance about the allocation of resources between the reduction of random sampling errors and the reduction of the other errors and to develop methods for computing standard errors and confidence limits that remain valid when the other errors are present.

13.2 EFFECTS OF NONRESPONSE

We will use the term *nonresponse* to refer to the failure to measure some of the units in the selected sample. In the study of nonresponse it is convienient to think of the population as divided into two "strata", the first consisting of all units for which measurements would be obtained if the units happened to fall in the sample, the second of the units for which no measurements would be obtained. The compositions of the two strata depend intimately on the methods used to find the units and obtain the data. A survey in which at least three calls are made, if necessary, on every house and in which a supervisor with exceptional powers of persuasion calls on all persons who refuse to give data will have a much smaller "nonresponse" stratum than one in which only a single attempt is made for every house.

TABLE 13.1

RESPONSES TO THREE REQUESTS IN A MAILED INQUIRY

	Number of Growers	% of Population	Average Number of Fruit Trees per Grower
Response to first mailing	300	10	456
Response to second mailing	543	17	382
Response to third mailing	434	14	340
Nonrespondents after 3 mailings	1839	59	290
Total population	3116	100	329

This division into two distinct strata is, of course, an oversimplification. Chance plays a part in determining whether a unit is found and measured in a given number of attempts. In a more complete specification of the problem we would attach to each unit a probability representing the chance that it would be measured by a given field method if it fell in the sample.

The sample provides no information about the nonresponse stratum 2. This would not matter if it could be assumed that the characteristics of stratum 2 are the same as those of stratum 1. Where checks have been made, however, it has often been found that units in the "nonresponse" stratum differ from units that are measurable. An illustration appears in Table 13.1. The data come from an experimental sampling of fruit orchards in North Carolina in 1946. Three successive mailings of the same questionnaire were sent to growers. For one of the questions—number of fruit trees—complete data were available for the population (Finkner, 1950).

The steady decline in the number of fruit trees per grower in the successive responses is evident, these numbers being 456 for respondents to the first mailing, 382 in the second mailing, 340 in the third, and 290 for the refusals to all three letters. The total response was poor, more than half the population failing to give data even after three attempts.

We now consider the effects of nonresponse on the sample estimate. Let N_1, N_2 be the numbers of units in the two strata and let $W_1 = N_1/N$, $W_2 = N_2/N$, so that W_2 is the proportion of nonresponse in the population. Assume that a simple random sample is drawn from the population. When the field work is completed, we have data for a simple random sample from stratum 1 but no data from stratum 2. Hence the amount of bias in the sample mean is

$$E(\bar{y}_1) - \bar{Y} = \bar{Y}_1 - \bar{Y} = \bar{Y}_1 - (W_1\bar{Y}_1 + W_2\bar{Y}_2)$$
$$= W_2(\bar{Y}_1 - \bar{Y}_2) \tag{13.1}$$

The amount of bias is the product of the proportion of nonresponse and the difference between the means in the two strata. Since the sample provides no information about $\bar{Y}_2$, the size of the bias is unknown unless bounds can be placed on $\bar{Y}_2$ from some source other than the sample data. With a continuous variate, the only bounds that can be assigned with certainty are often so wide as to be useless.

Consequently, with continuous data, any sizable proportion of nonresponse usually makes it impossible to assign useful confidence limits to $\bar{Y}$ from the sample results. We are left in the position of relying on some guess about the size of the bias, without data to substantiate the guess.

In sampling for proportions the situation is a little easier, since the unknown proportion P_2 in stratum 2 must lie between 0 and 1. If W_2 is known, these bounds for P_2 enable us to construct confidence limits for the population proportion P. Suppose that a simple random sample of n units is drawn and that measurements are obtained for n_1 of the units in the sample. Assuming n_1 large enough, 95% confidence limits for P_1 are given by

$$p_1 \pm 2\sqrt{p_1 q_1/n_1}$$

where p_1 is the sample proportion and the fpc is ignored.

When we try to derive a confidence statement about P, we are on safe ground if we assume $P_2 = 0$ when finding $\hat{P}_L$ and $P_2 = 1$ when finding $\hat{P}_U$. Thus we might take, for 95% limits,

$$\hat{P}_L = W_1(p_1 - 2\sqrt{p_1 q_1/n_1}) + W_2(0) \tag{13.2}$$

$$\hat{P}_U = W_1(p_1 + 2\sqrt{p_1 q_1/n_1}) + W_2(1) \tag{13.3}$$

It is easy to verify that these limits are conservative, that is, that

$$Pr(\hat{P}_L \le P \le \hat{P}_U) > 0.95$$

The limits can be narrowed a little by a more careful argument (Cochran, Mosteller, and Tukey, 1954, p. 280), since P_2 cannot be 0 and 1 simultaneously, as assumed above.

The limits are distressingly wide unless W_2 is very small. Table 13.2 shows the average limits for a sample size $n = 1000$ and a series of values of W_2 and p_1. Since the limits in (13.2) and (13.3) depend on the value of n_1 (number of respondents in the sample), we have taken $n_1 = nW_1$, its average value, in computing Table 13.2.

TABLE 13.2

95% CONFIDENCE LIMITS FOR P (%) WHEN $n = 1000$

% Nonresponse, $100W_2$	Sample Percentage, $100p_1$			
	5	10	20	50
0	(3.6, 6.4)	(8.1, 11.9)	(17.5, 22.5)	(46.7, 53.2)
5	(3.4, 11.1)	(7.6, 16.3)	(16.5, 26.5)	(44.4, 55.6)
10	(3.2, 15.8)	(7.2, 20.8)	(15.6, 30.4)	(42.0, 58.0)
15	(3.0, 20.5)	(6.8, 25.2)	(14.7, 34.3)	(39.6, 60.4)
20	(2.8, 25.2)	(6.3, 29.7)	(13.7, 38.3)	(37.2, 62.8)

The rapid increase in the width of the confidence interval with increasing W_2 is evident. It is of interest to examine what values of n would be needed to give the same widths of confidence interval if W_2 were zero. This is easily done when p_1 is 50%. For $W_2 = 5\%$, Table 13.2 shows that the half-width of the confidence interval is 5.6. The equivalent sample size n_e, assuming no nonresponse, is found from the equation

$$5.6 = 2\sqrt{(50)(50)/n_e}$$

$$n_e = 320$$

For $W_2 = 10, 15$, and 20%, the values of n_e are 155, 90, and 60, respectively. It is evidently worthwhile to devote a substantial proportion of the resources to the reduction of nonresponse.

If the population nonresponse rate W_2 is not known, as will usually be the case, conservative confidence limits can be calculated from the sample data by a method suggested by a student. In calculating the lower limit, assume that all sample nonrespondents would have given a negative response. In calculating the upper limit, assume that sample nonrespondents would have given a positive response. For example, suppose $n = 1000$, $n_1 = 800$, and $p_1 = 10\%$, so that 80 sample members give a positive response and the sample nonresponse rate is 20%. Then, in percents,

$$\hat{P}_L = 8 - 2\sqrt{(8)(92)/1000} = 6.3\%$$

$$\hat{P}_U = 28 + 2\sqrt{(28)(72)/1000} = 30.8\%$$

The limits are a little wider than those for $p_1 = 10\%$, $W_2 = 20\%$ in Table 13.2.

If W_2 is known from previous experience in the particular type of survey, Birnbaum and Sirken (1950a, 1950b) give a method of finding the sample size n that guarantees with risk α an absolute error in the sample proportion less than a specified amount d. No advance knowledge of P_1, P_2, or P is assumed. If there were no nonresponse, we would take (by section 4.4),

$$n = t_\alpha^2 PQ/d^2 \tag{13.4}$$

where t_α is the normal deviate corresponding to the risk that the error exceeds d. With no advance information about P, the least favorable case is $P = 0.5$, giving

$$n = \frac{t_\alpha^2}{4 d^2} \tag{13.5}$$

By taking the least favorable combination of the bias $W_2(P_1 - P_2)$ and the value of P_1, Birnbaum and Sirken show that a value of n that still guarantees an error less than d, with risk α, is approximately

$$n \doteq \frac{t_\alpha^2}{4d(d - W_2)W_1} - 1 \tag{13.6}$$

Note that no value of n suffices if $W_2 > d$. If $W_2 = 0$, this equation reduces to (13.5) apart from the term -1, which comes from an approximation in the analysis. Some values of n given by Birnbaum and Sirken's method are shown in Table 13.3.

This table tells the same sad story as Table 13.2. If we are content with a crude estimate ($d = 20$), amounts of nonresponse up to 10% can be handled by doubling the sample size. However, any sizable percentage of nonresponse makes it impossible or very costly to attain a highly guaranteed precision by increasing the sample size among the respondents.

TABLE 13.3

Smallest Value of n for Given Limit of Error d, with Risk $\alpha = 0.05$

% Nonresponse, $100 W_2$	d (%)			
	20	15	10	5
0	24	43	96	384
2	27	50	122	653
4	31	60	166	2000
6	36	75	255	....
8	43	99	521	....
10	53	142	...	....
15	112	...	...	....

13.3 TYPES OF NONRESPONSE

Some methods for handling the nonresponse problem are described in succeeding sections. A rough classification of the types of nonresponse is as follows.

1. *Noncoverage.* This is failure to locate or to visit some units in the sample.

This is a problem with areal sampling units, in which the interviewer must find and list all dwellings (according to some definition) in a city block. It arises also from the use of incomplete lists. Sometimes weather or poor transportation facilities make it impossible to reach certain units during the period of the survey.

2. *Not-at-homes.* This group contains persons who reside at home but are temporarily away from the house. Families in which both parents work and families without children are harder to reach than families with very young children or with old people confined to the house.

3. *Unable to answer.* The respondent may not have the information wanted in certain questions or may be unwilling to give it. Skillful wording and pretesting of the questionnaire are a safeguard.

4. *The "hard core."* Persons who adamantly refuse to be interviewed, who are incapacitated, or who are far from home during the whole time available for field work constitute this sector. It represents a source of bias that persists no matter how much effort is put into completeness of returns.

The detection and measurement of noncoverage are difficult. With areal sampling, one method is to revisit the primary units, making a careful listing that serves as a check. Comparisons of counts of numbers of people or dwellings with those in another survey sometimes give a warning that some have been missed. When the principal frame is a directory of street addresses, it may be supplemented by an area sample, the purpose of which is to sample parts of the town (new building) not adequately covered by the directory and, in parts in which the directory seems accurate, to search for addresses missing from the directory. Surveys using these methods, with a discussion of the problem of noncoverage, are described by Kish and Hess (1958) and Woolsey (1956).

In regard to the not-at-homes, the problem is easier in surveys in which *any* adult in the home is capable of answering the questions than in those in which a single adult, chosen at random, is to be interviewed. A single adult is usually preferred if the survey is one of individuals in which one person cannot report accurately for another or if intrahousehold correlations are expected to be high, so that the measurement of more than one person per family is uneconomical. In this connection, a useful method of selecting a single person from a household was developed by Kish (1949). The original plan was intended for households that may contain as many as six eligible persons, but the procedure can be adapted for smaller or larger households.

The interviewer lists on the schedule the eligible persons in the household and then numbers them: males first in order of decreasing age, then females in order of decreasing age. Each schedule has printed on it *one* of the sets of instructions in Table 13.4.

Each eligible person in a household of a given size has an equal chance of being selected, except that adults 3 and 5 in households of size 5 are slightly overrepresented. Since male respondents are concentrated in Tables A, B, and C, the interviewer can devote evening calls to households so designated.

TABLE 13.4

INSTRUCTIONS FOR SELECTING A SINGLE RESPONDENT

Relative Frequency of Use	Table Number	If Number of Adults in Household is					
		1	2	3	4	5	6
		Select Adult Numbered					
1/6	A	1	1	1	1	1	1
1/12	B1	1	1	1	1	2	2
1/12	B2	1	1	1	2	2	2
1/6	C	1	1	2	2	3	3
1/6	D	1	2	2	3	4	4
1/12	E1	1	2	3	3	3	5
1/12	E2	1	2	3	4	5	5
1/6	F	1	2	3	4	5	6

13.4 CALL-BACKS

A standard technique is to specify the number of call-backs, or a minimum number, that must be made on any unit before abandoning it as "unable to contact." Stephan and McCarthy (1958) give data from a number of surveys on the percentage of the total sample obtained at each call. Average results are shown in Table 13.5.

TABLE 13.5

NUMBER OF CALLS REQUIRED FOR COMPLETED INTERVIEWS

Respondent	% of Sample contacted on			Per Cent Nonresponse	Total
	First Call	Second Call	Third or Later Call		
Any adult*	70	17	8	5	100
Random adult	37	32	23	8	100

*Two surveys in which the respondent was a housewife and a farm operator, respectively, have been included in the "any adult" group.

In surveys in which any adult in the house could answer the questions, the first call obtained about 70% of the sample and the first two calls, 87%. The increased cost of sampling when a randomly chosen adult is to be interviewed is evident, the first call producing only 37% of the required interviews. The marked success of the second call reflects the work of the interviewer in finding out in advance when the desired respondent would be at home and available.

Little has been published on the relative costs of later calls to the first call. Later calls would be expected to be more expensive per completed interview, since the houses are more sparsely located in the area assigned to the interviewer and since the occupants are presumably people who spend more than an average amount of time away from home. From British experience, Durbin (1954) suggests that later calls may be less expensive than would be anticipated. The following figures (Table 13.6) show estimated relative costs per completed interview (i.e., money spent on ith calls divided by number of new interviews obtained) for each call up to the fifth in a special study reported by Durbin and Stuart (1954).

TABLE 13.6

RELATIVE COSTS PER NEW COMPLETED INTERVIEW AT THE iTH CALL

Call	1	2	3	4	5
Relative cost	100	112	127	151	250

The estimation of these costs requires care. If the desired respondent is not at home at the first call, the interviewer may spend time inquiring when this person will be at home and making a tentative appointment. In the costing such time should be assigned to the second call instead of to an unsuccessful first call.

A more useful measure is the average cost per completed interview *over all interviews obtained up to the ith call*. These figures give the relative costs of obtaining n completed interviews when we insist on i calls before the interviewer gives up. In order to compute these figures, we must know how many interviews are obtained at each call. In Table 13.7 these calculations are made under two sets of assumptions. The first simulates surveys in which any adult can answer the questions, the second those demanding a random adult. The data on numbers of interviews obtained were taken from Table 13.5.

The details of the calculation are shown only for the first assumption, the method being exactly the same for the second. The symbol n_o denotes the *original* sample size.

Insistence on up to three calls costs only 4% more *per completed interview* than single calls if any adult is a satisfactory respondent, and only 10% more if a random adult must be interviewed. How typical these results are is not known, but the method provides realistic estimates of the cost of insisting on call-backs if the necessary cost and sample size data have been collected. There is also the time factor: call-backs delay the final results.

TABLE 13.7

RELATIVE COSTS PER COMPLETED INTERVIEW UP TO THE ith CALL

Respondent = Any Adult

Call	Relative Cost	At ith Call		Up to ith Call			"Random" Adult	
		No. of Ints.*	Cost of Ints.	Total No. of Ints.	Total Cost	Cost per Int.	No. of Ints.	Cost per Int.
1	100	$0.70n_0$	$70n_0$	$0.70n_0$	$70n_0$	100	$0.37n_0$	100
2	112	$0.17n_0$	$19.04n_0$	$0.87n_0$	$89.04n_0$	102	$0.32n_0$	106
3	127	$0.07n_0$	$8.89n_0$	$0.94n_0$	$97.93n_0$	104	$0.16n_0$	110
4	151	$0.04n_0$	$6.04n_0$	$0.98n_0$	$103.97n_0$	106	$0.09n_0$	114
5	250	$0.02n_0$	$5.00n_0$	$1.00n_0$	$108.97n_0$	109	$0.06n_0$	122

* Interviews

13.5 A MATHEMATICAL MODEL OF THE EFFECTS OF CALL-BACKS

Deming (1953) developed a useful and flexible mathematical model for examining in more detail the consequences of different call-back policies. The population is divided into r classes, according to the probability that the respondent will be found at home. Let

w_{ij} = probability that a respondent in the jth class will be reached on or before the ith call

p_j = proportion of the population falling in the jth class

μ_j = item mean for the jth class

σ_j^2 = item variance for the jth class

For simplicity we assume $w_{ij} > 0$ for all classes, although the method is easily adapted to include persons impossible to reach. If $\bar{y}_{ij}$ is the mean for those in class j who were reached on or before the ith call, it is also assumed that $E(\bar{y}_{ij}) = \mu_j$.

The true population mean for the item is

$$\bar{\mu} = \sum_j p_j \mu_j \tag{13.7}$$

Consider the composition of the sample after i calls. The persons in the sample can be classified into $(r+1)$ classes as follows: in the first class and interviewed; in the second class and interviewed; and so on. The $(r+1)$th class consists of all those not yet interviewed after i calls. If the fpc is ignored, the numbers falling in these

$(r+1)$ classes are distributed according to the multinomial

$$[w_{i1}p_1 + w_{i2}p_2 + \cdots + w_{ir}p_r + (1 - \textstyle\sum w_{ij}p_j)]^{n_o}$$

where n_o is the initial size of the sample

It follows that the number n_i who have been interviewed in the course of i calls is binomially distributed with number of trials $= n_o$ and probability of success $\sum w_{ij}p_j$. Hence

$$E(n_i) = \text{expected number of interviews in } i \text{ calls} = n_o \sum_j^r w_{ij}p_j \qquad (13.8)$$

For fixed n_i, the numbers of interviews n_{ij} obtained $(j = 1, 2, \ldots, r)$ follow a multinomial with probabilities $w_{ij}p_j / \sum w_{ij}p_j$. It follows that

$$E(n_{ij}|n_i) = \frac{n_i w_{ij}p_j}{\sum w_{ij}p_j}$$

Hence, if $\bar{y}_i$ is the sample mean obtained after i calls,

$$E(\bar{y}_i|n_i) = E\left(\frac{\sum n_{ij}\bar{y}_{ij}}{n_i}\right) = \frac{\sum n_i w_{ij}p_j\mu_j}{n_i \sum w_{ij}p_j} = \frac{\sum w_{ij}p_j\mu_j}{\sum w_{ij}p_j} = \bar{\mu}_i \qquad (13.9)$$

Since this result does not depend on n_i, the unconditional mean of $\bar{y}_i$ is also $\bar{\mu}_i$. The bias in the estimate $\bar{y}$ is therefore $(\bar{\mu}_i - \bar{\mu})$.

The conditional variance of $\bar{y}_i$ for given n_i is found similarly to be

$$V(\bar{y}_i|n_i) = \frac{\sum_j^r w_{ij}p_j[\sigma_j^2 + (\mu_j - \bar{\mu}_i)^2]}{n_i \sum_j^r w_{ij}p_j} \qquad (13.10)$$

The unconditional variance, ignoring terms of order $1/n_i^2$, is given approximately by replacing n_i in (13.10) by its expected value form (13.8).

Finally, the mean square error of the estimate obtained after i calls is

$$\text{MSE}(\bar{y}_i|i) = V(\bar{y}_i|i) + (\bar{\mu}_i - \bar{\mu})^2 \qquad (13.11)$$

The cost of making i calls must also be considered. The expected number of *new* interviews obtained in the kth call is $\sum (w_{kj} - w_{k-1,j})p_j$. Hence, if c_k is the cost per completed interview at the kth call, the expected total cost of making i calls is $n_o C(i)$, where

$$C(i) = c_1 \sum w_{1j}p_j + c_2 \sum (w_{2j} - w_{1j})p_j + \cdots + c_i \sum (w_{ij} - w_{i-1,j})p_j$$

Example.   A population with three classes is shown in Table 13.8. The p_j and w_{ij} are intended to represent surveys in which a random adult is interviewed. At the first call the probabilities w_{1j} of obtaining an interview are taken as 0.6, 0.3, and 0.1 in the three classes. At the second and subsequent calls, the conditional probabilities of interviewing a person missed previously are 0.9, 0.5, and 0.2. These figures were made higher than the corresponding probabilities at the first call in order to represent the effect of intelligent inquiry by the interviewer.

TABLE 13.8

CHARACTERISTICS OF THE THREE CLASSES

	Class		
	1	2	3
p_j	0.45	0.50	0.05
w_{ij}	$0.6 + (0.4)[1 - (0.1)^{i-1}]$	$0.3 + (0.7)[1 - (0.5)^{i-1}]$	$0.1 + (0.9)[1 - (0.8)^{i-1}]$
I μ_j	55	50	45
II μ_j	60	50	40

TABLE 13.9

NUMBER OF INTERVIEWS, COSTS PER INTERVIEW AND BIASES

Number of Calls Required	Number of Interviews Obtained	Average Cost per Interview	I Bias	II Bias
1	$0.425n_o$	100	+1.118	+2.235
2	$0.771n_o$	105	+0.711	+1.421
3	$0.882n_o$	108	+0.421	+0.842
4	$0.933n_o$	110	+0.266	+0.532
5	$0.960n_o$	114	+0.180	+0.360

The item being estimated is a binomial percentage close to 50%. Two sets of μ_j are considered (I, II). For simplicity, the within-class variances $\sigma_j^2 = \mu_j(100 - \mu_j)$ were all taken as 2500. The relative costs per completed interview at successive calls were those given in Table 13.6.

Table 13.9 shows (a) expected total number of interviews obtained for a total of i calls, (b) the average cost of these calls per interview, and (c) the bias $(\bar{\mu}_i - \bar{\mu})$ in the estimate $\bar{y}$ under assumptions I and II about the μ_j.

In II, for example, the true population mean $\bar{\mu}$ is 54%. The mean $\bar{\mu}_1$ obtained from first calls is 56.235%, giving the bias of +2.235% shown in the table. A policy that requires three calls reduces this bias to +0.842%.

The values of MSE$(\bar{y})$ obtained from a given expenditure of money were compared for the different call-back policies. In the first comparisons the amount of money is sufficient to take $n_o = 500$ if only one call is made. From Table 13.9 the expected number of interviews obtained in the first call is $E(n_1) = (500)(0.425) = 212.5$. If two calls are made, this expected number must be reduced to $E(n_2) = 212.5/1.05 = 202.4$. to maintain the same cost, and similarly for 3, 4, and 5 call-backs. These values of $E(n_i)$ were substituted in equation (13.10) to give $V(\bar{y})$ and hence MSE$(\bar{y})$.

TABLE 13.10

VALUES OF MSE($\bar{y}$) FOR DIFFERENT CALL-BACK POLICIES
COSTING THE SAME AMOUNT

Number of Calls Required	$n_o = 500$ (for first calls only)			$n_o = 1000$		$n_o = 2000$	
	No Bias	I*	II*	I	II	I	II
1	**11.8**	13.0	16.9	7.1	10.9	4.2	8.0
2	12.4	12.9	14.6	6.7	8.3	3.6	5.2
3	12.7	**12.9**	13.6	**6.5**	7.1	3.4	3.9
4	13.0	13.1	**13.4**	6.6	**6.9**	**3.3**	3.6
5	13.5	13.5	13.8	6.8	6.9	3.4	**3.5**

* These represent populations with smaller (I) and greater (II) amounts of bias, as defined in Table 13.8.

Table 13.10 presents the resulting MSE's for three amounts of expenditure, corresponding to $n_o = 500$, 1000, 2000 for a single call. When $n_o = 500$, the values of MSE($\bar{y}$) are also given for the "no bias" situation in which every $\mu_j = 50$. This column shows the effect of call-backs when they are unnecessary, since no bias results from confining the survey to a single call.

The policies giving the lowest MSE's are shown in boldface type. Consider first the smallest sample size, $n_o = 500$. If call-backs are unnecessary, a policy demanding as many as four call-backs results in only a modest increase in the MSE. In I, involving the smaller amount of bias, the different policies produce about the same accuracy, although three is the optimum. In II, three to five call-backs are satisfactory, a single call giving a MSE about 25% above the minimum.

For the larger sample sizes the optimum number of call-backs increases to four or five, and the use of a single call results in more substantial losses of accuracy.

This is, of course, only an illustration. The importance of the method is that as information accumulates about costs and relative biases an economical policy can be worked out for any specific type of survey.

13.6 OPTIMUM SAMPLING FRACTION AMONG THE NONRESPONDENTS

After the first attempt to reach the persons in the sample has been made, another approach, due to Hansen and Hurwitz (1946), is to take a random subsample of the persons who have not been reached and make a major effort to interview everyone in the subsample. This technique was first developed for surveys in which the initial attempt was made by mail, a subsample of persons who did not return the completed questionnaire being approached by the more expensive method of a personal interview.

This method can be regarded as an application of the technique of double sampling for stratification presented in section 12.2 We first take a simple random sample of n' units. Let n_1' be the number of units in the sample that provide the data sought and n_2' the number in the nonresponse group. By intensive efforts, the data are later obtained from a random subsample $n_2 = v_2 n_2'$ out of the n_2'. Hansen and Hurwitz use the notation $v_2 = 1/k$, so that $n_2 = n_2'/k$.

In the framework of section 12.2 there are two strata. Stratum 1 consists of those who would respond to a first attempt, with a measured sample of size $n_1 = n_1'$, so that $v_1 = 1$. Stratum 2 consists of those who would respond to the second attempt, with $n_2 = n_2'/k$.

The cost of taking the sample is

$$c_0 n' + c_1 n_1' + \frac{c_2 n_2'}{k} \tag{13.12}$$

where the c's are the costs per unit: c_0 is the cost of making the first attempt, c_1 is the cost of processing the results from the first attempt, and c_2 is the cost of getting and processing the data in the second stratum. If W_1, W_2 are the population proportions in the two strata, the expected cost is

$$C = c_0 n' + c_1 W_1 n' + \frac{c_2 W_2 n'}{k} \tag{13.13}$$

As an estimate of $\bar{Y}$, we take

$$\bar{y}' = w_1 \bar{y}_1 + w_2 \bar{y}_2 = \frac{(n_1' \bar{y}_1 + n_2' \bar{y}_2)}{n'} \tag{13.14}$$

where $\bar{y}_1$, $\bar{y}_2$ are the means of the samples of sizes $n_1 = n_1'$ and $n_2 = n_2'/k$. By theorem 12.1, the estimate $\bar{y}'$ will be unbiased if responses are obtained from all the selected random subsample of size $n_2 = n_2'/k$.

By formula (12.3), the variance of $\bar{y}'$ is

$$V(\bar{y}') = \left(\frac{1}{n'} - \frac{1}{N}\right)S^2 + \frac{W_2 S_2^2}{n'}\left(\frac{1}{v_2} - 1\right)$$

$$= \left(\frac{1}{n'} - \frac{1}{N}\right)S^2 + \frac{(k-1)W_2 S_2^2}{n'} \tag{13.15}$$

The quantities n' and k are then chosen to minimize the product $C(V + S^2/N)$. From (13.15) and (13.13) we have

$$V + S^2/N = \frac{(S^2 - W_2 S_2^2)}{n'} + \frac{k W_2 S_2^2}{n'} \tag{13.16}$$

$$C = (c_0 + c_1 W_1)n' + \frac{c_2 W_2 n'}{k} \tag{13.17}$$

By Cauchy's inequality, the optimum k is

$$k_{opt} = \sqrt{\frac{c_2(S^2 - W_2 S_2{}^2)}{S_2{}^2(c_0 + c_1 W_1)}}$$ (13.18)

The initial sample size n' may be chosen either to minimize C for specified V or V for specified C by solving for n' from (13.16) or (13.17). If V is specified,

$$n'_{opt} = \frac{N[S^2 + (k-1)W_2 S_2{}^2]}{(NV + S^2)}$$ (13.19)

where V is the value specified for the variance of the estimated population mean.

The solutions require a knowledge of W_2: this can often be estimated from previous experience. In addition to S^2, whose value must be estimated in advance in any "sample size" problem, the solutions also involve $S_2{}^2$, the variance in the nonresponse stratum. The value of $S_2{}^2$ may be harder to predict; it will probably not be the same as S^2. For instance, in surveys made by mail of most kinds of economic enterprise, the respondents tend to be larger operators, with larger between-unit variances than the nonrespondents.

If W_2 is not well known, a satisfactory approximation is to work out the value of n'_{opt} from a provisional (13.18) and (13.19) for a range of assumed values of W_2 between 0 and a safe upper limit. The maximum n'_{opt} in this series is adopted as the initial sample size n'. When the replies to the mail survey have been received, the value of n_2' is known. In seeking the value of k to be used with this method, we use the variance $v_c(\bar{y}')$ conditional on the known values of n_2 and n'. This can be obtained from (12.4) and 12.7) as

$$V_c(\bar{y}') = S^2\left(\frac{1}{n'} - \frac{1}{N}\right) + \frac{(k-1)n_2' S_2{}^2}{n'^2}$$ (13.20)

Equation (13.20) is solved to find the k that gives the desired conditional variance. The cost for this method is usually only slightly higher than the optimum cost for known W_2.

Example.   This example is condensed from the paper by Hansen and Hurwitz (1946). The first sample is taken by mail and the response rate W_1 is expected to be 50%. The precision desired is that which would be given by a simple random sample of size 1000 if there were no nonresponse. The cost of mailing a questionnaire is 10 cents, and the cost of processing the completed questionnaire is 40 cents. To carry out a personal interview costs $4.10.

How many questionnaires should be sent out and what percentage of the nonrespondents should be interviewed?

In terms of the cost function (13.12) the unit costs in dollars are as follows.

c_0 = cost of first attempt   = 0.1
c_1 = cost of processing data for a respondent = 0.4
c_2 = cost of obtaining and processing data
      for a nonrespondent   = 4.5

The optimum n' and k can be found from (13.19) and (13.18). If the variances S^2 and S_2^2 are assumed equal and N is assumed to be large, then

$$k_{opt} = \sqrt{\frac{c_2(1-W_2)}{c_0+c_1 W_1}} = \sqrt{\frac{(4.5)(0.5)}{0.1+(0.4)(0.5)}} = \sqrt{7.5} = 2.739$$

$$n'_{opt} = \frac{S^2[1+(k-1)W_2]}{V} = 1000\{1+(1.739)(0.5)\}$$

$$= 1870$$

Note that we have put $S^2/V = 1000$, or $V = S^2/1000$, since this is the variance that the sample mean would have if a sample of 1000 were taken and complete response were obtained.

Consequently, 1870 questionnaires should be mailed. Of the 935 that are not returned, we interview a random subsample of 935/2.739, or 341. The cost is $2095.

As Durbin (1954) has pointed out, subsampling is unlikely to show a marked profit unless c_2 is large in relation to $(c_0+c_1 W_1)$. The two quantities are comparable, since $(c_0+c_1 W_1)$ is the expected cost per unit of making the first attempt and processing the results and c_2 has the same meaning for the second attempt. From the equations it can be shown that the ratio of the cost of obtaining a prescribed V with $k = 1$ (no subsampling) to the minimum cost for optimum k is

$$\frac{S^2(c_0+c_1 W_1+c_2 W_2)}{[\sqrt{(S^2-W_2 S_2^2)(c_0+c_1 W_1)}+\sqrt{c_2 W_2 S_2^2}]^2} = \frac{c_0+c_1 W_1+c_2 W_2}{[\sqrt{W_1(c_0+c_1 W_1)}+\sqrt{c_2 W_2}]^2}$$

if S^2 and S_2^2 are approximately equal. If r is the ratio of c_2 to $(c_0+c_1 W_1)$, the cost ratio becomes

$$\frac{1+rW_2}{(\sqrt{W_1}+\sqrt{rW_2})^2}$$

For instance, the cost ratio is 1.029 for $W_1 = 0.5$ if $r = 4$, 1.146 if $r = 10$, and 1.228 if $r = 16$. If, however, S^2 is substantially greater than S_2^2, there is more to be gained from subsampling.

With stratified sampling, the optimum values of the n_h' and the k_h in the individual strata are rather complex. A good approximation is to estimate first, by the methods in sections 5.5 and 5.9, the sample sizes n_{oh} that would be required in the strata if there were no nonresponse. Now, from (13.19), if $W_2 = 0$, we have

$$n_o = \frac{NS^2}{NV+S^2} \tag{13.21}$$

Hence (13.19) can be written as

$$n'_{opt} = n_o\left[1+\frac{(k-1)W_2 S_2^2}{S^2}\right] \tag{13.22}$$

This equation, applied separately to each stratum, gives an approximation to the optimum n_h'. The values of k_h are found by applying (13.18) in each stratum.

These techniques can be used with ratio or regression estimates. With the ratio estimate, the quantities S^2 and S_2^2 are replaced by S_d^2 and S_{2d}^2, where $d_i = y_i - Rx_i$. With a regression estimate, S^2 becomes $S^2(1-\rho^2)$ and S_2^2 becomes $S_2^2(1-\rho^2)$.

13.7 ADJUSTMENTS FOR BIAS WITHOUT CALL-BACKS

An ingenious method of diminishing the biases present in the results of the first call was suggested by Hartley (1946) and developed by Politz and Simmons (1949, 1950) and Simmons (1954). Suppose that all calls are made during the evening on the six week-nights. The respondent is asked whether he was at home, at the time of the interview, on each of the five preceding weeknights. If the respondent states that he was at home t nights out of five, the ratio $(t+1)/6$ is taken as an estimate of the frequency π with which he is at home during interviewing hours.

The results from the first call are sorted into six groups according to the value of t, (0, 1, 2, 3, 4, 5). In the tth group let n_t be the number of interviews obtained and $\bar{y}_t$, the item mean. The Politz-Simmons estimate of the population mean μ is

$$\bar{y}_{PS} = \frac{\sum_{t=0}^{5} 6n_t\bar{y}_t/(t+1)}{\sum_{t=0}^{5} 6n_t/(t+1)} \tag{13.23}$$

This approach recognizes that the first call results are unduly weighted with persons who are at home most of the time. Since a person who is at home, on the average, a proportion π of the time has a relative chance π of appearing in the sample, his response should receive a weight $1/\pi$. The quantity $6/(t+1)$ is used as an estimate of $1/\pi$. Thus $\bar{y}_{PS}$ is less biased than the sample mean $\bar{y}$ from the first call, but its variance is greater, because an unweighted mean is replaced by a weighted mean with estimated weights.

In presenting the mean and variance of $\bar{y}_{PS}$, we use the notation of section 13.5. The population is divided into classes, people in the jth class being at home a fraction π_j of the time. Note that the tth group (i.e., persons at home t nights out of the preceding five) will contain persons from various classes. Let n_{jt}, $\bar{y}_{jt}$ be the number and the item mean for those in class j and group t. Then $\bar{y}_{PS}$ may be written as follows.

$$\bar{y}_{PS} = \frac{\sum\sum 6n_{jt}\bar{y}_{jt}/(t+1)}{\sum\sum 6n_{jt}/(t+1)} = \frac{N}{D} \quad \text{(say)} \tag{13.24}$$

This is a ratio type of estimate. In large samples its mean is approximately $E(N)/E(D)$.

If n_o is the initial size of sample (responses plus not-at-homes) and n_j is the number from class j who are interviewed, the following assumptions are made.

(i) $\dfrac{n_j}{n_o}$ is a binomial estimate of $p_j \pi_j$

(ii) $E(n_{jt}|n_j) = n_j \dfrac{5!}{t!(5-t)!} \pi_j^t (1-\pi_j)^{5-t}$

(iii) $E(\bar{y}_{jt}) = \mu_j$, for any j and t

Assumption (ii) is open to question. Without giving a detailed discussion, it assumes that people report correctly whether they were at home.

For given j, using assumption (ii),

$$E \sum_{t=1}^{5} n_{jt}\left(\frac{6}{t+1}\right) = n_j \sum_{t=1}^{5} \left(\frac{6}{t+1}\right)\frac{5!}{t!(5-t)!} \pi_j^t (1-\pi_j)^{5-t} \qquad (13.25)$$

$$= \frac{n_j}{\pi_j}[1-(1-\pi_j)^6] \qquad (13.26)$$

Hence

$$E(D) = \sum_{j=1}^{r} \frac{E(n_j)}{\pi_j}[1-(1-\pi_j)^6] = n_o \sum_{j=1}^{r} p_j[1-(1-\pi_j)^6] \qquad (13.27)$$

using assumption (i). Furthermore, since $E(\bar{y}_{jt}) = \mu_j$ for any j and t, this gives the result

$$E(\bar{y}_{PS}) = \bar{\mu}_{PS} \doteq \frac{\displaystyle\sum_{j=1}^{r} p_j \mu_j[1-(1-\pi_j)^6]}{\displaystyle\sum_{j=1}^{r} p_j[1-(1-\pi_j)^6]} \qquad (13.28)$$

Since the true mean $\bar{\mu} = \sum p_j\mu_j$, some bias remains in $\bar{y}_{PS}$. In a certain sense, this estimate has the same bias as $\bar{y}_6$, the sample mean given by the call-back method with a requirement that as many as six calls be made if necessary. In section 13.5, equation (13.9), it was shown that the call-back method, with a total of i calls, gives an unbiased estimate of $\bar{\mu}_{ji} = \sum w_{ij}p_j\mu_j/\sum w_{ij}p_j$, where w_{ij} is the probability that a person in class j who falls in the sample will be interviewed. Now $w_{1j} = \pi_j$. If at subsequent calls the probability of finding at home a person not previously reached remains at π_j, then

$$w_{ij} = [1-(1-\pi_j)^i]$$

so that $\bar{\mu}_{PS} = \bar{\mu}_6$. However, with the call-back method the probability of an interview at a later call may be greater than π_j as a result of information obtained by the interviewer at the first or earlier calls. In this event the call-back method has less bias after six calls.

The variance of $\bar{y}_{PS}$ is rather complicated. With the usual approximation for a ratio estimate, it may be expressed, following Deming (1953), as

$$V(\bar{y}_{PS}) \doteq \frac{1}{n_o U}\{\sum \pi_j p_j B_j [\sigma_j^2 + (\mu_j - \bar{\mu}_{PS})^2]$$

$$+ (n_o - 1) \sum (\pi_j p_j)^2 (B_j - A_j^2)(\mu_j - \bar{\mu}_{PS})^2\}$$

where

$$U = 1 - \sum p_j (1 - \pi_j)^6$$

$$A_j = \frac{1}{\pi_j}[1 - (1 - \pi_j)^6]$$

$$B_j = \sum_{t=0}^{5} \left[\frac{6}{(1+t)}\right]^2 \frac{5!}{t!(5-t)!} \pi_j^t (1 - \pi_j)^{5-t}$$

Although this expression is difficult to appraise without applying it to specific populations, two comments can be made. If the μ_j do not differ greatly, that is, if the bias from first calls is moderate, the dominating term is the first.

$$\frac{1}{n_o U} \sum \pi_j p_j B_j \sigma_j^2$$

This expression tends to be 25 to 35% higher than the variance of the unweighted mean of the first calls. Also, $V(\bar{y}_{PS})$ contains a term that does not decrease as n_o increases and becomes important in very large samples.

To summarize, comparisons made on simulated populations by Deming (1953), Durbin (1954), and myself suggest that this method shows to best advantage, in relation to call-backs, when the biases from early calls are substantial and the sample is large. The reductions in MSE for the same outlay are small, however, unless call-backs cost substantially more than postulated here. The Politz-Simmons technique has the advantage of saving time. Errors and incompleteness in the values of t, not considered in the analysis, are a disadvantage. The method may also be applied, as suggested by Simmons (1954), in conjunction with several call-backs.

Several other methods for mitigating the "not-at-home" bias have been proposed. Bartholomew's (1961) applies to a survey with two calls. He supposes that, for those not at home on the first call, the interviewer, by careful inquiry, can make the probability of finding them on the second call approximately equal. If this is so, the n_2 persons interviewed at the second call are a random subsample of the $(n_o - n_1)$ persons missed at the first call. Hence $[n_1 \bar{y}_1 + (n_o - n_1)\bar{y}_2]/n_o$ is an unbiased estimate of the mean of the initial target sample. The method worked well on some British surveys to which Bartholomew applied it. In repeated surveys Kish and Hess (1959a) suggest that nonresponses from recent surveys may serve as a replacement for nonresponses in a current survey. Wherever the bias from early calls shows a systematic pattern, as in Table 13.1, Hendricks (1949) has outlined extrapolation methods to estimate the average results that would be given by nonrespondents.

13.8   A MATHEMATICAL MODEL FOR ERRORS OF MEASUREMENT

Conceptually, we can imagine that a large number of independent repetitions of the measurement on the ith unit are possible. Let $y_{i\alpha}$ be the value obtained in the αth repetition. Then

$$y_{i\alpha} = \mu_i + e_{i\alpha} \qquad (13.29)$$

where μ_i = correct value

$e_{i\alpha}$ = error of measurement.

The idea of a "correct value" requires a little discussion. With some items the concept is simple and concrete. For instance, in an inventory taken by sampling, the correct value may be the number of fan belts lying on a shelf at 12 noon on a specified day. In some cases the correct value can be defined operationally. A person's correct diastolic blood pressure at a specified time might be defined as the value obtained when it is measured by a certain standard instrument under carefully prescribed conditions. We may realize, however, that our standard instrument is itself subject to errors of measurement, and we may expect that in course of time a more precise instrument will be developed. With other items, for instance, some aspect of an employee's attitude toward his employer or of a person's feelings of ability to cope with his day-to-day problems, nobody may claim to have a satisfactory method of measuring the "correct value." Nevertheless, the concept is useful even in such cases.

Under repeated measurements of the same unit, the errors $e_{i\alpha}$ will follow a frequency distribution. For the ith unit, let $e_{i\alpha}$ have mean β_i and variance σ_i^2. The term β_i represents a bias in the measurements. The magnitudes of β_i and σ_i^2 will, of course, depend on the nature of the item being measured and on the measuring instrument. They may depend also on numerous other factors. With human populations the prevailing economic and political climate and the amount and type of advance publicity received by the survey may influence the responses to the questionnaire.

The next step is to consider how the errors of measurement change when we move from one unit to another. Various complications can arise.

For the bias component β_i, there may be a constant bias, say, $E(\beta_i) = \beta$, that affects all units in the population. There will also be a component $(\beta_i - \beta)$ that follows a frequency distribution over the population. This component may be correlated with the correct value μ_i; for instance, the measuring device may consistently underestimate high values of μ_i and overestimate low values.

There may be a correlation between the values of $e_{i\alpha}$ on different units in the same sample. The simplest example is the "interviewer bias." Dramatic differences are sometimes found in the mean values of $y_{i\alpha}$ obtained by different interviewers who are sampling comparable parts of the same population (see Lienau, 1941, Mahalanobis, 1946, and Barr, 1957).

A similar effect has appeared when samples of a growing crop are cut by different teams and when chemical or biological analyses are done in different laboratories. The human factor is not the only cause for correlations among units that are measured at about the same time. Many measuring processes are affected by the weather; some use raw materials whose quality varies from batch to batch. In estimating the current sale price of homes built some years ago, Hansen, Hurwitz, and Bershad (1961) point out that if some houses in the sample have been sold recently their prices establish a level that guides the interviewer and the householder in assigning values to houses that have not been sold for many years. In fact, the average price recorded for the sample may depend on the order in which the recently sold houses appear in the sample.

In order to handle these intrasample correlations in their most general terms, a more complex model than that presented here is required. In particular, the notation for $e_{i\alpha}$ and β_i would have to indicate that their values may depend on the other units present in the sample However, the types of correlation that are believed to be most common in practice can be represented by the present model or by simple extensions of it.

The components of the error of measurement are summarized in Table 13.11.

We have noted further that values of β_i and $d_{i\alpha}$ on different units in the same sample may be correlated with one another, where $d_{i\alpha} = e_{i\alpha} - \beta_i$.

TABLE 13.11

COMPONENTS OF THE ERROR OF MEASUREMENT ON THE ith UNIT

Symbol	Nature of Component
β	Constant bias over all units
$\beta_i - \beta$	Variable component of bias, which follows some frequency distribution with mean zero as i varies and may be correlated with the correct value μ_i
$d_{i\alpha} = e_{i\alpha} - \beta_i$	Fluctuating component of error, which follows some frequency distribution with mean zero and variance σ_i^2 as α varies for fixed i

Models in general similar to the preceding have been developed by Hansen et al. (1951), Sukhatme and Seth (1952), Hansen, Hurwitz, and Bershad (1961), and Hansen, Hurwitz, and Pritzker (1965). Fellegi (1964) has given a more extensive model that includes numerous cross correlations that may exist between different sources of measurement error in practical situations.

In their 1961 and 1965 papers, Hansen et al. expressed our model in slightly different terminology, including adding a subscript G to all variables as a reminder that the errors and their sizes may depend on the general (G) conditions of the survey. They express results in terms of the $d_{i\alpha}$ and of

$$\mu_i' = E(y_{i\alpha}|i) = \mu_i + \beta_i \tag{13.30}$$

The quantity μ_i' (which they denote by Y_i, or P_i with a proportion) is conceptually the average obtained from many repetitions of the measuring process on unit i. Hence

$$d_{i\alpha} = e_{i\alpha} - \beta_i = y_{i\alpha} - \mu_i' \tag{13.31}$$

They call $d_{i\alpha}$ the *response deviation* on unit i, whereas we have called it the fluctuating component of the measurement error. Thus

$$y_{i\alpha} - \mu = d_{i\alpha} + (\mu_i' - \mu') + (\mu' - \mu) \tag{13.32}$$

where μ' is the *population* mean of the μ_i'.

Averaging over the sample, we have

$$\bar{y}_\alpha - \mu = \bar{d}_\alpha + (\bar{\mu}' - \mu') + (\mu' - \mu) \tag{13.33}$$

This gives the formula

$$\text{MSE}(\bar{y}_\alpha) = V(\bar{d}_\alpha) + V(\bar{\mu}') + (\mu' - \mu)^2 + 2 \text{ cov}(\bar{d}_\alpha, \bar{\mu}') \tag{13.34}$$

For the sample mean $\bar{y}_\alpha$, the terms on the right are called, respectively, the response variance, the sampling variance, the square of the overall bias, and twice the covariance between the sample's average response deviation and sampling error. Since $E(d_{i\alpha}|i) = 0$ under our model, this covariance vanishes under repetitions of the measuring process on the same set of sample units in the same order. It might not vanish under repetitions over different samples or different orders if the $d_{i\alpha}$ are affected by the other units in the sample. Although this term will be ignored here for simplicity, Fellegi (1964) has shown in a Canadian study that it may materially bias some methods of estimating components of the response variance $V(\bar{d}_\alpha)$.

Using Cornfield's approach, as described in section 2.9, Koch (1973) has given a general decomposition of the MSE of the estimator in multivariate sample surveys, with applications to subclass means.

13.9 EFFECTS OF CONSTANT BIAS

Suppose that the measurements y_i on all units are subject to a constant bias β whose magnitude is unknown. Then the mean $\bar{y}$ of a simple random sample is also subject to bias β. In the estimated error variance, which we attach to the sample mean, the bias cancels out, since this estimate is derived from a sum of squares of terms $(y_i - \bar{y})^2$. Consequently, the usual computation of confidence limits for $\bar{Y}$ from the sample data takes no account of the bias. The same results hold in stratified random sampling.

The situation is essentially the same with regression and ratio estimates. Consider the regression estimate

$$\bar{y}_{lr} = \bar{y} + b(\bar{X} - \bar{x})$$

where both the y_i and the x_i may be subject to constant biases β_y and β_x, respectively. Since the least squares estimate b remains unchanged and since the bias β_x cancels out of the term $(\bar{X} - \bar{x})$, it follows that $\bar{y}_{lr}$ is subject to a bias β_y. It is easy to verify that the sample estimate of $V(\bar{y}_{lr})$ contains no contribution due to the biases.

With the ratio estimate

$$\bar{y}_R = \frac{\bar{y}}{\bar{x}} \bar{X}$$

the bias is also β_y, to a first approximation, since in large samples $E(\bar{X}/\bar{x})$ is approximately 1 even if the x_i are subject to a constant bias. In large samples the sample estimate of variance

$$v(\bar{y}_R) = \frac{(N-n)}{Nn} \frac{\sum (y_i - \hat{R} x_i)^2}{n-1} \tag{13.35}$$

will be almost free from bias as an estimate of

$$E(\bar{y}_R - \bar{Y})^2$$

that is, as an estimate of the variance about the *biased* mean $\bar{Y}$.

To summarize, a constant bias passes undetected by the sample data. As we have seen (section 1.8), the 95% confidence probabilities are almost unaffected if the ratio of β_y to the standard error of the estimated mean is less than 0.1 but, as the ratio increases beyond this value, the computation of confidence limits becomes misleading. Estimates of *change* from one time period to another, or from one stratum to another, remain unbiased, provided that the bias is constant throughout.

13.10 EFFECTS OF ERRORS THAT ARE UNCORRELATED WITHIN THE SAMPLE

In this section a few results about $V(\bar{y}_\alpha)$ and $v(\bar{y}_\alpha)$ are given for simple random sampling in the simplest situation in which errors of measurement are uncorrelated within the sample. This situation may apply in surveys taken from records, in self-filled questionnaires (as in mail surveys) in which members of the same sample do not consult one another, and in some surveys of inanimate populations in which the measurement is objective. The assumption of no correlation cannot be made lightly: intrasample correlation may enter, for example, in interviewing, editing, coding, and transferring the data to the computer if the same person handles a number of elements in the sample.

From (13.34) we get, for the variance,

$$V(\bar{y}_\alpha) = V(\bar{d}_\alpha) + V(\bar{\mu}')$$ (13.36)

In finding variances we average first over repetitions of the measuring process on a specific sample and then over different simple random samples. Now $V(d_{i\alpha}) = \sigma_i^2$ and errors are uncorrelated on different units within the sample. Hence, for simple random sampling, where S denotes a specific sample,

$$E(\bar{d}_\alpha^2 | S) = \frac{1}{n^2} \sum_i^n \sigma_i^2$$ (13.37)

When we average subsequently over all simple random samples, we have

$$V(\bar{y}_\alpha) = \frac{1}{nN} \sum_i^N \sigma_i^2 + \frac{1-f}{n} \frac{\sum_i^N (\mu_i' - \mu)^2}{N-1}$$ (13.38)

$$= \frac{1}{n} \sigma_d^2 + \frac{(1-f)}{n} S_\mu'^2$$ (13.39)

where σ_d^2 denotes the population average of the variances of the errors of measurement. In the Hansen et al. terminology, σ_d^2/n is the response variance of the sample estimate $\bar{y}_\alpha$.

With uncorrelated errors the same model can be applied in estimating a population proportion P (Hansen, Hurwitz, and Bershad, 1961). For any unit, let the correct value μ_i be 1 if the unit is in class C and zero otherwise. If errors of measurement occur, this implies that units are sometimes incorrectly classified. For the ith unit, the recorded value $y_{i\alpha}$ is sometimes 1, sometimes 0. Let P_i denote the proportion of measurements on the ith unit for which $y_{i\alpha} = 1$. Then, for given i, $y_{i\alpha}$ is a binomial variate in repeated measurement, with mean $\mu_i' = P_i$ while the variance of $d_{i\alpha}$ is $P_i Q_i$. Hence, if $p_\alpha = \bar{y}_\alpha$ is the sample estimate, (13.38) becomes

$$V(p_\alpha) = V(\bar{y}_\alpha) = \frac{1}{nN} \sum^N P_i Q_i + \frac{(1-f)}{n} \frac{\sum^N (P_i - P)^2}{N-1}$$ (13.40)

$$< \frac{1}{n(N-1)} \left(\sum^N P_i - \sum^N P_i^2 + \sum^N P_i^2 - NP^2 \right)$$

$$= \frac{N}{n(N-1)} PQ$$ (13.41)

where

$$P = \frac{\sum\limits_{N}^{N} P_i}{N}$$

As (13.41) shows, the sum of the response variance and the sampling variance has an upper limit $NPQ/n(N-1)$. This upper limit is also the variance of the sample mean of the *correct* measurements on the units if the fpc is ignored and there is no overall bias. For the correct measurement μ_i on any unit is then a binomial variate with mean P, so that

$$\frac{S_\mu{}^2}{n} = \frac{N}{n(N-1)} PQ \geq V(p_\alpha) \tag{13.42}$$

This rather puzzling result holds because (*i*) the sampling variance entering into (13.39) and (13.40) is that of $\mu_i' = (\mu_i + \beta_i)$ and (*ii*) in the estimation of a proportion, μ_i and β_i are always *negatively* correlated. When $\mu_i = 1$, $\beta_i \leq 0$, since $P_i = (\mu_i + \beta_i) \leq 1$, and similarly when $\mu_i = 0$, $\beta_i \geq 0$. Thus the term

$$\frac{S_{\mu'}{}^2}{n}$$

in (13.39) is always less than

$$\frac{S_\mu{}^2}{n}$$

by just about the response variance of $\bar{y}_\alpha$.

With uncorrelated errors, a useful result is that the usual formula for $v(\bar{y}_\alpha)$ in simple random sampling remains unbiased if the fpc is negligible. From theorem 2.4 corollary, this formula, developed by assuming *no* errors of measurement, is

$$v(\bar{y}_\alpha) = \frac{1-f}{n} s^2 = \frac{1-f}{n} \frac{\sum\limits^{n} (y_{i\alpha} - \bar{y}_\alpha)^2}{(n-1)} \tag{13.43}$$

Now

$$y_{i\alpha} - \bar{y}_\alpha = (d_{i\alpha} - \bar{d}_\alpha) + (\mu_i' - \bar{\mu}') \tag{13.44}$$

Squaring and averaging first over repeated measurements and then over repeated sample selections, we obtain

$$Ev(\bar{y}_\alpha) = \frac{1-f}{n} \sigma_d{}^2 + \frac{1-f}{n} S_{\mu'}{}^2 \tag{13.45}$$

while, from (13.39)

$$V(\bar{y}_\alpha) = \frac{1}{n} \sigma_d{}^2 + \frac{(1-f)}{n} S_{\mu'}{}^2 \tag{13.46}$$

Thus $Ev(\bar{y}_\alpha) = V(\bar{y}_\alpha)$ if f is negligible.

In the same way the formulas in preceding chapters for the sample estimates of sampling error variances can be shown to remain valid in stratified and multistage sampling, as do the large-sample formulas applicable to ratio and regression estimates, provided that errors of measurement in $y_{i\alpha}$ and $x_{i\alpha}$ are uncorrelated within the sample and that the fpc's are negligible.

13.11   EFFECTS OF INTRASAMPLE CORRELATION BETWEEN ERRORS OF MEASUREMENT

Suppose that some or all of the values of $d_{i\alpha}$ for units in the same sample are correlated. The term $\bar{d}_\alpha{}^2$ can be written

$$\bar{d}_\alpha{}^2 = \frac{1}{n^2}\left(\sum_i^n d_{i\alpha}{}^2 + 2\sum_{i<j}^n d_{i\alpha}d_{j\alpha}\right) \tag{13.47}$$

Hence, averaging over repeated measurements and simple random samples,

$$V(\bar{d}_\alpha) = E(\bar{d}_\alpha{}^2) = \frac{1}{n}\sigma_d{}^2 + \frac{2n(n-1)}{2n^2}E(d_{i\alpha}\, d_{j\alpha}) \tag{13.48}$$

where the products are averaged over all pairs of units in the same sample. By analogy with cluster sampling, the average intrasample correlation coefficient ρ_w may be defined by the equation

$$E(d_{i\alpha}\, d_{j\alpha}) = \rho_w\sigma_d{}^2 \tag{13.49}$$

This gives, from (13.48),

$$V(\bar{d}_\alpha) = \frac{\sigma_d{}^2}{n}[1 + (n-1)\rho_w] \tag{13.50}$$

Hansen, Hurwitz, and Bershad (1961) have called $V(\bar{d}_\alpha)$ the *total* response variance as it affects the sample mean. Its component $\sigma_d{}^2/n$ is called the *simple* response variance, while the term $(n-1)\rho_w\sigma_d{}^2/n$ is the *correlated component* of the total response variance.

From (13.34) we get, assuming cov $(\bar{d}_\alpha, \bar{\mu}') = 0$,

$$V(\bar{y}_\alpha) = \frac{\sigma_d{}^2}{n}[1 + (n-1)\rho_w] + \frac{1-f}{n}S_{\mu'}{}^2 \tag{13.51}$$

The average value of the usual $v(\bar{y}_\alpha)$ in (13.43) is found in the same way to be

$$Ev(\bar{y}_\alpha) = \frac{(1-f)}{n}[\sigma_d{}^2(1-\rho_w) + S_{\mu'}{}^2] \tag{13.52}$$

Since ρ_w is likely to be positive for many types of measurement error, the standard formula for $v(\bar{y}_\alpha)$ is usually an *underestimate* in this case and makes the sample estimate appear more precise than it is. This is true even when the fpc is negligible.

Perhaps the most frequent example of intrasample correlation between errors of measurement is the intrainterviewer correlation previously mentioned, particularly on questions involving opinions and judgment. Suppose that $n = mk$, and that each of the k interviewers obtains data from m respondents. If we can assume that there is no correlation between the errors of measurement of different interviewers,

$$V(\bar{d}_\alpha) = \frac{\sigma_d^2}{n}[1 + (m-1)\rho_w] \qquad (13.53)$$

where ρ_w is the average intrainterviewer correlation coefficient.

Even a small ρ_w may make a major contribution to $V(\bar{y}_\alpha)$, since it is multiplied by roughly the size m of the interviewer's assignment. There may also be some correlation between errors of measurement for different interviewers (e.g., if they have been trained or directed by the same supervisor).

This model represents only the simplest type of intrasample correlation. With stratified sampling, for instance, a coder may process results from several strata and through a misunderstanding of instructions may introduce correlated errors that extend over the strata. The mathematical model can be adapted to apply to situations of this type.

13.12   SUMMARY OF THE EFFECTS OF ERRORS OF MEASUREMENT

In terms of the model, the mean $\bar{y}$ of a simple random sample would be unbiased, with variance S_μ^2/n (ignoring the fpc), if all measurements were fully accurate. As a result of the types of errors of measurement discussed here, the mean may be subject to a bias of amount β, and its mean square error is

$$\text{MSE}(\bar{y}_\alpha) = \frac{1}{n}\{S_{\mu'}^2 + \sigma_d^2[1 + (n-1)\rho_w]\} + \beta^2 \qquad (13.54)$$

where $\mu_i' = \mu_i + \beta_i$.

Formula 13.54 contains two terms, $S_{\mu'}^2/n$ and $\sigma_d^2(1-\rho_w)/n$, that decrease as $1/n$. The remaining two terms, $\rho_w\sigma_d^2$ and β^2, appear at first sight to be independent of n. This is probably an oversimplification. Any material change in the size of sample may require a change in the field methods of measurement, and this may affect ρ_w and β^2. However, these two terms should change relatively slowly, if at all, with n. Thus, in large samples, the MSE is likely to be dominated by these two terms, the ordinary sampling variance becoming unimportant and misleading as a guide to the real accuracy of the results.

13.13   THE STUDY OF ERRORS OF MEASUREMENT

In recent years much of the research on sampling practice has been devoted to the study of errors of measurement. The objectives are to discover the components that make a large contribution to the MSE and to find ways of decreasing these contributions. Some of the principal methods are described in this and the following sections. It is already clear that progress will be slow and expensive. One reason is that, as already mentioned, the measurement errors depend intimately both on the items and on the measuring process. Results about measurement errors found in one survey can seldom be assumed to apply to other surveys.

Ideally, the best method of studying errors of measurement is to obtain the correct values μ_i. In practice, this approach is limited to items for which a feasible method of finding μ_i exists and by problems of expense and execution. Examples are given by Belloc (1954), who compared data on hospitalization as reported in household interviews with the hospital records for the individual, and by Gray (1955), who compared employees' statements of sick leave with the personnel office records. Checks of this type—sometimes called "record-checks"—are possible with items such as age, occupation, number of years of schooling, and price paid for car. One difficulty is that sometimes the records contain no exact match of the person interviewed.

Failing a method of determining the correct value, an alternative is to remeasure by an independent method that is considered more accurate. Kish and Lansing (1954) engaged professional appraisers to estimate the selling prices of homes that had already been reported by the home owners. In surveys of illness respondents' replies have been compared either with doctors' records on the respondents or with the results of a complete medical examination [Sagen, Dunham, and Simmons, 1959, Trussell and Elinson, 1959). The results of such comparisons are not easy to interpret in terms of the model, since the superior instrument is itself subject to measurement errors, but the comparisons will at least indicate the items for which the routine instrument agrees well with the superior instrument and those for which it does not.

In household interview surveys, a method with a similar purpose is to reinterview a subsample of the respondents by more expert interviewers, the questionnaire being more detailed and probing. After the reinterview, the expert discusses with the respondent any discrepancies between the original and the repeat answers, the objectives being to determine the most accurate answer and the reason for the discrepancy. Much useful information may be gained. In presenting some devices for the measurement of response errors, Madow (1965) has discussed the use of double sampling, with a difference estimator of the form $[\bar{y}' - (\bar{y} - \hat{\mu})]$ as a means of reducing response bias. Here, $\hat{\mu}$ is the mean of unbiased or less biased measurements made on the subsample, while $\bar{y}'$ and $\bar{y}$ are means of the original measurements on the sample and subsample.

Occasionally, overall comparisons between the results of two different surveys are feasible. For a number of items, the results of the U.S. Census can be compared with those given by the Current Population Survey taken at the same time. Since the Survey is considered more accurate, particularly for items difficult to measure, rough estimates of the measurement bias β in the Census data can be made (Hansen, Hurwitz, and Bershad, 1961). A number of comparisons between the results of quota samples and probability samples are discussed by Stephan and McCarthy (1958).

In the following sections we consider some methods designed to produce quantitative estimates of the components of the total variance (response variance plus sampling variance) of a sample estimate.

13.14 REPEATED MEASUREMENT OF SUBSAMPLES

In recent studies, interest has centered on estimating: (a) the *total* response variance and the relative sizes of the simple response variance and the correlated component as contributors to it, and (b) the relative sizes of the total response variance and the sampling variance. For (a) a common method is to select a subsample of the measuring agents (interviewers, coders, etc.) and remeasure their assignments by second agents presumably of the same skill. Suppose mk is the size of the subsample, m being the size of an agent's assignment and k the number of agents chosen from the original sample.

Let y_{i1}, y_{i2} be the two measurements on the ith unit of an agent's assignment. If (13.31) holds,

$$y_{i\alpha} = d_{i\alpha} + \mu_i'$$
(13.55)

Hence, averaging over the assignment,

$$\frac{E \sum_i^m (y_{i1} - y_{i2})^2}{2m} = \frac{\sigma_{d1}^2 + \sigma_{d2}^2}{2} - \text{cov}(d_{i1} d_{i2})$$
(13.56)

This expression estimates σ_{d1}^2, the simple response variance for agent 1 in the survey, if two conditions (A) hold: no correlation between response errors d_{i1}, d_{i2} on the same unit; and $\sigma_{d1}^2 = \sigma_{d2}^2$. Equation (13.56) applies to a single pair of agents and is averaged over the subsample of k pairs of agents.

Conditions A may hold when the measurement is *coding*, the agents being coders of similar skill trained by different supervisors, neither coder seeing the other's work. With interviewing, a positive cov $(d_{i1} d_{i2})$ is to be expected because some respondents repeat a first incorrect response from memory. In this event $\sum^m (y_{i1} - y_{i2})^2/2m$ underestimates σ_{d1}^2. It also underestimates σ_{d1}^2 if the second agent is more skilled than the first, as shown for a (0,1) measurement by Hansen, Hurwitz, and Pritzker (1965). Moreover, $\sum (y_{i1} - y_{i2})^2/2m$ has been found to decline if the second interviewer is given the responses obtained by the first interviewer, even if told *not* to look at them until the repeat interview is completed (Koons, 1973). These complexities illustrate why the realistic study of errors of measurement is difficult.

For the *total* response variance the relevant estimate from (13.55) for a single pair of agents is $(\bar{y}_{.1} - \bar{y}_{.2})^2/2$, where $\bar{y}_{.1}, \bar{y}_{.2}$ are the means of the m first and second measurements. Under conditions (A).

$$\frac{E(\bar{y}_{.1} - \bar{y}_{.2})^2}{2} = \frac{\sigma_d^2}{m}[1 + (m-1)\rho_w] \tag{13.57}$$

where ρ_w is the correlation between response errors on different units by the same agent. Equation (13.57) provides only a single degree of freedom, but is averaged over the k pairs of agents. Having an estimate of σ_d^2 from (13.56), we can estimate the relative sizes of the simple response variance and the correlated component. In interview surveys the correlated component is usually found much larger than the simple response variance except for basic items such as age, sex, and marital status (Fellegi, 1964). Partly for this reason the 1970 U.S. Census used self-enumeration by mail extensively (Hansen and Waksberg, 1970).

If (13.55) holds, we can also study the ratio of the simple response variance to the sampling variance that applies to the sample in an agent's assignment. For under conditions A,

$$E \sum_{\alpha}^{2} \sum_{i}^{m} (y_{i\alpha} - \bar{y}_{.\alpha})^2/2(m-1) = \sigma_d^2(1-\rho_w) + S_{\mu'}^2 \tag{13.58}$$

Pritzker and Hanson (1962) have called the ratio

$$I = \frac{\sigma_d^2}{(\sigma_d^2 + S_{\mu'}^2)} \tag{13.58)'}$$

the *index of inconsistency*. It is analogous to the quantity $(1 - \phi)$, where ϕ is the coefficient of reliability used in studying errors of measurement in psychology. If ρ_w is negligible, I can be estimated from the ratio of (13.56) to (13.58).

With a (0, 1) variate, equations (13.40) and 13.41) in section 13.10, putting $n = 1$, show that for a single measurement of a single unit, the sum of the simple response variance and the sampling variance is $V(y_{1\alpha}) \doteq PQ$, where $P = \sum P_i/N$, assuming f negligible. The set of joint values (y_{i1}, y_{i2}) obtained by the two agents can be summarized in the following 2×2 frequency table.

| | | Number of responses Second agent | | |
		1	0	Total
First	1	a	b	$a+b$
agent	0	c	d	$c+d$
		$a+c$	$b+d$	m

Thus, b is the number of units on which the first agent records 1, the second agent 0. Under conditions A

$$\hat{\sigma}_d^2 = \frac{\sum\limits_{i}^{m} (y_{i1} - y_{i2})^2}{2m} = (b+c)/2m \tag{13.59}$$

As an estimate of I from the subsample, one choice is

$$\hat{I} = \frac{\hat{\sigma}_d^2}{\hat{P}\hat{Q}} = \frac{m(b+c)}{(a+b)(c+d) + (a+c)(b+d)} \tag{13.60}$$

.which averages the PQ estimates from the two sets of measurements. Estimates of the index have been published for both Census and sample survey items by Pritzker and Hanson (1962), Fellegi (1964), and Koons (1973). These estimates are useful in comparing the relative unreliabilities of measurement for different items, in successive censuses, or as between different methods of measurement, thus providing some appraisal of new methods of measurement.

The interpretation of these comparisons is, of course, often clouded by doubts as to whether conditions A apply. More complex estimates based on more realistic assumptions are given by Fellegi (1964).

13.15 INTERPENETRATING SUBSAMPLES

This technique, particularly useful for the study of correlated errors, was proposed by Mahalanobis (1946). To present is in the simplest terms, a random sample of n units is divided *at random* into k subsamples, each subsample containing $m = n/k$ units. The field work and processing of the sample are planned so that there is no correlation between the errors of measurement of any two units in *different* subsamples. For instance, suppose that the correlation with which we have to deal arises solely from biases of the interviewers. If each of k interviewers is assigned to a different subsample and if there is no correlation between errors of measurement for different interviewers, we have an example of the technique.

With the same mathematical model, it is convenient to label the units by double subscripts. Let

$$y_{ij\alpha} = \mu'_{ij} + d_{ij\alpha} \tag{13.61}$$

where i denotes the subsample (interviewer) and j the member within the subsample. The fpc is ignored.

Since the ith subsample is a random subsample, it is itself a simple random sample of size m. Hence, by (13.51), the variance of its mean is

$$V(\bar{y}_{i\alpha}) = \frac{1}{m}\{S_{\mu'}^2 + \sigma_d^2[1 + (m-1)\rho_w]\} \tag{13.62}$$

where ρ_w is the correlation between the $d_{ij\alpha}$ obtained by the same interviewer.

Since errors are independent in the different subsamples,

$$V(\bar{y}_\alpha) = \frac{1}{k} V(\bar{y}_{i\alpha}) = \frac{1}{n}\{S_{\mu'}^2 + \sigma_d^2[1+(m-1)\rho_w]\} \qquad (13.63)$$

From the sample results we can compute an analysis of variance of the km observations into the components "Between interviewers (subsamples)" with $(k-1)$ degrees of freedom and "Within interviewers" with $k(m-1)$ degrees of freedom. It is easy to verify that the expected values of the mean squares, averaged over selections of the interviewers and of the random samples and subsamples, work out as in Table 13.12.

TABLE 13.12

EXPECTATIONS OF THE MEAN SQUARES (ON A SINGLE-UNIT BASIS)

	df	ms	$E(ms)$
Between interviewers (subsamples)	$k-1$	$s_b^2 = \dfrac{m \sum (\bar{y}_{i\alpha} - \bar{y}_\alpha)^2}{k-1}$	$S_{\mu'}^2 + \sigma_d^2[1+(m-1)\rho_w]$
Within interviewers	$k(m-1)$	$s_w^2 = \dfrac{\sum\sum (y_{ij\alpha} - \bar{y}_{i\alpha})^2}{k(m-1)}$	$S_{\mu'}^2 + \sigma_d^2(1-\rho_w)$

Table 13.12 contains two important results. By comparison with (13.63) we see that s_b^2/n is an unbiased estimate of $V(\bar{y}_\alpha)$, ignoring the fpc. Thus interpenetrating subsamples provide an estimate of $V(\bar{y}_\alpha)$ that takes proper account of both the simple response variance and the correlated component.

The analysis also enables us to estimate the correlated component, since

$$\frac{E(s_b^2 - s_w^2)}{m} = \rho_w \sigma_d^2 \qquad (13.64)$$

Consequently, comparison of $(m-1)(s_b^2 - s_w^2)/m$ with s_b^2 estimates the relative amount which the correlated component of the response variance contributes to the total variance of $\bar{y}_\alpha$. With measurements in which the correlated component is much larger than the simple response variance, the ratio $(m-1)(s_b^2 - s_w^2)/ms_b^2$ has been used alternatively as a measure of the relative contribution of the total response variance to the total variance of $\bar{y}_\alpha$. Tepping and Boland (1972) present estimates of this ratio for items in the Current Population Survey.

When the interpenetration method is applied in a multistage sample covering a wide geographic area, the most common practice is to have pairs of interviewers measure interpenetrating subsamples drawn from the *smallest* clusters among the successive stages. In this way the number of ultimate units in an interviewer's assignment is kept at its customary level in the survey, although the interviewer has to travel over twice the usual area. For a single cluster, Table 13.12 provides 1 df between interviewers and $2(m-1)$ df within interviewers: the corresponding mean squares are averaged over the c clusters chosen for the study. The sampling variance that is measured is, of course, only that within the last stage of clustering. As a reminder of this fact, the sampling variance term in $E(ms)$ is sometimes written $S_{\mu'}^2(1-\rho_s)$ instead of $S_{\mu'}^2$, where ρ_s is the intracluster correlation among the sampling errors for different subunits.

The interpenetration method was used in this form in Response Variance Study I by the U.S. Census Bureau (1968), designed to estimate the correlated components of the total response variances of items in the 1960 Census. The areas in this study were compact clusters of households, the clusters being scattered all over the U.S.A. In any sample cluster, two interpenetrating subsamples were formed, each subsample being assigned to a different interviewer.

In half the clusters the two interviewers had *different* crew leaders. In this half it was assumed, as seems reasonable, that the response errors of the two interviewers were uncorrelated. Thus $(s_b^2-s_w^2)/m$ estimates $\rho_w\sigma_d^2$, where s_b^2, is now the average mean square between interviewers in the same cluster. In the other half of the clusters the two interviewers had the *same* crew leader. The objective here was to measure the extent to which "crew leader effect" induced a covariance between $\bar{d}_{.1}$ and $\bar{d}_{.2}$ for the two interviewers in a cluster. If so, s_b^2 in Table 13.12 now estimates

$$S_{\mu'}^2(1-\rho_s)+\sigma_d^2[1+(m-1)\rho_w]-mE\,\mathrm{cov}\,(\bar{d}_{.1}\bar{d}_{.2})$$

and $(s_b^2-s_w^2)/m$ estimates

$$\rho_w\sigma_d^2-E(\mathrm{cov}\,\bar{d}_{.1}\bar{d}_{.2}) \tag{13.65}$$

Comparison of the two sets of values of $(s_b^2-s_w^2)/m$ reveals the presence of a "crew leader effect." Since differences between estimates of variance like s_b^2 nad s_w^2 are unstable, large numbers of clusters in the two halves of this type of study are necessary to measure "crew leader effect" with any precision.

The interpenetration technique extends to stratified and multistage sampling. If the primary interest is in an unbiased estimate of $V(\bar{y}_\alpha)$ that takes proper account of the effects of errors of measurement, all that is necessary is that the sample consist of a number of subsamples of the same structure in which we are sure that errors of measurement are independent in different subsamples. Strictly, this requires that different interviewing teams, supervisors, and data processors be used in different subsamples. If $\bar{y}_{i\alpha}$ is the mean of the ith subsample, the quantity $\sum(\bar{y}_{i\alpha}-\bar{y}_\alpha)^2/k(k-1)$ is an unbiased estimate of $V(\bar{y}_\alpha)$, with $(k-1)$ df. This result holds because the subsample can be regarded as a single complex sampling unit, the sample being in effect a simple random sample of these complex units, with uncorrelated errors of measurement between different complex units. Consequently, the results in section 13.10 apply.

Numerous applications of this method, sometimes called *replicated sampling*, are described by Deming (1960), who has used the method extensively. For other discussions of its advantages, see Jones (1955) and Koop (1960). Travel costs of interviewers are increased by interpenetration, but this can be mitigated if the sample is stratified into compact areas. For instance, each stratum might contain two random samples, assigned to a different interviewer. Each interviewer is required to travel over the whole stratum instead of over only half the stratum. Every stratum provides 1 df for the estimate of $V(\bar{y}_\alpha)$.

13.16 COMBINATION OF INTERPENETRATION AND REPEATED MEASUREMENT

As we have seen (Section 13.14), repeated measurement of the units in an agent's assignment by another agent of similar quality provides estimates of the simple response variance and the total response variance if conditions A apply, although it may underestimate in interview surveys if the respondent's errors on the two occasions are positively correlated. The interpenetration scheme (section 13.15) provides estimates of the correlated component of the response variance and its contribution to the total variance, response variance plus sampling variance.

Much more may be learned from an ingenious combination of interpenetration and repetition, as used by Fellegi (1964) in a study of response errors in the 1961 Canadian Census of Population. The study was conducted in 134 Enumeration Areas ($E.A.$'s), each containing about 150 households, the size of an interviewer's assignment. Contiguous E. A.'s were grouped into 67 pairs. Two interviewers were assigned to each pair, each interviewing a random half of the households in the pair. Thus each enumerator had the regular work load, but spread over twice the area. Then the assignments of the two interviewers were switched, giving the desired combination of interpenetration and repeated measurement.

If S_1, S_2 denote the two interpenetrating subsamples and I_1, I_2, the two interviewers, comparison of (I_1S_1) with (I_2S_1) or (I_1S_2) with (I_2S_2) gives the "repeated measurement" analysis, while comparison of (I_1S_1) with (I_2S_2) or (I_1S_2) with (I_2S_1) gives the "interpenetration" analysis. These comparisons lead to estimates of the simple response variance, the correlated component, the total response variance, and the index of inconsistency. The sampling variance involved is that between households within pairs of $E.A.$'s. More extensive analysis by Fellegi also estimates the covariance between sampling and response deviations for the same interviewer. This term has been neglected in the model presented here, but Fellegi shows that it may create sizable biases in the estimates of $\rho\sigma_d^2$.

A good exposition of the strengths and weaknesses of different variants of the repeated measurement and the interpenetration approaches has been given by

Bailar and Dalenius (1969). Hansen and Waksberg (1970) review the research work of the U.S. Census Bureau on measurement errors as they affect the Census and some of the most important sample surveys taken by the Census Bureau. Resulting changes in the 1970 Census included more widespread use of self-enumeration (omitting interviewer biases) in a Census by mail, further use of sampling as distinct from a complete Census, and advance computer selection of the sample to avoid some biases that had been detected in selection of the final-stage sample by the interviewer. Disturbing errors found in data on occupation, industry, and housing quality as well as recall problems in expenditure data are under continuing study.

13.17 SENSITIVE QUESTIONS: RANDOMIZED RESPONSES

A situation likely to lead either to refusals to answer or to evasive answers occurs when a question in a survey is sensitive or highly personal (e.g. does the respondent regularly engage in shoplifting or use drugs?). Consider first the estimation of a binomial proportion—the proportion π_A of respondents who belong to a certain class A or have committed a certain act. By ingenious use of a randomizing device, Warner (1965) showed that it is possible to estimate this proportion without the respondent revealing his or her personal status with respect to this question. The objective is to encourage truthful answers while fully preserving confidentiality.

The randomizing device, such as a spinning arrow or a box with red and white balls, selects one of two statements or questions, each requiring a "yes" or "no" response, to be presented to the respondent. The interviewer does *not* know which question any respondent has answered, but does know the relative probabilities P and $(1-P)$ with which the two statements are presented. The success of the method depends, of course, on the respondent's being convinced that by participating he or she will not be revealing personal status with regard to the sensitive issue.

In Warner's original proposal the two statements are:

"I am a member of class A."   (presented with probability P)
"I am not a member of class A."

With a random sample of n respondents the interviewer records a binomial estimate $\hat{\phi} = m/n$ of the proportion ϕ of "yes" answers. If the questions are answered truthfully, the relation between ϕ and π_A in the population is

$$\phi = P\pi_A + (1-P)(1-\pi_A) = (2P-1)\pi_A + (1-P) \qquad (13.66)$$

With known P, this relation suggests the estimate

$$\hat{\pi}_{AW} = \frac{[\hat{\phi} - (1-P)]}{(2P-1)} \qquad \left(P \neq \frac{1}{2}\right) \qquad (13.67)$$

This estimate turns out to be the maximum likelihood estimate of π_A. (The suffix W denotes "Warner".) The estimate is unbiased, with variance

$$V(\hat{\pi}_{AW}) = \frac{\phi(1-\phi)}{n(2P-1)^2} \tag{13.68}$$

Writing $(1-\phi)$ in the form

$$(1-\phi) = (2P-1)(1-\pi_A) + (1-P) \tag{13.69}$$

we find easily

$$V(\hat{\pi}_{AW}) = \frac{\pi_A(1-\pi_A)}{n} + \frac{P(1-P)}{n(2P-1)^2} \tag{13.70}$$

The first term in $V(\hat{\pi}_{AW})$ is the variance that $V(\hat{\pi}_A)$ would have if all n respondents answered truthfully a direct question about class A membership. Except for π_A near $\frac{1}{2}$ and $P > 0.85$, the second term is greater than the first, often much greater. The method is thus quite imprecise in general. This might be expected, since the interviewer does not know whether a "yes" answer implies membership in class A or the opposite. As Warner showed, however, his method may give a smaller MSE than a direct sensitive question would, if the latter produced numerous refusals or false answers.

13.18 THE UNRELATED SECOND QUESTION

As an alternative to the Warner method, Simmons suggested (Horvitz, Shah, and Simmons, 1967) that respondent cooperation might improve if the second statement was not in any way sensitive, being unrelated to the first. For example,

"I was born in the month of May."

The first statement remains unchanged. If all respond truthfully, the population proportion of "yes" answers is now

$$\phi = P\pi_A + (1-P)\pi_U \tag{13.71}$$

where π_U is the proportion in the sampled population who were born in May. If π_U is known, the obvious (and maximum likelihood) estimate of π_A is

$$\hat{\pi}_{AU} = \frac{[\hat{\phi} - (1-P)\pi_U]}{P} \tag{13.72}$$

with variance

$$V(\hat{\pi}_{AU}) = \frac{\phi(1-\phi)}{nP^2} \tag{13.73}$$

Morton (Greenberg et al., 1969, p. 532) has suggested how the case, π_U *known* can always be achieved. A box contains red, white, and blue balls in known proportions P_1, P_2, P_3. Drawing a red ball produces the sensitive statement. Drawing a white or a blue ball produces the statement: "The color of this ball is white." Thus, $\pi_U = P_2/(P_2 + P_3)$.

Dowling and Shachtman (1975) have shown that $V(\hat{\pi}_{AU}) < V(\hat{\pi}_{AW})$ for all π_A, π_U, provided that P exceeds about $\frac{1}{3}$. (The variance of $\hat{\pi}_{AW}$ is symmetrical about $P = \frac{1}{2}$ but that of $\hat{\pi}_{AU}$ is not, a small P providing few responses on the sensitive question with this method.)

If it is necessary to estimate both π_A and π_U we can have two random samples of sizes n_1, n_2, with different proportions P_1, P_2 for the sensitive question. With ϕ_1, ϕ_2 denoting the proportions of "Yes" answers in the populations defined by the choices P_1 and P_2,

$$\phi_1 = P_1\pi_A + (1 - P_1)\pi_U \tag{13.74}$$

$$\phi_2 = P_2\pi_A + (1 - P_2)\pi_U \tag{13.75}$$

These relations suggest the estimate

$$\hat{\pi}_{AU} = \frac{[\hat{\phi}_1(1 - P_2) - \hat{\phi}(1 - P_1)]}{(P_1 - P_2)} \tag{13.76}$$

with

$$V(\hat{\pi}_{AU}) = \frac{1}{(P_1 - P_2)^2}\left[\frac{\phi_1(1 - \phi_1)(1 - P_2)^2}{n_1} + \frac{\phi_2(1 - \phi_2)(1 - P_1)^2}{n_2}\right] \tag{13.77}$$

If $P_1 > \frac{1}{2}$. Greenberg et al. (1969) showed that this variance is minimized when $P_2 = 0$, that is, when all in the second (n_2) sample are asked the unrelated π_U question. Moors (1971) has recommended this procedure, but Greenberg et al. (1969) suggest $P_1 + P_2 = 1$ as a working rule, in case the choice $P_2 = 0$ might weaken cooperation by respondents. When $P_1 = 0.8$, for example, 80% of sample 1 and 20% of sample 2 would be asked the sensitive question on the average.

With the optimum n_1, n_2 for given $n = n_1 + n_2$, the Cauchy-Schwarz inequality shows that the resulting minimum variance of $\hat{\pi}_{AU}$ is

$$V_{min}(\hat{\pi}_{AU}) = \frac{1}{n(P_1 - P_2)^2}[(1 - P_2)\sqrt{\phi_1(1 - \phi_1)} + (1 - P_1)\sqrt{\phi_2(1 - \phi_2)}]^2 \tag{13.78}$$

The minimizing n_1/n_2 ratio is

$$\frac{n_1}{n_2} = \frac{(1 - P_2)}{(1 - P_1)}\sqrt{\frac{\phi_1(1 - \phi_1)}{\phi_2(1 - \phi_2)}} \tag{13.79}$$

This choice requires advance estimates of π_A and π_U, but the optimum is fairly flat. Greenberg et al. (1969) give recommendations about the choices of $P_1, P_2, n_1,$

and n_2. For preservation of confidentiality it helps to have π_U approximately equal to π_A.

Numerous variants of these methods have been studied (e.g., use of two unrelated questions), as well as the biases produced in $\hat{\pi}_{AW}$ and $\hat{\pi}_{AU}$ if a fraction of the respondents answer the question falsely. Greenberg et al. (1971) have applied the two-sample technique to estimate the mean μ_A for a sensitive discrete or continuous variable by methods analogous to those leading to equations (13.74) and (13.75). The unrelated question estimates the mean μ_U for a nonsensitive variable. Random subgroups of n_i subjects ($i = 1,2$) receive the sensitive question with probability P_i, the nonsensitive with probability $(1 - P_i)$. The recorded variable z_i for a subject therefore follows a mixture of two distributions in proportions P_i, $(1 - P_i)$, one distribution with mean μ_A, variance $\sigma_A{}^2$, the other with mean μ_U, variance $\sigma_U{}^2$. Hence

$$E(z_i) = P_i\mu_A + (1 - P_i)\mu_U \tag{13.80}$$

$$V(z_i) = P_i\sigma_A{}^2 + (1 - P_i)\sigma_U{}^2 + P_i(1 - P_i)(\mu_A - \mu_U)^2 \tag{13.81}$$

Analogous to (13.76), the estimate of μ_A is

$$\hat{\mu}_{AU} = \frac{[(1 - P_2)\bar{z}_1 - (1 - P_1)\bar{z}_2]}{(P_1 - P_2)} \tag{13.82}$$

For maximum efficiency the conditions $\mu_U = \mu_A$, $\sigma_U{}^2 = 0$ are required, while for preservation of anonymity, $\mu_U = \mu_A$, $\sigma_U{}^2 = \sigma_A{}^2$ are best.

Warner (1971) has given a theoretical framework for a broad class of randomized response models. As (13.74) and (13.75) suggest, the trick is to estimate certain linear functions of the sensitive and unrelated π's or μ's with as many equations as there are parameters to be estimated.

The method has been applied to obtain estimates of the proportions of illegitimate births, of induced abortions, of users of heroin, of persons having contact with organized crime, and of mean income and number of abortions as continuous-discrete applications. Since the method has attracted widespread attention, further applications are likely to appear. An excellent review is given by Horvitz, Greenberg, and Abernathy (1975).

There have recently been discussions of the degree of privacy that the respondents have in different versions of randomized interviewing. In some versions the interviewer may be able to guess the status of some respondents with regard to the sensitive issue with a fairly high probability of being correct—an undesirable feature for this method.

13.19 SUMMARY

In regard to their effects on the formulas given in preceding chapters, nonsampling errors may be classified as follows.

1. With noncoverage and nonresponse, the most important consequence is that estimates may become biased, because the part of the population that is not reached may differ from the part that is sampled. There is now ample evidence that these biases vary considerably from item to item and from survey to survey, being sometimes negligible and sometimes large. A second consequence is, of course, that the variances of estimates are increased because the sample actually obtained is smaller than the target sample. This factor can be allowed for, at least approximately, in selecting the size of the target sample.

2. Errors of measurement that are independent from unit to unit within the sample and average to zero over the whole population are properly taken into account in the usual formulas for computing the standard errors of the estimates, provided that fpc terms are negligible. Such errors decrease the precision of the estimates, and it is worthwhile to find out whether this decrease is serious.

3. If errors of measurement on different units in the sample are correlated, the usual formulas for the standard errors are biased. The standard errors are likely to be too small, since the correlations are mostly positive in practice. This type of disturbance is easily overlooked and may often have passed unnoticed.

4. A constant bias that affects all units alike is hardest of all to detect. No manipulations of the sample data will reveal this bias.

As this chapter has indicated, the study of these problems is slow and difficult. Nevertheless, a good beginning has been made. Much ingenuity has been shown in devising techniques for the assessment and control of nonsampling errors. Although this is a field in which broad generalizations are hard to attain, information is accumulating about the nature and magnitudes of errors of measurement in different types of survey. More is being learned also about what can be accomplished with good training and supervision of the interviewers, with pretesting, with mechanisms to control the quality of the field work, and with a postsurvey appraisal of the successes and weaknesses in the operation.

Under certain assumptions, a second measurement of a subsample of the units by another interviewer of similar skill provides estimates of the simple response variance and its ratio to the total response variance, as well as an approximate estimate of its ratio to the sampling variance to which this subsample is subject. A second device—interpenetrating subsamples—furnishes estimates of the total variance (sampling variance plus response variance) and an estimate of the correlated component of the response variance. Combination of interpenetration and repeat measurement is particularly fruitful.

EXERCISES

13.1 Suppose that, by field methods of different intensities, it is possible to make the "response" stratum consist of 60, 80, 90, or 95% of the whole population. For a percentage that is to be estimated, the true "response" stratum means are: 60% stratum, 40.7; 80% stratum, 43.5; 90% stratum, 44.8; 95% stratum, 45.4; last 5%, 59.0. (a) For a method that samples only the 60% stratum, show that the root mean square error of the estimated percentage for the whole population is

$$\sqrt{(2414/n)+28.94}$$

where n is the number of completed questionnaires obtained. (b) Show that a root mean square error of 5% cannot be achieved by a method with 60% response but can be obtained with slightly over 100 completed questionnaires for the methods that have a response of 80% or better. (c) If a root mean square error of 2% is prescribed, what methods can achieve it and what sample sizes are needed?

13.2   In 13.1 (c) suppose that it costs $5 per completed questionnaire for the field method that has a 90% response. To obtain a completed questionnaire from the *next* 5% of the population costs $20. For a root mean square error of 2%, is it cheaper to use the method with 90% response rate or that with 95% response rate?

13.3   A population consists of two strata of equal sizes. The probability of finding the respondent at home and willing to be interviewed at any call is 0.9 for persons in stratum 1 and 0.4 for persons in stratum 2. (a) In the notation of section 13.5 show that

$$w_{i1} = 1-(0.1)^i, \qquad w_{i2} = 1-(0.6)^i$$

(b) If the original sample size is n_o, compute the total expected number of interviews obtained for 1, 2, 3, 4 and 5 calls. (c) If the relative costs per completed interview at the ith call are 100, 120, 150, 200, and 300 for $i = 1, 2, 3, 4, 5$, respectively, compute the average cost per interview for all interviews obtained up to the ith call. (d) The money available for the survey is enough to pay for 300 completed first calls. If the policy is to insist on i calls, what are the expected total numbers of completed interviews that can be obtained for the same amount of money when $i = 1, 2, 3, 4, 5$?

13.4   In exercise 13.3 persons in stratum 1 have a mean of 40% for some binomial percentage that is being estimated and persons in stratum 2 have a mean of 60%. (a) Compute the bias in the sample mean for $i = 1, 2, 3, 4, 5$ calls. (b) Compute the variances of the sample means for the cost situation in part (d) of exercise 13.3. (To save computing, the variance may be taken as $2600/n_i$, where n_i is the expected total number of interviews obtained.) (c) Which policy gives the lowest MSE?

13.5   In section 13.6 (subsampling of the nonrespondents) verify the formula (p. 373) for the ratio of the expected cost of obtaining a specified V with no subsampling to the minimum expected cost,

$$\text{Ratio} = \frac{F(c_0+c_1W_1+c_2W_2)}{[\sqrt{(F-W_2)(c_0+c_1W_1)}+ W_2\sqrt{c_2}]^2}$$

where $F = S^2/S_2^2$. Let $c_0 = c_1 = 1$, $c_2 = 16$. (a) If $F = 1$, show that the cost ratio has a maximum 1.25 when $W_2 = 0.2$ or 0.25. (b) If $F = 1.5$, show that the maximum is 1.41 for $W_2 = 0.3$ or 0.35.

13.6   In a survey on poultry and pigs kept in gardens and certain small holdings (Gray, 1957) a postal inquiry with several reminders was followed by interviews of a subsample of nonrespondents. By advance judgment, $k = 2$ was chosen (i.e., a 50% subsample). The following data were available after the survey for one important item, in the notation of exercise 13.5.

$$\frac{c_1}{c_0} \doteq 0.15, \qquad \frac{c_2}{c_0} \doteq 9.5, \qquad W_1 \doteq 0.8, \qquad S^2 \doteq S_2^2$$

By finding VC for $k = 2$ and for the optimum k, determine whether $k = 2$ was a good choice.

13.7   In a survey by the Politz-Simmons method 390 respondents in an initial sample of 660 were found at home on the first call. The numbers who stated that they were at home on $0, 1, \cdots , 5$ of the five previous nights and the number answering yes to a question in the survey were as follows.

	0/5	1/5	2/5	3/5	4/5	5/5
Number	14	35	55	74	94	118
Yes answers	4	13	20	30	42	156

Compute the Politz-Simmons estimate of the proportion of "yes" answers in the population and compare it with the simple binomial estimate.

13.8   A population with $N = 6$ contains three units for which the correct answer to a question is yes and three for which it is no. Owing to errors of measurement, the probability of obtaining a "yes" response on a yes unit is 0.9. and the probability of obtaining a "no" response on a no unit is also 0.9. (*a*) By working out the distribution of all possible responses for samples of size 2, show that the probabilities are 0.218, 0.564, and 0.218 that the sample gives 0, 1, 2 "yes" responses. (*b*) Show that the variance of the estimated proportion of "yes" responses is 0.1090. Verify results (13.40) and (13.41) in section 13.10. (*c*) What would be the variance of the estimated proportion of "yes" responses if there were no errors of measurement?

13.9   In part of the 1942 Bengal Labour Enquiry (Mahalanobis, 1946) a random sample of about 175 families was taken in each of three strata. The sample in each stratum was divided into five random subsamples, each assigned to a different interviewer. The five interviewers worked in all three strata. For expenditure on food, the relevant part of the analysis of variance (on a *single-family* basis) is as follows.

	df	ms	$E(ms)$
Between interviewers	4	22.3	$\sigma_{\mu'}^2 + \sigma_d^2 + 35\sigma_{IS}^2 + 105\sigma_I^2$
Interviews × strata	8	9.6	$\sigma_{\mu'}^2 + \sigma_d^2 + 35\sigma_{IS}^2$
Within subsamples	510	9.9	$\sigma_{\mu'}^2 + \sigma_d^2$

If g_i, w_{hi} represent biases of interviewer i, the model for a single family is

$$y_{hij\alpha} = \bar{\mu}_h + g_i + w_{hi} + (\mu'_{hij} - \bar{\mu}_h) + d_{hij\alpha}$$

Variances: $\sigma_I^2 \quad \sigma_{IS}^2 \quad \sigma_{\mu'}^2 \qquad \sigma_d^2$

Verify the expressions given for $E(ms)$ and estimate the proportion of the total variance of the mean that may be ascribed to enumerator biases.

13.10   Consider an illegal act that 10% of the population have committed ($\pi_A = 0.1$). If all respondents answer truthfully, compare the $V(\hat{\pi}_A)$ for $n = 500$ given by (*a*) a direct sensitive question, (*b*) the Warner method with $P = 0.8$, (*c*) the unrelated question method with $\pi_u = 0.2$ known, (*d*) the two-sample unrelated question method with π_u actually 0.2 but unknown, when $P_1 = 0.8, P_2 = 0$ as recommended by Moors, (*e*) the same method when $P_1 = 0.8$, $P_2 = 1 - P_1$. Assume that you can use the optimal n_1/n_2 in (*d*) and (*e*). Some decimals are avoided by calculating $V(100\hat{\pi}_A) = 10^4 V(\hat{\pi}_A)$ for the methods.

13.11 In exercise 13.10, suppose that all respondents answer truthfully with any of the randomized response methods (b), (c), (d), or (e), but that under a single direct sensitive question, method (a), some respondents who have committed the act deny this. For $n = 500$, which of the randomized response methods give a smaller $\text{MSE}(\hat{\pi}_A)$ than method (a) if (i) 15%, (ii) 20%, (iii) 25% of those who have committed the act deny it under method (a)?

References

Armitage, P. (1947). A comparison of stratified with unrestricted random sampling from a finite population. *Biometrika*, **34,** 273–280.

Arvesen, J. N. (1969). Jackknifing U-Statistics. *Ann. Math. Stat.*, **40,** 2076–2100.

Avadhani, M. S. and Sukhatme, B. V. (1973). Controlled sampling with equal probabilities and without replacement. *Int. Stat. Rev.*, **41,** 175–183.

Bailar, B. A. (1975). The effects of rotation group bias on estimates from panel surveys. *Jour. Amer. Stat. Assoc.*, **70,** 23–30.

Bailar, B. A., and Dalenius, T. (1969). Estimating response variance components of the U.S. Bureau of the Census Survey Model. *Sankhya*, **B31,** 341–360.

Barr, A. (1957). Differences between experienced interviewers. *App. Stat.*, **6,** 180–188.

Bartholomew, D. J. (1961). A method of allowing for "not-at-home" bias in sample surveys, *App. Stat.*, **10,** 52–59.

Bartlett, M. S. (1949). Fitting a straight line when both variables are subject to error. *Biometrics*, **5,** 207–212.

Basu, D. (1958). On sampling with and without replacement. *Sankhya*, **20,** 287–294.

Bayless, D. L. (1968). Variance estimation in sampling from finite populations. Ph.D. Thesis, Texas A & M University.

Bayless, D. L., and Rao, J. N. K. (1970). An empirical study of stabilities of estimators and variance estimators in unequal probability sampling ($n = 3$ or 4). *Jour. Amer. Stat. Assoc.*, **65,** 1645–1667.

Beale, E. M. L. (1962). Some uses of computers in operational research. *Industrielle Organisation*, **31,** 51–2.

Bean, J. A. (1970). Estimation and sampling variance in the health interview survey. National Center for Health Statistics, Washington, D.C., Series 2, 38.

Bean, J. A. (1975). Distribution and properties of variance estimators for complex multistage probability samples. National Center for Health Statistics, Washington, D.C., Series 2, 65.

Beardwood, J., Halton, J. H., and Hammersley, J. M. (1959). The shortest path through many points, *Proc. Cambridge Phil. Soc.*, **55,** 299–327.

Bellhouse, D. R. and Rao, J. N. K. (1975). Systematic sampling in the presence of a trend. *Biometrika*, **62,** 694–697.

Belloc, N. B. (1954). Validation of morbidity survey data by comparison with hospital records. *Jour. Amer. Stat. Assoc.*, **49,** 832–846.

Birnbaum, Z. W., and Sirken, M. G. (1950a). Bias due to nonavailability in sampling surveys. *Jour. Amer. Stat. Assoc.*, **45,** 98–111.

Birnbaum, Z. W., and Sirken, M. G. (1950b). On the total error due to noninterview and to random sampling. *Int. Jour. Opinion and Attitude Res.*, **4,** 179–191.

Blythe, R. H. (1945). The economics of sample size applied to the scaling of saw-logs. *Biom. Bull.*, **1,** 67–70.

Booth, G., and Sedransk, J. (1969). Planning some two-factor comparative surveys. *Jour. Amer. Stat. Assoc.*, **64**, 560–573.

Bose, Chameli (1943). Note on the sampling error in the method of double sampling. *Sankhya*, **6**, 330.

Brewer, K. W. R. (1963a). A model of systematic sampling with unequal probabilities. *Australian Jour. Stat.*, **5**, 5–13.

Brewer, K. W. R. (1963b). Ratio estimation in finite populations: Some results deducible from the assumption of an underlying stochastic process. *Australian Jour. Stat.*, **5**, 93–105.

Brewer, K. W. R., and Hanif, M. (1969). Sampling without replacement and probability of inclusion proportional to size. I Methods using the Horvitz and Thompson estimator. II Methods using special estimators. Unpublished manuscript.

Brewer, K. W. R., and Hanif, M. (1970). Durbin's new multistage variance estimator. *Jour. Roy. Stat. Soc.*, **B32**, 302–311.

Brooks, S. (1955). The estimation of an optimum subsampling number. *Jour. Amer. Stat. Assoc.*, **50**, 398–415.

Bryant, E. C., Hartley, H. O., and Jessen, R. J. (1960). Design and estimation in two-way stratification. *Jour. Amer. Stat. Assoc.*, **55**, 105–124.

Buckland, W. R. (1951). A review of the literature of systematic sampling. *Jour. Roy. Stat. Soc.*, **B13**, 208–215.

Burstein, H. (1975). Finite population correction for binomial confidence limits. *Jour. Amer. Stat. Assoc.*, **70**, 67–69.

Cameron, J. M. (1951). Use of variance components in preparing schedules for the sampling of baled wool. *Biomerics*, **7**, 83–96.

Chatterjee, S. (1966). A programming algorithm and its statistical applications. O.N.R. Tech. Rept. 1, Department of Statistics, Harvard University, Cambridge.

Chatterjee, S. (1967). A note on optimum stratification. *Skand. Akt.*, **50**, 40–44.

Chatterjee, S. (1968). Multivariate stratified surveys. *Jour. Amer. Stat. Assoc.*, **63**, 530–534.

Chatterjee, S. (1972). A study of optimum allocation in multivariate stratified surveys. *Skand. Akt.*, **55**, 73–80.

Chung, J. H., and DeLury, D. B. (1950). *Confidence Limits for the Hypergeometric Distribution.* University of Toronto Press, Toronto, Canada.

Cochran, W. G. (1942). Sampling theory when the sampling units are of unequal sizes. *Jour. Amer. Stat. Assoc.*, **37**, 199–212.

Cochran, W. G. (1946). Relative accuracy of systematic and stratified random samples for a certain class of populations. *Ann. Math. Stat.*, **17**, 164–177.

Cochran, W. G. (1961). Comparison of methods for determining stratum boundaries. *Bull. Int. Stat. Inst.*, **38**, 2, 345–358.

Cochran, W. G., Mosteller, F., and Tukey, J. W. (1954). *Statistical Problems of the Kinsey Report. American Statistical Association, Washington, D.C.*, p. 280.

Coleman, J. S. (1966). *Equality of Educational Opportunity*, U.S. Government Printing Office, Washington, D.C.

Cornell, F. G. (1947). A stratified random sample of a small finite population. *Jour. Amer. Stat. Assoc.*, **42**, 523–532.

Cornfield, J. (1944). On samples from finite populations. *Jour. Amer. Stat. Assoc.*, **39**, 236–239.

Cornfield, J. (1951). The determination of sample size. *Amer. Jour. Pub. Health*, **41**, 654–661.

Cox, D. R. (1952). Estimation by double sampling. *Biometrika*, **39**, 217–227.

Dalenius, T. (1957). *Sampling in Sweden.* Contributions to the methods and theories of sample survey practice. Almqvist and Wicksell, Stockholm.

Dalenius, T., and Gurney, M. (1951). The problem of optimum stratification. II. *Skand. Akt.*, **34**, 133–148.

Dalenius, T., and Hodges, J. L., Jr. (1959). Minimum variance stratification. *Jour. Amer. Stat. Assoc.*, **54,** 88–101.

Das, A. C. (1950). Two-dimensional systematic sampling and the associated stratified and random sampling. *Sankhya*, **10,** 95–108.

David, F. N., and Neyman, J. (1938). Extension of the Markoff theorem of least squares. *Stat. Res. Mem.*, **2,** 105.

David, I. P., and Sukhatme, B. V. (1974). On the bias and mean square error of the ratio estimator. *Jour. Amer. Stat. Assoc.*, **69,** 464–466.

Deming, W. E. (1953). On a probability mechanism to attain an economic balance between the resultant error of non-response and the bias of non-response. *Jour. Amer. Stat. Assoc.*, **48,** 743–772.

Deming, W. E. (1956). On simplifications of sampling design through replication with equal probabilities and without stages. *Jour. Amer. Stat. Assoc.*, **51,** 24–53.

Deming, W. E. (1960). *Sample Design in Business Research.* John Wiley and Sons, New York.

Deming, W. E., and Simmons, W. R. (1946). On the design of a sample for dealer inventories. *Jour. Amer. Stat. Assoc.*, **41,** 16–33.

Des Raj (1954). On sampling with probabilities proportional to size. *Ganita*, **5,** 175–182.

Des Raj (1956a). Some estimators in sampling with varying probabilities without replacement. *Jour. Amer. Stat. Assoc.*, **51,** 269–284.

Des Raj (1956b). A note on the determination of optimum probabilities in sampling without replacement. *Sankhya*, **17,** 197–200.

Des Raj (1958). On the relative accuracy of some sampling techniques. *Jour. Amer. Stat. Assoc.*, **53,** 98–101.

Des Raj (1964). The use of systematic sampling with probability proportional to size in a large-scale survey. *Jour. Amer. Stat. Assoc.*, **59,** 251–255.

Des Raj (1966). Some remarks on a simple procedure of sampling without replacement. *Jour. Amer. Stat. Assoc.*, **61,** 391–396.

Des Raj, and Khamis, S. H. (1958). Some remarks on sampling with replacement. *Ann. Math. Stat.*, **29,** 550–557.

Dowling, T. A., and Shachtman, R. H. (1975). On the relative efficiency of randomized response models. *Jour. Amer. Stat. Assoc.*, **70,** 84–87.

Durbin, J. (1953). Some results in sampling theory when the units are selected with unequal probabilities. *Jour. Roy. Stat. Soc.*, **B15,** 262–269.

Durbin, J. (1954). Non-response and call-backs in surveys. *Bull. Int. Stat. Inst.*, **34,** 72–86.

Durbin, J. (1958). Sampling theory for estimates based on fewer individuals than the number selected. *Bull. Int. Stat. Inst.*, **36,** 3, 113–119.

Durbin, J. (1959). A note on the application of Quenouille's method of bias reduction to the estimation of ratios. *Biometrika*, **46,** 477–480.

Durbin, J. (1967). Design of multi-stage surveys for the estimation of sampling errors *App. Stat.*, **16,** 152–164.

Durbin, J., and Stuart, A. (1954). Callbacks and clustering in sample surveys: an experimental study. *Jour. Roy. Stat. Soc.*, **A117,** 387–428.

Eckler, A. R. (1955). Rotation sampling. *Ann. Math. Stat.*, **26,** 664–685.

Ekman, G. (1959). An approximation useful in univariate stratification. *Ann. Math. Stat.*, **30,** 219–229.

Erdös, P., and Rényi, A. (1959). On the central limit theorem for samples from a finite population. *Pub. Math. Inst. Hungarian Acad. Sci.*, **4,** 49–57.

Ericson, W. A. (1969). Subjective Bayesian models in sampling finite populations. *Jour. Roy. Stat. Soc.*, **B31,** 195–233.

Evans, W. D. (1951). On stratification and optimum allocations. *Jour. Amer. Stat. Assoc.*, **46,** 95–104.

Fellegi, I. (1963). Sampling with varying probabilities without replacement: rotating and non-rotating samples. *Jour. Amer. Stat. Assoc.*, **58**, 183–201.

Fellegi, I. (1964). Response variance and its estimation. *Jour. Amer. Stat. Assoc.*, **59**, 1016–1041.

Feller, W. (1957). *An Introduction to Probability Theory and Its Applications*, John Wiley and Sons, New York, second edition.

Fieller, E. C. (1932). The distribution of the index in a normal bivariate population. *Biometrika*, **24**, 428–440.

Finkner, A. L. (1950). Methods of sampling for estimating commercial peach production in North Carolina. *North Carolina Agr. Exp. Stat. Tech. Bull.*, **91**.

Finkner, A. L., Morgan, J. J., and Monroe, R. J. (1943). Methods of estimating farm employment from sample data in North Carolina. *N. C. Agr. Exp. Sta. Tech. Bull.*, **75**.

Finney, D. J. (1948). Random and systematic sampling in timber surveys. *Forestry*, **22**, 1–36.

Finney, D. J. (1949). On a method of estimating frequencies. *Biometrika*, **36**, 233–234.

Finney, D. J. (1950). An example of periodic variation in forest sampling. *Forestry*, **23**, 96–111.

Fisher, R. A. (1958). *Statistical Methods for Research Workers*. Oliver and Boyd, Edinburgh, thirteenth edition, section 21, fourth ed. (1932).

Fisher, R. A., and Mackenzie, W. A. (1922). The correlation of weekly rainfall. *Quart. Jour. Roy. Met. Soc.*, **48**, 234–245.

Fisher, R. A., and Yates, F. (1957). *Statistical Tables for Biological, Agricultural and Medical Research*. Oliver and Boyd, Edinburgh, fifth edition.

Foreman, E. K., and Brewer, K. W. R. (1971). The efficient use of supplementary information in standard sampling procedures. *Jour. Roy. Stat. Soc.* **B33**, 391–400.

Frankel, M. R. (1971). Inference from survey samples. Institute for Social Research, Ann Arbor, Mich.

Fuller, W. A. (1966). Estimation employing post strata. *Jour. Amer. Stat. Assoc.*, **61**, 1172–1183.

Fuller, W. A. (1970). Sampling with random stratum boundaries. *Jour. Roy. Stat. Soc.*, **B32**, 209–226.

Fuller, W. A., and Burmeister, L. F. (1972). Estimators for samples selected from two overlapping frames. *Proc. Soc. Stat. Sect. Amer. Stat. Assoc.*, 245–249.

Gallup, G. (1972). Opinion polling in a democracy. *Statistics, a Guide to the Unknown*. J. M. Tanur et al. (eds.), Holden-Day, Inc., San Francisco, 146–152.

Gautschi, W. (1957). Some remarks on systematic sampling. *Ann. Math. Stat.*, **28**, 385–394.

Godambe, V. P. (1955). A unified theory of sampling from finite populations. *Jour. Roy. Stat. Soc.*, **B17**, 269–278.

Goodman, L. A., and Hartley, H. O. (1958). The precision of unbiased ratio-type estimators. *Jour. Amer. Stat. Assoc.*, **53**, 491–508.

Goodman, R., and Kish, L. (1950). Controlled selection—a technique in probability sampling. *Jour. Amer. Stat. Assoc.*, **45**, 350–372.

Graham, J. E. (1973). Composite estimation in two cycle rotation sampling designs. *Comm. in Stat.*, **1**, 419–431.

Gray, P. G. (1955). The memory factor in social surveys. *Jour. Amer. Stat. Assoc.*, **50**, 344–363.

Gray, P. G. (1957). A sample survey with both a postal and an interview stage. *App. Stat.*, **6**, 139–153.

Gray, P. G., and Corlett, T. (1950). Sampling for the social survey. *Jour. Roy. Stat. Soc.*, **A113**, 150–206.

Greenberg, B. G., et al. (1969). The unrelated question randomized response model: Theoretical framework. *Jour. Amer. Stat. Assoc.*, **64**, 520–539.

Greenberg, B. G. et al. (1971). Application of the randomized response technique in obtaining quantitative data. *Jour. Amer. Stat. Assoc.*, **66**, 243–250.

Grundy, P. M., Healy, M. J. R., and Rees, D. H. (1954). Decision between two alternatives—how many experiments? *Biometrics*, **10**, 317–323.

Grundy, P. M., Healy, M. J. R., and Rees, D. H. (1956). Economic choice of the amount of experimentation. *Jour. Roy. Stat. Soc.*, **B18**, 32–55.

Hagood, M. J., and Bernert, E. H. (1945). Component indexes as a basis for stratification. *Jour. Amer. Stat. Assoc.*, **40**, 330–341.

Hájek, J. (1958). Some contributions to the theory of probability sampling. *Bull. Int. Stat. Inst.*, **36**, 3, 127–134.

Hájek, J. (1960). Limiting distributions in simple random sampling from a finite population. *Pub. Math. Inst. Hungarian Acad. Sci.*, **5**, 361–374.

Haldane, J. B. S. (1945). On a method of estimating frequencies. *Biometrika*, **33**, 222–225.

Hansen, M. H., et al. (1951). Response errors in surveys. *Jour. Amer. Stat. Assoc.*, **46**, 147–190.

Hansen, M. H., and Hurwitz, W. N. (1942). Relative efficiencies of various sampling units in population inquiries. *Jour. Amer. Stat. Assoc.*, **37**, 89–94.

Hansen, M. H., and Hurwitz, W. N. (1943). On the theory of sampling from finite populations. *Ann. Math. Stat.*, **14**, 333–362.

Hansen, M. H., and Hurwitz, W. N. (1946). The problem of nonresponse in sample surveys. *Jour. Amer. Stat. Assoc.*, **41**, 517–529.

Hansen, M. H., and Hurwitz, W. N. (1949). On the determination of the optimum probabilities in sampling. *Ann. Math. Stat.*, **20**, 426–432.

Hansen, M. H., Hurwitz, W. N., and Bershad, M. (1961). Measurement errors in censuses and surveys. *Bull. Int. Stat. Inst.*, **38**, 2, 359–374.

Hansen, M. H., Hurwitz, W. N., and Gurney, M. (1946). Problems and methods of the sample survey of business. *Jour. Amer. Stat. Assoc.*, **41**, 173–189.

Hansen, M. H., Hurwitz, W. N., and Jabine, T. B. (1963). The use of imperfect lists for probability sampling at the U.S. Bureau of the Census. *Bull. Int. Stat. Inst.*, **40**, 1, 497–517.

Hansen, M. H., Hurwitz, W. N., and Madow, W. G. (1953). *Sample Survey Methods and Theory*. John Wiley and Sons, New York, Vols. I and II.

Hansen, M. H., Hurwitz, W. N., Nisselson, H., and Steinberg, J. (1955). The redesign of the census current population survey. *Jour. Amer. Stat. Assoc.*, **50**, 701–719.

Hansen, M. H., Hurwitz, W. N., and Pritzker, L. (1965). The estimation and interpretation of gross differences and the simple response variance. *Contributions to statistics presented to Professor P. C. Mahalanobis*. Pergamon Press, Oxford, and Statistical Publishing Society, Calcutta, 111–136.

Hansen, M. H., and Waksberg, J. (1970). Research on non-sampling errors in censuses and surveys. *Rev. Int. Stat. Inst.*, **38**, 318–332.

Hanson, R. H., and Marks, E. S. (1958). Influence of the interviewer on the accuracy of survey results. *Jour. Amer. Stat. Assoc.*, **53**, 635–655.

Hartley, H. O. (1946). Discussion of paper by F. Yates. *Jour. Roy. Stat. Soc.*, **109**, 37.

Hartley, H. O. (1959). *Analytic Studies of Survey Data*. Istituto di Statistica, Rome, volume in honor of Corrado Gini.

Hartley, H. O. (1962). Multiple frame surveys. *Proc. Soc. Stat. Sect. Amer. Stat. Assoc.*, 203–206.

Hartley, H. O. (1974). Multiple frame methodology and selected applications. *Sankhya*, **C36**, 99–118.

Hartley, H. O., and Hocking, R. (1963). Convexing programming by tangential approximation. *Management Science*, **9**, 600–612.

Hartley, H. O., and Rao, J. N. K. (1962). Sampling with unequal probabilities and without replacement. *Ann. Math. Stat.*, **33**, 350–374.

Hartley, H. O., and Rao, J. N. K. (1968). A new estimation theory for sample surveys. *Biometrika*, **55**, 547–557.

Hartley, H. O., and Rao, J. N. K. (1969). A new estimation theory for sample surveys, II. In *New Developments in Survey Sampling*, N. L. Johnson and H. Smith (eds.), Wiley-Interscience, New York, 147–169.

Hartley, H. O., Rao, J. N. K., and Kiefer, G. (1969). Variance estimation with one unit per stratum. *Jour. Amer. Stat. Assoc.*, **64**, 841–851.

Hartley, H. O., and Ross, A. (1954). Unbiased ratio estimates. *Nature*, **174**, 270–271.

Harvard Computation Laboratory (1955). *Tables of the Cumulative Binomial Probability Distribution*. Harvard University Press, Cambridge, Mass.

Haynes, J. D. (1948). An empirical investigation of sampling methods for an area. M. S. thesis, University of North Carolina.

Hendricks, W. A. (1944). The relative efficiencies of groups of farms as sampling units. *Jour. Amer. Stat. Assoc.*, **39**, 367–376.

Hendricks, W. A. (1949). Adjustment for bias by non-response in mailed surveys. *Agr. Econ. Res.*, **1**, 52–56.

Hendricks, W. A. (1956). *The Mathematical Theory of Sampling*. Scarecrow Press, New Brunswick, N.J.

Hess, I., Riedel, D. C., and Fitzpatrick, T. B. (1976). *Probability Sampling of Hospitals and Patients.* University of Michigan, Ann Arbor, Mich., second edition.

Hess, I., Sethi, V. K., and Balakrishnan, T. R. (1966). Stratification: A practical investigation. *Jour. Amer. Stat. Assoc.*, **61**, 74–90.

Hoeffding, W. (1948). A class of statistics with asymptotically normal distribution. *Ann. Math. Stat.*, **19**, 293–325.

Homeyer, P. G., and Black, C. A. (1946). Sampling replicated field experiments on oats for yield determinations. *Proc. Soil Sci. Soc. America*, **11**, 341–344.

Horvitz, D. G. (1952). Sampling and field procedures of the Pittsburgh morbidity survey. *Pub. Health Reports*, **67**, 1003–1012.

Horvitz, D. G., Greenberg, B. G., and Abernathy, J. R. (1975). Recent developments in randomized response designs. *A survey of statistical design and linear models.* J. N. Srivastava (ed.), American Elsevier Publishing Co., New York, 271–285.

Horvitz, D. G., Shah, B. V., and Simmons, W. R. (1967). The unrelated randomized response model. *Proc. Soc. Stat. Sect. Amer. Stat. Assoc.*, 65–72.

Horvitz, D. G., and Thompson, D. J. (1952). A generalization of sampling without replacement from a finite universe. *Jour. Amer. Stat. Assoc.*, **47**, 663–685.

Huddleston, H. F., Claypool, P. L., and Hocking, R. R. (1970). Optimum sample allocation to strata using convex programming. *App. Stat.* **19**, 273–278.

Hutchinson, M. C. (1971). A Monte Carlo comparison of some ratio estimators. *Biometrika*, **58**, 313–321.

Hyman, H. H. (1954). *Interviewing in Social Research*, University of Chicago Press, Chicago, Ill.

James, A. T., Wilkinson, G. N., and Venables, W. N. (1975). Interval estimates for a ratio of means. *Sankhya* (in press).

Jebe, E. H. (1952). Estimation for sub-sampling designs employing the county as a primary sampling unit. *Jour. Amer. Stat. Assoc.*, **47**, 49–70.

Jensen, A. (1926). Report on the representative method in statistics. *Bull. Int. Stat. Inst.*, **22**, 359–377.

Jessen, R. J. (1942). Statistical investigation of a sample survey for obtaining farm facts. *Iowa Agr. Exp. Sta. Res. Bull.*, 304.

Jessen, R. J. (1955). Determining the fruit count on a tree by randomized branch sampling. *Biometrics*, **11**, 99–109.

Jessen, R. J., et al. (1947). On a population sample for Greece. *Jour. Amer. Stat. Assoc.*, **42**, 357–384.

Jessen, R. J., and Houseman, E. E. (1944). Statistical investigations of farm sample surveys taken in Iowa, Florida and California. *Iowa Agr. Exp. Sta. Res. Bull.*, 329.

Johnson, F. A. (1941). A statistical study of sampling methods for tree nursery inventories. M. S. thesis, Iowa State College.

Johnson, F. A. (1943). A statistical study of sampling methods for tree nursery inventories. *Jour. Forestry*, **41**, 674–689.

Jones, H. W. (1955). Investigating the properties of a sample mean by employing random subsample means. *Jour. Amer. Stat. Assoc.*, **51**, 54–83.

Kempthorne, O. (1969). Some remarks on inference in finite sampling. *New Developments in Survey Sampling*, N. L. Johnson and H. Smith, Jr. (eds.), John Wiley & Sons, New York, 671–695.

Kendall, M. G., and Smith, B. B. (1938). Randomness and random sampling numbers. *Jour. Roy. Stat. Soc.*, **101**, 147–166.

Keyfitz, N. (1957). Estimates of sampling variance where two units are selected from each stratum. *Jour. Amer. Stat. Assoc.*, **52**, 503–510.

Khan, S., and Tripathi, T. P. (1967). The use of multivariate auxiliary information in double-sampling. *J. Ind. Stat. Assoc.*, **5**, 42–48.

King, A. J., and McCarty, D. E. (1941). Application of sampling to agricultural statistics with emphasis on stratified samples. *Jour. Marketing*, April, 462–474.

Kiser, C. V., and Whelpton, P. K. (1953). Resume of the Indianapolis study of social and psychological factors affecting fertility. *Population Studies*, **7**, 95–110.

Kish, L. (1949). A procedure for objective respondent selection within the household. *Jour. Amer. Stat. Assoc.*, **44**, 380–387.

Kish, L. (1957). Confidence limits for clustered samples. *Amer. Soc. Rev.*, **22**, 154–165.

Kish, L. (1965). *Survey Sampling.* John Wiley & Sons, New York.

Kish, L., and Frankel, M. R. (1974). Inference from complex samples. *Jour. Roy. Stat. Soc.*, **B36**, 1–37.

Kish, L., and Hess, I. (1958). On noncoverage of sample dwellings. *Jour. Amer. Stat. Assoc.*, **53**, 509–524.

Kish, L., and Hess, I. (1959a). A "replacement" procedure for reducing the bias of nonresponse. *Amer. Statistician*, **13**, 4, 17–19.

Kish, L., and Hess, I. (1959b). On variances of ratios and their differences in multistage samples. *Jour. Amer. Stat. Assoc.*, **54**, 416–446.

Kish, L., and Lansing, J. B. (1954). Response errors in estimating the value of homes. *Jour. Amer. Stat. Assoc.*, **49**, 520–538.

Kish, L., Namboodiri, N. K., and Pillai, R. K. (1962). The ratio bias in surveys, *Jour. Amer. Stat. Assoc.*, **57**, 863–876.

Koch, G. (1973). An alternative approach to multivariate response error models for sample survey data with applications to estimators involving subclass means. *Jour. Amer. Stat. Ass.*, **68**, 906–913.

Kokan, A. R. (1963). Optimum allocation in multivariate surveys. *Jour. Roy. Stat. Soc.*, **A126**, 557–565.

Koons, D. A. (1973). Quality control and measurement of nonsampling error in the Health Interview Survey. *Nat. Center for Health Stat.*, Series 2, 54.

Koop, J. C. (1960). On theoretical questions underlying the technique of replicated or interpenetrating samples. *Proc. Soc. Stat. Sect. Amer. Stat. Assoc.*, 196–205.

Koop, J. C. (1968). An exercise in ratio estimation. *Amer. Statistician*, **22**, 1, 29–30.

Kulldorff, G. (1963). Some problems of optimum allocation for sampling on two occasions. *Rev. Int. Stat. Inst.*, **31**, 24–57.

Lahiri, D. B. (1951). A method for sample selection providing unbiased ratio estimates. *Bull. Int. Stat. Inst.*, **33**, 2, 133–140.

Lieberman, G. J., and Owen, D. B. (1961). *Tables of the Hypergeometric Probability Distribution.* Stanford University Press, Stanford, Calif.

Lienau, C. C. (1941). Selection, training and performance of the National Health Survey field staff. *Amer. Jour. Hygiene*, **34**, 110–132.

Lund, R. E. (1968). Estimators in multiple frame surveys. *Proc. Soc. Sci. Sect. Amer. Stat. Assoc.*, 282–288.

McCarthy, P. J. (1966). Replication: An approach to the analysis of data from complex surveys. National Center for Health Statistics, Washington, D. C., Series, 2, 14.

McCarthy, P. J. (1969). Pseudo-replication: Half-samples. *Rev. Int. Stat. Inst.*, **37**, 239–264.

McVay, F. E. (1947). Sampling methods applied to estimating numbers of commercial orchards in a commercial peach area. *Jour. Amer. Stat. Assoc.*, **42**, 533–540.

Madow, L. H. (1946). Systematic sampling and its relation to other sampling designs. *Jour. Amer. Stat. Assoc.*, **41**, 207–214.

Madow, L. H. (1950). On the use of the county as a primary sampling unit for state estimates. *Jour. Amer. Stat. Assoc.*, **45**, 30–47.

Madow, W. G. (1948). On the limiting distributions of estimates based on samples from finite universes. *Ann. Math. Stat.*, **19**, 535–545.

Madow, W. G. (1949). On the theory of systematic sampling, II. *Ann Math. Stat.*, **20**, 333–354.

Madow, W. G. (1953). On the theory of systematic sampling, III. *Ann. Math. Stat.*, **24**, 101–106.

Madow, W. G., and Madow, L. H. (1944). On the theory of systematic sampling. *Ann. Math. Stat.*, **15**, 1–24.

Madow, W. G. (1965). On some aspects of response error measurement. *Proc. Soc. Stat. Soc. Amer. Stat. Assoc.*, 182–192.

Mahalanobis, P. C. (1944). On large-scale sample surveys. *Phil. Trans. Roy. Soc. London*, **B231**, 329–451.

Mahalanobis, P. C. (1946). Recent experiments in statistical sampling in the Indian Statistical Institute. *Jour. Roy. Stat. Soc.*, **109**, 325–370.

Matérn, B. (1947). Methods of estimating the accuracy of line and sample plot surveys. *Medd. fr. Statens Skogsforsknings Institut.*, **36**, 1–138.

Matérn, B. (1960). Spatial variation. *Medd. fr. Statens Skogsforsknings Institut.*, **49**, 5, 1–144.

Mickey, M. R. (1959). Some finite population unbiased ratio and regression estimators. *Jour. Amer. Stat. Assoc.*, **54**, 594–612.

Midzuno, H. (1951). On the sampling system with probability proportionate to sum of sizes *Ann. Inst. Stat. Math.*, **2**, 99–108.

Milne, A. (1959). The centric systematic area sample treated as a random sample. *Biometrics*, **15**, 270–297.

Moors, J. J. A. (1971). Optimization of the unrelated question randomized response model. *Jour. Amer. Stat. Assoc.*, **66**, 627–629.

Murthy, M. N. (1957). Ordered and unordered estimators in sampling without replacement. *Sankhya*, **18**, 379–390.

Murthy, M. N. (1967). *Sampling Theory and Methods*. Statistical Publishing Society, Calcutta, India.

Narain, R. D. (1951). On sampling without replacement with varying probabilities. *Jour. Ind. Soc. Agric. Stat.*, **3**, 169–174.

National Bureau of Standards (1950). *Tables of the Binomial Probability Distribution*. U.S. Government Printing Office, Washington, D.C.

Neter, J. (1972). How accountants save money by sampling. *Statistics, A Guide to the Unknown*, J. M. Tanur et al. (eds.), Holden-Day, Inc., San Francisco, 203–211.

Neyman, J. (1934). On the two different aspects of the representative method: The method of stratified sampling and the method of purposive selection. *Jour. Roy. Stat. Soc.*, **97**, 558–606.

Neyman, J. (1938). Contribution to the theory of sampling human populations. *Jour. Amer. Stat. Assoc.*, **33**, 101–116.

Nordbotten, S. (1956). Allocation in stratified sampling by means of linear programming. *Skand. Akt. Tidskr.*, **39**, 1–6.

Nordin, J. A. (1944). Determining sample size. *Jour. Amer. Stat. Assoc.*, **39**, 497–506.

Olkin, I. (1958). Multivariate ratio estimation for finite populations. *Biometrika*, **45**, 154–165.

Osborne, J. G. (1942). Sampling errors of systematic and random surveys of cover-type areas. *Jour. Amer. Stat. Assoc.*, **37**, 256–264.

Patterson, H. D. (1950). Sampling on successive occasions with partial replacement of units. *Jour. Roy. Stat. Soc.*, **B12**, 241–255.

Patterson, H. D. (1954). The errors of lattice sampling. *Jour. Roy. Stat. Soc.*, **B16**, 140–149.

Paulson, E. (1942). A note on the estimation of some mean values for a bivariate distribution. *Ann. Math. Stat.*, **13**, 440–444.

Payne, S. L. (1951). *The Art of Asking Questions*. Princeton University Press, Princeton, N.J.

Plackett, R. L., and Burman, J. P. (1946). The design of optimum multifactorial experiments. *Biometrika*, **33**, 305–325.

Platek, R., and Singh, M. P. (1972). Some aspects of redesign of the Canadian Labor Force Survey. *Proc. Soc. Stat. Sect. Amer. Stat. Assoc.*, 397–402.

Politz, A. N., and Simmons, W. R. (1949, 1950). An attempt to get the "not at homes" into the sample without callbacks. *Jour. Amer. Stat. Assoc.*, **44**, 9–31, and **45**, 136–137.

Pritzker, L., and Hanson, R. (1962). Measurement errors in the 1960 Census of Population. *Proc. Soc. Stat. Sect. Amer. Stat. Assoc.*, 80–89.

Quenouille, M. H. (1949). Problems in plane sampling. *Ann. Math. Stat.*, **20**, 355–375.

Quenouille, M. H. (1956). Notes on bias in estimation. *Biometrika*, **43**, 353–360.

Raiffa, H., and Schlaifer, R. (1961). *Applied Statistical Decision Theory*. Harvard Business School, Cambridge, Mass.

Rand Corporation (1955). *A Million Random Digits*. Free Press, Glencoe, Ill.

Rao, C. R. (1971). Some aspects of statistical inference in problems of sampling from finite populations. *Foundations of Statistical Inference*. V. P. Godambe and D. A. Sprott, (eds.), Holt, Rinehart, and Winston, Toronto, Canada 177–202.

Rao, J. N. K. (1962). On the estimation of the relative efficiency of sampling procedures. *Ann. Inst. Stat. Math.*, **14**, 143–150.

Rao, J. N. K. (1965). On two simple schemes of unequal probability sampling without replacement. *Jour. Ind. Stat. Assoc.*, **3**, 173–180.

Rao, J. N. K. (1966). Alternative estimators in *pps* sampling for multiple characteristics. *Sankhya*, **A23**, 47–60.

Rao, J. N. K. (1968). Some small sample results in ratio and regression estimation. *Jour. Ind. Stat. Assoc.*, **6**, 160–168.

Rao, J. N. K. (1969). Ratio and regression estimators. *New Developments in Survey Sampling*, N. L. Johnson and H. Smith, Jr. (eds.), John Wiley & Sons, New York, 213–234.

Rao, J. N. K. (1973). On double sampling for stratification and analytical surveys. *Biometrika*, **60**, 125–133.

Rao, J. N. K. (1975a). On the foundations of survey sampling. In *A Survey of Statistical Design and Linear Models*, J. N. Srivastava (ed.), American Elsevier Publishing Co, New York, 489–505.

Rao, J. N. K. (1975b). Unbiased variance estimation for multistage designs. *Sankhya* (in press).

Rao, J. N. K., and Bayless, D. L. (1969). An empirical study of the stabilities of estimators and variance estimators in unequal probability sampling of two units per stratum. *Jour. Amer. Stat. Assoc.*, **64**, 540–559.

Rao, J. N. K., and Beegle, L. D. (1967). A Monte Carlo study of some ratio estimators. *Sankhya*, **B29**, 47–56.

Rao, J. N. K., and Graham, J. E. (1964). Rotation designs for sampling on repeated occasions. *Jour. Amer. Stat. Assoc.*, **59**, 492–509.

Rao, J. N. K., and Kuzik, R. A. (1974). Sampling errors in ratio estimation. *Indian Jour. Stat.* **36**, C, 43–58.

Rao, J. N. K., Hartley, H. O., and Cochran, W. G. (1962). A simple procedure of unequal probability sampling without replacement. *Jour. Roy. Stat. Soc.* **B24**, 482–491.

Rao, J. N. K., and Pereira, N. P. (1968). On double ratio estimators. *Sankhya*, **A30**, 83–90.

Rao, J. N. K., and Singh, M. P. (1973). On the choice of estimator in survey sampling. *Australian Jour. Stat.*, **15**, 95–104.

Rao, P. S. R. S., and Mudholkar, G. S. (1967). Generalized multivariate estimations for the mean of finite populations. *Jour. Amer. Stat. Assoc.*, **62**, 1008–1012.

Rao, P. S. R. S., and Rao, J. N. K. (1971). Small sample results for ratio estimators. *Biometrika*, **58**, 625–630.

Robson, D. S. (1952). Multiple sampling of attributes. *Jour. Amer. Stat. Assoc.*, **47**, 203–215.

Robson, D. S. (1957). Applications of multivariate polykays to the theory of unbiased ratio type estimation. *Jour. Amer. Stat. Assoc.*, **52**, 511–522.

Robson, D. S., and King, A. J. (1953). Double sampling and the Curtis impact survey. *Cornell Univ. Agr. Exp. Sta. Mem.*, 231.

Romig, H. G. (1952). *50–100 Binomial Tables.* John Wiley & Sons, New York.

Roy, J., and Chakravarti, I. M. (1960). Estimating the mean of a finite population. *Ann. Math. Stat.*, **31**, 392–398.

Royall, R. M. (1968). An old approach to finite population sampling theory. *Jour. Amer. Stat. Assoc.*, **63**, 1269–1279.

Royall, R. M. (1970a). On finite population sampling theory under certain linear regression models. *Biometrika*, **57**, 377–387.

Royall, R. M. (1970b). Finite population sampling—on labels in estimation., *Ann. Math. Stat.*, **41**, 1774–1779.

Royall, R. M. (1971). Linear regression models in finite population sampling theory. *Foundations of Statistical Inference*, V. P. Godambe, and D. A. Sprott (eds.), Holt, Rinehart, & Winston, Toronto, Canada, 259–279.

Royall, R. M., and Herson, J. (1973). Robust estimation in finite populations, I. *Jour. Amer. Stat. Assoc.*, **68**, 880–889.

Sagen, O. K., Dunham, R. E., and Simmons, W. R. (1959). Health statistics from record sources and household interviews compared. *Proc. Soc. Stat. Sect. Amer. Stat. Assoc.*, 6–15.

Sampford, M. R. (1967). On sampling without replacement with unequal probabilities of selection. *Biometrika*, **54**, 499–513.

Sandelius, M. (1951). Truncated inverse binomial sampling. *Skandinavisk Aktuarietidskrift*, **34**, 41–44.

Särndal, C. E. (1972). Sample survey theory vs. general statistical theory: Estimation of the population mean, *Rev. Int. Stat. Inst.*, **40**, 1–12.

Satterthwaite, F. E. (1946). An approximate distribution of estimates of variance components. *Biometrics*, **2**, 110–114.

Scott, A. J., and Smith, T. M. F. (1974). Analysis of repeated surveys using time series methods. *Jour. Amer. Stat. Assoc.*, **69**, 674–678.

Sedransk, J. (1965). A double sampling scheme for analytical surveys. *Jour. Amer. Stat. Assoc.*, **60**, 985–1004.

Sedransk, J. (1967). Designing some multi-factor analytical studies. *Jour. Amer. Stat. Assoc.*, **62**, 1121–1139.

Sen, A. R. (1953). On the estimate of variance in sampling with varying probabilities. *Jour. Ind. Soc. Agric. Stat.*, **5**, 119–127.

Sen, A. R. (1972). Successive sampling with p ($p \geq 1$) auxiliary variables. *Ann. Math. Stat.*, **43**, 2031–2034.

Sen, A. R. (1973a). Theory and application of sampling on repeated occasions with several auxiliary variables. *Biometrics*, **29**, 383–385.

Sen, A. R. (1973b). Some theory of sampling on successive occasions. *Australian Jour. Stat.*, **15**, 105–110.

Seth, G. R., and Rao, J. N. K. (1964). On the comparison between simple random sampling with and without replacement. *Sankhya*, **A26**, 85–86.

Sethi, V. K. (1963). A note on optimum stratification for estimating the population means. *Australian Jour. Stat.*, **5**, 20–33.

Sethi, V. K. (1965). On optimum pairing of units. *Sankhya*, **B27**, 315–320.

Simmons, W. R. (1954). A plan to account for "not-at-homes" by combining weighting and callbacks. *Jour. of Marketing*, **11**, 42–53.

Singh, D., Jindal, K. K., and Garg, J. N., (1968). On modified systematic sampling. *Biometrika*, **55**, 541–546.

Sittig, J. (1951). The economic choice of sampling system in acceptance sampling. *Bull. Int. Stat. Inst.*, **33**, V, 51–84.

Slonim, M. J. (1960). *Sampling in a Nutshell*. Simon & Schuster, New York.

Smith, H. F. (1938). An empirical law describing heterogeneity in the yields of agricultural crops. *Jour. Agric. Sci.*, **28**, 1–23.

Smith, T. M. F. (1976). The foundations of survey sampling: A Review. *Jour. Roy. Stat. Soc.*, **A139**, 183–204.

Snedecor, G. W., and Cochran, W. G. (1967). *Statistical Methods*. Iowa State University Press, Ames, Iowa, sixth edition.

Srinath, K. P. (1971). Multiphase sampling in nonresponse problems. *Jour. Amer. Stat. Assoc.*, **16**, 583–586.

Stein, C. (1945). A two-sample test for a linear hypothesis whose power is independent of the variance. *Ann. Math. Stat.*, **16**, 243–258.

Stephan, F. F. (1941). Stratification in representative sampling. *Jour. Marketing*, **6**, 38–46.

Stephan, F. F. (1945). The expected value and variance of the reciprocal and other negative powers of a positive Bernoulli variate. *Ann. Math. Stat.*, **16**, 50–61.

Stephan, F., and McCarthy, P. J. (1958). *Sampling Opinions*. John Wiley and Sons, New York, p. 243.

Stuart, A. (1954). A simple presentation of optimum sampling results. *Jour. Roy. Stat. Soc.*, **B16**, 239–241.

Sukhatme, P. V. (1935). Contribution to the theory of the representative method. *Supp. Jour. Roy. Stat. Soc.*, **2**, 253–268.

Sukhatme, P. V. (1947). The problem of plot size in large-scale yield surveys. *Jour. Amer. Stat. Assoc.*, **42**, 297–310.

Sukhatme, P. V. (1954). *Sampling Theory of Surveys, With Applications*. Iowa State College Press, Ames, Iowa.

Sukhatme, P. V., and Seth, G. R. (1952). Non-sampling errors in surveys. *Jour. Ind. Soc. Agr. Stat.*, **4**, 5–41.

Sukhatme, P. V., and Sukhatme, B. V. (1970). *Sampling Theory of Surveys With Applications*. Food and Agriculture Organization, Rome, second edition.

Tepping, B. J., and Boland, K. L. (1972). Response variance in the Current Population Survey. U.S. Bureau of the Census Working Paper No. 36, U. S. Government Printing Office, Washington, D.C.

Tin, M. (1965). Comparison of some ratio estimators. *Jour. Amer. Stat. Assoc.*, **60**, 294–307.

Trueblood, R. M., and Cyert, R. M. (1957). *Sampling Techniques in Accounting*, Prentice-Hall, Englewood Cliffs, N.J.

Trussell, R. E., and Elinson, J. (1959). *Chronic Illness in a Large City*. Harvard University Press, Cambridge, Mass., pp. 339–370.

Tschuprow, A. A. (1923). On the mathematical expectation of the moments of frequency distributions in the case of correlated observations. *Metron*, **2**, 461–493, 646–683.

Tukey, J. W. (1950). Some sampling simplified. *Jour. Amer. Stat. Assoc.*, **45**, 501–519.

Tukey, J. W. (1958). Bias and confidence in not-quite large samples. *Ann. Math. Stat.*, **29**, 614.

U. N. Statistical Office (1950). *The preparation of sample survey reports*. Stat. Papers Series C, No. 1.

U. N. Statistical Office (1960). *Sample Surveys of Current Interest*. Eighth Report.

U. S. Bureau of the Census. (1968). *Evaluation and Research Program of the U.S. Census of Population and Housing, 1960: Effects of Interviews and Crew Leaders*. Series ER 60, No. 7, Washington, D.C.

Warner, S. L. (1965). Randomized response: A survey technique for eliminating evasive answer bias. *Jour. Amer. Stat. Assoc.*, **60,** 63–69.

Warner, S. L. (1971). The linear randomized response model. *Jour. Amer. Stat. Assoc.*, **66,** 884–888.

Watson, D. J. (1937). The estimation of leaf areas. *Jour. Agr. Sci.*, **27,** 474.

West, Q. M. (1951). *The Results of Applying a Simple Random Sampling Process to Farm Management Data.* Agricultural Experiment Station, Cornell University.

Williams, W. H. (1963). The precision of some unbiased regression estimators. *Biometrika*, **17,** 267–274.

Wishart, J. (1952). Moment-coefficients of the k-statistics in samples from a finite population. *Biometrika*, **39,** 1–13.

Wold, H. O. A. (1954). *A Study in the Analysis of Stationary Time Series.* Almqvist and Wicksell, Stockholm, second edition.

Woodruff, R. S. (1959). The use of rotating samples in the Census Bureau's Monthly Surveys. *Proc. Soc. Stat. Sect. Amer. Stat. Assoc.*, 130–138.

Woodruff, R. S. (1971). A simple method for approximating the variance of a complicated estimate. *Jour. Amer. Stat. Assoc.*, **66,** 411–414.

Woolsey, T. D. (1956). Sampling methods for a small household survey. *Pub. Health Monographs*, No. 40.

Yates, F. (1948). Systematic sampling. *Phil. Trans. Roy. Soc. London*, **A241,** 345–377.

Yates, F. (1960). *Sampling Methods for Censuses and Surveys.* Charles Griffin and Co., London, third edition.

Yates, F., and Grundy, P. M. (1953). Selection without replacement from within strata with probability proportional to size. *Jour. Roy. Stat. Soc.*, **B15,** 253–261.

Zarkovic, S. S. (1960). On the efficiency of sampling with various probabilities and the selection of units with replacement. *Metrika*, **3,** 53–60.

Zukhovitsky, S. I., and Avdeyeva, L. I. (1966). *Linear and Convex Programming.* W. B. Saunders, Philadelphia.

Answers to Exercises

1.1 (*a*) Examples of problems of definition are decisions whether to count words in a preface or index and how mathematical symbols are treated as "words." In a book such as this one with many mathematical symbols, however, it seems unlikely that a count of words, either including or omitting symbols, would be wanted. (*b*) (*1*) The pages constitute a convenient frame. A disadvantage of the page as a sampling unit in which we count all words on any sample page is that with numerous illustrations the number of words per page may be quite variable because of the incomplete pages. It may be worthwhile first to list all the incomplete pages, forming two subpopulations or strata, one of incomplete pages and one of complete pages, that are sampled separately by the method of stratified sampling described in Chapter 5. (*2*) A problem with the line is that obtaining a listing of lines so that lines can be sampled directly is time-consuming. Also, there are incomplete lines at the end of paragraphs. Since words per line should be fairly stable, however, the solution may be to use two-stage sampling (Chapter 11), first drawing a sample of pages and then counting the number of lines on each selected page and drawing a subsample of lines on these pages.

1.2 This question supposes that we first draw a sample of cards with equal probability. (*a*) If sample names not in the target population are discarded, the only problem is that the size of the sample of names from the target population will generally be less than the number of cards and will be a random variable, depending on the cards that happen to be chosen. (*b*) The problem is that names appearing on several cards have higher probabilities of selection. One way of handling this is to count the number of cards on which a selected name appears and use this number in making the estimate by methods appropriate to selection with unequal probabilities (Chapter 9A). Another way that gives each name an equal chance but may involve many rejections is to retain a card only if it is the first of the set on which this name appears. (*c*) As in (*b*), names are being selected with unequal probabilities. I know of no easy method of giving each name an equal chance. If the number of cards on which each name appears has been recorded somewhere, an unequal-probability method can be used, as in (*b*).

1.3 Suggestions are: (*a*) a recent directory of department and luggage stores, (*b*) the repositories for lost articles maintained by the subway and bus companies, and (*c*) hospitals and private physicians in the geographical area in which snake bites occur, plus any public health organization to which reporting of bites is compulsory. Weaknesses in all three frames are likely to be incompleteness, plus high costs in (*c*) if snake bites are rare and not centrally reported. (*d*) A list of households is often used as a frame for selecting a sample of families. Although there will be some incompleteness (families who cannot be reached), the major problem may be errors in measurement.

1.4 A problem is incompleteness because of new construction. In a sample of addresses, new dwellings can usually be handled by the interviewer, who checks, for any sample address, whether there are new dwellings between this address and the next address in the directory and, if so, includes these new dwellings in the sample. Whole areas of new construction may not be mentioned in the directory and require development of a separate frame. Drawing a list of addresses is preferable to drawing a list of persons, since addresses are more permanent. For reasons of travel expense, however, the sampling unit may be a city block from which a subsample of dwellings is selected.

1.5 \$80,390 and \$82,970.

1.6 The confidence probability is about 0.054 (found from $t = -1.67$ with 25 degrees of freedom). This assumes that future receipts follow the same frequency distribution as the sample of 26 receipts, and that this distribution is normal.

1.7 When the MSE is due entirely to bias, the estimate is always wrong by $1\sqrt{MSE}$. The probability of an error $\geq 1\sqrt{MSE}$ is therefore unity and the probability of an error $\geq 1.96\sqrt{MSE}$ or $\geq 2.576\sqrt{MSE}$ is zero.

2.4 $\hat{Y} = 51{,}473$. Probability about 0.9.

2.5 Yes. $\sigma(\hat{Y})$ is 98.4.

2.6 $\hat{Y} = 20{,}238$, $s(\hat{Y}) = 849$.

2.7 (a) Public: $\hat{R} = 15.46$. Private: $\hat{R} = 12.75$. (b) Public: $s(\hat{R}) = 0.761$. Private: $s(\hat{R}) = 0.727$. For the fpc we take $f = 100/468$. (c) $14.2 < R < 16.7$.

2.8 Diff./s.e.$_{\text{diff}} = 2.71/1.186 = 2.28$. P about 0.023. Note that the fpc is not used in computing s.e.$_{\text{diff}}$.

2.9 (a) 9408, s.e. $= 780$; (b) 9472, s.e. $= 1104$.

2.10 S.e. (in 1000's) $= (a)$ 14,800; (b) 3900; (c) 3140.

2.11 9.2. (a) 2.7; (b) 2.4.

2.12 (a) $n = 60$, with 30 from each domain; (b) $n = 80$ will do if the number of owners in the sample lies anywhere between 20 and 60. With $n = 80$, the probability that this happens is only about 0.54 (from the binomial tables). With $n = 100$, any sample number of owners between 19 and 81 will suffice, the probability that this happens being about 0.94.

2.14 (a) 420; (b) 490; (c) both are unbiased; (d) estimate (b).

3.2 1066, 1334 as given by the normal approximation, equation 3.19.

3.3 Nearly conclusive.

3.6 (a) $76.2 \pm 3.6\%$; (b) 1738 ± 280 families.

3.7 1789 ± 268 families.

3.8 As an exact result,

$$\frac{V(\hat{A}_1)}{V(\hat{A}_1')} = \frac{N_1^2 n Q_1}{N^2 n_1 (1 - \pi)(Q_1 + P_1 \pi)}$$

Now $N_1 = N(1 - \pi)$, and in large samples $n_1 \doteq n(1 - \pi)$. These substitutions give the stated result. In order that $V(\hat{A}_1)V(\hat{A}_1')$ be small, we must have $\pi(1 - Q_1)/Q_1$ large. This means that Q_1 must be small: in other words, the proportion of domain 1 that lies in class C must be large. For given Q_1, π should be large.

3.9 All give $A_U = 13$. By the hypergeometric, the probability of no units in C in the sample is 0.0601 for $A_U = 12$ and 0.0434 for $A_U = 13$. By the binomial, $P_U = 0.4507$ and $\sqrt{1 - f}P_U = 0.4114$, giving $A_U = 12.3$. Page 59, Ex. 3 gives 0.061 for $A_U = 12$ and 0.044 for $A_U = 13$.

3.11 Estimate (b) seems more precise.

3.12 The highest value is PQ/n as compared with PQ/mn by the binomial formula. This occurs when every cluster consists entirely of 1's or entirely of 0's. The lowest value can be zero if every cluster gives the same proportion P. (This is possible only for certain values of P and m).

3.13 Variance is 0.00184 by the ratio method and 0.00160 by the binomial formula.

3.14 Average size of sample $= m/P$.

4.1 735 houses. This sample size is needed for two-car households if $P = 10\%$.

4.2 About 260 sheets.

4.3 (a) 2475; (b) 4950.

4.4 $n = 21$ (taking $t = 2$).

4.5 $n = 484$. For number of unemployed, the cv would be about 15%.

4.6 62 more.

4.7 (a) $n = 278$; (b) $n = 2315$; (c) $n = 3046$.

4.8 If a rectangular distribution is assumed within each class, we take $S^2 = 0.083h^2$ or $S = 0.29h$. This gives estimates of 230, 580, 2030, and 11,600 in the four classes. If the right-triangular distribution is used in the fourth class, we take $S = 0.24h$, giving 9600 for this class.

4.11 $n_{opt} = \left(\dfrac{UNS}{2c\sqrt{2\pi}} \right)^{2/3}$,

4.12 (a) $n = 1250$; (b) $n = 679$. In this part, we can give a dummy variate y_i the value $+100$ for a (Yes, No) answer, -100 for a (No, Yes) answer, and 0 otherwise. Then $\bar{Y} = P_1 - P_2$ in percentages. With an advance sample of $n_1 = 200$, formula (4.7) in Section 4.7 can be used, giving $n = 679$.

4.13 (a) The probability that a family of four persons has 1,2, and 3 females is approximately $1/4$, $1/2$, and $1/4$, respectively. For estimating the proportion P of females, a simple random sample of n families gives $V(\hat{P}) \doteq 0.03125/n$, as against $0.0625/n$ for a sample of $4n$ persons. The *deff* factor is about $1/2$. In the corresponding example in Table 3.5 with 30 households of unequal sizes, the estimated *deff* factor was 0.475. (b) The *deff* factor would be slightly raised by families with identical twins, since the proportions of families with 1 and 3 females would be slightly increased.

5.1 (a) Neyman allocation gives $n_1 = 0.87$, $n_2 = 3.13$. (b) There are three possible estimates under optimum allocation and nine under proportional allocation. $V_{opt}(\bar{y}_{st}) = \frac{1}{6} = 0.167$: $V_{prop}(\bar{y}_{st}) = \frac{7}{12} = 0.583$. (d) Formula 5.27 gives $V_{opt}(\bar{y}_{st}) = 0.159$.

5.2 (a) $n_1 = 375$, $n_2 = 625$; (b) $n_1 = 750$, $n_2 = 250$.

5.3 $RP = 181\%$ for proportional allocation and 214% for optimum allocation.

5.5 When $W_1 = W_2$ the relative increases equal 0.029 for $c_2/c_1 = 2$ and 0.111 for $c_2/c_1 = 4$.

5.6 (a) $n_1/n = \frac{1}{3}$, $n_2/n = \frac{2}{3}$, (b) $n = 264$, $n_1 = 88$, $n_2 = 176$; (c) \$1936.

5.7 (a) \$2288 against \$1936. (b) No. The minimum field cost to reduce V to 1 is \$2230.

5.8 (a) $n_1 = 384$, $n_2 = 192$; (b) $n_1 = 400$, $n_2 = 1600$; (c) $n_1 = 1200$, $n_2 = 2400$.

5.9 Fractional increase $= \frac{1}{9}$.

5.10 $n_1 = 541$, $n_2 = 313$, $n_3 = 146$.

5.12 In population 1, $V_{prop} = 0.143/n$; $V_{opt} = 0.134/n$. In population 2, $V_{prop} = 0.0491/n$, $V_{opt} = 0.0423/n$. The reduction in variance from optimum allocation is about 6% in population 1 as against 14% in population 2.

5.14 (a) If we guess $P_1 = 45\%$, $P_2 = 25\%$, $P_3 = 7.5\%$ as a compromise, this gives $n_1 = 268$, $n_2 = 116$, $n_3 = 16$; (b) s.e. $= 0.0225$; (c) s.e. $= 0.0241$.

5.15 As n approaches N, a stage is reached in which the standard formula $n_h \propto N_h S_h$ for Neyman optimum allocation is no longer applicable, since it would require $n_h > N_h$ in at least one stratum. As noted in Section 5.8, formula (5.27) then ceases to hold. The student is in error if he claims that (5.27) is always wrong; the formula has a limited range which, however, covers nearly all applications.

5A.4 No. In each of the worst cases $[\Sigma(w_h - W_h)\bar{Y}_h]^2$ is $(0.105)^2 = 0.0110$. Thus, with stratification, MSE$(\bar{y}_{st})$, as given by formula (5A.6), is $0.0108 + 0.0110 = 0.0218$. With simple random sampling, $V(\bar{y}) = 0.0177$.

5A.6 (a) $n = 1024$. The optimum allocation for the second variate (average amount invested) satisfies both requirements.

5A.7 $W_1 = 0.728$, $W_2 = 0.272$, $S_1 = 1.806$, $S_2 = 4.698$ (in the coded scale). (a) The optimum sample sizes are $n_1 = 0.507n$, $n_2 = 0.493n$. (b) $V(\bar{y}) = 31.95/n$, $V_{opt}(\bar{y}_{st}) = 6.72/n$.

5A.8 (b) $\displaystyle\int_0^a \sqrt{f(y)}\, dy = \int_0^a \sqrt{2(1-y)}\, dt = 2\sqrt{2}[1 - (1-a)^{3/2}]/3$. Hence we want $[1 - (1-a)^{3/2}] = \frac{1}{2}$.

5A.9 In Exercise 5A.7, $W_1 = 0.728$, as in the Dalenius–Hodges rule, comes closest to satisfying the Ekman rule. In Exercise 5A.8, the Ekman rule gives $a = (3 - \sqrt{5})/2 = 0.38$.

5A.10 The optima are $L = 7$ for $p = 0.95$, $L = 5$ for $p = 0.9$, and $L = 4$ for $p = 0.8$. Either $L = 5$ or $L = 6$ is a good compromise.

5A.11 (a) Gain in precision is about 110%. (b) Gain from proportional stratification over simple random sampling is about 90%.

5A.12 Increase n_1 as the hint suggests, leave $n_2 = 400$. We require $n_1 = 140$, giving $n = 540$.

6.1 For the ratio estimate $V(\hat{Y}_R) = N^2(1-f)S_d^2/n$ and for simple expansion $V(\hat{Y}) = N^2(1-f)S_y^2/n$, where $d = (y - Rx)$. For the sample of 21 households the estimates of S_d^2 and S_y^2 are as follows. Number of children, $s_d^2 = 0.49$, $s_y^2 = 1.61$; number of cars, $s_d^2 = 0.41$, $s_y^2 = 0.39$; number of TV sets, $s_d^2 = 0.51$, $s_y^2 = 0.45$. The ratio estimate appears superior for children.

6.2 Gain = 66%. At leat 11 units by the ratio method.

6.3 Quadratic limits (27,100, 29,870); normal limits (27,030, 29,700).

6.4 Apply theorem 6.3 to the estimation of $R = \bar{Y}/\bar{X}$. With large samples, use $\bar{y}/\bar{X}$ if $r \leq$ (cv of x)/2(cv of y), and use $\bar{y}/\bar{x}$ otherwise, where r and the cv's are sample estimates.

6.5 The MSE's are 46.5 for the separate ratio estimate and 40.6 for the combined ratio estimate. In both cases the contribution of bias to the MSE is negligible.

6.6 For Lahiri's method, $V(\hat{Y}_{Rs}) = 40.1$.

6.7 Estimated population total = 116.21 millions. The relative variance is 0.00111, so that the s.e. is $(0.0333)(116.21) = 3.87$ millions. The estimate is within 1 s.e. of the true total.

6.8 The estimates are (a) 1896, (b) 1660, (c) 1689. In (c) we find $w_1 = 2.38$, $w_2 = -1.38$. Estimated s.e.'s are (a) 256, (b) 36.9, (c) 18.6. For the s.e. in (b) I used the formula s.e. = $\hat{\bar{Y}}_R\sqrt{(1-f)(c_{yy} + c_{11} - 2c_{y1})/n}$, where $\hat{\bar{Y}}_R$ is the ratio estimate of $\bar{Y}$, that is, 1660. For the s.e. in (c) I used $\hat{\bar{Y}}_{MR} = 1689$.

7.1 Estimate = 11,080; s.e. = 152 (including the fpc).

7.2 No, since b is very close to 1.

7.3 $\hat{Y}_{lr} = 28,177 \pm 570$. The relative precision is 113%.

7.4 $27,751 \pm 694$.

7.6 For the difference estimate, $V(\bar{y}) = S_e^2/n$, for the linear regression estimate, $V(\bar{y}_{lr}) = S_e^2 S_y^2/n(S_e^2 + S_y^2)$. The regression estimate has the smaller variance, but its superiority is unimportant if S_e^2/S_y^2 is small.

7.7 $\text{MSE}(\hat{Y}_{lrs}) = 34.5$, $\text{Bias}^2 = 9.7$; $\text{MSE}(\hat{Y}_{lrc}) = 11.9$, $\text{Bias}^2 = 1.2$.

7.9 $\text{MSE}(\hat{Y}_{Rs}) = 8.9$, $\text{MSE}(\hat{Y}_{Rc}) = 6.7$.

8.1 Variances are 8.19 (systematic), 11.27 (simple random), 8.25 (stratified, 2), 7.46 (stratified, 1).

8.2 $V_{sys} = 0.00141$, $V_{ran} = 0.00340$.

8.3 The systematic sample should be superior for the proportion of people of Polish descent, since this variable exhibits geographical stratification. It is likely to be inferior for proportion of children because the sampling interval, 1 in 5, coincides with the average size of a household. The same is true, though to a smaller extent, for proportion of males.

8.4 The variances are as follows. Males, $V_{srs} = 0.0204$, $V_{sys} = 0.0216$; children, $V_{srs} = 0.0204$, $V_{sys} = 0.0776$; professional, $V_{srs} = 0.0192$, $V_{sys} = 0.0016$.

8.5 Actual variance = 8.19. Method (a) gives 11.29. For method (b) the estimated variance from a single sample is $(1-f)(\bar{y}_{i1} - \bar{y}_{i2})^2/4$, where $\bar{y}_{i1}$, $\bar{y}_{i2}$ are the means of the two halves. The average is 3.24. The serious underestimation is unexpected.

8.7 Both variances are $(k^2 - 1)/6$.

8.8 Simple random sampling is better unless $n = 1$ or $k = 1$.

8.9 Every kth unit, $V(\hat{\bar{Y}}) = 362.2$; Yates, MSE$(\hat{\bar{Y}}) = 7.3$; Sethi, $V(\hat{\bar{Y}}) = 21.0$; Singh et al., $V(\hat{\bar{Y}}) = 81.0$. The last two estimators are unbiased in this example.

9.1 Relative costs of using the four types of unit are as 100, 90.1, 79.7, and 77.8 (taking the first unit as a standard).

9.2 Relative precision of the household is 211% for the sex ratio and 38% for the proportion who had seen a doctor.

9.3 Relative precision of the large unit is 0.566 with simple random sampling and 0.625 with stratified random sampling.

9.5 (a) $M = 5$; (b) $M = 1$.

9.6 The optimum M should decrease because travel cost, which varies as $\sqrt{n}$, becomes relatively less important when n increases.

9A.1 (a) 34,242; (b) 5534; (c) 6493.

9A.3 (a) If the s.d. among large units in class $h \propto M_h$. (b) If probability $\propto \sqrt{M_h}$.

9A.4 (b) $V(\hat{Y}_{HT}) = 1.75$, $V(\hat{Y}_M) = 0.27$, $V(\hat{Y}_{RHC}) = 0.33$. In this problem y_i/z_i varies little, but $\hat{Y}_{HT}$ performs relatively poorly because with the method of sample selection $\pi_i \neq 2z_i$. Murthy's estimator is devised for this method of sample selection having $V(\hat{Y}_M) = 0$ if y_i/z_i is constant, as (9A.60) shows. (c) $V(\hat{Y}_M)/V(\hat{Y}_{ppz}) = 0.54$, while $(N-n)/(N-1) = \frac{1}{2}$.

9A.5 MSE$(\hat{Y}_{HT(A)}^*) = 7.06$, (Bias)^2/MSE $= 0.065$, $V(\hat{Y}_{SRS}) = 6.5$.

10.1 (a) 2.00; (b) 2.13.

10.3 (a) $165/n$; (b) $148.5/n$; (c) $132/n$.

10.4 (a) $n = 660$ fields; (b) $n = 530$ fields. Protein requires fewer fields than yield.

10.5 $c_1/c_2 = 8$.

10.7 (a) 0.93%; (b) 0.51%; (c) 0.36%.

10.8 (a) Either $m_0 = 7$ or $m_0 = 8$ is suitable; (b) 89% for $m_0 = 7$ and 93% for $m_0 = 8$; (c) 86% for $m_0 = 7$ and 89% for $m_0 = 8$.

11.2 The relative precision of III to II drops from 3.02 to 2.75. If two sampling plans differ primarily in their between-units contribution to the variance, the relative precision of the superior plan will in general decrease as the ratio of the within-units variance to the total variance increases.

11.3 The explanation is, roughly speaking, that with these data the Y_i/z_i are more stable than the Y_i/M_i. If we took $z_i = \frac{1}{33}, \frac{8}{33}$, and $\frac{24}{33}$, the between-units contribution in method IV would vanish.

11.4 Total variance: 0.00504 (Ia), 0.02358 (II), 0.00554 (III).

11.6 Estimated percentage 14.2 ± 2.16. Estimated number 3540 ± 540.

11.7 Estimated percentage 13.9 ± 2.49.

11.9 Total rooms, 29,400, total persons, 50,550, persons per room, 1.72; s.e.'s: total persons, 2,440, persons per room 0.066.

12.1 $n = 267$, $n' = 1320$ or $n = 268$, $n' = 1280$. $V(p_{st})$ with optimum allocation is 6.67 when p_{st} is in %'s. With single sampling, $V(p) = 8.33$.

12.2 $c_n/c_n' > 9$.

12.3 $n/n' = 1/19$.

12.4 $n' > 16n$.

12.5 By formula (12.67), s.e. $= 1.25$, ignoring $1/N$.

12.6 Per cent gains from the second to the sixth occasion are 50, 75, 91, 100, and 105, respectively.

12.8 The values of $nV(\bar{y}_2'')/S^2$ and $nV(\bar{y}_2')/S^2$ are as follows: $\mu = \frac{1}{4}, \rho = 0.8$: 0.885, 0.875; $\mu = \frac{1}{4}$; $\rho = 0.9$: 0.843, 0.840; $\mu = \frac{1}{2}, \rho = 0.8$: 0.824, 0.810; $\mu = \frac{1}{2}, \rho = 0.9$: 0.752, 0.746.

12.12 (a) Let $y_i = 1$ for any unit that has the first attribute and $y_i = 0$ otherwise and let stratum 1 be the stratum in which every unit has the second attribute. In the notation of theorems 12.1, 12.2, with $1/N$ negligible, $S^2 = P_1 Q_1$ and $S_1^2 = P_1 P_2/(P_1 + P_2)^2$. Results in (a) follow from the theorems; (b) if C^* is the expected cost, double sampling with $\nu_1 = 1/2$, $k = 2$ (the optimum) gives $V(\hat{Y}_{st}) = N^2(0.844)/C^*$, while a simple random sample gives $V(\hat{Y}) = N^2(1.875)/C^*$, over twice as large.

13.1 (c) 90% response with 1047 completed questionnaires or 95% response with 701 completed questionnaires.

13.2 The method with 90% response costs \$5235. That with 95% response costs \$5.7895 per completed questionnaire, or \$4058 total cost.

13.3 (b) $0.65n_0$, $0.815n_0$, $0.8915n_0$, $0.9351n_0$, $0.9611n_0$; (c) 100, 104, 108, 112, 117; (d) 300, 288, 277, 267, 256.

13.4 (a) Bias (in %) $= -3.85, -2.15, -1.21, -0.69, -0.40$; (b) variances are 8.67, 9.03, 9.39, 9.74, 10.16; (c) four calls.

13.6 Yes. VC for $k = 2$ is only about 2% over the minimum VC.

13.7 Politz-Simmons estimate, 39.7%; binomial, 42.3%.

13.8 (c) Variance $= 0.1$.

13.9 If each enumerator's error of measurement were independent from family to family, the variance of the sample mean would be $(\sigma_{\mu'}^2 + \sigma_d^2 + \sigma_{Is}^2 + \sigma_I^2)/525$ instead of $(\sigma_{\mu'}^2 + \sigma_d^2 + 35\sigma_{Is}^2 + 105\sigma_I^2)/525$. Enumerator biases contribute about 55% of the total variance.

13.10 $10^4 V(\hat{\pi}_A) = V(100\hat{\pi}_A) = (a)$ 1.80 (Direct); (b) 10.69 (Warner); (c) 3.30 (π_u known); (d) 5.12 (Moors); (e) 6.30 ($P_2 = 1 - P_1$).

13.11 (a) $10^4 \text{MSE}(\hat{\pi}_A) = 3.81$; $\hat{\pi}_{AU}$ is superior if π_U is known. (b) $10^4 \text{MSE}(\hat{\pi}_A) = 5.47$; the two-question $\hat{\pi}_{AU}$ is also superior if $P_2 = 0$. (c) $10^4 \text{MSE}(\hat{\pi}_A) = 7.64$. All methods are superior except Warner's original method.

Author Index

Abernathy, J.R., 395
Arvesen, J.N., 179
Avadhani, M.S., 48, 127
Avdeyeva, L.I., 122

Bailar, B.A., 392
Balakrishnan, T.R., 130, 134
Barr, A., 378
Bartholomew, D.J., 376
Basu, D., 30
Bayless, D.L., 265, 267, 269, 270
Beale, E.M.L., 176, 178
Bean, J.A., 135, 171, 324
Beardwood, J., 96
Beegle, L.D., 180
Bellhouse, D.R., 216, 217
Belloc, N.B., 385
Bernert, E.H., 131
Bershad, M., 354, 378, 381, 383, 386
Birnbaum, Z.W., 363
Black, C.A., 228, 233
Blythe, R.H., 84, 88
Boland, K.L., 389
Booth, G., 122, 123, 335, 336
Bose, Chameli, 340
Brewer, K.W.R., 158, 256, 261, 310
Brooks, S., 282, 285
Bryant, E.C., 124
Burman, J.P., 320
Burmeister, L.F., 146
Burstein, H., 59

Cameron, J.M., 281
Chakravarti, I.M., 48
Chatterjee, S., 121, 122, 147
Chung, J.H., 57, 59
Cochran, W.G., 19, 130, 131, 132, 197, 220, 226, 256, 266, 361
Coleman, J.S., 4
Cornell, F.G., 106
Cornfield, J., 28, 80, 247, 379
Corlett, T., 292

Cox, D.R., 79
Cyert, R.M., 12

Dalenius, T., 122, 127, 129, 130, 131, 134, 140, 392
Das, A.C., 228
David, I.P., 161
DeLury, D.B., 57, 59
Deming, W.E., 4, 81, 148, 321, 367, 376, 391
Des Raj, 30, 256, 263, 300, 309
Dowling, T.A., 394
Dunham, R.E., 385
Durbin, J., 142, 144, 175, 261, 262, 263, 300, 310, 325, 366, 373, 376

Eckler, A.R., 355
Ekman, G., 130
Elinson J., 385
Erdös, P., 39
Evans, W.D., 117

Fellegi, I., 263, 279, 387, 388, 391
Feller, W., 39
Fieller, E.C., 156
Finkner, A.L., 172, 239, 360
Finney, D.J., 71, 222
Fisher, R.A., 42, 59, 221
Fitzpatrick, T.B., 126, 127
Foreman, E.K., 256
Frankel, M.R., 321, 322
Fuller, W.A., 140, 146, 149

Gallup, G., 4
Garg, J.N., 217
Gautschi, W., 226
Godambe, V.P., 45, 48
Goodman, R., 126
Graham, J.E., 354
Gray, P.G., 385
Greenberg, B.G., 394, 395
Grundy, P.M., 84
Gurney, M., 130, 165, 172, 261

Hagood, M., 131
Hájek, J., 39
Haldane, J.B.S., 77
Halton, J.H., 400
Hammersley, J.M., 400
Hanif, M., 261, 310
Hansen, M.H., 22, 68, 139, 145, 155, 165, 172, 243, 246, 250, 283, 297, 314, 354, 355, 370, 372, 378, 381, 383, 386, 392
Hanson, R., 387, 388
Hartley, H.O., 45, 122, 124, 139, 142, 145, 146, 162, 174, 266, 374
Haynes, J.D., 227
Hendricks, W.E., 243, 244, 377
Herson, J., 160
Hess, I., 126, 127, 134, 183, 364, 377
Hocking, R.R., 122
Hodges, J.L., 129
Hoeffding, W., 179
Homeyer, P.G., 228, 233
Horvitz, D.G., 44, 259, 272, 393, 395
Houseman, E.E., 38, 102
Huddleston, H.F., 122
Hurwitz, W.N., 22, 68, 139, 145, 165, 172, 243, 246, 250, 283, 297, 314, 354, 355, 370, 372, 378, 381, 383, 386
Hutchinson, M.C., 178
Hyman, H.H., 6

Jabine, T.B., 145
James, A.T., 157
Jebe, E.H., 297
Jensen, A., 11
Jessen, R.J., 6, 38, 102, 119, 124, 160, 237, 243, 244, 346
Jindal, K.K., 217
Johnson, F.A., 222, 225, 235
Jones, H.W., 391

Kempthorne, O., 45
Kendall, M.G., 19
Keyfitz, N., 169, 170, 184
Khamis, S.H., 30
Khan, S., 340
Kiefer, G., 139
King, A.J., 131, 330
Kiser, C.V., 4
Kish, L., 68, 126, 162, 183, 242, 321, 322, 325, 364, 377, 383
Koch, G., 379
Kokan, A.R., 122, 149
Koons, D.A., 386, 388

Koop, J.C., 163, 391
Kulldorf, G., 348
Kuzik, R.A., 178

Lahiri, D.B., 175, 206, 251
Lansing, J.B., 383
Lieberman, G.J., 57
Lienau, C.C., 378
Lund, R.E., 146

McCarthy, P.J., 136, 320, 321, 365, 386
McCarty, D.E., 131
McVay, F.E., 243, 247
Mackenzie, W.A., 221
Madow, L.H., 210, 213, 218
Madow, W.G., 39, 139, 145, 206, 210, 213, 243, 246, 265, 283, 354, 355, 385
Mahalanobis, P.C., 243, 274, 321, 378, 388
Matérn, B., 219, 221, 226, 227
Mickey, M.R., 175, 203
Midzuno, H., 175
Milne, A., 218, 227
Monroe, R.J., 239
Moors, J.J.A., 394
Morgan, J.J., 239
Morton, R., 394
Mosteller, F., 361
Mudholkar, G.S., 186
Murthy, M.N., 22, 30, 130, 206, 217, 263, 265

Namboodiri, N.K., 162
Neter, J., 4
Neyman, J., 99, 101, 327
Nordbotten, S., 149
Nordin, J.A., 84

Olkin, I., 184, 186
Osborne, J.G., 221, 222, 226
Owen, D. B., 57

Patterson, H.D., 348
Paulson, E., 156
Payne, S.L., 6
Pereira, P., 183
Pillai, R.K., 162
Plackett, R.L., 320
Platek, R., 309
Politz, A.N., 374
Pritzker, L., 379, 386, 387, 388

Quenouille, M.H., 175, 178, 221, 228

Raiffa, H., 85
Rao, C.R., 45
Rao, J.N.K., 45, 136, 139, 155, 163, 178, 179,
    180, 183, 197, 203, 216, 217, 261, 262, 266,
    267, 269, 270, 300, 311, 330, 337, 338, 354
Rao, P.S.R.S., 155, 178, 186
Rényi, A., 39
Riedel, D.C., 126, 127
Robson, D.S., 330
Romig, H.G., 55, 59
Ross, A., 162, 174
Roy, J., 48
Royall, R.M., 45, 48, 158, 160, 200, 339

Sagen, O.K., 385
Sampford, M.R., 262, 263
Sandelius, M., 71
Särndal, C.E., 49
Satterthwaite, F.E., 96
Schlaifer, R., 85
Scott, A.J., 355
Sedransk, J., 122, 123, 335, 336
Sen, A.R., 261, 348
Seth, G.R., 30, 378
Sethi, V.K., 130, 134, 217
Shachtman, R.H., 394
Shah, B.V., 393
Simmons, W.R., 148, 374, 385, 393
Singh, D., 217
Singh, M.P., 261, 309
Sirken, M.G., 363
Sittig, J., 85
Slonim, M.J., 4
Smith, B.B., 19
Smith, H.F., 244
Smith, T.M.F., 45, 355

Snedecor, G.W., 19
Srinath, K.P., 330
Stein, C., 78
Stephan, F., 118, 131, 135, 136, 365, 386
Stuart, A., 97, 366
Sukhatme, B.V., 48, 117, 127, 161
Sukhatme, P.V., 117, 163, 308, 326, 378

Tepping, B.J., 389
Thompson, D.J., 44, 259, 272
Tin, M., 176, 178
Tripathi, T.P., 340'
Trueblood, R.M., 12
Trussell, R.E., 385
Tschuprow, A.A., 99
Tukey, J.W., 29, 178, 361

Venables, W.N., 157

Waksberg, J., 387, 392
Warner, S.L., 392, 393, 395
Watson, D.J., 189
West, Q.M., 44
Whelpton, P.K., 4
Wilkinson, G., 157
Williams, W.A., 203
Wishart, J., 29
Wold, H.O.A., 221
Woodruff, R.S., 170, 319, 355
Woolsey, T.D., 364

Yates, F., 59, 84, 121, 122, 142, 184, 189, 216,
    217, 222, 226, 228, 256, 261, 348

Zarcovic, S.S., 256
Zukhovitskay, S.L., 122

Subject Index

Acceptance sampling, 3
Advantages of sampling, 1–2
Aligned systematic sample, 228
Analytical surveys, 4. *See also* Domains of study
Area sample, 89
Attributes, *see* Proportions
Autocorrelated populations, 219

Balanced repeated replications, 320
Best linear unbiased estimate, 158
Bias:
  definition, 13
  deliberate use of biased procedures, 12, 15
  due to errors in stratum weights, 117
  due to nonresponse, 359–361
  of interviewer, 378
  of ratio estimator, 160
  of regression estimator, 198
  in small areal units, 233
Binomial distribution, 59
  confidence limits, 59
  erroneous use in cluster sampling, 64–68
  references to tables, 55
  for sample proportions, 55
Boundaries of strata, rule for determining, 127–130

Call-backs, 365
  effects on response ratio, 365
  mathematical model, 367
  optimum policy, 369–370
  relative cost, 366–367
Cauchy-Schwarz inequality, 97
Census, use of sampling, 2
Centrally-located systematic samples, 205
Change, estimates of, 345, 352–353
Cluster sampling (single-stage):
  clusters of equal sizes, 233
    compared with SRS of elements, 242
    estimation of proportions, 64–65, 246
    optimum size and shape of cluster, 234–240, 246
    variance as function of size of cluster, 243
  variance of mean per element, 240
  clusters of unequal sizes, 249
  comparison of $\hat{Y}_u$, $\hat{Y}_R$, $\hat{Y}_{pps}$, 255–258
    equal-probability selection, unbiased estimator $\hat{Y}_u$, 249
    ratio estimator $\hat{Y}_R$, 250
  Horvitz-Thompson estimator, 259–261
    optimum selection probabilities, 255
    ratio estimators, 271
    selection:
      methods related to systematic sampling, 265
      without replacement (WOR), 258–259
    selection WOR, Brewer's method, 261
      comparison of methods, 270
      Murthy's method, 263
      Rao, Hartley, Cochran method, 266
    in stratified sampling, 270
    unequal-probability selection with replacement (WR), $\hat{Y}_{ppz}$, 251–255
Cluster sampling (with subsampling), *see* Two-stage sampling
Coefficient of variation(cv), 54
Collapsed strata method, 139, 227
Combined ratio estimate, 165
  compared with separate ratio estimate, 167
  estimated variance, 167
  in estimating domain means, 143
  optimum allocation for, 172
  in stratified two-stage sampling, 318
  upper bound to relative bias, 166
  variance, 166
Combined regression estimate, 202
  compared with separate regression estimate, 203
  estimated variance, 203
  variance, 202
Composite estimate, in repeated sampling, 354
Conditional distribution, of proportions, 61
Confidence limits, 12
  conditional, 60
  definition for attributes, 57
  effect of bias on, 14
  effect of nonresponse on, 361–363

for means in simple random sampling, 27
for proportions or percentages, 57–60
for ratio estimates, 156
in stratified random sampling, 95
validity of normal approximation, 39–44
Consistency:
definition, 21
of mean of simple random sample, 21
of ratio estimate, 153
of regression estimate, 190
Controlled selection, 124
Correction for continuity, 58, 64
"Correct value," 377
Correlation coefficient:
in finite populations, 154
intracluster, 209
within a systematic sample, 209
Correlogram, 219
Cost function:
for number of strata, 134
in determining optimum probabilities of
selecting primary units, 311–318
in determining optimum sampling fraction
for nonrespondents, 371
in determining optimum size of cluster unit,
244
in determining optimum subsampling
fraction, 280, 313–316
in determining sample size, 84
in double sampling:
for analytical comparisons, 335
for regression estimates, 341
with stratification, 331
in stratified random sampling, 96
in two-frame sampling, 145
Covariance of sample means, 25
Cumulative $\sqrt{f}$ rule, 129

Degrees of freedom, effective number in stratified
sampling, 96
Descriptive surveys, 4
Design effect (Deff), 21
Domains of study (subpopulations), 34
comparisons between means, 39
comparisons between proportions and totals, 64
domains cutting across strata, 142–144
estimates of means and totals (continuous
variate), 34–35
estimates of proportions and totals (0-1 variate),
63
sample size needed, 82–83
strata as domains, 140

Double sampling, 327
ratio estimates, 343–344
Double sampling with regression estimates, 335
comparison with simple random sampling, 341
estimated variance, 343
optimum sample sizes, 341
variance, 339–340
Double sampling with stratification, 327
comparison with simple random sampling, 332
estimated variance, 333
estimation of proportions, 330
optimum sample sizes, 331
variance, 328

E, average over all possible samples, 11
E_1, E_2, averages over first and second stages, 275
Elements, 31
End corrections, 216
Errors in surveys, types of, 359
Errors of measurement, 377
effects of constant bias, 379
effects of correlation between errors, 383
effects of errors uncorrelated within sample,
380
mathematical model, 377–379
summary of effects, 384
techniques for studying, 384–392
use of interpenetrating subsamples, 389, 391
Estimates of population variances:
effects of errors in S_h on precision of stratified
sampling, 115
for determining sample size, 78
Expansion factor, 21
Eye estimates, 189

Finite population correction (fpc), 24
rule for ignoring, 25
in stratified random sampling, 92
in two-stage sampling, 277
Frame, 6
estimates when frame has units not in target
population, 37–38
sampling from two frames, 144–146

Geographic stratification, 102
construction of strata, 131
gains in precision, 102–103

Hard core in nonresponse, 364
Horvitz-Thompson estimator, 259–261
Hypergeometric distribution, 55
as conditional distribution, 61

confidence limits, 57
    reference to charts and tables, 57

Index of inconsistency, 387
Inflation factor, 21
Interpenetrating subsamples, 388–392
Interviewer bias, 378, 384
    mathematical models, 377–379
Intracluster correlation, 209
Inverse sample (Haldane's method), 77
Item, definition, 20

Jacknife method, 174
    for estimating variance of a ratio, 178
    for variance of nonlinear function, 321

Keyfitz, short-cut method of variance estimation,
        169–171
Knight's move latin square, 229

Labels, 44
Lattice sampling, 228
Linear regression estimator, *see* Regression
        estimator
Listing of primary units, effect of cost on
        optimum selection probability, 316
Loss function, 83, 121

Mail surveys, 145, 360, 370
Matching in repeated sampling of same
        population, 364–365
Mean of sample, optimum properties, 44–45
Mean square error, definition, 15
    justification for use of, 15
    relation to variance and bias, 15
Measure of homogeneity, 248
Model-unbiased, 158
Multivariate ratio estimate, 185

Neyman allocation, 99
    best stratum boundaries for, 127
Noncoverage, 364
Nonnormality:
    effect of stratification on, 44
    effect on confidence limits, 41
    frequently encountered in sampling practice, 40
Nonresponse, 359
    bias produced by, 359–361
    effect of call-backs, 365–370
    effect on confidence limits, 361–363
    effect on variance in stratified sampling, 144
    optimum sampling fraction among
        nonrespondents, 370–374

Politz and Simmons' method, 374–377
    reasons for, 364
Normal distribution, 8
    as approximation to hypergeometric, 57
    as limiting distribution of sample means, 39
    use in surveys, 11
    validity with means from continuous data,
        39–44
Notation:
    for errors of measurement, 377–379
    for proportions, 50, 63
    for ratio estimates, 151
    for simple random sampling, 20
    for stratified sampling, 90
    for two-stage sampling, 276
    for variances of estimates, 27

Optimum allocation in stratified sampling:
    comparison with proportional allocation, 99,
        109
    comparison with simple random sampling, 99,
        109
    with double sampling, 330
    effect of deviations from the optimum, 115
    effect of errors:
        in S_h, 115–117
        in stratum sizes, 117
    estimation from previous data, 102, 131
    for fixed sample size, 98–99
    for fixed total cost, 96–98
    with more than one item, 119–123
    with ratio estimates, 172
    requiring more than 100% sampling, 104
    in sampling for proportions, 108
Optimum allocation in stratified two-stage
        sampling, 280
Optimum per cent matched, in repeated sampling,
        347, 350
Optimum size of subsample:
    primary units of equal size, 280
    primary units of unequal size, 313–316

Percentages, estimation of, *see* Proportions,
        estimation of
Periodic variation, effect on systematic sampling,
        217
Pilot survey:
    use in estimating optimum sampling and
        subsampling fractions, 283–285
    use in estimating sample size, 78
Politz and Simmons' method for handling
        nonresponse, 374–377

Population, 5
  sampled, 5
  target, 5
Poststratification, 134
  compared with proportional allocation, 134
Precision contrasted with accuracy, 16
  relative, 99
Pretest, 7
Primary sampling units (primary units), 274
Probability proportional to size, pps, (one-stage
      sampling), 250
  compared with equal probabilities, 255–258
  method of drawing sample, 250
  selection without replacement, 258
Pps (two-stage sampling), 295, 308
  compared with equal probabilities, 299, 310
  selection without replacement, 308
Probability sampling, definition and properties, 9
  alternatives to, 10
Product estimator, 186
Proportional allocation in stratified sampling, 91
  advice on use, 103, 110
  comparison with optimum allocation, 99, 103,
      109
  comparison with poststratification, 134
  comparison with simple random sampling, 99,
      109
  in estimating proportions, 109
  sample self-weighting, 91
  variance formulas, 92–95
Proportions, estimation of, 50
  in cluster sampling, 64–68, 246
  in double sampling, 330
  effect of nonresponse, 361
  effect of population P on precision, 53
  with more than two classes, 60, 61
  in simple random sampling, 50–68
  size of sample for, 74
  in stratified random sampling, 107–111
  in two-stage sampling (units of equal size), 279
  in two-stage sampling (units of unequal size),
      311
Purposive selection, 10

Quadratic confidence limits for ratio estimate, 156
Qualitative characteristics, 50. *See also*
      Proportions, estimation of
Quota sampling, 135

Raising factor, 21
Random member of household, method of
      selecting, 365

Random numbers, 19
Randomized response method, 392
  other variants, 395
  unrelated second question, 393
  Warner's original method, 392
Random sampling, *see* Simple random sampling
Rare items, Haldane's method, 77
Ratio estimator, 30, 150
  accuracy of approximate variance, 162
  adjustments to decrease bias, 174
  bias, 160–162
  in cluster sampling, 31, 66
  compared with mean per unit, 157
  compared with regression estimate, 195
  compared with stratified sampling, 169
  conditions under which unbiased, 158
  confidence limits, 156
  consistency, 153
  in double sampling, 343
  estimated variance, 32, 155, 178
  Hartley-Ross estimate, 174
  jackknife estimate of variance, 178
  Lahiri's method, 175
  multivariate, 185
  optimum allocation in stratified sampling, 172
  optimum conditions for, 160
  as special case of regression estimate, 190
  standard error for comparison of two ratios, 180
  in stratified random sampling, 164
  in stratified two-stage sampling, 317
  unbiased ratio-type estimates, 174
  upper bound to relative bias, 162
  variance in large samples, 30, 153
  see also Combined ratio estimate; Separate ratio
      estimate
Ratio of two ratio estimators, 183
Record checks, 385
Regression coefficient in finite populations, 191
Regression estimator, 189
  accuracy of large-sample variance, 197
  bias, 198
  compared with mean per unit, 195
  compared with ratio estimate, 195
  conditions under which unbiased, 199
  in double sampling, 338
  effect of error in slope, 192
  estimated variance, 195
  with preassigned slope, 190
  in repeated sampling of same population
      346–355
  in stratified random sampling, 201, 202
  uses, 189

variance in large samples, 194
see also Combined regression estimate;
　　Separate regression estimate
Reinterviews in study of errors of measurement,
　　385, 386
Relative precision, 99
　method of calculating, 103
Repeated measurements, 386, 391
Repeated sampling of population, 344
　composite estimate, 354
　current estimates, 346–355
　estimates of change, 345, 352
　optimum percent matched, 347, 350
　rotation policies, 354
　use of nonrespondents from previous surveys,
　　377
Response deviation, 379
Response variance, 379
　correlated component, 383
　simple, 383
　total, 383

Sampled population, 5
Sampling fraction, 21
　first-stage, 277
　second-stage, 277
Sampling unit, definition, 6
Sampling with replacement (WR), 18, 29
Sampling without replacement (WOR), 18
Satterthwaite's approximation to number of df, 96
Self-weighting estimate, 91
　in two-stage sampling, 303, 304, 307, 317
Sensitive questions, 392
　randomized response method, 392–395
Separate ratio estimate, 164
　compared with combined ratio estimate, 167
　estimated variance, 167
　liability to bias, 165
　optimum allocation, 172
　variance, 164
Separate regression estimate, 201
　compared with combined regression estimate,
　　203
　estimated variance, 202
　liability to bias, 202
　variance, 202
Short-cuts, in computation of variance of ratio
　　estimate, 173
Simple expansion, 169
Simple random sampling, 18
　for classification into more than two classes, 60
　confidence limits:

for sample mean, 27
for sample proportion, 57
distribution of sample proportion, 55–57
estimated variance, 202
　of sample mean, 26
　of sample proportion, 52
method of drawing, 19
optimum linear estimator of population mean,
　44
sample size:
　needed for means, 77
　sample size needed for proportions, 75
Simple response variance, 383
Size of sample for specified limits of error:
　analysis of problem, 73
　for comparisons between domains, 83
　with continuous data, 77
　Cox's method of two-step sampling, 78–80
　for means over domains, 82
　by minimizing cost plus loss due to errors, 83
　with more than one item, 81
　for normal approximation to confidence limits
　　for continuous data, 41–44
　for normal approximation to confidence limits of
　　proportions, 58
　with proportions, 75
　with rare items, 76
　in stratified random sampling, 105, 110
Skewness:
　coefficient of, 42, 197
　effect of stratification on, 44
　effect on confidence limits, 41
Square grid sample, 227
Standard deviation in finite population, 23
Standard error:
　of difference between domain means, 39
　of difference between two ratios, 180
　of domain mean, 33–34
　　in stratified sampling, 145–147
　of domain total, 35
　　in stratified sampling, 143
　of estimated population total from simple
　　random sample, 24
　of mean:
　　in cluster sampling, 240
　　of simple random sample, 24, 25–27
　　in stratified sampling, 91–98
　　of systematic sample, 207–212
　　in three-stage sampling, 286–287
　of proportion:
　　in cluster sampling, 64–68
　　in simple random sampling, 51–55

in stratified sampling, 107–109
in two-stage sampling, 279
over a domain, 62–63
of ratio estimate, 31–32, 153–156
in stratified sampling, 164–167
of ratio in two-stage sampling, 311
of regression estimate, 190, 194, 195
of regression in stratified sampling, 200–203
of sample standard deviation, 43
of total in population possessing some attribute, 52
Standard error (approximate) of nonlinear estimators, 318
balanced repeated replications, 320
comparison of methods, 322–324
jackknife method, 321
Taylor series method, 319
Steps in a sample survey, 4
Strata, 89
construction, 127–131
optimum number, 132–134
Stratification, 89
best variable for, 101
effect of number of strata on precision, 132
effect on normality of variate, 44
two-way, 124
Stratified random sampling, 89
compared with simple random sampling, 99–101
compared with ratio estimate, 169
compared with systematic sampling, 209–223
construction of strata, 127–131
estimate p st, 107–108
estimate ȳ st, 91–92
estimate of gain in precision, 136
estimated variance of ȳ pt, 95
estimated variance of p st, 108
estimates for domains of study, 142–144
Neyman allocation, 98–99
with one unit per stratum, 138
optimum allocation for fixed cost, 96–98
with ratio estimates, 164
with regression estimates, 200
size of sample, 105, 110
type of population giving large gains, 101
Subpopulations, see Domains of study
Subunits (elements), 233
Superpopulation, 158
Systematic sampling, 205
in autocorrelated populations, 219
compared with simple random sampling, 208–221
compared with stratified sampling, 209–223
effect of periodic variation, 217
end corrections, 216
estimation of variance, 223–226
method of drawing, 206
in natural populations, 221
in populations:
in "random" order, 212
with linear trend, 214–217
in single-stage cluster sampling, 265
stratified systematic sampling, 226
in two dimensions, 227
in two-stage sampling, 279
recommendations about use, 229
relation to cluster sampling, 207
variance of mean, 207–212

Target population, 5
Taylor series method, 319
Theory, in sample surveys, 8
Three-stage sampling, 285
optimum sampling and subsampling fractions, 288
variance of mean per third-stage unit, 286–287
Total response variance, 383
Travel costs, formula, 96
Two-phase sampling, see Double sampling
Two-stage sampling (units of equal size):
advantage, 274
optimum sampling and subsampling fractions, 280
stratified sampling of the primary units, 288–289
table for selecting optimum size of subsample, 283
use of pilot survey, 283
variance:
of estimated mean, 276
of estimated proportions, 279
Two-stage sampling (units of unequal size), 292
comparison of methods, 310
methods with one unit per stratum, 293–299
optimum sampling and subsampling fractions, 313–316
ratio estimators, 311
units chosen:
with equal probabilities, 293–295, 303–305
with unequal probabilities WOR, 308–310
with unequal probabilities WR, 306–308

Unaligned systematic sample, 228
Unbiased estimate, definition, 11

Unit (sampling unit), definition, 6
United States Census, use of sampling, 2–3
Uses of sample surveys, 2–4
  by business and industry, 3
  in decennial censuses, 2–3
  in market research, 3
  by members of United Nations, 2

  in opinion polls, 4

Variance, definition of S^2 and σ^2, 22, 25
Variance function, 243
Variance of population, advance estimates, 78–81
Variance of sample estimates, *see* Standard error